STUDY OF TRAVELLING INTERPLANETARY PHENOMENA/1977

ASTROPHYSICS AND SPACE SCIENCE LIBRARY

A SERIES OF BOOKS ON THE RECENT DEVELOPMENTS
OF SPACE SCIENCE AND OF GENERAL GEOPHYSICS AND ASTROPHYSICS
PUBLISHED IN CONNECTION WITH THE JOURNAL
SPACE SCIENCE REVIEWS

VOLUME 71
PROCEEDINGS

STUDY OF TRAVELLING INTERPLANETARY PHENOMENA 1977

PROCEEDINGS OF THE L. D. DE FEITER MEMORIAL SYMPOSIUM
HELD IN TEL AVIV, ISRAEL, JUNE 7–10, 1977

Sponsored by

SPECIAL COMMITTEE ON SOLAR TERRESTRIAL PHYSICS (SCOSTEP)
COMMITTEE ON SPACE RESEARCH (COSPAR)
INTERNATIONAL ASTRONOMICAL UNION (IAU)

Edited by

M. A. SHEA and D. F. SMART
Air Force Geophysics Laboratory, Bedford, Massachusetts, U.S.A.

and

S. T. WU
University of Alabama, Huntsville, Alabama, U.S.A.

D. REIDEL PUBLISHING COMPANY
DORDRECHT-HOLLAND / BOSTON-U.S.A.

Library of Congress Cataloging in Publication Data

L. D. de Feiter Memorial Symposium on the Study of Travelling Interplanetary Phenomena, Tel-Aviv, Israel, 1977.
Study of travelling interplanetary phenomena/1977.

(Astrophysics and space science library ; v. 71)
Includes bibliographies and index.
1. Sun – Congresses. 2. Solar wind – Congresses. 3. Solar activity – Congresses. 4. Cosmic rays – Congresses. I. Feiter, Leendert Dirk de. II. Shea, M. A., 1937 - III. Smart, D. F., 1935 - IV. Wu, Shi Tsan, 1933 - V. International Council of Scientific Unions. Special Committee on Solar–Terrestrial Physics. VI. International Council of Scientific Unions. Committee on Space Research. VII. International Astronomical Union. VIII. Title. IX. Series.
QB521.L18 1977 523.7 77–17121
ISBN-13: 978-94-010-1281-2 e-ISBN-13: 978-94-010-1279-9
DOI: 10.1007/978-94-010-1279-9

Published by D. Reidel Publishing Company,
P.O. Box 17, Dordrecht, Holland

Sold and distributed in the U.S.A., Canada and Mexico
by D. Reidel Publishing Company, Inc.
Lincoln Building, 160 Old Derby Street, Hingham,
Mass. 02043, U.S.A.

TABLE OF CONTENTS

Foreword IX

In Memoriam XI

I. SOLAR PHYSICS AS IT CONCERNS
THE STUDY OF TRAVELLING INTERPLANETARY PHENOMENA (STIP)

Mass Ejections from the Solar Corona into Interplanetary Space
E. Hildner 3

Energetic Solar Flare Particles and Interplanetary Shock Waves
R. P. Lin 23

Dynamic Modeling of Coronal and Interplanetary Responses to Solar Events
S. T. Wu, Y. Nakagawa, and M. Dryer 43

II. SOLAR RADIO ASTRONOMY AND
INTERPLANETARY SCINTILLATIONS

Solar Stereo Radioastronomy
J.-L. Steinberg 65

Measurements of the Solar Wind Using Spacecraft Radio Scattering Observations
R. Woo 81

Observations of Interplanetary Scintillation: Solar Wind Velocity Measurements
T. Kakinuma 101

Scintillation Observations of the Interplanetary Plasma
Z. Houminer 119

III. SOLAR WIND THEORY AND OBSERVATIONS

The Three-Dimensional Structure of the Interplanetary Medium
W. I. Axford 145

Theoretical Contributions to Solar Wind Research - A Review
S. Cuperman 165

The Large Scale and Long Term Evolution of the Solar Wind Speed Distribution and High Speed Streams
Devrie S. Intriligator 195

Pioneer 10, 11 Observations of Evolving Solar Wind Streams and Shocks Beyond 1 AU
E. J. Smith and J. H. Wolfe 227

Wave-Particle Interaction Phenomena Associated with Shocks in the Solar Wind
F. L. Scarf 259

IV. COMETARY TOPICS

Comets in the STIP Context
Max K. Wallis 279

The Comet-Solar Wind Interaction
D. A. Mendis 291

Solar Wind Interaction with Type-1 Comet Tails
A. I. Ershkovich 305

V. SOLAR COSMIC RAY PROPAGATION, MODULATION, AND INTERPLANETARY ACCELERATION

Signatures of Solar Cosmic Ray Events and Their Relation to Propagation and Acceleration Processes
G. Wibberenz 323

Solar Energetic Particles Below 10 MeV
E. C. Roelof and S. M. Krimigis 343

Acceleration and Modulation of Electrons and Ions by Propagating Interplanetary Shocks
T. P. Armstrong, G. Chen, E. T. Sarris, and S. M. Krimigis 367

VI. STUDY OF TRAVELLING INTERPLANETARY PHENOMENA - STIP INTERVAL I AND STIP INTERVAL II

A Summary of Significant Solar-Initiated Events During STIP Intervals I and II
M. A. Shea, D. F. Smart, and H. E. Coffey 393

Overview of the HELIOS 1 and HELIOS 2 Missions and Their Participation in STIP Intervals I and II
H. Porsche 421

Index of Authors Cited 431

LEEN DE FEITER

1927 - 1975

FOREWORD

The Study of Travelling Interplanetary Phenomena (STIP) was formally established by the International Council of Scientific Unions' Special Committee on Solar-Terrestrial Physics (SCOSTEP) in August 1973 with M. Dryer as Convenor and M. A. Shea as Secretary. The scientific objectives of STIP are the study and search for understanding of quiet (i.e. normal or background) and active periods in the interplanetary medium. The concepts of informal, extemporaneous interdisciplinary research is continuously emphasised, and these concepts have proved to be extremely successful in conducting the very productive studies undertaken by the members.

About 200 scientists are actively participating in STIP, their interests ranging from solar physics (insofar as it concerns the initiation of phenomena which move out from the Sun) to the observation and study of comets and planetary magnetospheres and ionospheres. Solar wind plasma and fields, solar and galactic cosmic rays, interstellar interactions, solar radio astronomy and interplanetary scintillations of discrete radio sources are among the topics of interest.

A number of subgroups within STIP have been organized and are headed by coordinators from several different countries. Almost all of the coordinators conduct and carry out informal studies, quickly and efficiently within their groups, with a minimum of formalities. Both the informal character of the entire STIP program and the enthusiastic response to the interdisciplinary approach of the various studies, have contributed to the productivity and success of this group over the past four years. For example, two scientific sessions devoted to travelling interplanetary phenomena studies were held in conjunction with the XVIII COSPAR Meeting in Varna, Bulgaria in 1975; a special issue of *Space Science Reviews* was published in 1976 presenting review papers on the August 1972 solar-terrestrial phenomena; and three special periods for scientific studies, STIP Intervals I, II, and III, have been designated, with selection of these intervals based upon unique opportunities generated by interplanetary spacecraft configurations. So far unique observations have been possible during both STIP Intervals II and III, primarily because of the solar activity (and hence related interplanetary effects) that occurred at those times.

The L. D. de Feiter Memorial Symposium on the Study of Travelling Interplanetary Phenomena was still another successful activity conducted by STIP participants. Held in conjunction with the XX COSPAR Meeting and co-sponsored by SCOSTEP, COSPAR, and the IAU, the symposium was named in honor of the memory of Leen Dirk de Feiter, who actively encouraged and participated in the organization and initial activities of STIP.

The Proceedings of the L. D. de Feiter Memorial Symposium on the Study of Travelling Interplanetary Phenomena are being published in two parts. The invited papers are all contained in this volume; the contributed papers are being published by the Air Force Geophysics Laboratory in a report *Contributed Papers to the Study of Travelling Interplanetary Phenomena/1977* also edited by M. A. Shea, D. F. Smart, and S. T. Wu. In an effort to produce a completed proceedings in a timely fashion, all authors were requested to prepare their manuscripts in a camera ready format, following general guidelines provided by the publisher and the editors. Furthermore, the authors were requested to provide these completed manuscripts by the end of the conference. The editors gratefully acknowledge the cooperation given to them by all authors in complying with these various requests. In the process of reviewing and editing these proceedings, changes in some manuscripts were found to be necessary, and in a few cases, complete manuscripts were retyped. In this regard the typing services of Ms. Carol Holladay of the University of Alabama are gratefully acknowledged.

The papers in this volume are generally of a review nature and the authors have found it necessary to reproduce figures previously published in scientific journals. On behalf of the authors, we wish to acknowledge the following for permission to use previously published material: American Astronomical Society (*Astrophysical Journal*); American Geophysical Union (*Geophysical Research Letters, Journal of Geophysical Research*); Pergamon Press (*Planetary and Space Science*); and the D. Reidel Publishing Company (*Solar Physics*).

M. A. Shea
D. F. Smart
S. T. Wu

July 1977

L. D. de FEITER - IN MEMORIAM

The excellent initiative to organize a symposium on Travelling Interplanetary Phenomena was already conceived a couple of years ago, in the framework of SCOSTEP, the Special Committee for Solar-Terrestrial Physics. Further discussions took place at several occasions, particularly also during the COSPAR Meeting in Varna, Bulgaria, in 1975. Sponsorship was accepted by the COSPAR plenary in its Philadelphia Meeting in June 1976 and a first general announcement was circulated in August 1976. In the meantime it had been decided that this SCOSTEP-COSPAR symposium would also be cosponsored by the International Astronomical Union.

Valuable contributions to the first discussions, in Varna and even earlier, at the Sao Paulo Plenary of COSPAR, were given by the late Dr. Leendert Dirk de Feiter, and it is with a deep feeling of sorrow that we miss him here. The suggestion to dedicate this Symposium to his memory originated among several of those who contributed to the first discussions of this Symposium, and was not only thankfully accepted by the COSPAR Plenary, during the Philadelphia Meeting, but also by the other sponsoring organizations.

Leendert Dirk de Feiter was born on August 17, 1927 in Renesse, The Netherlands, a village behind the sand dunes that protect the island of Schouwen, in the province of Zeeland, from the North Sea. The family was far from wealthy; on the contrary from his young years on, Leen de Feiter was confronted with the fact that the struggle for life demands hard and continuous labour. It took sacrifices from the family to send their gifted son to secondary school, but university study was beyond their reach. Thus Leen de Feiter had to find his own way, through a job as an assistant at the Royal Meteorological Institute at De Bilt in 1947, from where he could make the jump to University study in 1951, a study which he could only assume as a part-time activity. Nevertheless he passed his Doctoral Examination in 1957, cum laude, a judicium that is only rarely given. In the meantime he had assumed a job at the Dutch Postal, Telephone and Telegraph Service (in the Department of Ionospheric and Solar Radio Science); in 1964 he became a staff member of the Astronomical Observatory in Utrecht. In October 1966 he successfully defended a thesis on "The Analysis of the Balmer Spectrum of Solar Flares". Thereafter he spent a year at the Sacramento Peak Observatory in the U. S. A., and this was followed, upon his return in 1967, by an appointment as Head of the Section for Interpretation and Theory at the Space Research Laboratory at Utrecht. Once again a period of hard labour started, since the establishment of a new section to a young laboratory was demanding, initially not very rewarding, and left very little time for purely scientific activities.

Only after a number of years could Leen de Feiter successfully return to scientific research, and there were all indications that a period of fertile scientific productivity had started, when he was most untimely taken away, from heart failure on September 2, 1975, while at his home in Bosch en Duin near Utrecht. Just earlier that year the University had appointed him as an Associate Professor, a rather late appreciation for a life of devoted scientific labour.

Dr. de Feiter was strongly involved in matters of international scientific cooperation. From 1961 until 1966 he was Secretary of the International Ursigram and World Days Service (IUWDS); during the period 1969-1973 he was President of SPARMO (now SBARMO - Scientific Ballooning and Radiations Monitoring Organization) and from 1973 to the time of his death he was Vice-President of that institution. He was a member of the International Astronomical Union and at the time of his death he was serving in the Organizing Committee of its Commission 10, on Solar Activity.

For many years Dr. de Feiter gave important contributions to the activities of COSPAR. He was the official representative of the Royal Netherlands Academy of Sciences on COSPAR, the Vice-Chairman of Panel 3.B on Solar Flares and Forecasts, and a member of Panel 4.C on the Polar Ionosphere. In addition, he served as an advisor to the COSPAR Finance Committee. He was the author or co-author of a good number of scientific papers, many of which were presented during COSPAR Meetings.

To me, personally, Leen de Feiter was not only one of my closest co-workers, but also an honest and sincere friend with whom it was good to cooperate. To his dynamic, sympathetic and open personality he added sincerity in human relations, and devotion to work and science. His death was, also for these reasons, a great loss to the Space Research Laboratory in Utrecht. It is therefore also in the name of our Laboratory, as well as for his wife Jose de Feiter, and their children, that I would like to thank the Programme Committee of this Symposium for having proposed that this Symposium be called the "Leendert Dirk de Feiter Memorial Symposium on Study of Travelling Interplanetary Phenomena".

Cornelis de Jager
President of COSPAR

PART I

SOLAR PHYSICS AS IT CONCERNS THE STUDY OF TRAVELLING INTERPLANETARY PHENOMENA (STIP)

MASS EJECTIONS FROM THE SOLAR CORONA INTO INTERPLANETARY SPACE

E. HILDNER
High Altitude Observatory, National Center for Atmospheric Research*, Boulder, Colorado 80307, U.S.A.

*The National Center for Atmospheric Research is sponsored by the National Science Foundation

ABSTRACT

Mass ejections from the corona are common occurrences, as observations with the High Altitude Observatory's white light coronagraph aboard Skylab showed. During 227 days of operation in 1973 and 1974 at least 77 mass ejections were observed, and as many more probably occurred unobserved. It is suggested that the frequency of ejections varies with the solar cycle and that ejections may contribute 10 percent or more of the total solar mass efflux to the interplanetary medium at solar maximum. Since ejections are confined to relatively low latitudes, their fractional mass flux contribution is greater near the ecliptic than far from it. From the behavior of ejecta, we can estimate the magnitude of the force driving them through the corona. It is also suggested that loop-shaped ejections -- the largest fraction of ejections -- are driven, primarily, by magnetic forces. By comparison, gas pressure forces are negligible, and forces due to wave pressure are completely inadequate. That magnetic forces are important is consistent with observations that ejections seem to come, primarily, from regions where the magnetic field is more intense and more complex than elsewhere. Indeed, ejections are associated with phenomena (flares and eruptive prominences) which occur over lines separating regions of opposite polarities.

I. INTRODUCTION

Mass ejections from the solar corona were seen at naturally occurring eclipses at least three times between 1860 and 1941 (Eddy, 1974; MacQueen, 1977; Shklovskii, 1965), but were not recognized for what they were until the advent of satellite-borne coronagraphs. The Naval

M. A. Shea et al. (eds.), Study of Travelling Interplanetary Phenomena/1977, 3–21.

Research Laboratory's coronagraph (Tousey, 1973) on board the OSO-7 satellite obtained images of several ejection events (Howard et al., 1976), and the High Altitude Observatory's coronagraph (MacQueen et al., 1976) on board the Skylab satellite obtained many more. This paper reviews recent results arising, primarily, from analysis of Skylab coronagraph images. An example of the Skylab observations is shown in Figure 1; such an event typically consists of 6×10^{15} g of material with energy 12×10^{30} erg. Gosling (1976) has reviewed several aspects of mass ejections not touched upon here, and the reader is referred to that work for additional information.

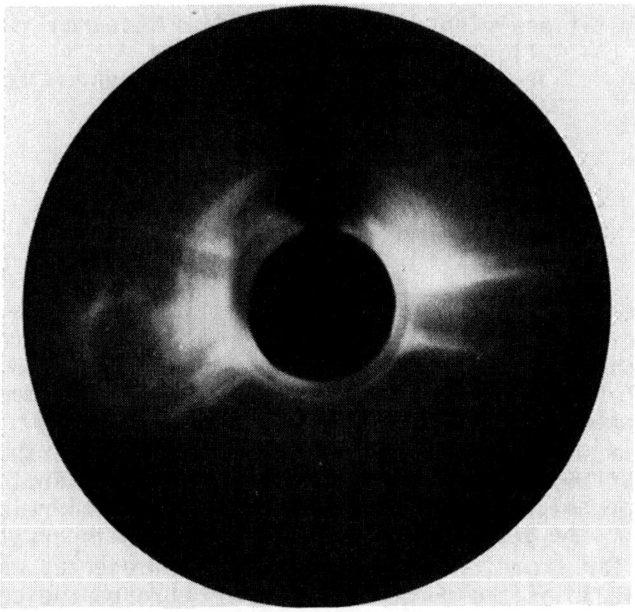

Figure 1
A coronal mass ejection photographed at 1448 GMT on 10 August 1973. The High Altitude Observatory's white light coronagraph aboard Skylab had an annular field of view ranging from the occulting disk at the center ($\sim$1.5 $R_\odot$ in radius) to approximately 6 $R_\odot$ at the outer edge. The strong gradient of Thomson-scattered coronal radiance has been suppressed in this image by instrumental vignetting and by dodging the print. From Gosling et al. (1975a).

This necessarily limited review is organized around two major themes. First, results are presented which allow us to determine that mass ejection coronal transients are important to our understanding of the interplanetary medium. Second, results and ideas are presented which contribute to current belief that magnetic forces, primarily, drive mass ejections from the corona. The final section of the paper is a brief summary.

II. MASS EJECTIONS' IMPORTANCE TO THE INTERPLANETARY MEDIUM

The average mass and energy, coupled with the frequency of occurrence, determine whether and to what degree transient mass ejections are important to our understanding of interplanetary space and the medium which fills it. In this section, we present results which indicate that mass ejections from the solar corona contribute a non-negligible fraction of the mass flux from the sun, especially towards the heliographic equator near solar activity cycle maximum.

A. Mass and Energy

Subtraction, point by point, of a pre-event image from one which shows an ejection results in a digital image which contains only contributions due to the ejection event. We convert the resulting brightness differences to columnar mass densities using the known properties of Thomson scattering of photospheric light. Integrations over the subtracted image then yield 1) the total amount of additional mass brought into the coronagraph's field of view by the ejection and 2) its potential energy (Hildner et al., 1975a). The speed of the disturbance is determined by measuring the height of the leading edge as a function of time. Since there appears to be a distribution of speeds within an ejection, we take one-half the speed of the leading edge to be appropriate for the calculation of kinetic energy. Table I presents, for 13 large ejection events, the mass, the leading edge, speed, and the total (potential plus kinetic) energy.

One surprising feature of this table is the energy expended in mass ejections associated with eruptive prominences. Prior to observations from orbiting coronagraphs that mass ejections often accompany eruptive prominences, the energy expended to cause a prominence eruption was taken to be just the energy required to lift the Hα-emitting, cool material (e.g., Tandberg-Hanssen, 1974). Since cool prominence material makes up only a small fraction (approximately 10%) of the total mass expelled (Hildner et al., 1975b; Schmahl and Hildner, 1978), total energy expenditures in eruptive events were vastly underestimated by this partial approach. Additionally, we note Table I indicates that the average energy of non-flare associated ejections studied to date is 6.5×10^{30} erg; this energy is about equal to the total liberated in an "average" flare, according to Švestka (1976). The largest, flare associated mass ejections can be quite energetic; from solar wind measurements at 1 AU, it can be inferred that they carry away approximately half the total energy released in a large flare (Hundhausen, 1974).

The mass given in Table I for each event is the additional material in the coronagraph's field of view at the time of a picture during the event. Subsequent outward flow of material is difficult to measure. We expect that outward flow continues for some time after the leading edge of an ejection leaves the field of view, so that the mass values in Table I represent lower limits on the amount of mass delivered to the

Table 1

TRANSIENT COMPARISON

ASSOC*	DATE 1973/74	MASS† (10^{15}g)	SPEED (km s^{-1})	ENERGY° (10^{30} erg)	REFERENCE
FLA**	11 JAN	17	620	10	Stewart et al., 1974a
FLA**	11 JAN	8	750	?	Stewart et al., 1974b
EPL	10 JUN	5	500	9.4	Hildner et al., 1975a
EPL	10 AUG	5	400	8.0	Gosling et al., 1975a
EPL	13 AUG	2	∼175	2.3	
EPL	21 AUG	8	530	14.3	Poland and Munro, 1976
EPL	26 AUG	1	100	2.0	Hildner et al., 1976
FLA	7 SEP	24	≥960	69.8	Gosling et al., 1975b
FLA?	15 SEP	2	∼720	3.8	Dulk et al., 1976
EPL	19 DEC	2	140	2.9	Schmahl and Hildner, 1978
FLA?	10 JAN	1	400	1.6	Munro et al., 1977
FLA	12 JAN	3	∼400	5.4	
FLA	17 JAN	3	850	12.8	
AVERAGES					
FLA	(inc 15 Sep and 10 Jan)	8.3	671	17.2	
EPL		3.8	308	6.5	
ALL		6.2	503	11.9	

Notes: *FLA = flare; EPL = eruptive prominence at the limb.
†Total excess (event - pre-event) mass in the corona within the coronagraph's field of view.
°Total = potential plus kinetic energies.
**Observed with the OSO-7 coronagraph of the Naval Research Laboratory.

interplanetary medium by each event.

B. Frequency of Ejections

If coronal mass ejections were rare, they would be only local perturbations of marginal importance to the global, long term interplanetary medium. But mass ejections are not rare. At least 115 coronal transients, among them 77 definite mass ejections, were imaged during the Skylab coronagraph's 227 observing days in late 1973 and early 1974 near the minimum of Solar Cycle 20 (Munro et al., 1978). From geometry alone, we infer that about twice as many transients occurred as were seen; any correction for the intermittent nature of the coronagraph's observations would further increase the number of inferred, unobserved transients.

Hildner et al. (1976) have shown that the frequency of occurrence of coronal transients from bands of longitude during Skylab was well correlated with the sunspot number and plage area times brightness index appropriate to each band. That is, the more active regions present, the more frequent are mass ejections from an area. If the relationship between transients and sunspots is independent of phase in the solar cycle, then transients occur in proportion to the sunspot number; extrapolation from Skylab data implies that the entire sun would produce >100 transients per month near solar activity maximum.

Radio bursts may provide another clue to the variation of mass ejection frequency over the solar cycle. Although only a small percentage of Skylab period transients were accompanied by metric radio bursts, nearly all (21 of 23) metric Type II and/or IV radio bursts arising from near the limbs of the sun were accompanied by mass ejections (Munro et al., 1978). Thus a solar cycle dependence of the occurrence of metric radio bursts might imply a similar fluctuation in the occurrence of mass ejections. Whether the proportion of ejections with radio bursts to those without radio bursts remains constant during the solar cycle is unknown; no data exist which enable this constancy to be evaluated.

Moving Type IV (or IVm) bursts have long been thought to be diagnostic of ejected plasmoids, and Type II bursts seem to be diagnostic of shock waves. (See Wild and Smerd (1972) for descriptions of burst types and their probable sources.) Figure 2 shows II and IVm burst occurrences during Solar Cycle 20. A solar cycle dependence is strongly evident in Type II and IVm occurrences, tending to support our speculation that the frequency of mass ejections -- or at least those associated with radio bursts -- depends upon phase in the solar cycle.

C. Events' Sizes and Trajectories

The tendency for Skylab transients 1) to occur at low latitudes, rather than uniformly over the sun, and 2) to rise through the corona on paths inclined slightly equatorward from vertical, means that mass

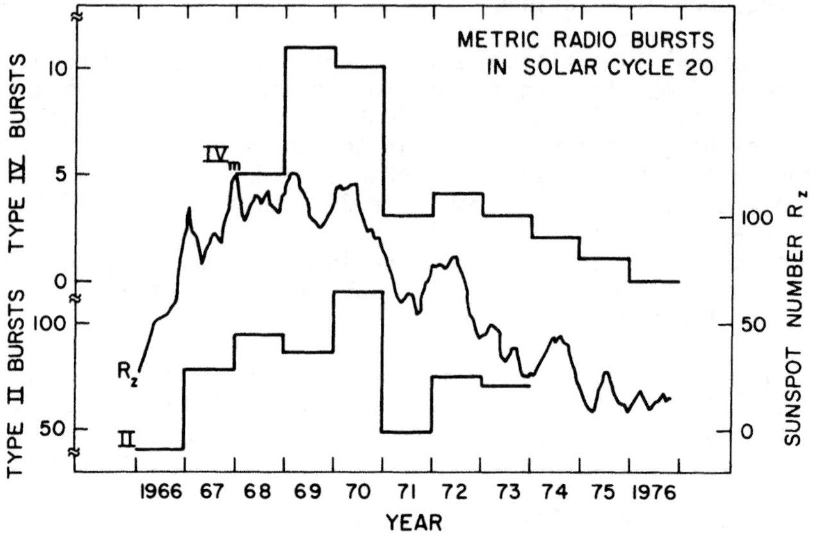

Figure 2
Solar cycle dependence of radio burst occurrence. The histograms show that the number of bursts of each type observed per year is well-correlated with the curve, labeled R_Z, of the 3-month running mean of monthly sunspot numbers. The type II data are from Dodge (1974); the type IVm data include only those events identified with the Culgoora radioheliograph of C.S.I.R.O. (Stewart and McLean, 1977).

ejections are even more important to the near-ecliptic interplanetary medium than the numbers presented in the preceding sections would indicate.

Somewhat surprisingly, as seen from sun center, mass ejection events during Skylab were relatively low-latitude phenomena and did not subtend increasingly larger solid angles as they progressed from 2 to 6 $R_\odot$ -- quite the converse. Figure 3 shows how the latitudes of the north and south edges of transients were measured. From these measurements, the latitudinal extent and mid-latitude at 2 and 4 $R_\odot$ can be determined.

The distribution of latitudinal extents shows that very few events were larger than 65° in apparent extent. The mid-latitudes of coronal transients were confined below 50° latitude, with most below 40°, and were distributed approximately symmetrically about a peak at the solar equator.

Figure 4 shows the results of taking the radio of latitudinal extents of each transient as measured at 2 and 4 $R_\odot$. As Figures 3 and 4 indicate, the typically straight northern and southern edges of a

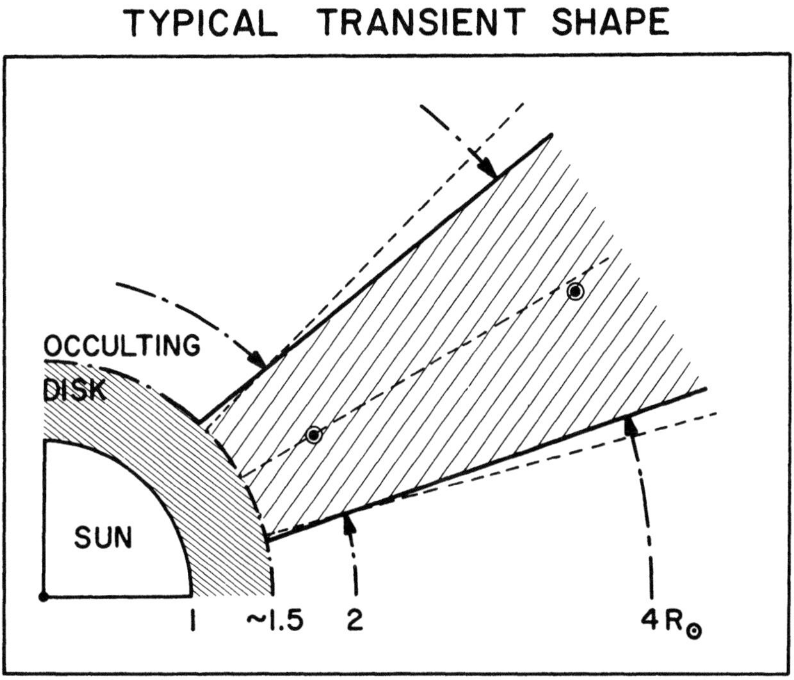

Figure 3
Schematic to show how angles were measured. The apparent latitudes of the northern and southern edges of each transient were measured at heights of 2 and 4 $R_\odot$. At each height, the difference in apparent latitudes is the latitudinal extent, and the mid-latitude of the transient at that height is represented by a circled dot. The dashed lines extend radially from sun center. As shown in this drawing, a transient's edges tend to be non-radial, and the mid-latitude tends to decrease with height. In most cases, very little lateral (Latitudinal) movement of transient edges is observed until many hours after the leading edge of the transient leaves the Skylab coronagraph's field of view.

visible transient tended to be closer together than if they were radial.

The mid-latitude of each transient tended to vary as a function of height. However, the variation of a fully formed ejection's mid-latitude as a function of height may either indicate simple, non-radial, straightline propagation or it may indicate a deflection, a curvature, of the ejection's outward path. For whatever reason, transients' mid-latitudes during Skylab tended to incline from local vertical toward the equator, as the histogram of Figure 5 shows. This tendency for the axes of transients to depart from the vertical is in addition to the observed departure of the legs or edges of transients from vertical (Figure 4).

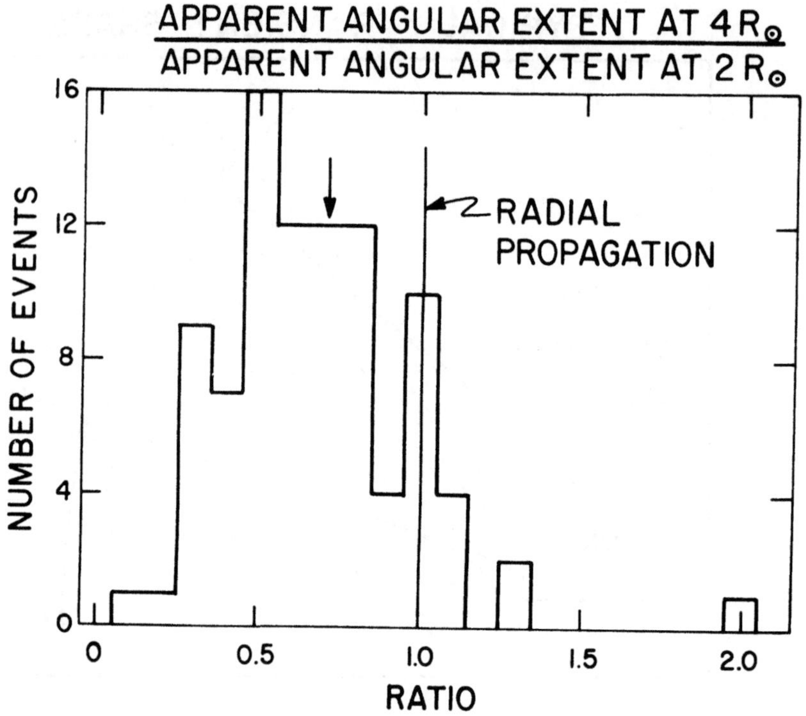

Figure 4
Ratio of transients' extents at two heights. The histogram shows that most transients were narrower in latitudinal extent at 4 $R_\odot$ than at 2 $R_\odot$. The arrow points to the average value. Thus, the northern and southern edges of almost every transient were inclined toward each other from the radial. See Figure 3 for an example.

It is tempting to speculate that the overall magnetic configuration of the solar field produced lateral pressure forces responsible for the equatorward average deflection of transients during Skylab. With large unipolar magnetic regions (coronal holes) of opposite polarities around north and south poles of the sun during the entire 9-month period (Nolte et al., 1976; Bohlin and Rubenstein, 1975), perhaps transients felt a push toward the "symmetry plane" of the large-scale magnetic field.

D. Fractional Contribution to the Interplanetary Medium

With the values presented above, we can estimate the contribution that mass ejections make to the interplanetary mass flux. Let us compare the interplanetary mass contribution of ejections to the average mass flux through the 1 A U sphere ($\sim 1.9 \times 10^{17}$ g day^{-1} if a particle

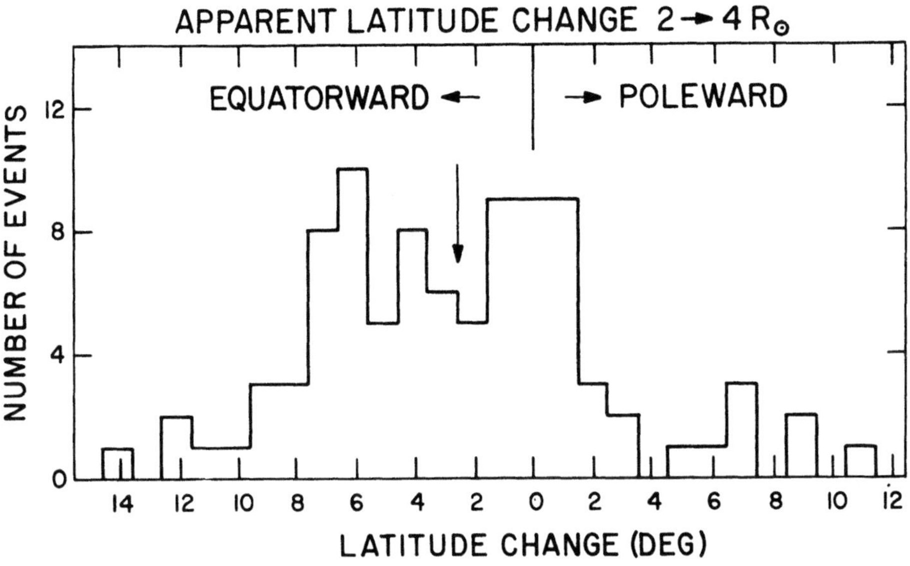

Figure 5
Inclination of transients' axes. The 4 $R_\odot$ mid-latitude, measured as shown in Figure 4, was compared to the 2 $R_\odot$ mid-latitude for each transient. As the arrow shows, on average, the mid-latitude was $\sim 2.5^0$ closer to the equator at 4 $R_\odot$ than at 2 $R_\odot$.

flux of 3.8×10^8 cm^{-2} sec^{-1} (Feldman et al., 1977) is appropriate to the entire 1 A U sphere). A very conservative estimate is that twice as many transients occurred during Skylab as were observed, or approximately 1 per day around the sun. If each of these transients expelled the average 6×10^{15} g of coronal material to the interplanetary medium, then mass ejections contributed 3% of the mass escaping through the 1 A U sphere during Skylab. Near solar maximum, ejections might contribute $\gtrsim 10\%$ of the outflowing mass. And, if mass ejections are confined to latitudes below 40^0, then their fractional contribution to the solar wind mass flux past Earth should be 16%.

III. FORCES DRIVING MASS EJECTIONS FROM THE CORONA

Any one or a combination of gas pressure, magnetic forces, or pressure due to waves might serve to drive coronal mass ejections outward. In this section, we present observations and their interpretations which lead to the belief that magnetic forces are primary, at least for loop-shaped transients, as ejections traverse the corona from 2 to 6 $R_\odot$. Among other things, the Skylab observations show that transients were associated with regions of increased magnetic field and with separations of unipolar regions of the field; that the forces driving transients outward acted to great heights long after the onsets of the events; and, for some events, we can infer that the magnetic energy density within the ejecta far exceeded the thermal energy density, that is, the behavior of the ejecta was magnetically controlled. The following subsections treat these topics in more detail.

A. Ejections' Associations With the Near-Surface Magnetic Field

Soon after the mission ended, it was noted that the source regions of coronal transients observed during the latter part of the Skylab era were not uniformly distributed in solar longitude, as shown in Figure 6. Indeed, 74% of these transients came from the hemisphere centered on Carrington longitude 180°; this hemisphere exhibited many more active regions than did the opposite hemisphere, as the sunspot curve in Figure 6 indicates. The extremely good correspondence between the number of transients arising from and the number of sunspots or active regions existing in each band of longitude implies that most transients are associated with active regions. That is, few coronal mass ejections originate far from active regions.

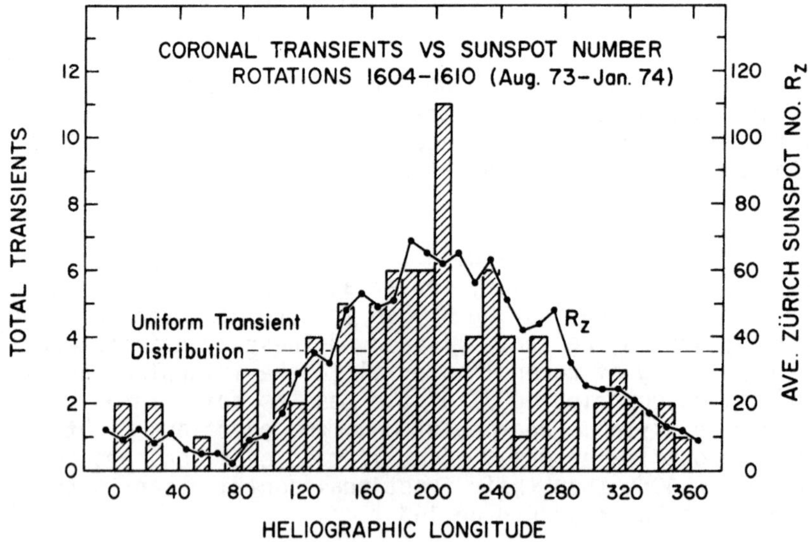

Figure 6
Longitudes of occurrence of transients during seven solar rotations. The broken curve represents the Zürich sunspot number for the hemisphere centered at each longitude, and the histogram shows the number of transients emanating from each 10° band of longitude during seven solar rotations. From Hildner et al. (1976).

As mentioned earlier, a recently completed study by Munro et al. (1978), identifies 77 definite mass ejections among the 115 transient events detected during the Skylab coronagraph's 227 observing days. The fainter events not classified as definite ejections might have been ejections which were weak or far from the plane of the sky, or perhaps they were only lateral rearrangements of material. In this paper "ejection" refers to one or more of the 77 definite ejections, "transient" applies to all transient events whether or not they are classified as ejections. Of the definite ejections, about half (those originating from the front side of the sun?) could be associated with events classically considered solar activity such as flares, eruptive prominences, etc. About twice as many ejections were associated with eruptive prominences as with flares; there was a one to one correspondence between flare-associated coronal mass ejections and flares with ejections of Hα material. On the basis of the ejections associated with flares and eruptive prominences, it appears that three-quarters of mass ejections originated in or above active regions. Although the exact connection between X-ray transient brightenings and white light ejections is not clear, we note in passing that three-quarters of the coronal X-ray transients observed with the American Science and Engineering X-ray telescope aboard Skylab (Vaiana et al., 1976) appeared to originate in active regions (Rust and Webb, 1977), while only one-quarter originated outside active regions (Webb et al., 1976).

The close association between active regions, rather than polar crown filaments, and mass ejections is consistent with our finding that transients during Skylab were largely confined to low latitudes. From this close association, we infer that more rapid evolution of the more intense, complex, and large-scale magnetic fields presumed to occur above active regions is important in causing ejections.

Since ejections are associated with classical forms of solar activity which take place there, we suppose that ejections originate near the lines which bound relatively large-scale, photospheric, unipolar, magnetic regions. That is, above so-called neutral lines. The region from which the ejected mass arises is below an orbiting coronagraph's field of view, as depletions of matter from the low corona show. These depletions have been observed with K-coronameter (Stewart et al., 1974a, b; Hansen et al., 1974), X-ray (Rust and Hildner, 1976), and green-line (Fisher, 1977) telescopes. The magnetic field in the region of depletion can not yet be observed. However, the Hα and X-ray manifestations accompanying ejections always occur near photospheric neutral lines (Munro et al., 1978; Webb et al., 1976; Rust and Webb, 1977).

B. Events' Shapes

The "typical" structure of a phenomenon often yields clues about the physics at work; as in a rainbow, for example. Unfortunately, we do not yet understand how to interpret or even define a "typical" morphology for coronal mass ejections. Of the 77 definite mass ejections observed with the Skylab coronagraph, about 1/3 appear to be loops;

that is, these are obvious, bright, relatively thin loops of emission rising through the corona. Loop-like ejections tend to be associated with eruptive prominences (Munro, 1977). About 1/4 of the ejections can be classified as clouds or amorphous blobs, and the remaining events display miscellaneous forms (Munro, 1977). An ejection could be comprised of a collection of loops of different orientations; such an event would not appear as and would not be classified as a loop. The apparent structure of a mass ejection depends somewhat upon the time at which the event was imaged. During some well-observed events, an initially obscure loop structure becomes more obvious as the material rises.

One mass ejection, observed 10 Jan 1974, appears to be a loop seen edge-on (Munro, 1977). That is, polarization and brightness measurements of this event rule out a shape which is cylindrically symmetric about an axis standing approximately vertical in the solar atmosphere; the measurements imply a near-planar loop seen in projection so that the two legs overlap.

The inference that many, if not all, mass ejections have the character of loops is consistent with our idea that the shape and ejecting force are usually magnetically controlled. X-ray photographs show that the lower corona is densely populated by the loops; these loops are thought to take their shape from the pervasive magnetic field. It is likely that one or more of these low coronal loops rises to become the loop-like transients seen in the outer corona with white light coronagraphs (Rust and Hildner, 1976).

Although it seems improbable that loop-shaped ejections could arise from a point source explosion (flare) in the low corona, some ejections do not appear as loops. The work of Dryer, Nakagawa, Steinolfson, Wu and their collaborators (see papers by Wu et al. in this volume and by Steinolfson et al., 1977 for references) may be appropriate to these non-loop transients.

C. Speed vs. Height

Gosling et al. (1976) recently published a study of the speeds of mass ejections. With an improved technique to measure the maximum height of the leading edges, it has been possible to determine speed as a function of height for a few more events and, in some cases, to apply slight corrections to the already published speeds. The results are shown in Figure 7.

In this plot, only the curve for 17 January indicates deceleration; the 17 January deceleration is approximately that expected for a particle which starts with the initial speed and at the height of the transient, and is acted upon thereafter by gravity alone. The remaining events show constant or accelerating speeds. Note that the initial accelerations must have been quite large to enable originally stationary material to get from near the surface to the first observed speed and height plotted in Figure 7. For a rising mass to exhibit constant or increasing speed requires that the net force upon it be zero or outward directed and increasing, respectively.

D. Driving Forces

Generally speaking, by the time ejecta rise into the field of view of an orbiting coronagraph, a great deal of the most interesting physics has already occurred and the major portion of kinetic and potential energy is already imparted. That is, most of the speed has been acquired and the climb out of the sun's gravitational well has been largely accomplished. Nonetheless, the curves of Figure 7 imply the existence of an outward force acting at great heights above and long times after the near-surface initiations of ejections. Perhaps these later forces, acting at 2 - 6 $R_{\odot}$, tell us something about the forces acting earlier near the surface. Gas pressure, magnetic forces, and wave pressure have been suggested to account for the propagation of loop-shaped ejections.

The loop-shaped mass ejection event of 15 September 1973 was well observed with the Skylab coronagraph, and its accompanying Type IV radio bursts were recorded at Culgoora, Australia and Boulder, Colorado. Dulk et al. (1976) have combined the radio and white light data to analyse this event. They suggest that the ejected mass was driven outward because an inward gradient of magnetic field strength existed in and around the transient. To account for the near-constancy of the width of the leading loop, they invoke a component of the magnetic field circumferential around the loop, giving rise to a helical field in the loop. When the loop was at 3.1 $R_{\odot}$, these considerations imply a helical field of strength $\sim$1.7 G and pitch angle 30° within the loop. The ratio of thermal to magnetic energy density within the loop at that height was $\beta \simeq 0.007$, indicating that gas pressure is certainly negligible in comparison with magnetic pressure, and the event was magnetically controlled.

A somewhat different proposal has been put forward by Mouschovias and Poland (1977) to explain the non-ballistic outward motion of loop transients. In their model, as in the Dulk et al. picture, a transient loop carrying ejected mass contains a helical field. However, Mouschovias and Poland suggest that the outward force arises from the circumferential rather than the axial field of the loop. The ambient field around the loop is considered unimportant. To explain the observed behavior of the event shown in Figure 1, with its slowly widening loop, Mouschovias and Poland find that a field strength of 1.0 G with pitch angle of 50° when the top of the loop is at 2 $R_{\odot}$ is appropriate. At this height, their model and the observed density suggest that the ratio of thermal to magnetic energy density is $\beta \simeq 0.2$. Thus, they conclude, in agreement with Dulk et al. (1976), that the gas pressure is unimportant in mass ejections driven outward by magnetic forces.

The third of the suggested forces, that due to the presence of waves which might deposit their energy or momentum in the ejecta is

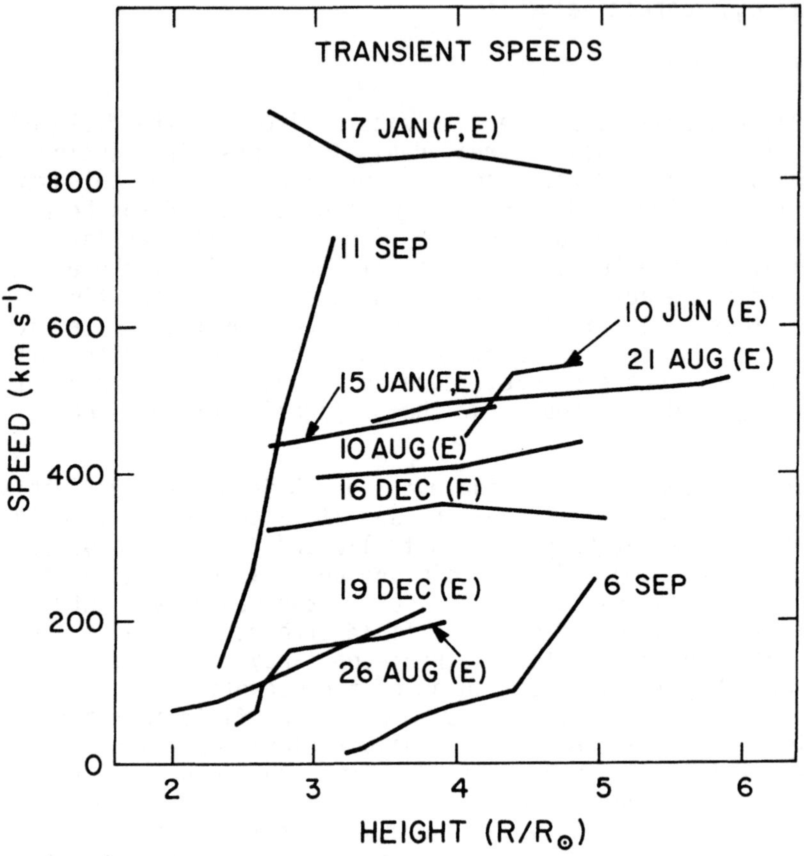

Figure 7
Speeds of loop transients' leading edges versus distance from sun center. Each point is plotted at the mid-height of the interval used to determine it. The points for each event are connected to guide the eye; breaks in the trajectory for an event undoubtedly arise from measurement uncertainties, which range from ±50 to ±100 km s^{-1}. All but the January (1974) events occurred in 1973. An F or E after the date indicates that the ejection was associated with a flare or eruptive prominence, respectively. After Gosling et al. (1976), as revised by Jackson (1977).

easily shown to be inadequate. An upper limit on the wave flux coming through the transition region into the corona is 10^5 erg cm^{-2} s^{-1} if our understanding of currently available observations is correct (Hollweg, 1972). Let us suppose that the footpoints of the loop at the leading edge of the ejection seen in Figure 1 intersect the transition region in an area 0.1 x 0.1 $R_\odot^2$, surely an overestimate. Then the energy available to drive the ejection by wave dissipation is $<5 \times 10^{24}$ erg s^{-1}.

Ejection events require energy inputs several orders of magnitude larger than this to exhibit the behavior shown in Figure 7. It has also been suggested (Hundhausen, 1976) that non-dissipative waves could exert a force on ejected material. The idea here is that, e.g., Alfvén waves enter a loop at one foot, are guided around the top of the loop, and exit at the other footpoint; as the wave vector is turned, the wave imparts a force to the loop. This force is also orders of magnitude too small to drive the ejections as observed through the Skylab coronagraph's field of view.

In sum, it appears that of the three suggested forces -- gas pressure, magnetic forces, and wave pressure -- only magnetic forces are adequate to explain the observations of loop transients.

IV. DISCUSSION

We have shown that, during the Skylab era, mass ejections from the corona probably contributed at least 3% of the mass flux from the sun. The frequency of transients is expected to increase with overall activity in the solar cycle, so more than 10% of the sun's mass efflux will be in the form of ejections at solar cycle maximum. The relatively low latitudes of origin and the tendency for trajectories to lie equatorward of radial means that ejections affect the interplanetary medium even more strongly near the ecliptic.

Our demonstration that ejections contribute appreciably to the interplanetary mass budget is convincing, but to date conclusive observations of earlier-identified mass ejections past plasma detectors at 1 A U are not available. Unfortunately, coronagraphs by their nature see only events near the solar limbs; most solar wind plasma detectors are in orbit around Earth. Only one ejection has been observed departing the corona and again at its 1 A U passage (Gosling et al. 1975b); that event was associated with a big flare. Using data acquired near Earth, Gosling et al. (1977) have identified non-compressive density enhancement (NCDE) events in the solar wind, apparently discrete clouds of denser, often cooler plasma at times of constant or declining solar wind speeds. They suggest that NCDEs are 1 A U manifestations of coronal mass ejections directed Earthward; the frequency of occurrence, estimated mass, estimated energy, and duration are all about the same for NCDEs as for coronal mass ejections.

Our earlier discussion of forces concentrated upon loop-shaped ejections as they rise through the corona. The initial acceleration near the surface, which may well be impulsive for flare associated events (see the 17 Jan curve in Figure 7), has not been considered, nor have non-loop-shaped ejections. Numerical modeling of coronal relaxation after impulsive perturbations has been carried out by Dryer, Nakagawa, Steinolfson, Wu and their collaborators (see Wu et al., 1977 and Steinolfson et al., 1977 for refs.). Observations are inadequate to specify computational conditions (initial, boundary, perturbation size), but some observed ejection characteristics are successfully reproduced.

Our belief that the most common, loop-shaped ejections are magnetically driven through the corona arises from several observations. 1) At least three-quarters of ejections arise from active regions where the large-scale magnetic field is more intense, more complex, and probably more rapidly varying than elsewhere. 2) Within these magnetic configurations, ejections are associated with lower atmospheric phenomena (flares, eruptive prominences) which only take place near a particular part of the magnetic field, viz., near lines which separate regions of uni-polar, photospheric magnetic field. 3) The loop shape itself is suggestive of a magnetic flux tube. 4) The observed behavior of ejecta as they traverse the corona enables us to estimate the net forces applied, and the magnetic nature of the outward-directed force can be inferred by elimination. Radio observations of one event imply that the magnetic energy density in the event was ∿140 times greater than the thermal energy density (Dulk et al., 1976). Mouschovias and Poland (1977) have proposed a magnetically driven model and applied it to explain the observed behavior of transients; in their model the estimated magnetic energy density is ∿5 times the thermal energy density deduced from the observed mass density. Thus, gas pressure forces are negligible in comparison with the magnetic forces which are entirely adequate to propel loop-shaped events outward. Forces due to wave pressure can, at most, contribute only a few percent of the gross outward forces.

ACKNOWLEDGMENTS

This review would not be possible without the author's close collaboration with the past and present members of the Skylab Coronagraph Experiment Team -- R. M. MacQueen, J. T. Gosling, T. Ch. Mouschovias, R. H. Munro, A. I. Poland, and C. L. Ross. The assistance of U. Anzer, G. A. Dulk, T. Holzer, B. V. Jackson, D. L. McLean and D. Webb is also appreciated. This work was supported by the National Aeronautics and Space Administration Contract NAS5-3950 to the University Corporation for Atmospheric Research.

REFERENCES

Bohlin, J. D. and Rubenstein, D. M.: 1975, NOAA World Data Center A for Solar Terrestrial Phys., Report UAG-51.

Dodge, J. C.: 1974, Report No. SN-5, Radio Astron. Obsvy. of U. of Colorado, Boulder, CO, USA. (For a discussion of these data, see Dodge, J. C.: 1975, Solar Phys. 42, 445-459.)

Dulk, G. A., Smerd, S. F., MacQueen, R. M., Gosling, J. T., Magun, A., Stewart, R. T., Sheridan, K. V., Robinson, R. D., Jacques, S.: 1976, Solar Phys. 49, 369.

Eddy, J. A.: 1974, Astron and Astrophys. 34, 235-240.

Feldman, W. C., Asbridge, J. R., Bame, S. J., Gosling, J. T.: 1977 chapter in The Solar Output and Its Variation, O. R. White, ed., Associated Univ. Press, Boulder, Colorado USA.
Fisher, R. R.: 1977, Solar Phys., submitted.
Gosling, J. T., Hildner, E., MacQueen, R. M., Munro, R. H., Poland, A. I., and Ross, C. L.: 1975a, J. Geophys. Res. 79, 4581.
Gosling, J. T., Hildner, E., MacQueen, R. M., Munro, R. H., Poland, A. I., and Ross, C. L.: 1975b, Solar Phys. 40, 439.
Gosling, J. T.: 1976, Proc. Int'l. Symp. Solar-Terrestrial Physics held 7-18 June, Boulder, CO, D. J. Williams, ed., publ. by Amer. Geophys. Union.
Gosling, J. T., Hildner, E., MacQueen, R. M., Munro, R. H., Poland, A. I., and Ross, C. L.: 1976, Solar Phys. 48, 389-398.
Gosling, J. T., Hildner, E., Asbridge, J. R., Bame, S. J., Feldman, W. C.: 1977, J. Geophys. Res., in press.
Hansen, R. T., Garcia, C. J., Hansen, S. F., Yasukawa, E.: 1974, Publ. Astron. Soc. Pac. 86, 500-575.
Hildner, E., Gosling, J. T., MacQueen, R. M., Munro, R. H., and Poland, A. I.: 1975a, Solar Phys. 42, 163.
Hildner, E., Gosling, J. T., Hansen, R. T., Bohlin, J. D.: 1975b, Solar Phys. 45, 363-376.
Hildner, E., Gosling, J. T., MacQueen, R. M., Munro, R. H., Poland, A. I., and Ross, C. L.: 1976, Solar Phys. 48, 127-135.
Hollweg, J. V.: 1972, Astrophys. J. 177, 255-259.
Howard, R. A., Koomen, M. J., Michels, D. J., Tousey, R., Detwiler, C. R., Roberts, D. E., Seal, R. T., and Whitney, J. D.: 1976, NOAA World Data Center A for Solar-Terrestrial Phys., Report UAG-48A.
Hundhausen, A. J.: 1974, Coronal Expansion and Solar Wind, Springer-Verlag, New York.
Hundhausen, A. J.: 1976, personal communication.
Jackson, B. V.: 1977, personal communication.
MacQueen, R. M., Gosling, J. T., Hildner, E., Munro, R. H., Poland, A. I., and Ross, C. L.: 1976, Phil. Trans. Roy. Soc. London 281, 405.
MacQueen, R. M.: 1977, personal communication - an arch appears in images obtained by the 1937 Nat'l. Geographic Soc./U.S. Navy Eclipse Expedition to Canton Island.
Mouschovias, T. Ch. and Poland, A. I.: 1977, Astrophys. J., submitted.
Munro, R. H.: 1977, personal communication.
Munro, R. H., Gosling, J. T., Hildner, E., MacQueen, R. M., Poland, A. I., and Ross, C. L.: 1978, Solar Phys., submitted.
Nolte, J. T., Krieger, A. S., Timothy, A. F., Vaiana, G. S., Zombeck, M. V.: 1976, Solar Phys. 46, 291-301.
Poland, A. I. and Munro, R. H.: 1976, Astrophys. J. 209, 927.
Rust, D. M. and Hildner, E.: 1976, Solar Phys. 48, 381-387.
Rust, D. M. and Webb, D. F.: 1977, Solar Phys., in press.
Schmahl, E. J. and Hildner, E.: 1978, Solar Phys., submitted.
Shklovskii, I. S.: 1965, Physics of the Solar Corona, 2nd ed., Addison-Wesley Publ. Co., Reading, Mass, (see Figure 5, which shows a 1941 eclipse).

Steinolfson, R. S., Wu, S. T., Dryer, M., and Tandberg-Hanssen, E.: 1977, in M. A. Shea, D. F. Smart, S. T. Wu (eds.), Contributed Papers to the Study of Travelling Interplanetary Phenomena/1977; Air Force Geophys. Lab., Bedford, Massachusetts, U.S.A.
Stewart, R. T., McCabe, M. K., Koomen, M. J., Hansen, R. T., and Dulk, G. A.: 1974a, Solar Phys. 36, 203.
Stewart, R. T., Howard, R. A., Hansen, S. F., Gergely, T., and Kundu, M.: 1974b, Solar Phys. 36, 219.
Stewart, R. T. and McLean, D. J.: 1977, personal communication.
Švestka, Z.: 1976, Solar Flares, D. Reidel, Dordrecht, p. 306.
Tandberg-Hanssen, E.: 1974, Solar Prominences, D. Reidel, Dordrecht.
Tousey, R.: 1973, Space Res. 13, 713.
Vaiana, G. S., Krieger, A. S., Timothy, A. F., and Zombeck, M.: 1976, Astrophys. and Space Sci. 39, 75.
Webb, D. F., Krieger, A. S., and Rust, D. M.: 1976, Solar Phys. 48, 159.
Wild, J. P. and Smerd, S.: 1972, Ann. Rev. Astron. and Astrophys. 10, 159.
Wu, S. T., Nakagawa, Y., and Dryer, M.: 1977, in M. A. Shea, D. F. Smart, S. T. Wu (eds.), Study of Travelling Interplanetary Phenomena/1977, in the series Astrophys. and Space Sci. Library, D. Reidel, Dordrecht, this volume.

DISCUSSION

Jockers: How much energy is necessary to drive an average coronal transient and how does this relate to the magnetic energy available in an average active region?

Hildner: For events (1 from OSO-7 and 11 from Skylab) analyzed to date, the average energy of 6 flare-related ejections is 17 x 10^{30} erg and the average energy of 6 eruptive prominence-associated events is 6 x 10^{30} erg. These energies are totals of kinetic and potential energies estimated while the event is within a coronograph's field of view and probably represent lower limits to the total energy expended.

Bird: Most of the observed coronal transients are associated with eruptive prominences rather than flares. Are there any examples of large solar flares near the limbs which definitely did not produce coronal mass ejection events?

Hildner: According to Munro et al. (1978) during Skylab each of the Importance 2 or larger flares within 45° of the limb were associated with coronal ejections. However, only 2 of the 18 Importance 1 flares within 10° of the limb were associated with ejections.

Wibberenz: What kind of solar observations does one have in the post-Skylab era; e.g., integral soft X-rays, to observe coronal transients?

Hildner: So far as I am aware, there are no post-Skylab instruments capable of observing coronal transients. However, extrapolation from the Skylab era permits us to infer the presence of a coronal mass ejection when we see a meter-wave radio burst of Type II and/or Type IV, or when we see a flare with Hα ejection, or when we see an eruptive prominence rise to a height greater than 1.3 $R_\odot$ from the sun's center.

Wibberenz: Are there any ideas about the deceleration of the coronal transients if they move further out?

Hildner: Not really. As Gosling et al. (1976) showed, the most common ejection events have average speeds at a few $R_\odot$ which are quiet similar to the average solar wind speed at Earth. Thus, one doesn't really need to decelerate most ejections; one only needs to cease the acceleration to agree with 1 AU observations.

Axford: On occasions when the effects of coronal transients appear to have been observed at 1 AU, is there any evidence for low temperatures (especially for the electrons) and composition anomalies?

Hildner: The interplanetary characteristics of mass ejections associated with IP shocks and large flares have been well reported and are well known. Let us suppose the non-compressive density enhancements (NCDEs) reported by Gosling et al. (1978) are 1 AU manifestations of slow, non-shocking mass ejections. Characteristically, the proton and/or electron temperature within NCDEs vary inversely as the density. Thus, as the density increases at the onset of an NCDE, the temperatures diminish. The alpha/proton ratio and its variability in NCDEs are typical of the low speed solar wind in which they are embedded.

Jockers: How do transients correlate with metric type II bursts? I ask because you mentioned that some transients produce shock waves and others do not.

Hildner: Munro et al. (1978) indicate that 23 reported meter-wave Type II and Type IV radio bursts during Skylab probably originated within 45° of a solar limb. Only 2 of these 23 events were not associated with observed mass ejections. Gosling et al. (1976) in a study of the speeds of ejections suggest that most events move too slowly to produce shocks and thus to produce Type II bursts. This, it seems fair to say that Type II bursts are well correlated with transients, but the converse is not true.

ENERGETIC SOLAR FLARE PARTICLES AND INTERPLANETARY SHOCK WAVES

R. P. LIN
Space Sciences Laboratory, University of California,
Berkeley, California 94720 USA

ABSTRACT

Estimates from hard X-ray measurements show that for many flares the bulk of the flare energy is released in the form of $\sim 10-10^2$ keV energy electrons. The interaction of these electrons with the solar atmosphere can produce the optical, UV, EUV, and radio emissions observed during the flare impulsive phase. In addition, explosive heating and evaporation of the chromosphere by these electrons can produce the $\sim 10^7\ ^\circ$K soft X-ray plasma. For the large solar flares which produce interplanetary shock waves, the accelerated $\sim 10-10^2$ keV electron population may produce the heating and mass motion required for mass ejection and the formation of the shock wave. The shock wave can in turn accelerate ions and electrons to higher energy as it travels through the corona and interplanetary medium.

INTRODUCTION

In many small solar flares (total energy release of $\sim 10^{28-30}$ ergs) the $\sim 10-100$ keV electrons accelerated during the initial phase may constitute the bulk of the total flare energy (Lin and Hudson, 1971). Thus the basic energy release mechanism for these flares essentially converts the available flare energy into fast electrons. These electrons can produce much of the observed flare X-ray, EUV, optical and radio emissions through various interactions with the solar atmosphere (e.g., Kane, 1974). Large solar flares may release as much as 10^{32} to 10^{33} ergs in a period of order 10^3 seconds. These flares emit electromagnetic radiation over the entire spectrum from γ-rays to low frequency radio bursts, but these emissions account for only a minor fraction of the total flare energy release. Ions and electrons are also accelerated up to relativistic energies by these

M. A. Shea et al. (eds.), Study of Travelling Interplanetary Phenomena/1977, 23–42.

flares. Most of the flare energy, however, eventually ends up contained in matter ejected at high velocity into the interplanetary medium to form a shock wave (see Fig. 1).

Here we consider the possibility that the ~10-100 keV electron population accelerated in the initial phase of large flares also provides the energy for the ejection of matter and formation of a shock wave which travels into the interplanetary medium. If we assume that the hard X-ray emission observed early in these large flares is due to bremsstrahlung from fast electrons colliding with a relatively cold plasma (kT ≪ fast electron energy), then we find that as in the case for small flares the bulk of the flare energy was initially contained in

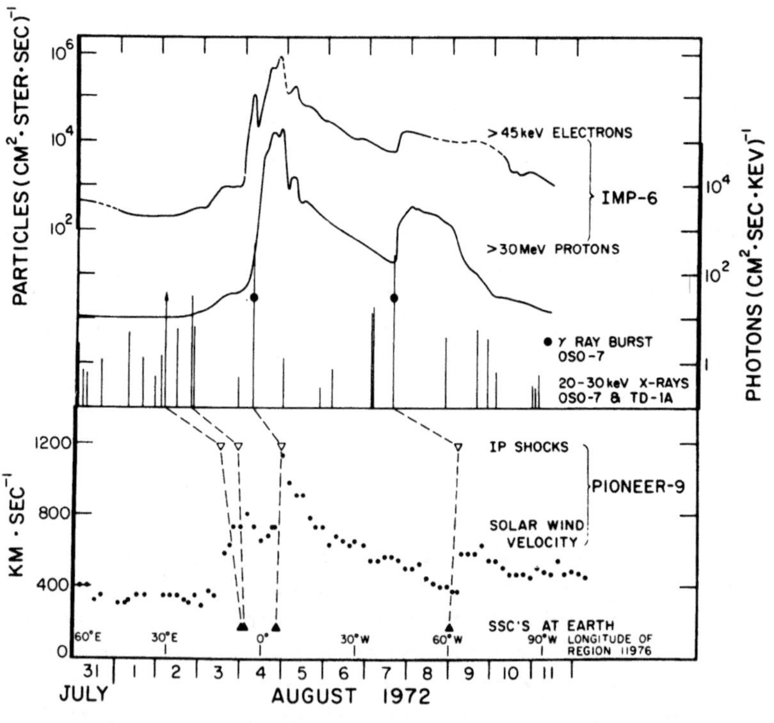

Fig. 1. A summary of solar non-thermal phenomena for the 31 July-11 August 1972 period, taken from the World Data Center "A" report UAG-28 and Solar Geophysical Data monthly reports. Except for the Pioneer 9 plasma observations at 0.78 AU all the data are from Earth-orbiting spacecraft. Note the correspondence between energetic particle injections, shock waves, γ-ray bursts, and intense hard X-ray bursts.

10-100 keV electrons (Lin and Hudson, 1976). Because of the remarkably large numbers and densities of fast electrons required the above assumption, however, must be examined with care. Finally we examine the evidence for subsequent acceleration of ions and electrons to high energies by the shock wave.

THE ENERGETIC PARTICLE POPULATIONS IN LARGE SOLAR FLARES

The physical processes of emission and propagation of hard X-rays and gamma-rays by fast particles are well and quantitatively understood. Thus observations of solar hard X-rays and gamma-rays can provide quantitative estimates of the energetic electron and ion populations at the sun. Suppose at a given instant there is a population, $dn/dE(cm^3\,keV)^{-1}$, of energetic particles contained in a volume V with ambient ion density n_i. Then the instantaneous photon emission at distance R is given by

$$\frac{dJ}{d(h\nu)} = \frac{1}{4\pi R^2}\iint \frac{dn}{dE}\,\frac{d\sigma}{d(h\nu)}\,n_i w\,dE\,dV \qquad (1)$$

where E and $h\nu$ are the particle and photon energy, respectively, $\frac{d\sigma}{d(h\nu)}(E, h\nu)$ the differential cross section, and w the particle velocity.

For hard X-rays from bremsstrahlung of electrons this equation can be numerically inverted to obtain dn/dE in terms of $dJ/d(h\nu)$ (Lin, 1977). Using the Beithe-Heitler approximation to $d\sigma/d(h\nu)$ for non-relativistic electrons (Jackson, 1962),

$$\begin{aligned}\frac{d\sigma}{d(h\nu)}(E,\ h\nu) &= \frac{1.58\times 10^{-24}}{E\ h\nu}\,\ell n\left[\left(\frac{E}{h\nu}\right)^{\frac{1}{2}} + \left(\frac{E}{h\nu}-1\right)^{\frac{1}{2}}\right] cm^2\,ion^{-1}\,keV^{-1} \\ &\qquad \text{for } E \geq h\nu \qquad (2)\\ &= 0, \text{ for } E < h\nu\end{aligned}$$

Equation (1) can be inverted analytically (Brown, 1971).

For a power law photon spectrum $dJ/d(h\nu) = A(h\nu)^{-\gamma}$ we obtain a power law electron spectrum

$$\frac{dn}{dE} = 1.2\times 10^{42}\ \gamma\,(\gamma-1)^2\ B(\gamma-\tfrac{1}{2},\ \tfrac{3}{2})\frac{AE^{-\gamma+\frac{1}{2}}}{n_i V} \qquad (3)$$

where B(x,y) is the beta function. It is important to note that these relationships [Equations (1) and (3)] are between the instantaneous number and spectrum of energetic electrons in the X-ray region and the instantaneous X-ray emission. The instantaneous electron population [dn/dE, denoted below by N(E, t)] must be related to the source of electrons F(E, t) (the accelerated electrons) by considering the electron loss mechanisms and using the continuity equation

$$\frac{\partial N(E,t)}{\partial t} = F(E,t) - L(E,t) - \frac{\partial}{\partial E}\left[N(E,t)\frac{dE}{dt}\right] \quad (4)$$

Here L(E, t) refers to electron escape from the X-ray emitting region and the last term takes into account energy change processes.

Suppose that accelerated electrons run into a cold hydrogen plasma ($T < 10^8$ °K) where they eventually lose all their energy via collisions with the gas, producing bremsstrahlung X-rays in the process. Then the energy loss term is given by (Trubnikov, 1965)

$$\frac{dE}{dt} = -4.9 \times 10^{-9}\, n_i E^{-\frac{1}{2}}\ (\mathrm{keV\ s^{-1}}) \quad (5)$$

Assuming no electrons escape (L = 0) then in a steady state situation ($\partial N/\partial t = 0$) we obtain $F(E) = \frac{\partial}{\partial E}\left[N(E)\frac{dE}{dt}\right]$. For a power law X-ray spectrum and a total duration Δt of the X-ray burst, we obtain for the source

$$F(E) = 5.9 \times 10^{33}\ \gamma^2(\gamma-1)^2\, B(\gamma - \tfrac{1}{2}, \tfrac{3}{2})\, A\Delta t E^{-\gamma-1}\ \mathrm{keV}^{-1} \quad (6)$$

This case is commonly called thick target bremsstrahlung. The energy in fast electrons for the thick target case can be shown to be the minimum needed for the observed bremsstrahlung X-ray emission, assuming that the electrons are stopped in a cold ($kT \ll E$) medium.

Chubb (1970), Brown (1975), Kahler (1975), Colgate et al. (1976), and others have considered the case where the X-rays may be produced in a volume where all the electrons have approximately the same energy, i.e., an extremely hot, $kT \sim 200$ keV plasma. Then electron-electron Coulomb collisions will only result in exchanges of energy among the electrons without any net change in the average energy of the population. The only energy loss process will be from electron-ion collisions, and if the ions are at the same temperature then only bremsstrahlung producing collisions result in energy escaping the volume. In steady state ($\partial N/\partial t = 0$) the source has to supply only the bremsstrahlung energy losses. These losses are $\sim 10^{-4}$ to 10^{-5} the collision losses computed for the thick target case. However, it is extremely difficult to contain such a hot plasma (as CTR research has shown).

Colgate et al. (1976) argue that directed fast electron beams cannot exist because the large current that will result would create a magnetic field whose total energy would be many orders of magnitude greater than the flare energy. This argument is fallacious in that there can be directed energy flux in the form of a fast electron beam without or with only a very small net current. Such situations commonly occur in nature (for example, auroral electron precipitation) and in laboratory plasmas. Consider the situation where some fast electrons begin to move in one direction through a plasma. Right away electric fields are produced by the charge separation and magnetic induction. These electric fields will accelerate positively charged particles in the direction of the fast electrons and negatively charged particles in the opposite direction, i.e. produce a reverse current. Because ambient plasma electrons have the lowest mass they will be accelerated the most by the electric field, and barring the presence of more complicated phenomena, are likely to be the major carriers of the reverse current. The strength of this reverse current will adjust to the fast electron beam so that the resultant net electric field is just enough to drive the reverse current. Thus there is never a large net current of the type envisioned by Colgate et al. (1976). We emphasize that the generation of a reverse current is the natural consequence of a fast electron beam. Although the plasma physics of each specific situation must be investigated in detail (as various plasma waves may be driven by the beam and reverse current), there is absolutely no a priori reason for ruling out fast electron beams. As we shall see later, the presence of fast electron beams fits naturally into the overall flare picture.

Quantitative hard X-ray flux and spectrum observations for large flares are limited because of instrumental considerations. The best observations were made by the ESRO TD-1 satellite during the August 1972 series of large flares. Figure 2 shows X-ray (Hoyng, 1975) and γ-ray (Chupp et al., 1973) observations of the 4 August 1972 large flare, and Table 1 lists the X-ray burst and computed thick target electron parameters for that and other flares of that series (from Lin and Hudson, 1976). We have also included computations from interplanetary observations of the number of escaping electrons. These typically constitute only on the order of 10^{-3} of the total number of electrons.

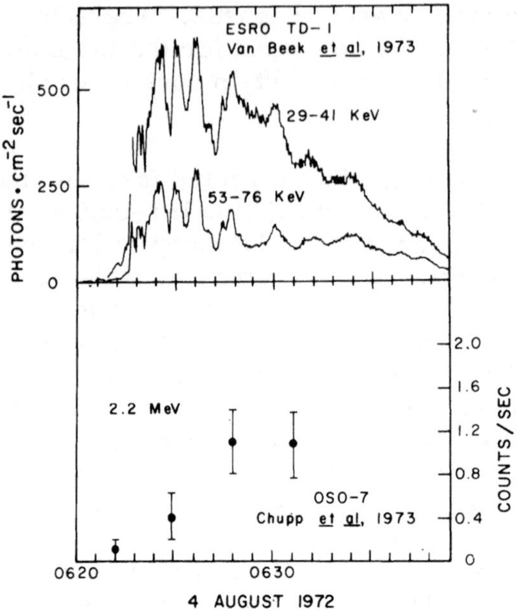

Fig. 2. The temporal evolution of the hard X-ray and 2.2 MeV γ-ray bursts for the 4 August 1972 flare is shown here. Note that the maximum of the γ-ray emission lags the X-ray maximum by several minutes.

Similar computations can be made for the energetic ion population from the observed γ-ray emission (see review by Ramaty and Lingenfelter, 1975). Only the flare of 4 August had γ-ray observations of sufficient quality to obtain quantitative estimates. These are given along with the estimates from interplanetary observations of escaping protons. Note that the number of escaping protons far exceeds the computed thick target number required for the γ-ray production.

THE ENERGY INPUT TO THE SOLAR ATMOSPHERE

Under the thick target model for the hard X-ray emission, the electrons must lose their energy through collisions in the solar atmosphere. For the thick target case the energy input rate from energetic particle collisional losses as a function of column density can be computed under the simplifying assumption that the energetic particles are precipitated from high in the corona, vertically downward into the solar atmosphere in a column of constant cross-sectional area. More complex particle injection models, i.e., models with non-vertical magnetic configurations, with various injection heights or with various initial pitch angle distributions, could be incorporated into the

TABLE 1. FLARE EVENTS: MCMATH PLAGE 11976 AUGUST 2-7, 1976

1. H_α flare						
Date	2 Aug	2 Aug	2 Aug	4 Aug	7 Aug	7 Aug
Start (UT)	0316	1839	1958	0617	0252	1449
Max	(0327 - 0515)	1844	2058	0640	0255	1534
End	0800	1857	2336	0855	0320	1721
Location	N14 E34	N14 E26	N14 E28	N14 E08	N15 W30	N14 W37
2. Hard X-rays Burst Parameter[1]						
A_m $(cm^2 \text{ sec keV})^{-1}$	7×10^6	3.5×10^5	4.5×10^4	1.2×10^6	2×10^5	4.5×10^5
γ_m	4.1	2.8	2.5	2.7	2.9	2.3
Δt	$\geqslant 260$	32	~ 600	500	43	$\geqslant 310$
3. Energetic Electrons						
Number >20 keV						
At the flare[2]	2.5×10^{38}	3×10^{37}	1.5×10^{38}	2×10^{39}	2×10^{37}	1×10^{39}
Escaping to the IP medium[3]	$>5 \times 10^{33}$	---	$>3 \times 10^{34}$	2.5×10^{36}	---	1×10^{35}
Total Energy >20 keV (ergs)	1×10^{31}	1.5×10^{30}	8×10^{30}	1.1×10^{32}	9×10^{29}	6×10^{31}
4. Energetic Protons						
Number >10 MeV						
At the flare[4]	NA	NA	NA	6×10^{33}	NA	$>8 \times 10^{32}$
Escaping to the IP medium[3]	1×10^{34}	---	5×10^{34}	5×10^{35}	---	2×10^{34}
Total Energy >10 MeV (ergs)	2×10^{29}	---	10^{30}	8×10^{30}	---	3×10^{29}
5. Interplanetary Shock Wave Arrival UT						
At Earth	0119 (8/4)		0220 (8/4)	2354 (8/4)		0037 (8/9)
At 0.78 AU (Pioneer 9)	0440 (8/3)		117 (8/3)	2323 (8/4)		0707 (8/9)

1. The X-ray spectrum at burst maximum is $\frac{dJ}{d(h\nu)} = A_m (h\nu)^{-\gamma m}$, and Δt is the effective duration at maximum flux. Except for OSO-7 observations of 2142 UT 2 August all X-ray observations are from ESRO TD-1.
2. The electron spectrum can be obtained from the X-ray parameters with Equation (6).
3. Because the 2 August flares are from E30 longitude, the electron estimates given here from near-earth observations are low. The protons are estimated from Pioneer 9, at $\sim 45°$ east of the sun-earth line.
4. For the 1521 UT 7 August event only the end of the γ-ray event was observed by OSO-7.

computation; but the vertical case illustrates the general features well.

For the fast electrons we follow the method of Brown (1973), which takes into account the scattering of the electrons in pitch angle. The rate of energy input per unit volume is then given by

$$I_B(N, t) = 1.0 \times 10^7 \, n(N, t) \, [x(N, t) + 0.55] \, \frac{A}{a} \, C(\gamma)$$

$$\times \, (1.1 \times 10^{-17} N)^{-\left(\frac{\gamma + 1}{2}\right)} \text{ ergs cm}^{-3} \text{ sec}^{-1} \qquad (7)$$

where $n(N, t)$ is the hydrogen number density at column density N and time t; $x(N, t)$ is the fractional ionization; a is the cross-sectional area in cm^2; $dJ/d(h\nu)^{-\gamma}$ is the photon spectrum at time t, photons $(cm^2 \text{ sec keV})^{-1}$; and $C(\gamma) = \gamma^2 \, (\gamma - 1)^2 \, B(\gamma - \frac{1}{2}, \frac{3}{2}) \, B(\frac{1}{3}, \frac{\gamma + 1}{2})$. This expression differs from that of Brown (1973) due to an error in his evaluation of the integral in his equation (7). The total energy dumped at a given height z in the solar atmosphere is independent of the cross-sectional area a.

$$d\mathcal{E} = I_B(z) \Delta t \, a(z) \, dz$$

$$d\mathcal{E} = -1 \times 10^7 \, AC(\gamma) \Delta t [X(N) + 0.55] \, (1.1 \times 10^{-17} N)^{-\left(\frac{\gamma + 1}{2}\right)} dN \qquad (8)$$

The total energy dumped below a given N was obtained by integration of equation (8) and is shown in Fig. 3 for each flare event of the August series. Except for the 4 August flare, a single power law has been fitted to the X-ray observations.

Similar computations can be performed for the energetic protons where γ-ray observations are available. The energy loss function for the energetic protons in the 4 August 1973 event is shown in Fig. 3. We note here that even under the most favorable thick target approximation, the energy lost by the protons is substantially less than the energy lost by the electrons.

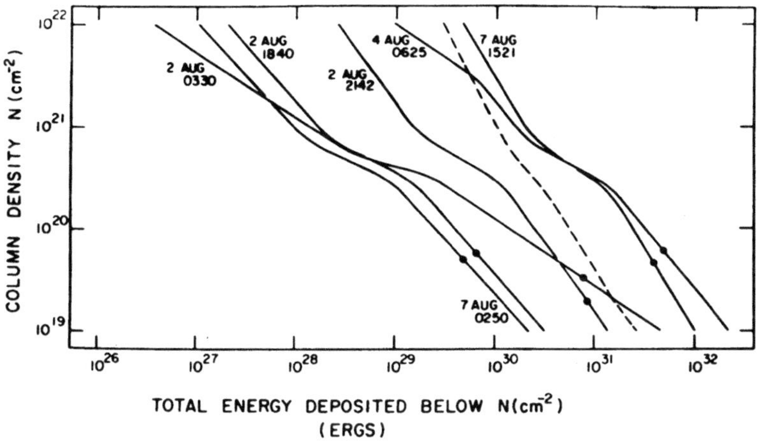

Fig. 3. The total energy deposited below a given column density by energetic electrons is plotted here for six flares of the August 1972 series. Also plotted is the proton energy deposition for the 4 August flare (dashed line). The dots indicate the dividing line between the radiative quasi-equilibrium region and the flare explosion region, i.e., the energy deposited below the dot goes primarily into radiation, and above to heating and mass ejection.

THE FLARE EXPLOSION

Because the emissivity as a function of temperature has a peak and because the energy dumped into the solar atmosphere by the energetic particles increases with height, below some critical level, N_c, the plasma will be able to maintain a quasi-equilibrium by radiating away the input energy. Above N_c the energy input rate will be too rapid for radiative equilibrium; the ambient gas will be explosively heated because the emissivity now decreases with increasing temperature. At even high levels of the atmosphere the energy gained by the ambient plasma may exceed that needed to escape the restraining forces of the sun's gravitational and magnetic fields. This material will then be ejected from the sun. If the velocity of the ejected material is sufficiently high, a shock wave will form in front. This shock wave can give rise to type II radio emission, and may be observed later as an interplanetary shock wave.

We can estimate lower limits to the energy and mass involved in this flare explosion process. The critical level N_c is obtained by equating the rate of energy input (equation 7) with the maximum plasma emissivity. For constant pressure the temperature for maximum emissivity is $T_c \approx 6 \times 10^4$ °K (Brown, 1973). The radiation loss rate is

$$\frac{dE_{rad}}{dt} \approx 1.9 \times 10^{-30}\, n N_c \text{ ergs cm}^{-3} \text{ sec}^{-1} \tag{9}$$

This simple treatment assumes that the ionization level follows the temperature as in a hot, optically thin gas. Equating dE_{rad}/dt to I_B we find that

$$N_c = 9.1 \times 10^{16} \left[9.1 \times 10^{19} \frac{A}{a} C(\gamma)\right]^{\frac{2}{\gamma+3}} \text{cm}^{-2} \tag{10}$$

The values of N_c are indicated in Figure 3 for the maximum rate of electron energy input (at the hard X-ray maximum) using the reported H_α flare area nearest the time of X-ray maximum.

The total energy available above N_c depends on the shape of the particle spectrum at low energies. Present hard X-ray observations are almost completely limited to $\gtrsim 20$ keV. On the other hand, the few available <20 keV observations indicate that the electron spectrum extends to ≤ 5 keV (Kahler and Kreplin, 1971; Peterson et al., 1973; Lin, 1973), so that a very conservative lower limit to the flare explosion energy would be $\mathcal{E}(>20 \text{ keV}) - \mathcal{E}_r$, where $\mathcal{E}_r$ is the energy dumped below N_c. ($\mathcal{E}_r$ presumably is radiated away.)

The total mass of material in the flare explosion is $M_c = N_c\, a/\eta$ in our simple model. We argue later that M_c is a lower limit to the mass of that material.

THE INTERPLANETARY SHOCK WAVE

The mass and energy involved in the flare explosion go into creating the hot, $\sim 10^7$ °K flare plasma, observed by its soft X-ray emission, and into the ejection of matter and formation of a shock wave.

Both the energy and mass involved in the hot flare plasma can be estimated from the soft X-ray observations. For the August flares the emission measure, $n_e n_i V$, and temperature, T, are available from soft X-ray measurements (Dere et al., 1973; Neupert et al., 1974). Assuming a typical density of $\sim 3 \times 10^{10}$ cm^{-3} for the X-ray flare we obtain a total plasma energy

$$\epsilon_{plasma} \approx 3\, n_i VkT = \frac{3(n_e n_i V)kT}{n_e} \tag{11}$$

ranging from 6×10^{30} to 4×10^{31} ergs (Table 2). Radiation losses during the initial phase are negligible compared to the total plasma energy content. The amount of material involved is just $M_{plasma} = \overline{m}\, n_e n_i V/n_e$ where $\overline{m}$ is the average ion mass.

The amount of energy involved in the interplanetary shock wave can be estimated from the plasma density ρ, velocity v, temperature T, and magnetic field B observed near 1 AU.

$$\mathcal{E}_{shock} = \Omega R^2 \int [\, (\tfrac{1}{2}\rho v^3 - \rho_b v_b^{\,3}) + \epsilon_g (\rho v - \rho_b v_b)$$

$$+ \tfrac{3}{2} k(\rho vT - \rho_b v_b T_b) + \frac{1}{8\pi}(B^2 v - B_b^{\,2} v_b)]\, dt \tag{12}$$

where the terms represent the flow, gravitational, thermal, and magnetic energy, respectively (the subscript b denotes the pre-shock background levels). $\epsilon_g \approx 2$ keV is the energy per particle associated with the gravitational potential and ΩR^2 is the area subtended by the shock. The mass of ejected material in these interplanetary events is

$$M_{shock} = \Omega R^2 \int \left[\rho v - \rho_b v_b\right] dt \tag{13}$$

Computations by Dryer et al. (1976), using Pioneer 9 plasma and assuming $\Omega = 2\pi$ steradians at 0.78 AU, show that $\mathcal{E}_{shock}$ ranges from ~ 1 to 5×10^{32} ergs and the M_{shock} from $\sim 10^{16}$ to 10^{17} gms (Table 2). The total shock wave energy can be supplied by the fast electrons if the electron spectrum extends down to 10 keV, 14 keV, and 5 keV for the 0330 UT 2 August, 4 August and 1500 UT 7 August flares, respectively. Hence the energy requirements of the shock wave could easily be satisfied by the non-thermal electrons; however,

TABLE 2. PARAMETERS OF THE FLARE EXPLOSION REGION FOR THE AUGUST 2-7, 1972, LARGE PROTON FLARES*

		0330 Aug 2	2058 Aug 2+	0625 Aug 4	1522 Aug 7
I. Observed Output					
a. Soft X-ray flare plasma	energy $\mathcal{E}_{plasma}$	6.0×10^{30}	6.0×10^{30}	3.5×10^{31}	1.5×10^{31}
	mass M_{plasma}	1.2×10^{15}	1.4×10^{15}	8.5×10^{15}	3.0×10^{15}
b. Interplanetary shock wave	energy $\mathcal{E}_{shock}$	8.8×10^{31}	4.6×10^{32}	1.1×10^{31}	3.8×10^{32}
	mass M_{shock}	3.4×10^{16}	1.1×10^{17}	1.6×10^{16}	9.4×10^{16}
c. Totals	energy	1.0×10^{32}	4.7×10^{32}	1.5×10^{32}	4.0×10^{32}
	mass	3.5×10^{16}	1.1×10^{17}	2.5×10^{16}	1.0×10^{17}
II. Computed Electron Input					
a. Minimum flare explosion energy [$\mathcal{E}(>20$ keV) $- \mathcal{E}_r$]		2.0×10^{30}	---	6.0×10^{31}	10^{31}
b. Low energy cutoff to supply total flare explosion		9.5 keV	---	16 keV	5 keV
c. Minimum mass M_c		1.5×10^{15}	---	1.5×10^{15}	2.0×10^{15}

*All units cgs unless otherwise labelled.

the decisive observations of $\lessdot$20 keV non-thermal X-rays are not available.

The calculated M_c are 18-50 times smaller than the sum of M_{plasma} and M_{schock}. Several factors may be responsible for this:

1. The electrons may be accelerated low in the solar atmosphere. Since there is sufficinet energy in the electrons, essentially all the material above the acceleration region will be ejected.
2. The flux tubes are not vertical but flare out with height. For magnetic fields with their sources in the photosphere or below, the cross-sectional area, a, of a loop would increase as a function of height. Since $M_c \propto a^{(\gamma+1)/(\gamma+3)}$, an increase of a factor of 10 in the area would yield a factor of ~5 increase in M_c.
3. The configuration may be a loop structure in which case all the mass in the loop above N_c would be ejected. There is strong evidence from the ATM white-light coronograph pictures for loop configurations in the ejected material (Gosling et al., 1974). Loop structures are also found in soft X-ray and EUV observations of flares (Neupert et al., 1974).
4. The ejected material may drag considerable quantities of the surrounding atmosphere with it, as in atmospheric nuclear explosions (Brown, 1973; Colgate, 1965).
5. Soft X-ray absorption (Somov, 1975) may heat the chromospheric gas and cause some further evaporation.

Detailed calculations of more realistic models are required to show whether these effects are sufficient to increase M_c to values comparable to the sum of the observed M_{plasma} and M_{shock}. However, it is clear that the ~10-100 keV electrons accelerated during the initial phase of large flares could be the energy source for the interplanetary shock wave. Note that in the August 1972 series only the flares with total >20 keV electron energy greater than $\approx 10^{31}$ ergs produced shock waves.

THE ACCELERATION OF IONS AND ELECTRONS TO RELATIVISTIC ENERGIES

A unique feature of these large energetic flares is the

acceleration of ions to $\gtrsim$ 10 MeV energies and electrons to relativistic ($\gtrsim$ 0.5 MeV) energies. These particles appear to be accelerated in a second stage which follows and is distinct from the initial phase acceleration of ~10-100 keV electrons. Direct measurements of the arrival times of these particles at 1 AU in the interplanetary medium show that energetic ions and relativistic electrons are likely to have been released at least a few minutes later than the 10-100 keV flash phase electrons (Lin, 1970; Sullivan, 1970; Simnett, 1974). Hard X-ray observations for these large flares often show a distinct second burst (Frost and Dennis, 1971), beginning after the initial flash phase impulsive burst (approximately at the start time of the type II radio burst) and lasting for tens of minutes. In the example published by Frost and Dennis (1971) the X-ray spectrum of the second stage burst extends in a single power law to the high energy limit of the detector ~200 keV, while the initial impulsive burst has a sharp downturn at ~100 keV, implying that additional acceleration of electrons to higher energies occurred in the second phase. For the 4 August 1972 large flare, an analysis of the time dependence of the 2.2 MeV gamma ray line flux indicates that the energetic, > 10 MeV protons which produce this line (via neutron production and capture into an excited state of deuterium) were accelerated several minutes later than the initial phase 10-100 keV electrons (Fig. 2) (Bai and Ramaty, 1976). Many years ago it was found that a distinct second phase of non-thermal radio emission occurred in large flares (Wild et al., 1963). This phase is characterized by type II and IV emission, generated by a shock wave passing through the solar atmosphere and gyrosynchrotron emission from mildly relativistic electrons, respectively.

We believe that the shock wave, as it traverses the solar corona, produces this second stage of acceleration. Shock waves in the interplanetary medium and the earth's bow shock accelerate particles, particularly ions, to energies up to MeV's (Anderson, 1965; Lin et al., 1974; Palmeira et al., 1971). There is evidence that on at least one occasion (following the 4 August flare) interplanetary shocks in a favorable configuration may have accelerated ions to energies above 1 GeV (Pomerantz and Duggal, 1974). In the stronger and more ordered coronal magnetic fields near the flare site configurations favorable to strong acceleration are likely to occur. Some trains of periodic radio oscillations have been observed which suggest that shock acceleration is occurring in a loop configuration (McLean et al., 1971). Acceleration by shock waves traversing outward through the corona provide a natural interpretation for the fact that most of the energetic ions in the 4 August 1972 flare escape the sun. Recently, measurements of accelerated ~100 keV/nucleon ions from flares have shown that these particles have ionization states typical of the 1.6×10^6 °K corona, rather than that expected for the $\geq 10^7$ °K flare plasma or ≤ 5 °K flare chromosphere (Gloeckler et al., 1976). Although those authors interpret this result in terms of storage of these ions in the corona after acceleration, a natural interpretation would be that these ions are accelerated in the low density corona without any change in their ionization state. Finally, recent comparisons of energetic proton phenomena with Skylab data indicate that all the energetic particle events are associated with fast coronal transients observed by the white-light coronagraph (E. Hildner, private communication).

DISCUSSION

A close relationship clearly exists between the particle acceleration processes in flares and the flare shock wave. Figure 4 illustrates a very simple but consistent flare model. Early in the flare electrons are energized to ~10-100 keV energies as the major initial energy

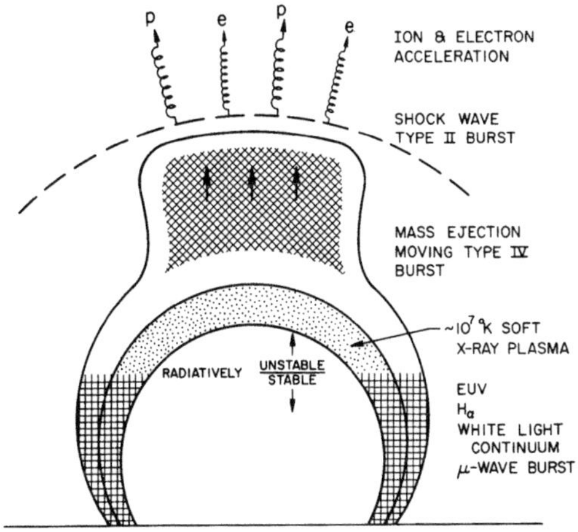

Fig. 4. This figure shows a schematic representation of a large flare. In the cross-hatched region the solar atmosphere is able to radiate away the input energy. Above that is the flare explosion region, divided into flare plasma and ejected material, depending on whether enough energy is dumped in fast electrons to overcome the restraining gravitational and magnetic forces.

release mechanism. These electrons collide with the cold solar atmosphere to produce heating and electromagnetic radiation at the deepest levels of the flare. Above the critical point, explosive heating takes place to form the hot $\sim 10^7$ °K flare plasma. If sufficient energy is dumped into the atmosphere to overcome the gravitational and magnetic restraining forces, then matter is ejected with the formation of a shock wave. This shock wave then accelerates ions and electrons up to high energies.

One prediction of this model is that very intense hard X-ray bursts should be followed by shock waves and ion and electron acceleration to high energies. Figure 5 shows the maximum count rate for solar hard X-ray burst observations of the OGO-1 and -3 (Arnoldy et al., 1968a, b; Kane and Winckler, 1969), AIMP-1 and -2, and IMP-4 ionization chamber experiments (normalized to the OGO-3 ionization chamber) plotted as a function of solar longitude. These experiments

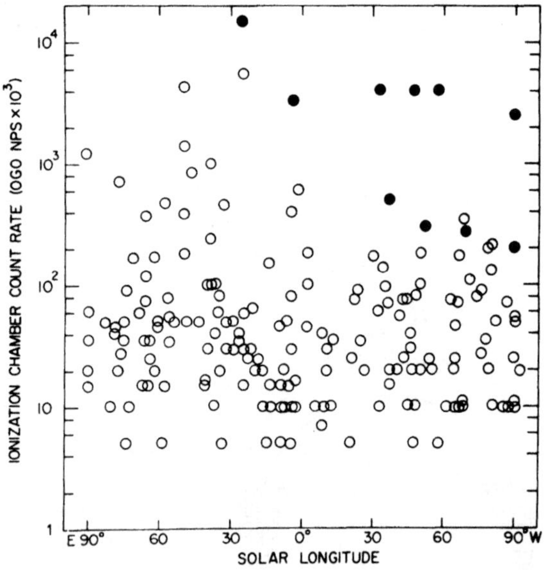

Fig. 5. The peak ionization chamber count rates for solar hard X-ray bursts observed from 1964 through 1968 are plotted here vs. the solar longitude of the associated flare. Flares which gave rise to proton events with >10 MeV fluxes above 10 cm^{-2} sec^{-1} at 1 A U are indicated by the solid dots. Taking into account the E-W propagation effect, essentially all the flares with hard X-ray intensity above 250 NPS x 10^{-3} gave rise to such a proton event at 1 A U. Approximately 10^{31} ergs in >20 keV electron deposition is required for that count rate.

can measure intense hard X-ray bursts without saturation, and over the period September 1964 to the end of 1968, 178 solar X-ray bursts were observed. Defining large proton events as those which emit $\geq$10 MeV protons with a maximum flux at 1 A U greater than 10 cm^{-2} sec^{-1}, we find that essentially every flare west of central meridian with an X-ray burst above ~250 NPS x 10^{-3} gave rise to a large proton event. This count rate corresponds to a total >20 keV electron energy deposition of 10^{31} ergs. Thus, the events above this threshold are comparable to the August 1972 large proton flares. Although many intense X-ray events were observed from flares east of central meridian, only two (E 04 and E25) were followed by large proton events. This east-west effect is well understood and is due to the limited angular extent of the propagation of the protons.

Reversing the correlation we find that of the twenty large proton events observed over the same period, ten were preceded by intense X-ray bursts; five were unassociated with either a flare on the visible solar disc or a hard X-ray burst, and thus presumably were from flares behind the west limb of the sun; and five occurred when there were no hard X-ray observations. This one-to-one correlation indicates that large proton event flares are highly non-thermal in character, with the bulk of the total flare energy going primarily into the acceleration of electrons to 10-100 keV energies.

Using type II radio bursts as an indicator of shock waves traveling through the solar atmosphere we find that eleven of the large proton events have associated type II emission, six had no radio observations. Of the remaining three, two were from over the limb flares, and one apparently showed no type II emission. Thus there is a consistent pattern of intense hard X-ray bursts, shock waves, and energetic particle acceleration for large flares.

We have only outlined the plausibility and energetic feasibility of these relationships between accelerated particles and shock waves in large flares. Many critical observations are yet to be made; particularly, measurements of the spectrum of electrons below 20 keV; improved γ-ray and ionization state measurements to define the energetic ion population; and spatially resolved hard X-ray measurements to locate the electron acceleration region. Detailed calculations of the flare explosion and shock wave formation processes would also be very useful.

ACKNOWLEDGMENTS

I wish to acknowledge many discussions with H. S. Hudson. This work was supported in part by NASA grant NGL-05-003-017.

REFERENCES

Anderson, K. A.: 1965, Proc. Ninth Int'l. Conf. on Cosmic Rays, London, p. 520.

Arnoldy, R. L., Kane, S. R. and Winckler, J. R.: 1968a, Astrophys. J., 151, 711.

Arnoldy, R. L., Kane, S. R. and Winckler, J. R.: 1968b, Univ. of Minnesota, Cosmic Ray Technical Report CR-108.

Bai, T. and Ramaty, R.: 1976, Solar Physics, 49, 343.

Brown, J. C.: 1971, Solar Phys., 18, 489.

Brown, J. C.: 1973, Solar Phys., 31, 143.

Brown, J. C.: 1975, in S. R. Kane (ed.), Solar Gamma, X- and EUV Radiation, IAU Symposium 68, Reidel, p. 211.

Chubb, T. A.: 1970, in E. R. Dyer (ed.), Proc. Leningrad Symp. Solar Terrestrial Physics/1970, part I, Reidel, p. 99.

Chupp, E. L., Forrest, D. J. and Suri, A. N.: 1973, in R. Ramaty and R. G. Stone (eds.), High Energy Phenomena on the Sun, NASA SP-342, p. 285.

Colgate, S. A.: 1965, J. Geophys. Res., 70, 3161.

Colgate, S. A., Audouze, J., and Fowler, W. A.: 1976, Orange Aid Preprint OAP-458.

Dere, K. P., Horan, D. M. and Kreplin, R. W.: 1973, Collected Data Reports on August 1972 Solar Terrestrial Events, World Data Center A, UAG-28, part II, p. 298.

Dryer, M., Smith, Z. K., Steinolfson, R. S., Mihalov, J. D., Wolfe, J. H. and Chao, J. K.: 1976, J. Geophys. Res., 81, 4651.

Frost, K. J. and Dennis, B. K.: 1971, Astrophys. J., 165, 655.

Gloeckler, G., Sciambi, R. K., Fan, C. Y., and Hovestadt, D., 1976: Astrophys. J., 209, L93, 97-A9.

Gosling, J. T., Hildner, E., MacQueen, R. M., Munro, R. H., Poland, A. I. and Ross, C. L.: 1974, J. Geophys. Res., 79, 4581.

Hoyng, P.: 1975, Doctoral Thesis, University of Utrecht, The Netherlands.

Jackson, J. D.: 1962, Classical Electrodynamics, J. Wiley and Sons, N. Y.

Kahler, S.: 1975, in S. R. Kane (ed.), Solar Gamma-, X-, and EUV Radiation, IAU Symposium 68, Reidel, p. 211.

Kahler, S. W. and Kreplin, R. W.: 1971, Astrophys. J., 168, 531.
Kane, S. R.: 1974, in Gordon Newkirk, Jr. (ed.), Coronal Disturbances, IAU Symposium No. 57, p. 105.
Kane, S. R. and Winckler, J. R.: 1969, Univ. of Minnesota Cosmic Ray Technical Report, CR-135.
Lin, R. P.: 1970, Solar Physics, 12, 209.
Lin, R. P.: 1973, in R. Ramaty and R. G. Stone (eds.), High Energy Phenomena on the Sun, NASA SP-342, p. 439.
Lin, R. P.: 1977, in preparation.
Lin, R. P. and Hudson, H. S.: 1971, Solar Phys., 17, 412.
Lin, R. P. and Hudson, H. S.: 1976, Solar Phys., 50, 153.
Lin, R. P., Meng, C.-I., and Anderson, K. A.: 1974, J. Geophys. Res., 79, 489.
McLean, D. J., Sheridan, K. V., Stewart, R. T., and Wild, J. P.: 1971, Nature, 234, 140.
Neupert, W. M., Thomas, R. J. and Chapman, R. D.: 1974, Solar Phys., 34, 349.
Palmeira, P. A. R., Allum, F. R., and Rao, U. R.: 1971, Solar Phys., 21, 204.
Peterson, L. D., Datlowe, D. W. and McKenzie, D. L.: 1973, in R. Ramaty and R. G. Stone (eds.), High Energy Phenomena on the Sun, NASA SP-342, p. 132.
Pomerantz and Duggal: 1974, J. Geophys. Res., 79, 913.
Ramaty, R. and Lingenfelter, R. E.: 1975, in S. R. Kane (ed.), Solar Gamma-, X-, and EUV Radiation, IAU Symposium 68, Reidel, p. 363.
Simnett, G. M.: 1974, Space Science Review, 16, 257.
Somov, B. V.: 1975, Solar Phys., 42, 235.
Sullivan, J.: 1970, "Two Solar Flare Electron Components and Their Relationship to the Proton Component," Ph.D. Thesis, Univ. of Chicago.
Trubnikov, B. A.: 1965, Rev. Plasma Phys. 1, 105.
Van Beek, H. F., Hoyng, P. and Stevens, G. A.: 1973, Collected Data Reports on August 1972 Solar Terrestrial Events, World Data Center A, UAG-28, part II, p. 319.
Wild, J. P., Smerd, S. F., and Weiss, A. A.: 1963, Ann. Rev. Astron. Astrophys., 1, 291.

DISCUSSION

Kaufmann: The 30 GHz burst associated with the 4 August 1972 event has apparently the same time structure as the hard X-ray burst. But it is nearly 2 min. in delay. If we believe the morphological similarity as physically meaningful, would this time displacement be too high and rather unusual?

Lin: The 30 GHz burst agrees very well with the 0.35-8 MeV X-ray burst, but is delayed from the lower energy 29-41 keV X-ray burst. This behavior is consistent if the radio emission is produced by gyrosynchrotron emission since that process is much more efficient above ~ 100 keV.

Krimigis: Observations published by Neupert et al. (1967) show Fe XXIV and XXV X-ray lines. You have indicated that ions are shock-accelerated through the cold plasma and retain their quiet-time charge state. In what phase of the flare do these X-ray lines arise, and can these be produced by thermal ionization? (Neupert, W. M., Gates, W. J., Swartz, M., and Young, R. M.: 1967, *Astrophys. J. 149*, L79).

Lin: The high ionization states of Fe are observed from the hot 1-2 x 10^7 K flare plasma. They can be explained as thermally excited. The fact that the accelerated, 100 keV/nucleon ions have ionization states typical of the quiet, ~ 10^6 K corona suggests that these ions are not accelerated in the flare plasma, but rather are accelerated in the quiet corona, possibly by the shock wave.

Hudson (Additional comment to Krimigis' question): Thermal processes seem adequate to explain Fe XXIV and Fe XXV. Fe XXVI does not appear strongly in the X-ray spectra, and this would be consistent with thermal models also, because of the longer ionization time to the hydrogen-like ion and the transient nature of the hot plasma.

Jockers: I wonder if there is (magnetic?) energy input required through the second stage of particle accelerati in the shock wave? Did you check the energy balance?

Lin: I have not studied this question in any detail. However the energy in accelerated > 10 MeV ions and > 20 keV electrons which escape is only a small fraction of the shock wave energy. We do not know how much energy is contained in < 10 MeV ions and < 20 keV electrons.

DYNAMIC MODELING OF CORONAL AND INTERPLANETARY RESPONSES TO SOLAR EVENTS

S. T. WU
The University of Alabama in Huntsville
Huntsville, Alabama 35807, U.S.A.

Y. NAKAGAWA
High Altitude Observatory
National Center for Atmospheric Research
Boulder, Colorado 80302, U.S.A.

M. DRYER
Space Environment Laboratory
NOAA/Environmental Research Laboratories
Boulder, Colorado 80302, U.S.A.

ABSTRACT

Recent progress in the dynamic modeling of responses of the corona and interplanetary medium to solar events (such as surges, eruptive prominences, flares, etc.) is reviewed. In particular, coronal transients and wave phenomena are discussed in some detail including pertinent mathematical requirements. Within the context of hydrodynamics and magnetohydrodynamics, a summary of both one- and two-dimensional time-dependent models is presented. A comparison of theoretical results with ground-based optical, radio, ATM Skylab observations, Pioneer 9 and Pioneer 10 solar wind observations is also presented.

It is illustrated that (i) substantial progress has been made in the theoretical (i.e., numerical) modeling of coronal and interplanetary responses within the last few years; (ii) two-dimensional, time-dependent modeling is needed to examine the details of nonlinear wave coupling, and (iii) theoretical results of modeling appear to reproduce physical consequences successfully. We suggest, therefore, that dynamic modeling is a promising and fruitful direction for the study of solar-interplanetary physics.

M. A. Shea et al. (eds.), Study of Travelling Interplanetary Phenomena/1977, 43–62.

I. INTRODUCTION

Recent progress in observational techniques and instrumentation has revealed many heretofore unknown dynamic phenomena on the sun as well as in the interplanetary environment. In order to provide a rational basis for interpretation of these phenomena, theoretical studies which can handle such non-linear phenomena have been pursued with numerical modeling (or simulation). It may be noted that this study of dynamics represents progress beyond the steady-state analyses used in the past.

At present, the study of dynamics is still in the state of infancy in comparison with the traditional steady-state analyses. However, it is apparent that dynamic studies are indispensable in the interpretation of time-dependent phenomena. In consideration of such values of dynamic studies, the current state-of-art of the dynamic modeling of coronal and interplanetary phenomena is briefly reviewed. The presentation is presented sequentially on the basis of distance from the sun so that the coronal responses are presented first in Section II, followed by the interplanetary responses in Section III.

In the final Section IV, an assessment of current research and the future direction of this line of research are discussed.

II. CORONAL RESPONSES

The observation of the faint solar corona, other than the visual observation of eclipses, started with the invention of the coronagraph. The instantaneous static picture of the corona indicating the gross outline of global solar magnetic fields and the implied solar wind flows along such field lines has gradually been replaced by recent observations of their temporal variations. However, not until the recent observations of "coronal transients" from space (ATM/Skylab experiments; for a recent summary, see the book edited by Kent et al., 1976) has the real dynamic nature of the solar corona been revealed. The common occurrence (several times per month during the 1973 Skylab period) of such transients, which travel up to speeds of approximately one thousand km s^{-1}, indicates that the traditional model of the quiet steady-state corona is at best suitable for the statistical averaged state. In addition, other dynamic phenomena in the corona have been observed in association with flares such as surges, sprays and eruptive prominences. Therefore, the theoretical modeling of the evolution of these phenomena represented the first challenge to the numerical technique. The first study of this nature, within the approximation of spherical symmetry (hereafter called the one-dimensional study), has been the natural consequence of the development of hydrodynamic and magnetohydrodynamic models. Subsequently, studies which considered other dimensional dependence (i.e., two- and three-dimensional models) have been developed. However, the only three-dimensional model presently available is the analysis of ray-tracing discussed later. In the following subsections, we shall discuss these models with emphasis on their physical significance.

A. One-dimensional, non-linear, hydrodynamic models

This type of model is designed to study basic physical processes responsible for observed non-linear phenomena which are apparently governed by or guided along flux tubes. Examples of such phenomena are surges, sprays, and eruptive prominences (Tandberg-Hanssen, 1966), as well as some coronal transients (McQueen et al., 1974). The first study of surges, sprays and eruptive prominences was made by Nakagawa et al. (1975) who examined the response of an isothermal static corona to finite amplitude perturbations of different physical quantities at the base of the corona. They succeeded in showing that the typical velocity evolution characterized by an initial rapid velocity increase followed by a slow down of the sprays and fast ejections (Valniček, 1964) can be correlated with large sudden enhancements of either temperature or velocity at the base of the corona. They also showed that the velocity evolution of eruptive prominences (a gradual increase of velocity followed by a large acceleration) can be understood in terms of the uplifting of large density; i.e., large density perturbations at the base of the corona. Some of their computational results are illustrated in Figure 1. It may be noted that sprays are produced with a temperature enhancement of long duration, while a fast ejection can result only with an initial velocity enhancement. The latter may be a consequence of the eruption of reconnected magnetic field lines. Surges are found to be the result of either velocity or temperature pulses, i.e., enhancements of short duration. Wu et al. (1975) have applied a similar model to the flare-prominence of 30 April 1974. In this case, in order to simulate part of the H_α-observed events, it was necessary to introduce a combined density-temperature pulse (i.e., a pressure pulse) as well as a frictional force proportional to local velocity.

For the modeling of coronal phenomena some distance from the sun, the effects of the solar wind cannot be ignored. Thus, in the subsequent study, the presence of the solar wind has been taken into account. The model of this nature has been developed by Nakagawa and Steinolfson (1976) who have examined the nature of the basic mixed initial-boundary value problem and provided the basic mathematical foundations for the model computation. An example of the results in which such a method is utilized (Steinolfson and Nakagawa, 1977) is shown in Figure 2. It can be seen from the figure that the 10 June 1973 coronal transient observed by the white light coronagraph on Skylab is successfully reproduced by a pressure pulse. Steinolfson and Nakagawa (1976) also showed that violation of the proper initial-boundary value problem leads to the formation of a non-physical shock.

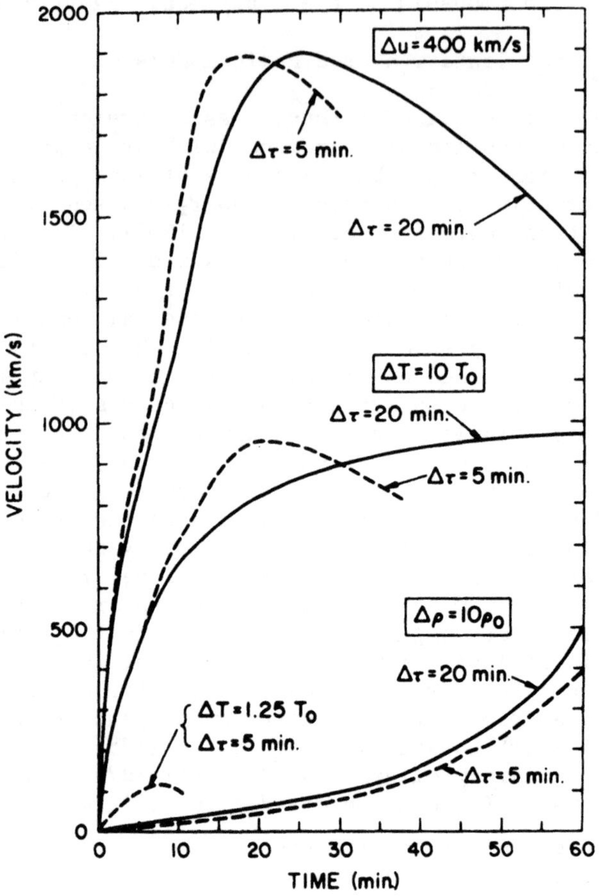

Figure 1. Evolution of velocity of coronal disturbances due to different physical causes for pulse durations of τ = 5 min and τ = 20 min. Note that the surge-like velocity curve results from a smaller temperature pulse $\Delta T = 1.25\ T_o$ of duration τ = 5 min. (Nakagawa et al., 1975).

B. Two-dimensional, non-linear, magnetohydrodynamic (MHD) models

The confinement of flow along magnetic flux tubes (namely, the flow which may be represented by the one-dimensional hydrodynamic model) may be a reasonable representation for the solar plasma when the magnetic field is dominant. However, it is apparent for energetic events that the magnetic field can also be modified. In such circumstances, two-dimensional MHD modeling is necessary. The non-linear, time-dependent, two-dimensional MHD model has been studied recently by Nakagawa et al. (1978) and Wu et al. (1978). These authors used the approximation of a plane analysis in which all physical variables are confined within a

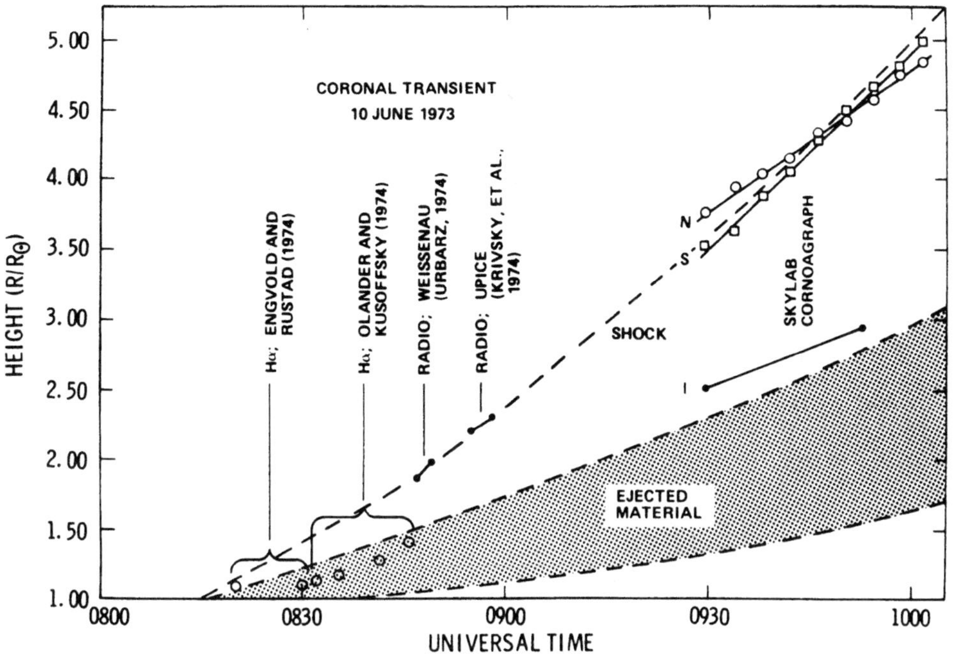

Figure 2. Overlay of the simulated shock and ejected material trajectories on the observations for a pressure pulse, $\delta p = 1.6$ (density enhancement being 0.4 and temperature enhancement being 4.0). The shaded region represents the outward progression of all the material ejected during the solar event. The shock fits the radio and white light observations; the density compression followed by a rarefaction is consistent with observations; and the hot, ejected material may be responsible for the bright knot. (Steinolfson and Nakagawa, 1977).

plane so that the non-compressible transverse (Alfvén) waves are excluded, but the fast and slow modes are included. Nakagawa et al. (1978) demonstrated the response of the solar corona to a sudden temperature perturbation (simulating a sudden local release of energy in a flare) on the configurations of the magnetic fields as well as on the field strengths. In particular, they showed that "bubble-like" density enhancements, similar to those seen as coronal transients, propagate in open (mostly radial) magnetic field configurations. In the closed (mostly azimuthal) magnetic field configuration, a density enhancement develops near the foot-point of field lines in the form of a pair of "horns" seen sometimes by the K-coronagraph after flares (Hansen et al., 1974). A typical evolution of the density enhancements for those two field configurations is shown in Figure 3 in which β, the ratio of gas pressure to magnetic pressure, is initially 0.7 at the

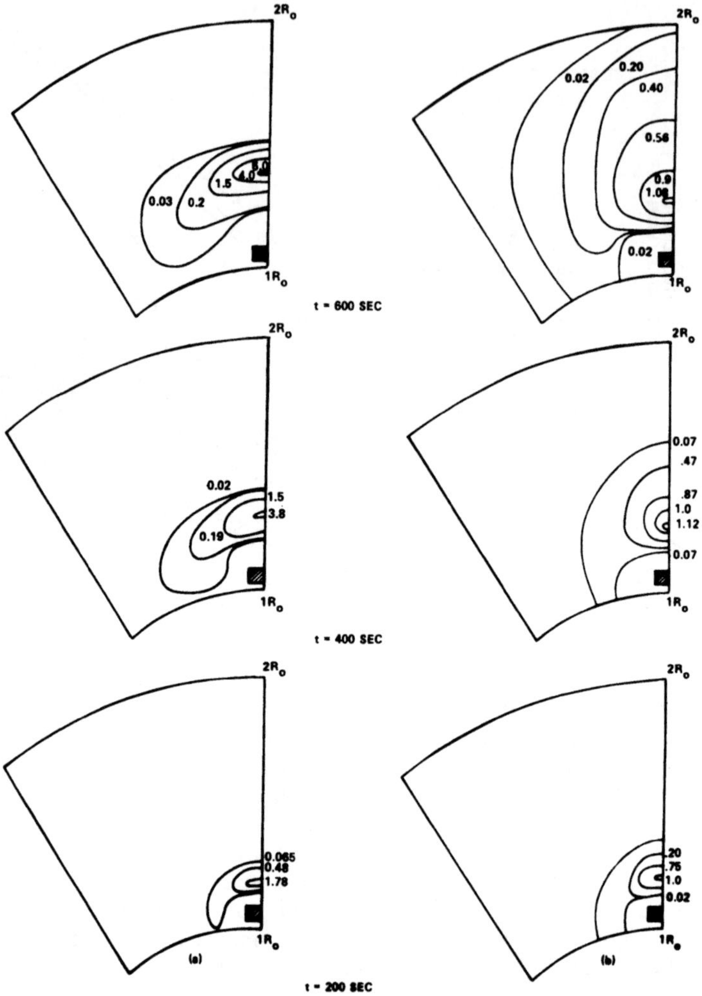

Figure 3. The evolution of density variation $\left(\frac{\Delta\rho}{\rho_0}\right)$ in (a) the open and (b) the closed magnetic field configurations for $\beta = 0.7$. (Nakagawa, Wu and Han, 1978).

bottom of the atmosphere on the axis of symmetry. It may be noted that β increases azimuthally away from the center in the closed field and decreases in the open field while it always decreases with height close to the sun. In their analysis (Nakagawa et al., 1978), it was noted that the basic characteristics of the density enhancement can be explained in terms of the anisotropic propagation and coupling of the fast- and slow-MHD waves. The details of the numerical results were also compared with the non-magnetic case by Wu et al., (1978) who demonstrated the channeling as well as the blocking effects of the magnetic field to the plasma motion, including the evolution of the fast- and slow-MHD shock waves, as shown in Figure 4. This figure shows the velocity vectors and the evolution of the magnetic topology for the open and closed field configurations (β = 0.7) as well as the non-magnetic case. The plasma flow is channeled mostly along the magnetic field lines immediately behind the weak parallel MHD shock. Away from the axis of symmetry, however, the development of a slow MHD shock can be noticed by the appearance of azimuthal velocity. The anisotropic expansion of the perturbation can be recognized by comparing the flow pattern with the non-magnetic case (Figure 4a) which clearly shows a uniform expansion from the location of the initial perturbation. The deformation of the magnetic field is illustrated on the right hand side of Figure 4b, in which the initial (dashed) and perturbed (solid) field lines are shown for comparison. It is clear that the field lines are pushed away by the plasma flow. Comparison of the velocity vectors with the deformed magnetic field clearly shows the separation of the fast and slow waves discussed by Nakagawa et al., (1978). In the case of the closed field configuration, again with β = 0.7 and the same temperature pulse, the results are shown in Figure 4c. The development of a very strong combination of radial and azimuthal flow can be seen. This flow pattern is attributed directly to the nature of the propagation of the fast and slow waves. The plasma flow is almost always perpendicular to the field lines, thereby giving maximum local effects of the Lorentz force. Again, the combination of the radial development and azimuthal decay of the perpendicular (almost everywhere) MHD shock is clearly indicated. On the right hand side of Figure 4c, we can clearly see the loop-type field lines being lifted up by the outward expansion of the infinitely-conducting plasma flow. Thermal conduction and radiative losses are expected (but not yet proved) to have minor effects on the bulk motion shown here.

Another plane, two-dimensional, MHD numerical model was examined by Nakagawa et al., (1976) in a Cartesian coordinate system in relation to the possible energy build-up due to the temporal change of magnetic fields in the solar atmosphere. For a given initial, two-dimensional, dipole-like field with an increase of the field strength, plasma flow developed and lifted the central part of the field lines, while the plasma flow was downwards near the foot points of the field lines. The general characteristics shown in Figure 5, therefore, resemble somewhat the case of closed field lines in the previous case. The results were found to be strongly dependent on the value of β. For example, representative cases of small β (0.03) and intermediate β (3.0) displayed a fountain-like behavior with fluid moving upward near the loop edges. However, for the case of the intermediate β, the fluid eventually (after ~ 15s) flowed downward throughout the lower portion of the loop carrying the magnetic field with it and, thereby, resulted in a rapid decrease in the magnetic energy and an increase in the kinetic energy. The behavior is thought to resemble the rapid disappearance of a prominence by flowing down into the chromosphere.

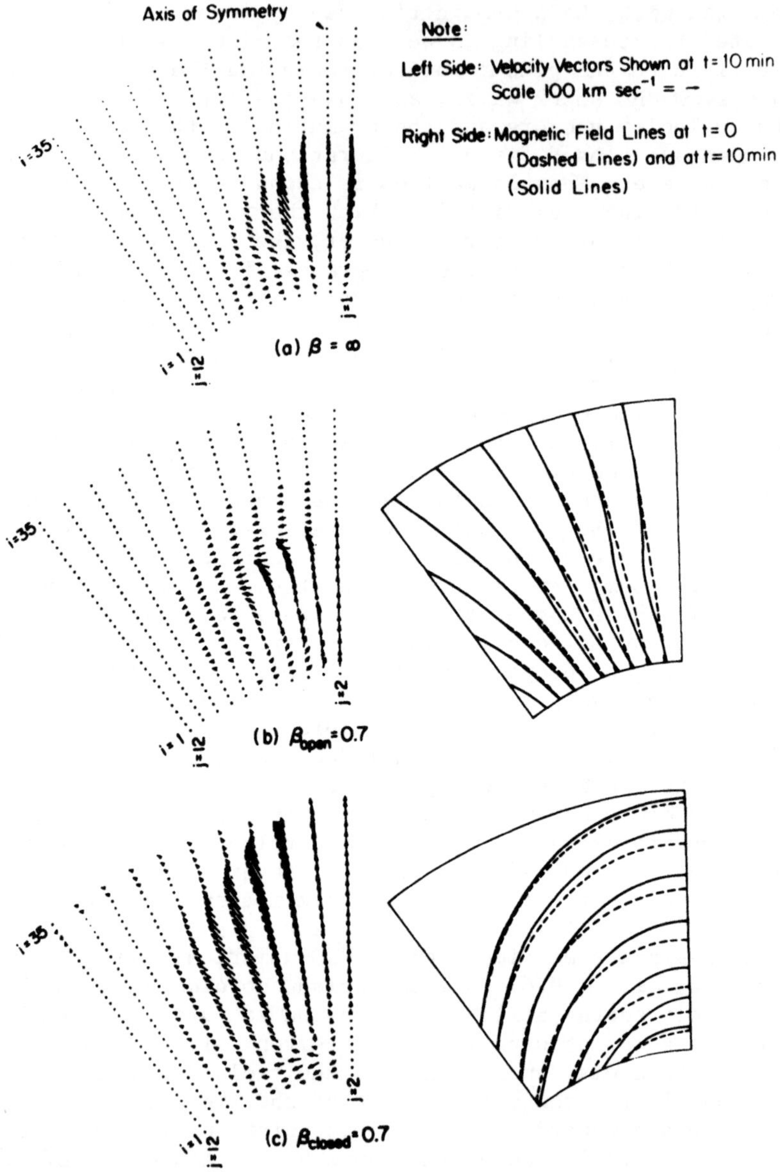

Figure 4. The velocity and magnetic fields resulting from an impulsive temperature enhancement ($T = 10\ T_o$) of 5 sec duration. (a) Velocity vectors in a nonmagnetic flow at t = 10 min. (b) Velocity vectors and magnetic field lines in an open magnetic configuration with $\beta = 0.7$ (computed at i = 1, j = 2). Magnetic field lines are plotted for t = 0 (dashed) lines and for t = 10 min (solid) lines. The fast parallel MHD shock wave decays into a fast transverse wave away from the axis of symmetry.

Figure 4 (continued)
(c) Velocity vectors and magnetic field lines in a closed magnetic configuration with $\beta = 0.7$ (computed at $i = 1$, $j = 2$). Magnetic field lines are plotted for $t = 0$ (dashed) lines and for $t = 10$ min (solid) lines. Note the development of the fast perpendicular MHD shock wave as indicated by the envelope of the sudden velocity vector enhancements at $j \approx 1$-7 and the gradual decay at $j \gtrsim 8$. The asymmetrical shock structure is shown by the varying radial position $22 \lesssim i \lesssim 33$. (Wu et al., 1978).

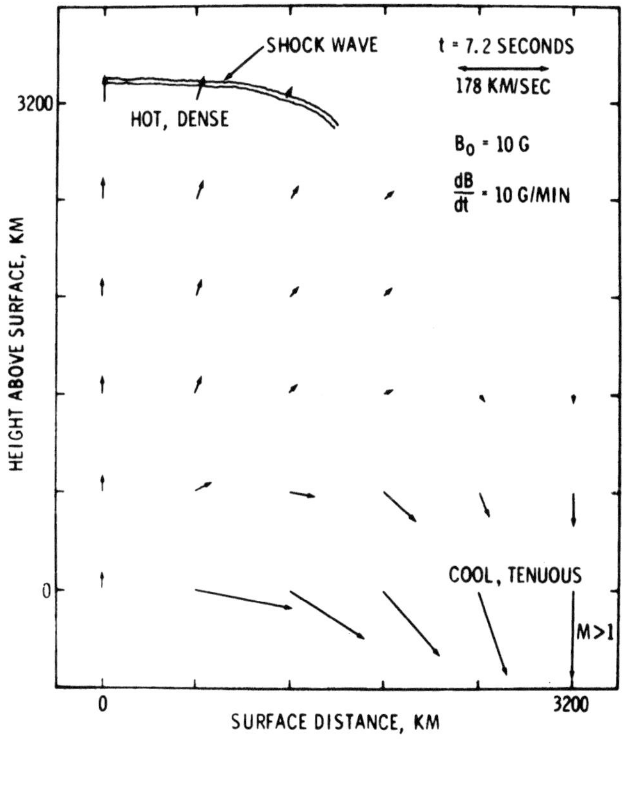

Figure 5. The velocity field for $B_o = 10$ G and $\frac{dB}{dt} = 10$ G min^{-1} ($\beta = 0.03$) after 7.2 s. The arrows represent the relative magnitude and direction of the velocity at the origin of the arrows. Only velocities greater than 15 km s^{-1} are indicated. (Nakagawa, et al., 1976).

C. Three-dimensional, linear, MHD models

The three-dimensional MHD study has only been performed within the linearized approximation because of the complexity of the non-linear problem which is still beyond our present capability. The linearized analysis is partially justified because of the likelihood of development of fast MHD wave velocities which could reduce the amplitude of disturbances rapidly with distance from the site of the initial disturbance. In other words, for the study of large scale propagation, the linear ray-tracing analysis can provide insight into the general behavior. On such grounds, Uchida et al., (1973) performed a general three-dimensional ray-tracing of the fast MHD waves to show that the flare-produced Moreton waves can be explained as the trace of the MHD fast wave within the chromosphere. Uchida (1974) extended the analysis to show that type II radio emission associated with flares can be considered as the consequence of such fast MHD waves impacting upon high density regions in the outer corona. The basic physical argument in the latter case was based on the formation of MHD shocks in the region of high density where the phase velocity of the fast MHD waves would exceed the local Alfvén wave speed. The detailed mechanisms of radio emissions through the formation of MHD shocks, however, requires further consideration of plasma turbulence. Uchida (1974) also suggested that these escaping masses behind the finite-amplitude MHD waves may act as a piston to generate the interplanetary shock observed near the earth (Hundhausen, 1972; Dryer, 1974, 1975). However, no computation has ever been performed.

III. INTERPLANETARY RESPONSES

On the basis of observation, it has been recognized (Scarf and Wolfe, 1974; Smith and Wolfe, 197) that solar disturbances propagating through the corona produce the interplanetary disturbances. This recognition is now firmly established by the successful observational programs initiated since the launch of interplanetary space probes more than a decade ago. However, somewhat similar to the dynamic responses of the solar corona, a systematic theoretical study to provide a rational basis of physical interpretation of the observations is still lacking. Nevertheless, a number of authors (Dryer et al., 1976; Wu et al., 1976, etc.) have recently begun to report the results of theoretical model studies of this problem within the context of hydrodynamics and magnetohydrodynamics.

A summary of observational and theoretical studies of interplanetary shocks due to solar flares has been given by Hundhausen (1972) and Dryer (1974, 1975). Therefore, in this section we present a brief review of recent developments of the theoretical models of the interplanetary responses.

A. Hydrodynamic models

With the assumption that either the effect of interplanetary magnetic field is negligible or that disturbances propagate along field lines, we can adopt the hydrodynamic model. Hundhausen and Gentry (1969a, b) were the first to show that the duration of flare-generated disturbances is important in determining the nature of disturbances in the solar wind at 1 AU. In their spherically symmetric analysis, the perturbations were introduced in the supersonic solar wind, i.e., beyond the critical point. Their results illustrated that, for the duration of the perturbation in a time that is short compared to the transient time to 1 AU, blast-wave like disturbances result, while for longer perturbations, the response resembles piston driven shocks. Recently, Hundhausen (1973a, b) has extended this model analysis to the evolution of stream-generated shocks in the heliocentric range from 0.4 to 6 AU. Using a triangular temperature pulse of several hours' duration (related to shock transient time to 1 AU), it was demonstrated that a double shock ensemble can be formed in interplanetary space. More recently, Wu et al., (1976) used a similar one-fluid hydrodynamic model of radially-expanding and polytropic flow (i.e., one-dimensional hydrodynamic model) to compute the propagation (initiated outside the critical point) of disturbances beyond 1 AU for separate perturbations of density, temperature and velocity. Specifically, these separate pulses of 20 min duration were taken as the manifestation of different solar activities such as the eruptive prominence, surge and spray as well as the solar flare. The correlation between density pulses and eruptive prominences, velocity pulses and sprays as well as flare-generated shocks, and between temperature pulses and flares has been discussed previously (Nakagawa et al., 1975). In particular, Wu et al. (1976) have shown for the June 1972 flare as an example that the specific physical disturbances can be discriminated by their model. Figure 6 illustrates the comparison of observations and calculated trajectories of the interplanetary shock waves resulting from separate density, temperature and velocity pulses. It can be seen that agreement between the theoretical results and observations by Pioneer 10 (1.6 AU) for the shock trajectory time is obtained by the velocity pulse. In addition, reasonably good agreement with Pioneer 10 observations was found for their calculated plasma properties (i.e., variation in density, temperature and velocity).

Dryer et al., (1976) used a similar model to examine the August 1972 flares in detail. They used rather complex time-dependent pulses to simulate successive flares at 1.01 solar radii (i.e., inside the critical point) and followed the disturbances through the critical point beyond 1 AU. Figure 7 shows the resulting variations of plasma properties between 0 to 1 AU at 25 hrs (a) and 30 hrs (b). This figure also clearly indicates the shock-shock interactions and the evolution which results from such complex flares. For example, in Figure 7a, it can be seen that the first flare on 0316 UT August 2 produced a shock, F_1, while the second flare on 1959 UT August 2 produced a forward-

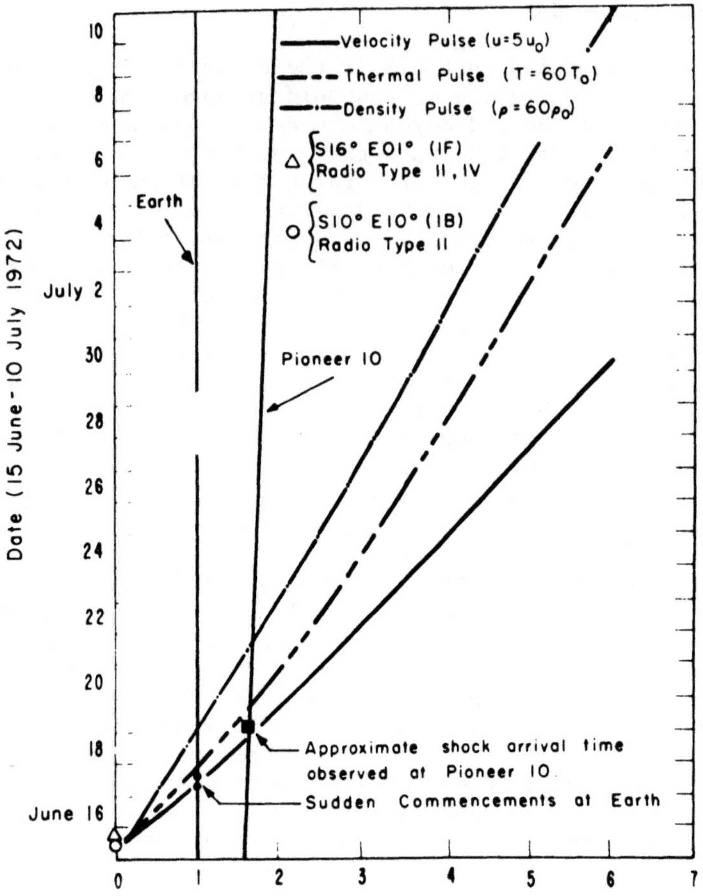

Figure 6. Calculated trajectories of the interplanetary shock waves resulting from density, temperature and velocity pulses initiated at ~ 1200 UT on 15 June 1972 and 0.08 AU for a duration of 20 min. Several flares (including position, optical importance and relevant radio burst) are indicated on 15 June 1972. Also indicated are sudden commencements at Earth and initiation of shock disturbance at Pioneer 10. (Wu et al., 1976).

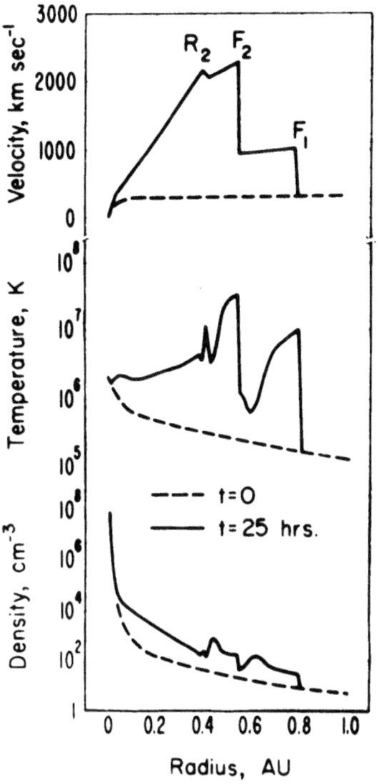

Figure 7a

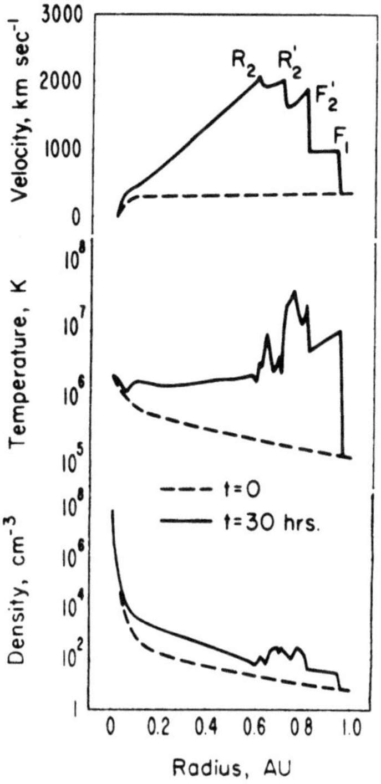

Figure 7b

Figure 7a. Simulated plasma parameters between 0 and 1 AU at t = 25 hours. Note that the disturbance from the second flare is overtaking that from the first flare. In particular, note that F_2 is about to overtake the density enhancement produced by the first flare's disturbance. The interaction of F_2 with this expansion produces F_2' and R_2' noted in 7b (Dryer et al., 1976)

Figure 7b. Simulated plasma parameters between 0 and 1 AU at t = 30 hours. Note that the second forward shock has developed into a new forward and a new reverse shock wave ensemble. This splitting occurred when F_2 overtook the density enhancement noted in Figure 7a. (Dryer et al., 1976).

reverse shock pair (F_2 and R_2). At 30 hrs (Figure 7b), F_2 has split into F_2' and R_2'; R_2' is the new reverse shock which is stronger than R_2 as can be seen by the large velocity, density, and temperature increases at 0.7 AU. It should be noted that the entire disturbance from the second flare is moving faster than that of the first flare, so that the heliocentric distance between F_1 and F_2' continues to diminish. This is expected from gasdynamics as the second disturbance is propagating into a higher temperature and velocity region.

Apart from these studies, a two-dimensional (i.e., the ecliptic plane) hydrodynamic model study has been made by Hirshberg et al., (1974). These authors showed that the solar wind response to a flare depends on the preflare solar modulations. Thus, in the presence of large azimuthal modulations, the flare response may be modulated but still detected at 1 AU in agreement with observational results.

B. Magnetohydrodynamic (MHD) models

The one-dimensional, time-dependent, non-linear MHD model has been used by Steinolfson et al., (1975) to study the comparison of interplanetary theoretical shock pair disturbances with space probe data. The comparison of MHD and OFD (ordinary fluid dynamic, i.e., hydrodynamic) models with space probe data, is shown in Figure 8a, 8b. It is apparent that the MHD results appear to agree better with the fluid properties observed by HEOS 1 than do the OFD results. This result, therefore, indicates the importance of the interaction between the plasma and magnetic field. Dryer and Steinolfson (1976) applied the same model to simulate the interplanetary disturbances caused by high speed streams starting from coronal holes. In this calculation they used two consecutive velocity pulses at 0.3 AU and the range of computation was extended to 10 AU. The result of this simulation in terms of the azimuthal interplanetary magnetic field is shown in Figure 9. The evolution of shock-shock interactions can be seen in the figure.

With reference to two-dimensional MHD models, the first study was made by Nakagawa and Wellck (1973) who confirmed most of the theoretical results inferred analytically by Matsuda and Sakurai (1972). The typical development and decay of azimuthal shocks and magnetic field obtained by Nakagawa and Wellck (1973) is shown in Figure 10. Another recent two-dimensional MHD model has been developed by Wu et al. (1977) for the examination of stream-stream interactions which start near the sun (~ 0.08 AU) and extend to interplanetary distances.

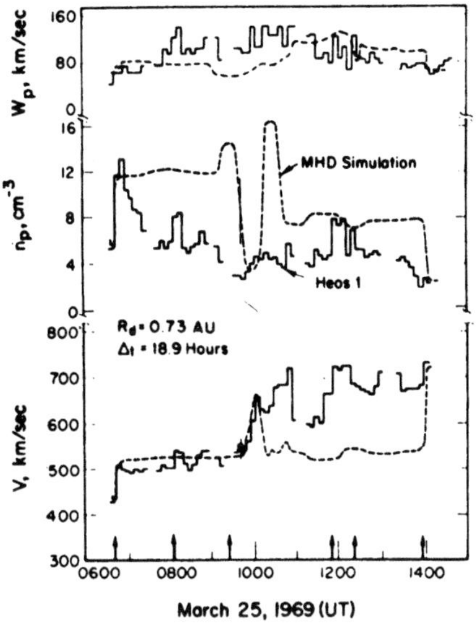

Figure 8a. Comparison of MHD simulation with 25 March 1969 event from HEOS 1. (Steinolfson et al., 1975).

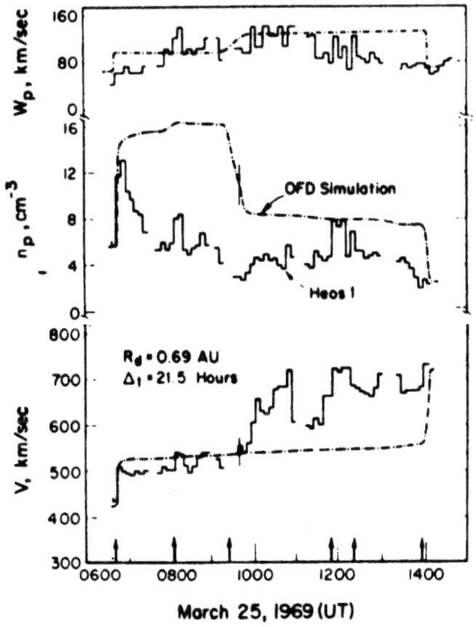

Figure 8b. Comparison of OFD simulation with 25 March 1969 event from HEOS 1. (Steinolfson et al., 1975).

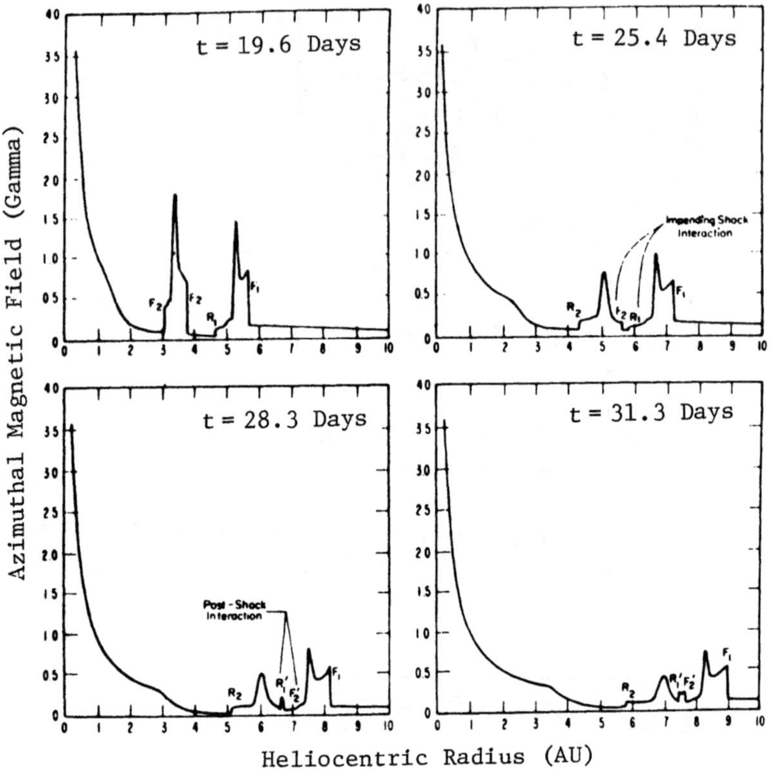

Figure 9. Profiles of azimuthal magnetic field between the inner boundary (0.3 AU) and 10 AU. The time t is the elapsed period since the initiation of the velocity pulses (Dryer and Steinolfson, 1976).

IV. CONCLUDING REMARKS

With the emphasis on numerical simulations, some recent developments in the study of non-linear dynamic phenomena in the solar corona and interplanetary medium are presented in this brief review. Even in the simplest case of one-dimensional, one-fluid hydrodynamics, it is apparent that such numerical simulations are valuable for the illucidation of basic physical processes. The subsequent more complex two-dimensional MHD models clarify the important roles of the magnetic field as well as the effects of additional geometric degrees of freedom.

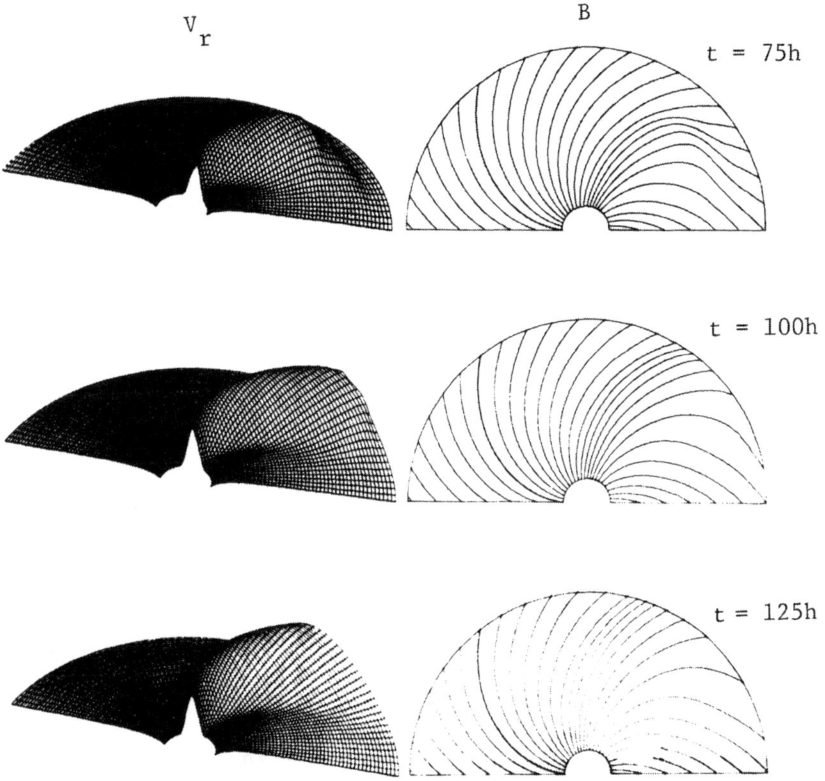

Figure 10. The evolutions of the radial velocity V_r and the magnetic field configuration denoted by B in response to stationary azimuthal modulations with respect to the Sun introduced at 30 R at $t \geq 0$; t refers to the lapsed time. (Nakagawa and Wellck, 1973).

Although the present review is confined to the single-fluid approach, the multi-fluid model is a natural consequence of the extension of such studies. In particular, in the multi-fluid approach additional physical processes such as the role of an electron gas in determining macroscopic transport properties becomes important. Also, it is necessary to consider different types of physical processes depending upon the circumstances. For example, in the examination of the chromospheric response to a solar flare, the effects of optical as well as X-ray radiation, electron-beam bombardment and heat conduction become important. These problems have been examined by a number of authors (Bessey and Kuperus, 1970; Craig and McClymont, 1976; Henoux and Nakagawa, 1977a, b; Kneer and Nakagawa, 1976; Kostyuk,

1975; Kostyuk and Pikelńer, 1975; Nakagawa et al., 1973; Somov and Syrovatsky, 1974; and Somov, 1975). These developments clearly indicate that the dynamic modeling of non-linear phenomena is one of the most promising and fruitful tools for the extraction of physically meaningful information from observational data. However, at present this line of study is still in the state of infancy; therefore, the authors hope this brief review will encourage future efforts along this line.

ACKNOWLEDGMENT

The work done by STW is supported by NASA/Marshall Space Flight Center under Contract NAS8-28097 and NOAA-Space Environment Lab./NOAA. The National Center for Atmospheric Research is sponsored by the National Science Foundation.

REFERENCES

Bessey, R. J. and Kuperus, M.: 1970, *Solar Phys. 12*, 216.
Craig, I. J. D., and McClymont, A. N.: 1976, *Solar Phys. 50*, 133.
Dryer, M.: 1974, *Space Science Reviews 15*, 403.
Dryer, M.: 1975, *Space Science Reviews 17*, 277.
Dryer, M. and Steinolfson, R. S.: 1976, *J. Geophys. Res. 81*, 5413.
Dryer, M., Smith, Z. K., Steinolfson, R. S., Mihalov, J. D., Wolfe, J. H. and Chao, J. K.: 1976, *J. Geophys. Res. 81*, 4651.
Hansen, R. T., Garcia, C. J., Hansen, S. F. and Yasukawa, E.: 1974, *Pub. Astron. Soc. Pacific 86*, 500.
Hénoux, J. C. and Nakagawa, Y.: 1977a, *Astron. and Astrophys. 57*, 105.
Hénoux, J. C. and Nakagawa, Y.: 1977b, *Astron. and Astrophys.* (submitted)
Hirshberg, J., Nakagawa, Y. and Wellck, R. E.: 1974, *J. Geophys. Res. 79*, 3726.
Hundhausen, A. J.: 1972, *Coronal Expansion and Solar Wind*, Springer-Verlag, New York.
Hundhausen, A. J.: 1973a, *J. Geophys. Res. 78*, 1528.
Hundhausen, A. J.: 1973b, *J. Geophys. Res. 78*, 2035.
Hundhausen, A. J. and Gentry, R. A.: 1969a, *J. Geophys. Res. 74*, 2908.
Hundhausen, A. J. and Gentry, R. A.: 1969b, *J. Geophys. Res. 74*, 6229.
Kent, M. I., Stuhlinger, E., and Wu, S. T., ed: 1976, *Scientific Investigations on the Skylab Satellite*, Progress in Astronautics and Aeroanutics, Vol. 48, American Institute of Aeronautics and Astronautics, New York.
Kneer, F. and Nakagawa, Y.: 1976, *Astron. and Astrophys. 47*, 65.
Kostyuk, N. D.: 1975, *Sov. Astron. 19*, 458.
Kostyuk, N. D. and Pikelńer, S. B.: 1975, *Sov. Astron. 18*, 590.
Matsuda, T. and Sakurai, T.: 1972, *Cosmic Electrodyn. 3*, 97.
MacQueen, R. M., Eddy, J. A., Gosling, J. T., Hildner, E., Munro, R. H., Newkirk, G. A., Jr., Poland, A. I. and Ross, C. L.: 1974, *Ap. J. (Letters) 187*, L85.

Nakagawa, Y., Wu, S. T. and Han, S. M.: 1973, *Solar Phys.* *30*, 111.
Nakagawa, Y., and Wellck, R. E.: 1973, *Solar Phys.* *32*, 257.
Nakagawa, Y., Wu, S. T. and Tandberg-Hanssen, E.: 1975, *Solar Phys.* *41*, 387.
Nakagawa, Y. and Steinolfson, R. S.: 1976, *Ap. J.* *207*, 296.
Nakagawa, Y., Steinolfson, R. S. and Wu, S. T.: 1976, *Solar Phys.* *47*, 193.
Nakagawa, Y., Wu, S. T. and Han, S. M.: 1978, *Ap. J.* (in press).
Scarf, F. L. and Wolfe, J. H.: 1974, *J. Geophys. Res.* *79*, 4179.
Smith, E. J. and Wolfe, J. H.: 1976, in M. Rycroft (ed.), *Space Research XVI*, Akademie-Verlag, Berlin.
Somov, B. V. and Syrovatsky, S. I.: 1974, *Solar Phys.* *39*, 415.
Somov, B. V.: 1975, *Solar Phys.* *42*, 235.
Steinolfson, R. S. and Nakagawa, Y.: 1976, *Ap. J.* *207*, 300.
Steinolfson, R. S. and Nakagawa, Y.: 1977, *Ap. J.* (in press).
Steinolfson, R. S., Dryer, M., and Nakagawa, Y.: 1975, *J. Geophys. Res.* *80*, 1989.
Tandberg-Hanssen, E.: 1966, *Solar Activity*, Blaisdell Publ. Co., Waltham, Mass.
Uchida, Y.: 1974, *Solar Phys.* *39*, 431.
Uchida, Y., Altschuler, M. D. and Newkirk, G., Jr.: 1973, *Solar Phys.* *28*, 495.
Valniček, D.: 1964, *Bull. Astron. Inst. Czech.* *15*, 207.
Wu, S. T., Dryer, M., McIntosh, P. S. and Reichmann, E.: 1975, *Solar Phys.* *44*, 117.
Wu, S. T., Dryer, M., and Han, S. M.: 1976, *Solar Phys.* *49*, 187.
Wu, S. T., Han, S. M., and Dryer, M.: 1977 (abstract) *EOS Trans. AGU*, *58*, 485.
Wu, S. T., Dryer, M., Nakagawa, Y., and Han, S. M.: 1978, *Ap. J.* (in press).

DISCUSSION

Jockers: Does your two-dimensional MHD code allow for interaction of the magnetic field with the plasma? In particular, can the topology of the field change from closed to open?

Wu: Yes, the code allows for self-consistent interaction of the magnetic field with the infinitely-conducting plasma. The magnetic diffusion mechanism is not included in the model, thus the model could not provide for the transition of magnetic field topology from an open to closed configuration or vice versa.

Jockers: Do your calculations allow conclusions with regard to the HAO coronal transients?

Wu: Yes. We have explicitly solved a self-consistent, time-dependent model for MHD flows in the corona. Basic flow responses under several representative conditions of plasma Beta and magnetic topologies are demonstrated. We have shown that plasma density increases occur in a spatial and temporal form which bears a strong resemblance to white light coronal transients resulting from density enhancements. We differ with Dr. Hildner's view that these enhancements are contained within expanding magnetic flux tubes. Alternatively, we found that the major characteristics of coronal transients are found in our "open", rather than "closed" magnetic topologies.

Armstrong: What is the nature of the equilibrium used to start the 2D, MHD model? Is it static? How accurate is it in equilibrium?

Wu: We first establish a magnetohydrostatic, isothermal atmosphere. This equilibrium solution is absolutely accurate within the context of the model; it is changed only by a perturbation in any of the physical parameters. With respect to empirical models (such as Newkirk's), the density compares to within 5% or less.

Wallis: I question the feasibility of realistically calculating the evolution of disturbances over a large distance range, to several AU. 1-dimensional modeling is poor; 2-dimensional, "limited to the equatorial plane", is theoretically suspect. In the absence of a 3-dimensional model one might just as well use a one-dimensional expansion $\sim r^{-n}$ with index n exceeding 2 (e.g., 3 or 4).

Wu: If only a one-dimensional model with expansion $\sim r^{-n}$ is used, then the details of nonlinear MHD waves and their interactions discussed by Nakagawa et al. (1978), Wu et al. (1978) and Steinolfson et al. (1977) would not be able to be demonstrated. If the one-dimensional approach which you suggested was used, then these essential two-dimensional characteristics would not be shown no matter what value is taken for n.

Steinolfson, R. S., Wu, S. T., Dryer, M., and Tandberg-Hanssen, E.: 1977, in M. A. Shea, D. F. Smart, S. T. Wu (eds.), Contributed Papers to the Study of Travelling Interplanetary Phenomena/1977; Air Force Geophys. Lab., Bedford, Massachusetts, U.S.A.

PART II

SOLAR RADIO ASTRONOMY AND INTERPLANETARY SCINTILLATIONS

SOLAR STEREO RADIOASTRONOMY

J.-L. STEINBERG
Département de Recherche Spatiale
Observatoire de Paris, F- 92190 MEUDON, France

ABSTRACT

Solar radio bursts can be used to probe the corona (bursts of types I and III) and the interplanetary medium (type III). Simultaneous observations in two widely different directions can yield the angular distribution of some burst characteristics or their position in space.

Type I bursts stereo observations were made in cooperation between the USSR and France, at 169 MHz (STEREO-1 experiment). They have shown that type I often radiate into a beam of less than 25° half-power width. When compared to other statistical data on size and space-time properties of type I bursts, these observations show that :

- the source may be smaller than seen from the ground (as required by theoreticians) and located in overdense fibers or arches which extend up to 0.8 solar radius altitude.
- it radiates at an angle to the magnetic field smaller than 40°.
- the corona above active regions is made of rather closely packed fibers or arches ; random scattering takes place in that medium and can explain most type I observations.
- the measured directivity is only compatible with some of the many proposed mechanisms.

Type III bursts when observed on meter waves during the STEREO mission are much less directive than type I bursts but more than anticipated from ray tracing in a quasi spherical corona. There is evidence of large scale dense coronal structures which are probably streamers. Models of the occultation of a point source by such structures have been built which can fit the observations. On hectometer waves, the directivity of type III bursts has been measured using two spacecraft.

On hecto- and kilometer wavelengths, multi-spacecraft observations using IMP-6 and 8, RAE-2, HELIOS-A and B and HAWKEYE-1 allowed 3-dimensional tracking of the 10-100 keV electrons which produce the radio events. Some of these observations are not model-dependent ; they yield maps of interplanetary magnetic field lines even out of the ecliptic

M. A. Shea et al. (eds.), Study of Travelling Interplanetary Phenomena/1977, 65–80.

and electron density models.

Although solar stereoradioastronomy is still very young, it holds great promise as one of the few techniques which can produce a 3-dimensional picture of the corona and interplanetary medium which is essentially free of the effects of integration along the line of sight.

INTRODUCTION

Associated with flares and other less spectacular transient phenomena which occur in active centers, particles are ejected from the sun and shock waves produced which later travel through the upper corona and the interplanetary medium. Shock waves can be observed with radioheliographs and space coronagraphs up to several solar radii altitudes while shocks and particles can be detected in situ providing the proper space-borne instruments are at the right place at the right time.

It is very fortunate that, over large portions of their trajectory, high energy electrons as well as shocks also trigger radio emissions known as radio bursts.

Solar radio bursts are intense and localized events which are produced by non-thermal mechanisms. The necessary external energy is brought in the plasma by fast electrons and shock waves and, through conversion processes which are not yet fully understood, electromagnetic waves are radiated. For any conversion mechanism to be efficient enough, the frequency of the radio waves must be close to one of the resonance frequencies of the plasma or its harmonics ; these particular frequencies are functions of the electron density N_e and the magnetic field B which vary over the corona. Therefore, when observed in a limited frequency band, the source of these emissions must be found where the plasma parameters obey some definite relationship and its size must be limited. Moreover close to the resonance frequencies, the index of refraction of the coronal plasma goes to zero or takes very large values ; on their path to the observer, the radio waves soon reach conditions where that index goes back to unity so that peculiar propagation effects including focusing can be expected. Strong refraction or even occultation of the radio source can be produced by relatively small variations of the plasma parameters N_e and/or B which could not be detected by conventional methods.

Being intense and localized, radio burst sources can be used as tracers in two different ways :

- If we can get some information on the angular distribution of their radiation, we can learn somthing on the radiation mechanism and on the medium where the propagation takes place towards us : scattering by small scale inhomogeneities or refraction and/or occultation by large structures.
- When we can localize these events at several frequencies, we can reach the trajectory of the exciter and its velocity, plus the value of

some plasma parameters at every observed point along the trajectory. To a large extent, the data obtained is free from the effects of line-of-sight integration.

In the last few years, extensive ground-based and space observations of these natural tracers have been carried out. Although this was just the beginning of stereo radioastronomy, some interesting results have been obtained which we shall now review.

THE DIRECTIVITY OF TYPE I BURSTS AND THE CORONAL STRUCTURE ABOVE ACTIVE CENTERS.

Type I bursts are very frequently observed phenomena in the meter-wavelength range. They are very short (some tenths of second), circularly polarized in general and narrow-banded (relative frequency bandwidth : about 2%). When observed at frequencies between 150 and 200 MHz, their altitude above the photosphere ranges from 0.2 to 0.8 solar radius, much higher than the critical level in the quiet corona ; they are mostly observed above active centers.

It has been known for a long time that some of the characteristics of type I bursts vary with the direction of observation : their probability of occurrence and their polarization ratio are maximum near the center of the disc ; some type I seem to move over their lifetime and these are more numerous when seen on the limbs. However center-limb statistical studies can yield no definite information on the angular distribution of any observable parameter of individual events unless some hypothesis is made beforehand : for instance that the axis of the beam of radiation is radially oriented. If we cannot make such an assumption, we must carry simultaneous observations from two widely different directions and, only in that way, can we provide unambiguous evidence of directivity (Steinberg and Caroubalos, 1970).

The first experiment of that kind called STEREO-1 was carried out in 1971-1972 as a cooperative program between the USSR and France. A receiver operating at 169 MHz was flown on the Soviet Mars-3 probe (Epstein, 1974). Observations on Earth were made in the USSR and Nançay (France). This space mission provided striking evidence of a strong beaming of type I radiation(fig.1 from Caroubalos and Steinberg, 1974). Indeed a detailed analysis of the data yielded an upper limit of 25° for the half power beamwidth of the bursts observed on October 18, 1971 (Steinberg et al., 1974). The orientation of the beam axis was found to change in a random way from event to event by no more than a few degrees rms.

Before discussing the implications of these observations, the limitations of the STEREO technique should be pointed out : the use of two lines of sight does not provide a complete 3-dimensional picture of the burst polar diagram but only some information on the directivity of a section of that diagram by the ecliptic plane. On the other hand, such an experiment yields only two sets of data on a phenomenon which can be

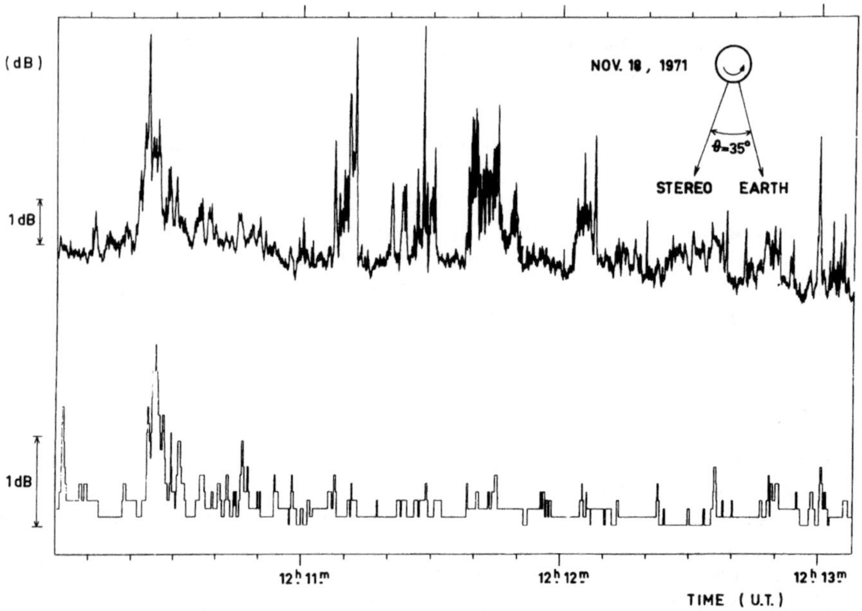

Figure 1 : Uncorrelated records of type I bursts obtained simultaneously on November 18, 1971 on the Earth (top) and on board Mars-3. (Caroubalos and Steinberg, 1974).

characterized by several parameters, some of them being random variables; for instance, in the STEREO experiment, the intensities of the bursts measured simultaneously from the Earth and Mars-3 are two random variables which are more or less correlated ; so that the model used to interpret the data should not contain more than two random variables and it might therefore be an oversimplification of the real situation. Nevertheless the STEREO-1 data contain interesting information and we shall now see how it can be used.

The size of type I sources as observed with ground-based radioheliographs ranges from the resolution limit of these instruments (about 50 000 km at 169 MHz) to more than 150 000 km in some instances. The scale height of the variations of the main plasma parameters, electron density N_e and magnetic field intensity B in the regions where type I are observed is smaller than the source size ; the expected relative bandwidth of the radiation should therefore be close to 50% while observations yield 2%. Therefore the primary source must be much smaller than the object we see. Indeed scattering mechanisms have been considered since 1960 as necessary ingredients of any type I theory. Models based on the extrapolation to type I altitudes of the inhomogeneous model deduced from interplanetary scintillation data have been used extensively (Fokker, 1965 ; Steinberg et al., 1971 ; Riddle, 1974) even though there was little justification for it.

In such models, the building up of the broadened scattered image of a point source is produced by the random refraction of the radio rays which are, at the same time, spread angularly ; and there is a definite relationship between the scattered image size and the final beamwidth of the observed radiation. This relationship is weakly dependent upon the model parameters as long as there are no reflecting inhomogeneities. For a maximum beamwidth of 25°, these models cannot account for an observed source size larger than about 0.7' or 35 000 km.

Moreover, if most of the scattering takes place in the corona over a long path length, the average size of type I bursts should increase at the limbs (since the total path length through the corona increases with the source longitude) and all sources should be about the same size at the same longitude. This is not the case (Kerdraon, 1973 ; Bougeret, 1977a).

It should also be noted that, within a given storm center, a small number of localized bursts sources is found (Bougeret, 1973), each of them radiating bursts with definite size and space-time properties. Therefore most of the scattering must take place in or very close to the source or that source must be made of a number of discrete sources each of them with a diameter smaller than 0.7', all of them triggered about simultaneously.

In the first case, a scattering process which preserves the primary source directivity must be found since the old weak-inhomogeneity model does not satisfy that requirement. There is evidence from high resolution eclipse photographs and from Skylab X-ray and EUV observations that above active centers the corona is highly inhomogeneous and contains arches and loops whose electron density is very different from that of the ambient corona. We suggest (Bougeret and Steinberg, 1977) that there are many more loops than the ones we see and which are the brightest ones. So that scattering by these overdense fibers may provide the required mechanism which broadens the small sized primary source to make it look as large as actually seen while preserving its directivity in planes containing the fibers.

If we accept the second interpretation, we must explain the existence of the discrete sources. It seems hard to explain it if they are located in the homogeneous background corona where the loops are embedded. It is much more likely that these discrete sources lie in different overdense fibers and, in any case, their altitude is so large that they have to be in overdense inhomogeneities. These discrete sources must be triggered by electrons injected into the structure from a common source, low in the corona ; there is not much difficulty in this simultaneous triggering since the duration of type I bursts (0.2 to 0.5 s) is comparable to the time taken by light to travel across them. But, if this picture is the correct one, we must again consider the propagation of radio radiation in bundles of fibers.

Therefore, it seems difficult to escape the conclusion that, above active centers, the corona is highly structured and made of dense fibers most of which are not bright enough to be seen in the X-ray or EUV. This result seems valid in a range of altitudes from 0.2 to 0.8 solar radius above the photosphere where only radio waves can bring us information. Lower in the corona, optical observations led to the same conclusions that most of the material is concentrated into inhomogeneities which occupy between 0.2 to 0.05 of the volume (Allen, 1975 ; Leroy and Trellis 1974).

Another result is that type I beamed emission is most probably oriented along the magnetic field and produced in the fibers. Indeed, we should mention that some type I sources have been found to radiate 60° away from the solar vertical (Bougeret, 1977b). If confirmed by the complete processing of all STEREO-1 data, these results should give us an important clue to the understanding of the radiation mechanism ; it is clear that only some of the proposed mechanisms can produce such a beam out of a fiber.

THE DIRECTIVITY OF TYPE III BURSTS

Type III solar radio emission.

As a function of time, the spectrum of type III bursts drifts

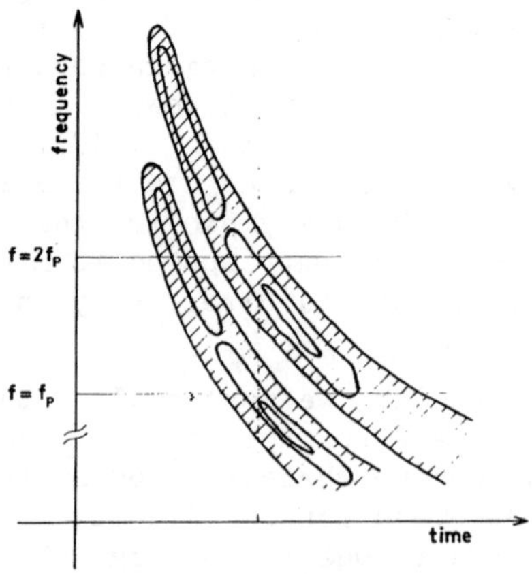

Figure 2 : Dynamic spectrum of a type III burst with two components : fundamental and harmonic.

towards the lower frequencies ; it is sometimes made of two bands, showing about the same detailed features at the same time and appearing in harmonic relation (fig. 2). As early as 1950, the mechanism of these events was supposed to involve an exciter ejected from the sun at a velocity of about c/3 which could trigger plasma waves along its path ; these plasma waves are converted into electromagnetic waves at the local plasma frequency $f_p = 0.009 \sqrt{N_e}$ (f_p in MHz, $N_e = cm^{-3}$) and its first harmonic at $2f_p$.

Although the detailed mechanism of production of the electromagnetic waves is not yet fully understood, the plasma hypothesis has been worked out in details and it can account for most type III properties. However, two of them require, to be explained, additional hypothesis :
i) If the fundamental component is radiated at the plasma frequency, the index of refraction for electromagnetic waves is very small but rises rapidly to unity as the waves leave the source region. Focusing is predictable which is not observed since type III fundamentals are seen on the limbs. One can eliminate this difficulty by assuming the existence of small scale inhomogeneities in electron density. Ray tracing computations (Steinberg et al., 1971 ; Riddle, 1974) confirm that scattering from weak inhomogeneities can make the fundamental components visible on the limbs while producing a scattered image of the primary point source which is not larger than observed.
ii) If the plasma mechanism is correct, the fundamental and harmonic components observed successively at a given frequency (fig. 2) should be found at different positions, those where the observing frequency is equal to the local plasma frequency or to twice the local plasma frequency. This is not the case : these two components are found at the same position. Here, again, scattering can help : it produces a scattered image of the fundamental which is displaced away from the sun center as compared to the primary source position so that the observations can be explained (Riddle, 1974 ; Leblanc, 1973). Computations were carried out in a spherical corona and in a Newkirk's streamer (Newkirk, 1967).

Meter waves stereo observations of type III bursts.

The STEREO experiment was designed to measure the variation with the direction of observation of the intensity of type III bursts. It has produced the following results :
i) the directivity of the fundamental component is larger than that of the harmonic ; this was the most direct evidence that the plasma hypothesis is correct until plasma waves were observed in situ at the same time as a type III (Gurnett and Anderson, 1976).
ii) for a source at about 70° longitude to the East, the directivity of the bursts was found to increase with the distance of the source to the central meridian. Since a variation of the altitude of the source along a given solar diameter would produce, by refraction, the opposite result, it is most likely that the directivity variations are due to longitude changes from an event to the next one. However, computations show that the observed effect cannot be explained by scattering in a spherical

corona at the observed longitude.

To explain the second result, it has been necessary to build strongly inhomogeneous models of the corona. Axisa et al. (1971) showed that the slowly varying component of solar radiation at 169 MHz is due to the continuum emission of overdense macrostructures based on filament chains

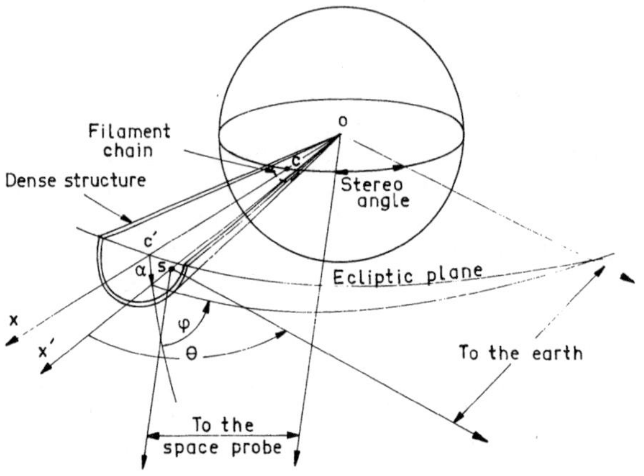

Figure 3. Geometry of the model of coronal overdense sheet used to interpret the measurements of type III directivity at 169 MHz on Nov. 14, 1971.(Hoang et al., 1977)

They suggested that these structures should be identified with streamers and that their geometry was that of "lames coronales" or vertical sheets of overdense plasma. This model has been adapted to the solar conditions encountered during the STEREO-1 mission (fig. 3). It was found necessary that the type III electrons travel out of the coronal sheet but close to it. The results (Hoang et al., 1977) show that most observations can be explained in this way : the fundamental component is more directive than the harmonic, the positions of the fundamental and harmonic components appear to be the same and the harmonic directivity increases with the source distance to the central meridian.

Hectometer waves stereo observations of type III bursts.

Weber et al. (1977) have measured the directivity of type III bursts on hectometer wavelengths using the HELIOS-A and B solar probes. Only one complex event has been analysed up to now. The ratio of the intensities measured at two very different directions is in the same range as found with STEREO-1 at 169 Mhz but the longitude of the flare as seen from HELIOS and from STEREO-1 is different. As a function of frequency, the directivity factor goes through a maximum (fig. 4) ; the results are interpreted as if the burst polar diagram rotated in the ecliptic plane as the type III electron beam travelled along an Archimidean spiral

The implicit assumption is, of course that the diagram is otherwise independent of frequency.

If we accept these hypotheses, we find that the type III radiation is beamed forward. The question which remains open is to know whether the beaming is due to the emission mechanism itself or to focusing ónly: the index of refraction near a source radiating at twice the local plasma frequency is still different from unity. When more events will have

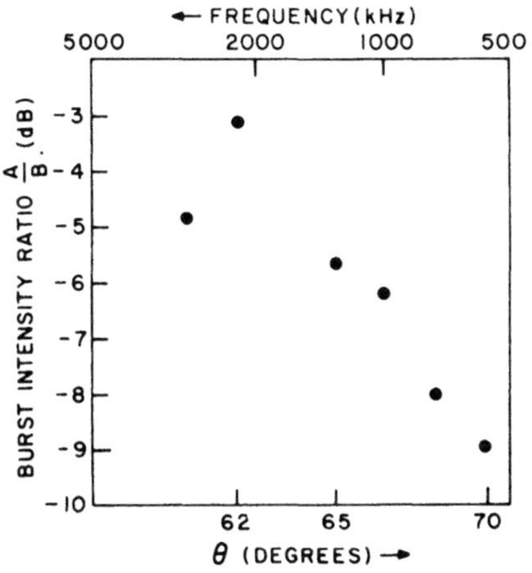

Figure 4. Burst intensity observed at HELIOS-A compared to HELIOS-B for different frequencies. θ is the angle between HELIOS-A and B measured from the source location. Consequently these data are a measure of the radiation pattern of the burst in the interplanetary medium. (Weber et al., 1977).

been studied and more ray tracing calculations performed, it will be possible to answer that question. There are some indications (Fitzenreiter et al., 1975) that the beam axis lies between the solar diameter through the source (grad N_e) and the magnetic field vector. This result is not yet explained.

Summary of type III burst directivity.

To sum up, the stereo observations of type III bursts directivity have shown that :

On meter waves, the observed directivity of fundamentals is larger than that of the harmonics, thus providing a new and strong argument for the plasma hypothesis. This result has also provided a new criterion for separating fundamental and harmonic components.

In the same wavelength range, observations can be quantitatively accounted for if some dense coronal structure is included in the model which can occult the source in some way ; the type III electrons must travel

on the side of an overdense plasma sheet and not inside it. This is an interesting implication of the model since there is no agreement on the relative positions of type III electron trajectories and streamers as measured using other methods.
In the hecto- and kilometer wavelength range, directivity has been measured on one event. As more observations will become available, we can hope to learn how much of the beaming comes from the emission mechanism itself and how much from propagation.

THREE DIMENSIONAL MAPPING OF THE HELIOSPHERE.

We have seen that type III bursts are produced by electrons which are guided along magnetic lines of force. In the interplanetary medium, the radiated frequency is that of the harmonic component, twice the local plasma frequency and ranges between a few MHz and about 30 kHz at the Earth orbit.

Thus if we can localize a type III burst at several frequencies, we shall follow the electron beam which produces it at various altitudes and map the line of force which guides it as well as the electron density along that line of force. This method is extremely powerful assuming we can localize type III events in the hecto- and kilometer wavelengths range where space-borne antennas are electrically very short and, therefore, not very directive.

Direction finding from a spinning spacecraft.

It is possible to localize a source of radio radiation using a spinning dipole since such an antenna is not sensitive in the direction of its axis. A spin stabilized satellite is used which carries a dipole antenna perpendicular to its axis of rotation. Then the received signal is amplitude modulated by the spacecraft rotation and the phase of that modulation yields the azimuth of the centroid of the source projected on the spin plane. No information is obtained on the elevation angle above or below that plane since a large angular diameter source in the spin plane will give the same depth of modulation as a narrow source at a higher elevation. Fainberg et al. (1972) have successfully used that technique and showed that the type III electrons do follow Archimedean spirals in the ecliptic. It is possible to obtain the elevation angle by using a tilted dipole (which is neither perpendicular nor parallel to the spin axis, Slysh, 1967) or correlation techniques (Kellogg, 1976).

In any case, the 3-dimensional position of a type III burst at a given frequency cannot be obtained from one direction only ; it is necessary to use an electron density model or two suitably instrumented spacecraft. One of these should determine a plane where the source lies, using a spin plane dipole ; the other a complete direction using, for instance, a tilted dipole. Such a scheme has not yet been put into operation but other methods have been used which have the potentiality to produce a complete 3-dimensional maps.

2-dimensional position measurements using two spacecraft.

Using two spinning spacecraft whose spin axis were nearly perpendicular to each other, Baumback et al. (1975) could determine a complete direction (fig. 5). Fitzenreiter et al. (1976) used IMP-6 and RAE-2 which was in a lunar orbit. Both of these teams used an electron density model, that deduced from RAE-1 measurements (Fainberg and Stone, 1971) or OGO-5 data (Alvarez and Haddock, 1973). These observations provide

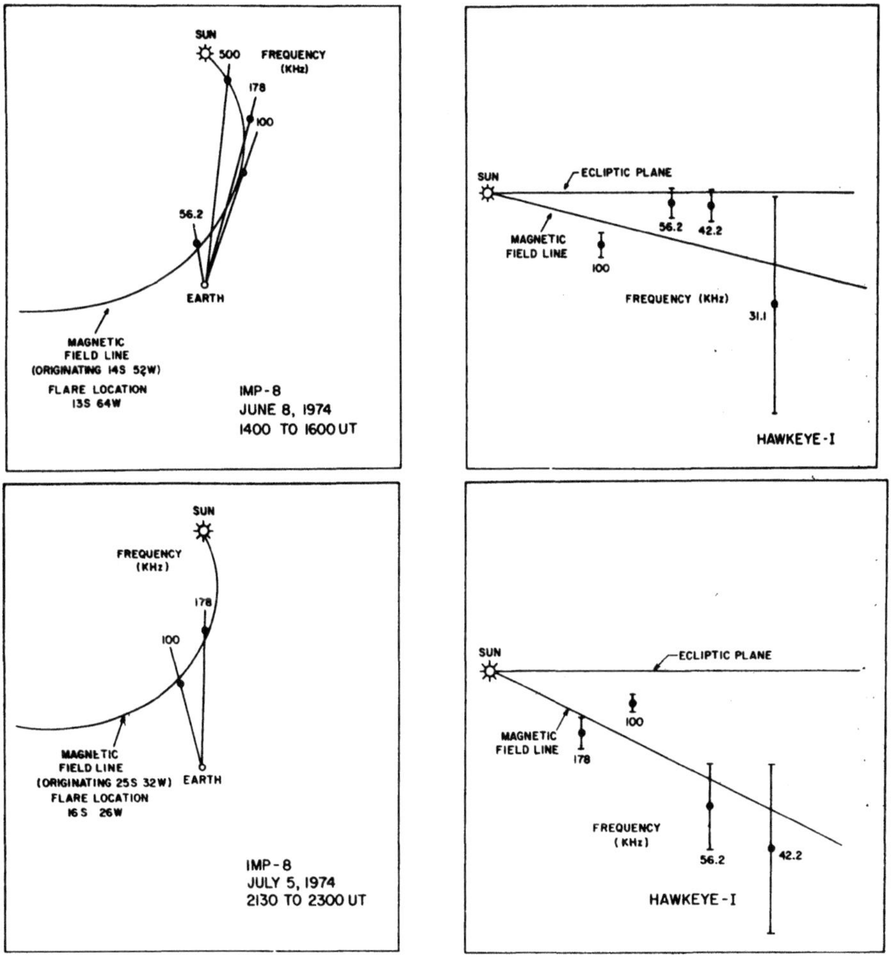

Figure 5. Source locations of a type III event on June 8, 1974 and July 5, 1974. Left as projected onto the ecliptic, right: solar latitude as a function of heliocentric distance (31.1 kHz corresponds to 1.4 AU). On June 8 the heliocentric latitude of the source seems to decrease with increasing distance to the sun; on July 5, the magnetic field line appears closer to a constant latitude model. (Baumback et al. 1976)

evidence for noticeable components of the interplanetary magnetic field perpendicular to the ecliptic. The most dramatic result is that of Fitzenreiter et al. who mapped a line of force (fig. 6) which was rooted in the northern hemisphere of the sun, later bent towards the ecliptic which it crossed at about 0.8 AU from the sun and reached a point south of the ecliptic at 20° heliocentric latitude. Such a configuration was not easily predictable from any previous measurements made within 7° from the solar equator using spacecraft in the ecliptic. However, the use of an electron density model is a weak point in the interpretation of the data : the RAE-1 model was obtained from the analysis of thousands of events but it is well known that the solar wind density can vary near the Earth orbit and, therefore elsewhere as well, by a factor

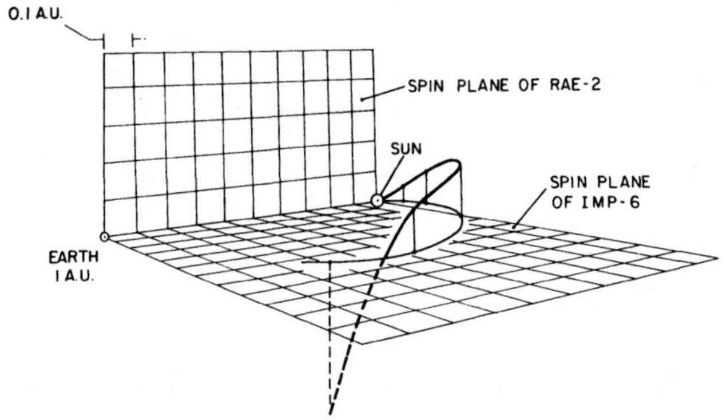

Figure 6. Out-of-ecliptic observation of a type III burst trajectory on June 23, 1973. (Fitzenreiter et al., 1976)

of 2 to 5 or more from day to day.

Towards 3-dimensional mapping of the interplanetary medium.

Very recently, Weber et al. (1977) used simultaneous observations from 3 spacecraft : HELIOS-A and -B and RAE-B to determine the positions of type III bursts at several frequencies. They also made use of the difference in the times of propagation from the source to the two spacecraft as a third piece of data to get a 3-dimensional result (fig.7). However they could not reach that goal because none of the spacecraft had any resolution across the ecliptic so that the accuracy on the distance to the ecliptic is poor and only 2-dimensional results are reported ; this is quite acceptable since the observations were obtained in 1976, in the low activity part of the solar cycle so that the latitude of active centers is low. The results (fig.8) are quite impressive,; for the first time, 2-dimensional positions are determined without any assumption on the electron density model ; the only assumption is that the harmonic component of type III bursts is observed but this is hardly an assumption any more in that range of frequencies.

An electron density model has thus been derived (fig.9) which appears to be very close to Newkirk's model (Newkirk, 1967) for the quiet sun near the minimum of activity. The RAE-1 scale was based on the assumption that the fundamental component was radiated but even taking that into account, it was between 3 and 5 times denser at a given altitude than the new one ; this difference can be explained by the variation of the level of activity and by day to day variations of the electron density near the Earth orbit.

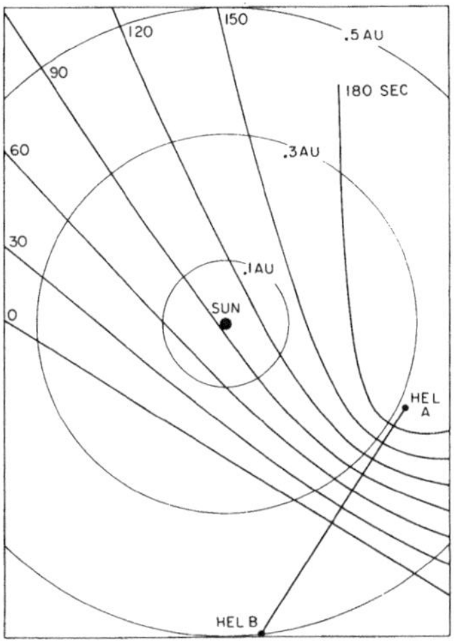

Figure 7. Source loci for time differences observed by HELIOS-A and -B on March 1976. (Weber et al.,1977)

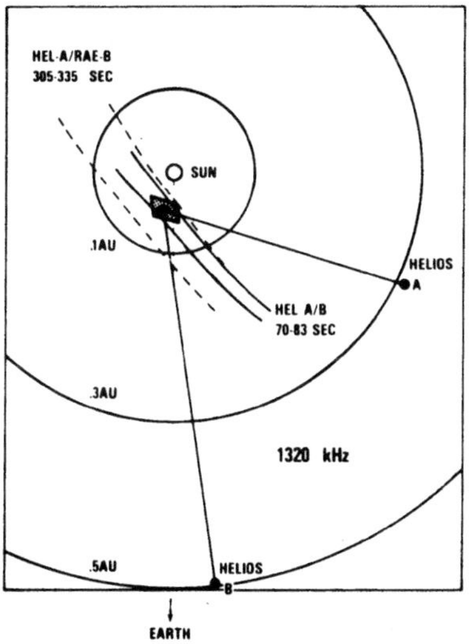

Figure 8. Location of burst source region given by intersections of the directions measured at HELIOS-A and -B and by arrival time differences at HELIOS-A and -B as well as HELIOS-A and RAE-2. Direction finding and arrival time results are in agreement. (Weber et al., 1977)

It is remarkable that the positions obtained from direction finding from the spin modulation at HELIOS-A and -B do agree with the arrival time determinations. This shows that propagation effects such as refraction and scattering do not require large corrections to what is deduced from simple geometry.

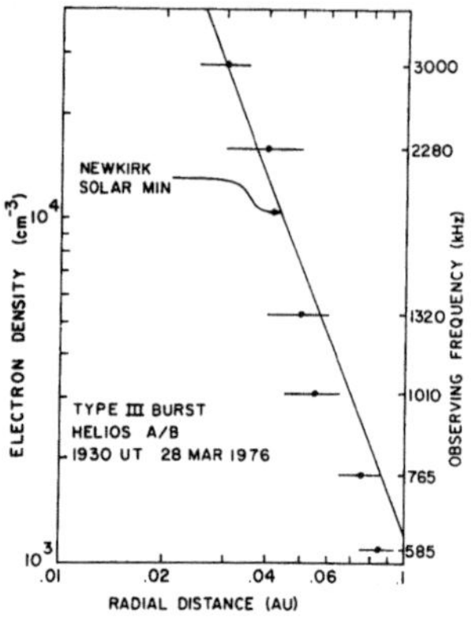

Figure 9. Electron density determined from type III burst trajectory using the measured source locations at each observing frequency and assuming emission occurs at twice the local plasma frequency.

The use of two satellites also allows the measurement of the size of the type III source (Baumback et al., 1976 ; Weber et al., 1977) which can only be estimated from one-satellite data (Kellogg, 1976). This size is controlled by two factors ; the electron beam extent and scattering in the medium (Steinberg, 1972). The source is found large, commonly between 30° and 70° as seen from the Sun and its size increases with decreasing frequency ; this size tends to increase when the source distance to the Earth decreases so that scattering is probably not the dominant factor.

CONCLUSION

Low frequency stereo radioastronomy has demonstrated its capability to provide a 3-dimensional picture of a large part of the heliosphere : a plot of some magnetic lines of force and an electron density distribution over a considerable range of distances to the Sun. It also provides data on the propagation of electrons, their scattering and deceleration if any. We may also note that type II bursts can be used to track shock waves but this is outside the scope of the present review.

All this information is obtained via a remote sensing technique and it has been shown that the results are not seriously affected by propagation effects in the interplanetary medium.

There are few published results yet and this is due to the fact that few events have been fully reduced and still less observed with the required number of properly instrumented spacecraft. In particular, there is a lack of data on the distance of type III sources to the ecliptic. We can hope that the ISEE-C mission will produce more information of that kind since there will be a joint Meudon-GSFC experiment on-board with 2-dimensional capability. We may also expect a higher level of solar activity in 1978-1981. This kind of data is badly needed if we want to learn something on the structure of the magnetic field out of the ecliptic and from it the velocity of the solar wind in these regions.

A 3-dimensional description is also necessary to improve our understanding of such phenomena as the solar modulation of cosmic rays. It will also be essential to the interpretation of most in situ measurements made out of the ecliptic in the next decade.

ACKNOWLEDGEMENTS

I wish to thank the colleagues who sent the data and papers in advance of publication particularly J. L. Bougeret from this laboratory and R. Weber and his co-workers from Goddard Space Flight Center. The camera-ready manuscript could not have been produced in time without the expertise and care of Mrs F. Axisa.

LITERATURE CITED

Allen,C.W. : 1975, Monthly Notices Roy. Astron. Soc. 172, 159.
Alvarez,H. and Haddock, F.T. : 1973, Solar Phys. 29, 197.
Axisa, F., Avignon, Y., Martres, M.J., Pick, M. and Simon, P. : 1971, Solar Phys. 19, 110.
Baumback,M.N., Kurth,W.S. and Gurnett, D.A. : 1976, Solar Phys. 46, 433.
Bougeret, J.L.: 1973, Astron. Astrophys. 24, 53.
Bougeret, J. L.: 1977a, Astron. Astrophys. (in press).
Bougeret, J. L.: 1977b (in preparation).
Bougeret, J.L. and Steinberg, J.L.: 1977, submitted to Astron. Astrophys.
Caroubalos,C. and Steinberg, J.L. : 1974, Astron. Astrophys. 32, 245.
Caroubalos,C., Poquerusse, M.,and Steinberg, J.L. : 1974, Astron. Astrophys. 32, 255.
Epstein,G. : 1974, Onde Electrique 54, 281.
Fainberg,J. and Stone, R.G. : 1971, Solar Phys. 17, 392.
Fainberg, J., Evans, L.G. and Stone, R. G. : 1972, Science 178, 743.
Fitzenreiter, R.J., Fainberg, J. and Bundy, R.B. : 1975, NASA-GSFC Document X-693-75-284.
Fitzenreiter, R.J., Fainberg, J., Weber, R.R., Alvarez, H., Haddock, F. T. and Potter, W.H. : 1976, NASA-GSFC Document X-692-76-244.
Fokker, A.D. : 1965, Bull. Astron. Inst. Netherlands 11, 118.
Gurnett, D.A. and Anderson, R.R. : 1976, Science 194, 1159.
Hoang, S., Poquerusse, M. and Steinberg, J.L. : 1977, Astron. Astrophys. 56, 283.
Kellogg, P.J. : 1976, Solar Phys. 46, 449.

Kerdraon, A. : 1973, Astron. Astrophys. 27, 361.
Leblanc, Y. : 1973, Astrophs. Letters 14, 41.
Leroy, J.L. and Trellis, M. : 1974, Astron. Astrophys. 35, 298.
Newkirk, G. : 1967, Ann. Rev. Astron. Astrophys. 5, 213.
Riddle, A.C. : 1974, Solar Phys. 35, 153.
Slysh, V.I. : 1967, Kosm. Issled. 5, 897 and Cos. Res. 5, 759.
Steinberg, J.L. and Caroubalos, C. : 1970, Astron. Astrophys. 9, 329.
Steinberg, J.L., Aubier-Giraud, M., Leblanc, Y. and Baoischot, A. : 1971, Astron. Astrophys. 10, 362.
Steinberg, J.L. : 1972, Astron. Astrophys. 18, 382.
Steinberg, J.L., Caroubalos, C. and Bougeret, J.L. : 1974, Astron. Astrophys. 37, 109.
Weber, R.R., Fitzenreiter, R.J., Novaco, J.C. and Fainberg, J. : 1977, paper submitted to Solar Phys.

DISCUSSION

Alurkar: The hypothesis of 2nd harmonic in the case of Type III bursts, is it supported by the observations of the Culgoora radio heliograph?

Steinberg: In the frequency range 300 to ≃ 30 MHz, both fundamental and harmonic seem to be observed, but lower than ≃ 10 MHz (from spacecraft) there is no trace of fundamentals (see for instance Alvarez, H., Haddock, F. T., and Potter, W. H.: 1974, *Solar Phys.* *34*, 413). Ray tracing computations in a scattering corona and the interplanetary medium (Hoang and Steinberg, to be published) show that the fundamental cannot get out because the scattering power gradient is low. There are other reasons such as the increasing role of Landau damping.

MEASUREMENTS OF THE SOLAR WIND USING SPACECRAFT RADIO SCATTERING OBSERVATIONS

R. WOO
Jet Propulsion Laboratory, California Institute of Technology, Pasadena, California 91103, U.S.A.

ABSTRACT

This paper reviews radio scattering measurements of the solar wind carried out with coherent, monochromatic, and point-source spacecraft signals. The observed phenomena which include spectral and angular broadening, and phase as well as intensity scintillations, have provided measurements of the solar wind previously not available from radio astronomical observations. These cover a wide range of heliocentric distances (as close as 1.7 $R_{\odot}$), and large- as well as small-scale electron density fluctuations.

1. INTRODUCTION

When a radio wave propagates through the interplanetary medium it is scattered by the electron density inhomogeneities which exist over an extensive range of scale sizes throughout the solar wind. This scattering gives rise to a number of radio phenomena that can be observed with natural radio sources as well as spacecraft signals. Although for many years radio astronomers have been using radio scattering phenomena such as angular broadening and intensity scintillations to obtain important information on the solar wind, it has not been until recent years that the advantages of spacecraft signals have been recognized and demonstrated. The extensive observations of natural radio sources have recently been reviewed (Hewish 1972; Jokipii 1973; Coles et al. 1974; Lotova 1975), and this paper summarizes the spacecraft radio scattering results. Spacecraft radio measurements such as range or time delay (Koehler 1968; Croft 1973; Landt 1974; Muhleman et al. 1977) and Faraday rotation (Levy et al. 1969; Stelzried et al. 1970), which have been used to study the mean density and magnetic field (as opposed to fluctuations in the case of scattering observations), will not be discussed in this paper.

M. A. Shea et al. (eds.), Study of Travelling Interplanetary Phenomena/1977, 81–100. All Rights Reserved.

2. SPECTRAL AND ANGULAR BROADENING

One radio scattering measurement that can be conducted with monochromatic spacecraft signals and not wideband natural radio sources is spectral broadening. The spectral broadening observations first demonstrated by Goldstein (1969) with Mariner 4 and Pioneer 6 also represent the first radio scattering measurements of the solar wind using a spacecraft signal. Further data have been obtained with Mars 2 (Yakovlev et al. 1974), Mariner 9 and Pioneer 10 (Cannon 1976) and Helios 1/2 (Woo 1977a). The potential of spectral broadening for probing the solar wind has only recently been realized with the Helios measurements, and this has come about as a result of the progress made in the important area of theoretical analysis (Woo et al. 1976a, 1977), the improvement in the spacecraft signal-to-noise ratio (SNR), and the favorable trajectory of the two Helios spacecraft. Earlier theoretical interpretations of spectral broadening were given by Hollweg and Harrington (1968) and Yakovlev et al. (1974).

A typical spectrogram of the received radio signal at 2.3 GHz measured at a closest approach distance $R = 2.18\ R_\odot$ is shown in Figure 1. The bandwidth is 500 Hz and the spectrogram is the result of 20 minutes of observation. Despite the close proximity to the sun, it is remarkably smooth, a direct consequence of the excellent SNR (~40 dB). Each spectrogram was processed to determine the bandwidth B defined in the following way

$$\int_0^{B/2} P(f)\ df = \frac{1}{2}\int_0^{\infty} P(f)\ df \tag{1}$$

where P(f) is the spectrogram of the received signal and f is the fluctuation frequency. The variation of B with heliocentric distance in the equatorial region of the solar wind is shown in Figure 2. The 1968 Pioneer 6 data were obtained by Goldstein (1969) and the "solar events" have been excluded (Woo et al. 1976a), while the 1975/1976 Helios data, which cover several solar rotations, were obtained by Woo (1977a). The scatter in the data represents changes in the solar corona rather than measurement error, and amounts to a factor of two. The bandwidth B is analogous to the scintillation index m for intensity scintillations (m is the rms intensity fluctuation about the mean, relative to the mean) in that it characterizes the strength of the density fluctuations (turbulence). There are some major differences between spectral broadening and intensity scintillations.

First, because of strong scattering, intensity scintillations saturate with increasing turbulence and, therefore, decreasing solar elongation. In other words, the scintillation index m reaches a maximum at approximately unity, and any further decrease in elongation is not accompanied by an increase in m (Gurvich and Tatarski 1975; Chang 1976). In fact, with natural radio sources, the finite source size

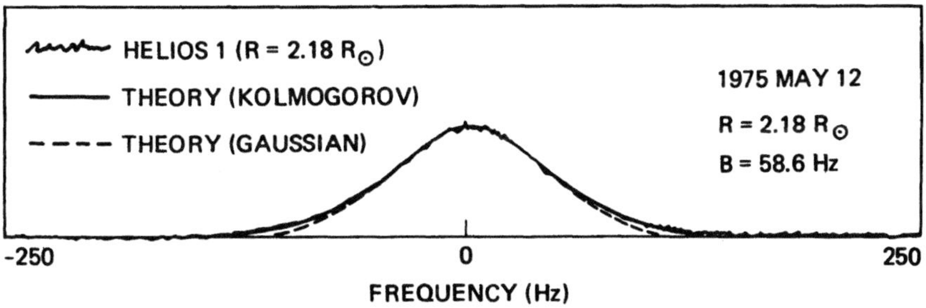

Figure 1. Typical spectrogram at 2.3 GHz.

causes the scintillation index to turn over and decrease (Coles et al. 1974). As can be seen from Figure 2, spectral broadening does not saturate but continues to increase with decreasing elongation and, therefore, permits the study of radial variation near the sun. The wide dynamic range also makes spectral broadening excellent for detecting changes that may be due to transients, events or streams close to the sun. The usefulness of this feature was well demonstrated in the measurements of Pioneer 6 which took place during the solar cycle maximum when several "solar events" marked by increased bandwidths by factors as high as 8 were observed (Goldstein 1969). Schatten (1970a, 1970b) and Pintér (1973) have related some of these events to solar flares. Such events were not seen at least during the times when the solar cycle minimum Helios measurements were conducted (Woo 1977a).

Second, while the analysis of intensity scintillations requires knowledge of the fourth-order moment of the field (Gurvich and Tatarski 1975; Rumsey 1975), spectral (as well as angular) broadening requires information only on the second moment (Woo et al. 1977). This simplifies the analysis considerably because the second-moment theory is better developed and better understood than the fourth (Gurvich and Tatarski 1975). This aspect is important because a reliable theory is essential for proper interpretation of radio scattering observations.

Third, for weak scattering, the scintillation index m is given by (Woo et al. 1976a)

$$m \cong 2k^{7/12} \, (a_1 R)^{1/2} \, (L_1 L_2/L)^{5/12} \, c_{no} \tag{2}$$

where k is the wavenumber, $a_1 = 0.85$, R, L_1, L_2, and L are defined in Figure 3, and the spatial wavenumber spectrum of the refractive index (and density) fluctuations has been assumed to be isotropic and Kolmogorov (power-law spectrum whose spectral index p = 11/3 and whose magnitude is characterized by the structure constant c_{no}). The structure constant c_{no} is related to the rms electron density fluctuation σ_{ne} by (Woo et al. 1976a):

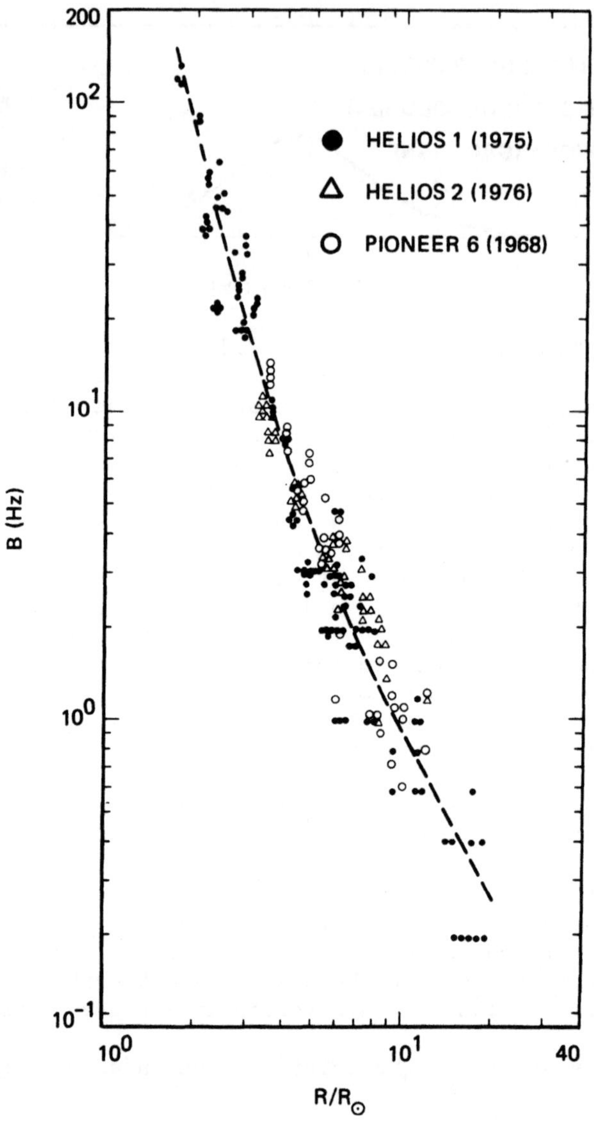

Figure 2. Heliocentric dependence of spectral broadening.

$$\sigma_{ne} \cong 4.1 \times 10^{13}\, c_{no}\, k^2\, L_o^{1/3} \tag{3}$$

where L_o is the outer scale of turbulence. On the other hand, the bandwidth B for both weak and strong scattering is given by (Woo et al. 1976a):

$$B = 0.542\ (c_{no}k)^{6/5}\ (a_1 R)^{3/5}\ v \tag{4}$$

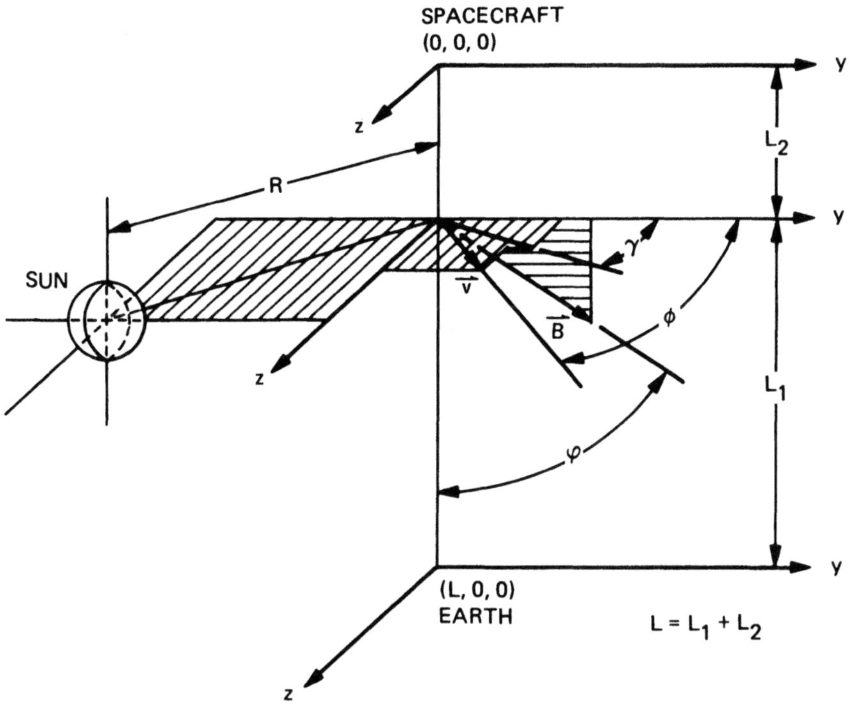

Figure 3. Relevant geometry.

where v is the solar wind velocity transverse to the line-of-sight path (see Figure 3). In comparing (2) and (4) we note that whereas the scintillation index measures the electron density fluctuations, spectral broadening measures a combination of electron density fluctuations and solar wind velocity. Also, B does not depend on the distances L_1 and L_2. Equation (4) is for the case when the density spectrum is isotropic and there are no velocity fluctuations. Anisotropy and wind fluctuations tend to decrease the bandwidth (Woo et al. 1977).

Spectral broadening is useful only near the sun where the induced broadening due to the solar wind exceeds the inherent line-width of the radio system which is typically 0.2 Hz. Since spectral broadening measures frequencies higher than 0.2 Hz, it provides information only on the smaller scales. In this respect it is similar to intensity scintillations which respond only to small scales because of Fresnel filtering (see, e.g., Jokipii 1973). At the high-frequency end both spectral broadening and intensity scintillations are limited by SNR. The ranges of heliocentric distance and scale size over which spectral broadening and intensity scintillations provide information are summarized in Figure 4, which has been drawn for 2.3 GHz assuming a SNR of 40 dB. The spatial wavenumber $\kappa = 2\pi f/v$ and is related to the scale size ℓ by $\kappa = 2\pi/\ell$.

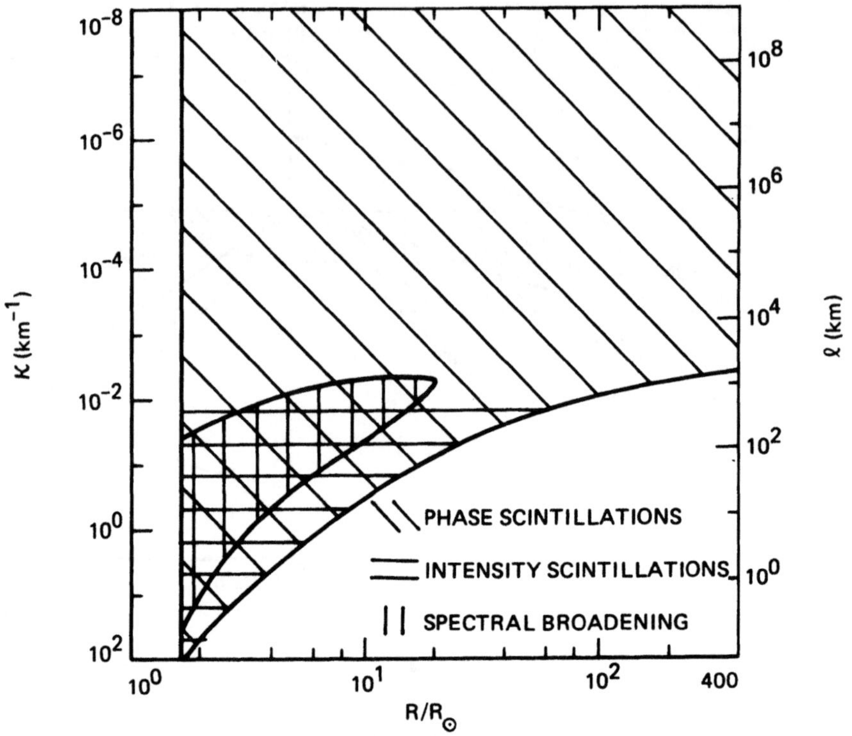

Figure 4. Range of heliocentric distances and spatial wavenumbers for spacecraft radio scattering observations.

The shape of the spectrogram depends on the spectral index p of the electron density spectrum, and can, therefore, be used to determine it (Woo et al. 1976a). Superimposed on the observed spectrogram in Figure 1 is the theoretical curve corresponding to the Kolmogorov spectrum ($p = 11/3$). As can be seen, the agreement is so good that the measured and theoretical curves are almost indistinguishable. For the sake of comparison, we have also fit the measured spectrogram with a gaussian curve (which corresponds to a gaussian density spectrum) by matching the spectrograms at half maximum. While the agreement is equally good for the lower frequencies, it is not as good at the high frequency end. It should be noted that the determination of the spectral index from the shape of the spectrogram is not affected by the anisotropy of the density fluctuations or the velocity fluctuations in the solar wind (Woo et al. 1977).

The closest spectral broadening measurement obtained with Helios was at 1.7 $R_{\odot}$ when the bandwidth B was 119.1 Hz. A careful examination of the data indicates that angular broadening, to the extent that the Helios point source was larger than the beamwidth of the 64-m receiving antenna, had occurred and that this resulted in a 4.52-dB deterioration in SNR at 1.7 $R_{\odot}$. Angular broadening is, therefore, the limiting factor for probing close to the sun, and with the 64-m antenna and a SNR of 40 dB this occurs at approximately 1.7 $R_{\odot}$, a result that is indicated in Figure 4. Two dimensional brightness distribution observations conducted near the solar cycle minimum in 1973 and 1974 (Ward 1975) indicate that in the heliocentric distance range of 5 to 15 $R_\odot$ the axial ratio Q is always less than 2 and that it exhibits little or no dependence on heliocentric distance. It is likely that this situation persists in 1975 even as close in as 1.7 $R_\odot$, and for this reason we will assume that the angular power spectrum of the scattered spacecraft signal has Q = 2. By also taking into account the defocussing attenuation and the pattern of the 64-m receiving antenna (angular halfwidth to e^{-1} of 0.08475^o at 2.3 GHz), and assuming that the absorption of the solar corona is negligible, Woo (1977a) has used the observed reduction in received signal power to infer a tangential scattering angular halfwidth to e^{-1} (semi-major axis of the ellipse representing the contour of equal intensity) of $8.82'$.

This Helios angular broadening measurement at 1.7 $R_\odot$ represents the first angular broadening measurement of a spacecraft signal, and it would be interesting to compare it with similar observations of natural radio sources. Ward's (1975) summary of the variation of the tangential scattering has been scaled to 2.3 GHz according to k^{-2}, and the results are combined with the plane-wave converted (Woo et al. 1977) Helios measurement in Figure 5. Data prior to 1968 are from Okoye and Hewish (1967), the 1969 to 1971 data from Blesing and Dennison (1972), and the 1973 to 1975 data from Ward (1975). As can be seen, the Helios result is consistent with the natural radio source observations further out. In comparing the 1975 Helios angular broadening measurement at 1.7 $R_\odot$ with that of the 1974 Crab Nebula at 5 $R_\odot$, we find, as might be expected from previous studies (Erickson 1964), that the scattering is very nearly proportional to the mean electron density (Newkirk 1967; Saito 1970).

Since angular broadening is a measure of density fluctuations, and spectral broadening of density fluctuations and solar wind velocity, simultaneous observations provide a means for measuring the solar wind velocity. For a density spectrum that is anisotropic, Woo et al. (1977) have found that

$$v = 5.77 \frac{B}{\theta} \frac{L}{kL_2} \left(\sin^2\phi + \frac{\cos^2\phi}{Q^2}\right)^{-1/2} \left(\sin^2\gamma + Q^2\cos^2\gamma\right)^{-1/2} \quad (5)$$

where θ is the angular halfwidth to e^{-1} measured in the plane which makes an angle γ with $\vec{B}$, the projection of the magnetic field on the plane transverse to the raypath (y-axis in Figure 3); ϕ is the angle $\vec{v}$ makes with $\vec{B}$; $v = |\vec{v}|$; and Q is the axial ratio in the yz plane. The solar wind is assumed to be radial and parallel to the magnetic field so that $\phi = 0$. This seems reasonable since the wind is radially directed and the magnetic field is too, even at distances as close as 1.7 $R_\odot$ when the sun is quiet (Waldmeier 1977). The angular broadening data (Ward 1975) show that the magnetic field is indeed radial in the heliocentric distance range of 5 to 15 $R_\odot$. From the trajectory of the Helios spacecraft, L_1 = 1.0 AU and L_2 = 0.84 AU, and substituting these results along with B = 119.1 Hz into (5) we obtain the solar wind velocity v at

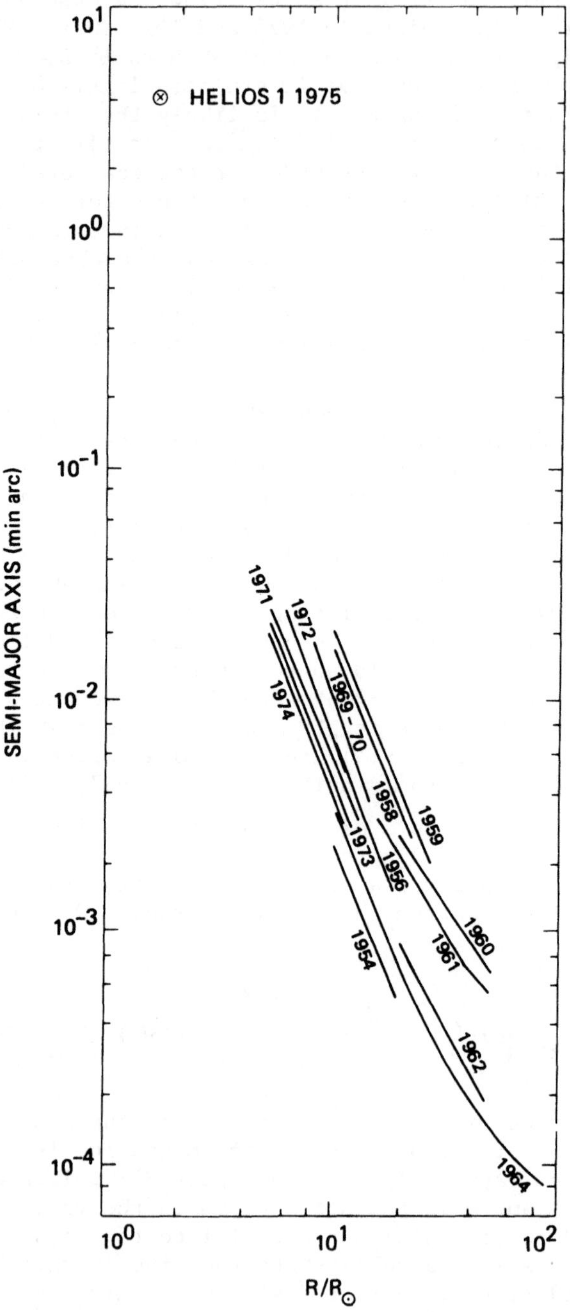

Figure 5. Heliocentric dependence of angular broadening at 2.3 GHz.

1.7 $R_\odot$ as 24.4 km s^{-1}. This result should be accurate to within a factor of 2 to 2.5 (Woo 1977a).

It is interesting to note that despite the disparity in time with respect to the solar cycle, the 1968 Pioneer 6 (solar cycle maximum) data appear to be within the scatter of the 1975/1976 Helios (solar cycle minimum) data. Of course, measurements corresponding to "solar events" as defined by Goldstein (1969) have been removed from the Pioneer 6 data and this would tend to lower the data set. It is clear that additional observations particularly during the solar cycle maximum need to be conducted before firm conclusions can be drawn about the solar cycle dependence. Although bandwidth increases with turbulence, it decreases (Woo et al. 1977) with the increase in axial ratio which appears to accompany the increase in turbulence (Ward 1975). Still, angular broadening is affected in a similar way (Woo et al. 1977), and observations of this phenomenon show a definite correlation with the solar cycle (Okoye and Hewish 1967; Blesing and Dennison 1972; Ward 1975). Of course, all this means is that the effect of the increased turbulence exceeds that due to the increased anisotropy. One point of difference that may also be important is the fact that the angular broadening observations contain the effects of "solar events", whereas the Pioneer 6 results do not.

To interpret the spectral broadening results shown in Figure 2, a dashed line representing an approximation of the radial dependence of the spectral broadening data has been drawn. The approximation in the 10 to 20 $R_\odot$ region is based on both spectral broadening and doppler scintillations, which will be discussed in the next section. To derive the solar wind profile from this result, it is assumed that the rms electron density fluctuation σ_{ne} is proportional to the mean electron density n_e, a result that has been used in interpreting angular broadening data (Erickson 1964; Newkirk 1967), and observed with small-scale intensity scintillations (Houminer and Hewish 1974; Chang 1976) and larger-scale spacecraft measurements (Sullivan et al. 1975). Furthermore, the radial profile of electron density given by Saito (1970) will be assumed. Within 5 $R_\odot$, Saito's model is almost identical to that of Newkirk (1967), and both models represent a widely accepted set of observed coronal densities. Then, since according to (4) bandwidth B is proportional to $\sigma_{ne}^{6/5}$ v, a wind velocity profile can be deduced from Figure 2 and the result, which is anchored to the velocity measurement at 1.7 $R_\odot$, is shown as a solid line in Figure 6. The important aspect of the coronal density profile in obtaining this result is its shape. Wilson (1976) has shown that the shape given by the Newkirk and, therefore, the Saito model agrees very well within 5 to 6 $R_\odot$ with the recent Skylab photographs taken by the HAO coronograph, even in streamers. Because the photometric observations are more reliable closer to the sun, we feel that the deduced wind profile is likewise. For the sake of comparison, the radar backscattering measurement made by James (1968) at 38 MHz and the 2.3-GHz two-station intensity scintillation results obtained by Ekers and Little (1971) at 6.6, 6.8, and 12 $R_\odot$ are included. As can be seen, the combined results in Figure 6 appear to form a

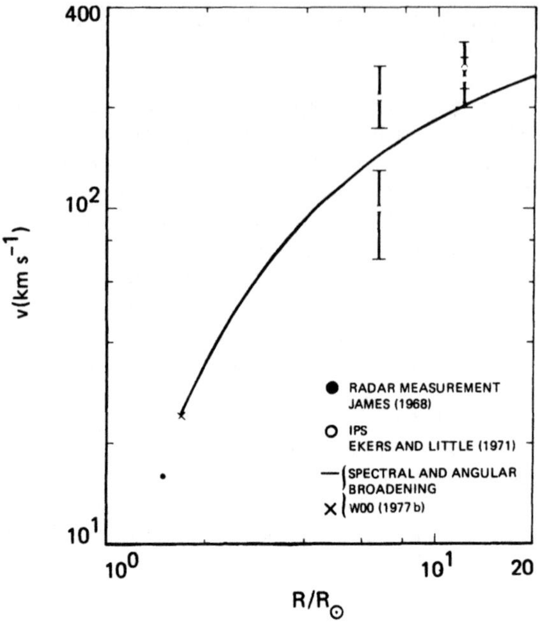

Figure 6. Radial dependence of solar wind velocity.

consistent wind profile in the acceleration region of the sun. Moreover, this profile is in reasonable agreement with current theoretical solar wind models (Hartle and Barnes 1970; Wolff et al. 1971; Hundhausen 1972).

Let us compare the measurements of the density spectrum deduced from the Helios spectral broadening observation at 1.7 $R_\odot$ to in situ spacecraft observations at 1 AU. Assuming $c_{no} \propto \sigma_{ne} \propto n_e$, Saito's density model and Q = 2, and using v = 24.4 km s^{-1} and the conversion to the one-dimensional density spectrum derived by Woo et al. (1976b), we have extrapolated the Helios measurement at 1.7 $R_\odot$ to 1 AU, and the result is shown in Figure 7, which assumes that the solar wind velocity is 350 km s^{-1} at 1 AU. As can be seen, it is consistent with the general discussion given by Jokipii and Hollweg (1970) and Cronyn (1972a), the direct spacecraft measurements of Intriligator and Wolfe (1970) and Neugebauer (1975), and the intensity scintillation results of Lovelace et al. (1970) and Chang (1976). In other words, the density spectrum is close to Kolmogorov and extends over a wide range of scale sizes.

3. PHASE SCINTILLATIONS

The ability to measure phase scintillations with coherent spacecraft signals represents a significant breakthrough in the probing of the solar wind with radio waves. The reason is that unlike intensity scintillations and spectral broadening, which are both limited to measuring smaller scales, phase scintillations provide information on the

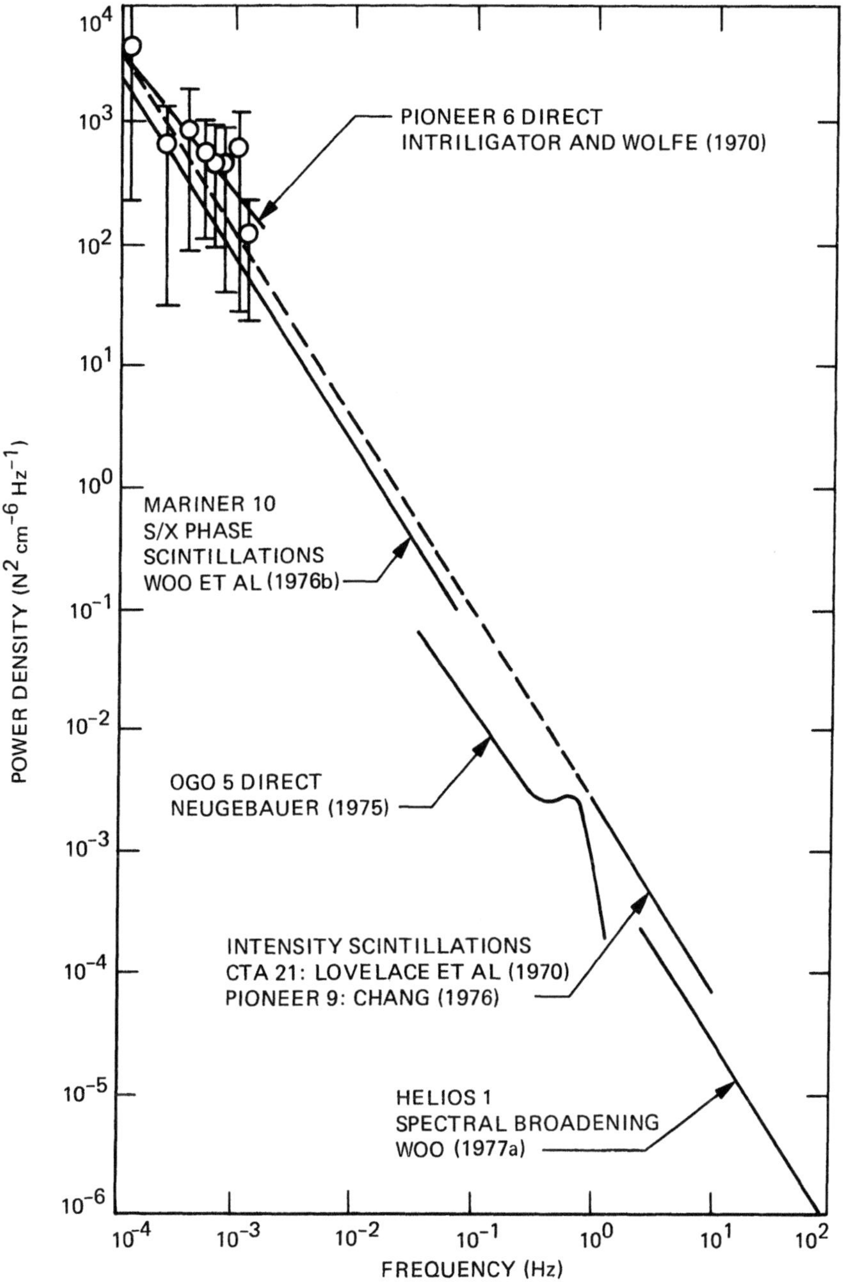

Figure 7. Density spectrum at 1 AU.

full range of scale sizes. With incoherent natural radio sources phase scintillations can be measured with very long baseline interferometry (VLBI) (Knight et al. 1971; Cronyn 1972b; Mutel 1975). Unfortunately, Cronyn (1972b) has shown that VLBI observations yield information only on scale sizes smaller than $\sigma/1.5$, where σ is the baseline distance. Moreover, the measurements are very noisy (Mutel 1975).

The measurement of phase scintillations at a single receiving antenna can be carried out only with coherent spacecraft signals. One problem with spacecraft phase scintillations is that the phase measurements include, in addition to the scintillations, the effect of the relative motion between the spacecraft and the receiving antenna, and this must be removed before an analysis of the scintillations can begin. Callahan (1975) has used a group-phase velocity technique called Differenced Range Versus Integrated Doppler (DRVID) to study the phase scintillations of Mariners 6, 7, and 9. In DRVID, measurements of range or time-delay are combined with those of phase, so that the spacecraft trajectory effect, which is common to both, is eliminated. However, because of the lower resolution of the ranging data (see, e.g., Koehler 1968), the DRVID technique is limited to probing close to the sun and to studying slow fluctuations (closest approach distance $R \lesssim 40\ R_{\odot}$, and fluctuation frequency $f < 10^{-3}$ Hz in the case of Mariners 6, 7, and 9).

The spacecraft trajectory is also eliminated in a phase differencing technique utilizing two coherently related radio frequencies. This important technique has recently been demonstrated by Woo et al. (1976b) with the Mariner 10 S/X (2.3 and 8.4 GHz) radio system, and is currently being applied to the extensive Viking observations. Because this technique uses the more highly resolved phase measurements, it is superior to DRVID.

For missions employing a single radio frequency (e.g., Helios 1/2), phase scintillations can be measured by subtracting from the observed phase the predicted value due to the spacecraft trajectory. The resulting residuals essentially represent the phase scintillations because the errors in predicting the trajectory are generally considerably lower than the phase scintillations (Woo 1977a). Since the residuals are based on phase measurements, they are also superior to DRVID. Because the residuals are generally used for navigation purposes, the quantity routinely computed is doppler rather than phase. Of course, conversion from one to the other is a simple matter since doppler is just the derivative of the phase.

Ranging residuals can also be computed and used to study phase scintillations (Muhleman et al. 1977), but, as in the case of DRVID, this data type is noisier, and, therefore, more limited in range of heliocentric distance and scale size.

Shown in Figure 8 is a typical frequency spectrum of phase scintillations. Figure 8 is the frequency spectrum $W_{\phi d}$ of the phase-difference scintillations obtained with the Mariner 10 dual-frequency S/X radio

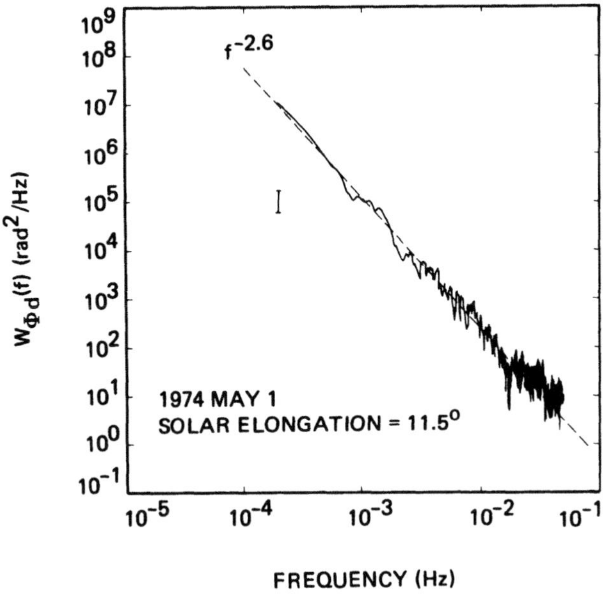

Figure 8. Typical frequency spectrum of phase scintillations.

signals when R = 43 $R_\odot$ (Woo et al. 1976b). The vertical bar represents the 90-percent confidence interval, and the dashed line an approximation to the measured spectrum. As expected (Woo 1975), the phase scintillation spectrum is roughly power-law with a slope of about -2.6. This result has been related to the one-dimensional density spectrum at 1 AU and the result is shown in Figure 7 (Woo et al. 1976b). Thus, large-scale turbulence can be measured with phase scintillations. The Mariner 10 measurements were limited to frequencies less than 5×10^{-1} Hz by the sampling rate. By increasing the sampling rate, the range of scale sizes could be extended substantially with SNR being the final limiting factor. On the other hand, there are no limitations at the low-frequency end. The range of scale sizes and heliocentric distances over which dual-frequency S/X phase scintillations provide information (assuming a 40-dB SNR) is summarized in Figure 4, and, as can be seen, the advantages over intensity scintillations and spectral broadening are considerable. Extensive phase scintillation observations will provide much needed measurements of the density spectrum in the vicinity of the proton gyroradius wavenumber, a region of much current interest (Jokipii 1973; Coles et al. 1974; Lotova 1975).

To study the variation of the phase scintillations with heliocentric distance, it is convenient to use the doppler residuals. In analogy to the scintillation index for intensity scintillations, we define the rms doppler scintillations σ_D over a 10-minute time span using data sampled at one per minute. Then, Woo (1977a) has found that for so-called two-way doppler,

$$\sigma_D = 0.088\ c_{no}\ k\ v^{5/6}\ (a_1 R)^{1/2} \tag{6}$$

In comparing (6) with (2), we note the σ_D, contrary to the scintillation index, depends on v and not on the distances L_1 and L_2. This difference stems merely from the fact that the range of f is fixed and just a segment of the frequency spectrum of the doppler scintillation is being studied. Equation (6) is similar to the result for spectral broadening as given by (4), in that σ_D responds to density fluctuations as well as solar wind velocity. Ishimaru (1977) has shown that for phase scintillations Rytov's approximation which was used to derive (6) is a good approximation even when the scattering is strong as is the case near the sun. The spectral amplitude studied by Callahan (1975) behaves the same way as σ_D and one must be careful about deducing information on turbulence from it (Woo 1977a). Another important point is that c_{no} for (4) and (6) are identical only if the small-scale (spectral broadening) and large-scale (doppler scintillations) density fluctuations form a continuous spatial spectrum, a result that is certainly strongly suggested by the measurements summarized in Figure 7. Assuming that this is the case, σ_D is related to B by

$$\sigma_D = 0.147\ B^{5/6} \tag{7}$$

Using (7), the spectral broadening data in Figure 2 have been combined with the measurements of σ_D by Helios 1 and Pioneers 10/11 in the equatorial region of the solar wind, and the results are shown in Figure 9.

The results in Figure 9 demonstrate the remarkable dynamic range of spectral broadening and doppler (phase) scintillations. This dynamic range is possible because neither saturates as do intensity scintillations when they are strong. Far from the sun, the two-way doppler residuals are limited by the error of the predicted trajectory and the contributions due to the Earth's ionosphere, which are both of the order of a few millihertz for a ten-minute integration time. For this reason the Pioneer data have been limited to $\sim 180\ R_\odot$. With dual-frequency S/X measurements such as those currently being made with the Viking spacecraft, measurements of phase and doppler scintillations are most likely possible to 1 AU and beyond.

The scatter in the data shown in Figure 9 represents structure such as streams which dominated the solar wind in 1975 (see for example Sheely, Harvey, and Feldman 1976). Comparison with direct spacecraft measurements of velocity and density is possible and is currently underway. It is clear from Figure 9 that the structure grows with radial distance and that these measurements provide important information on the evolution of the stream structure. Drawn in Figure 9 is a line representing the radial dependence $R^{-1.3}$ and, as can be seen, it is a fairly good approximation to the data for distances greater than $10\ R_\odot$. From (6) we see that the $R^{-1.3}$ variation corresponds to $\sigma_{ne}\ v^{5/6} \propto R^{-1.8}$, so that within the assumption that $\sigma_{ne} \propto n_e$ and $v^{5/6} \sim v$, mass flux is

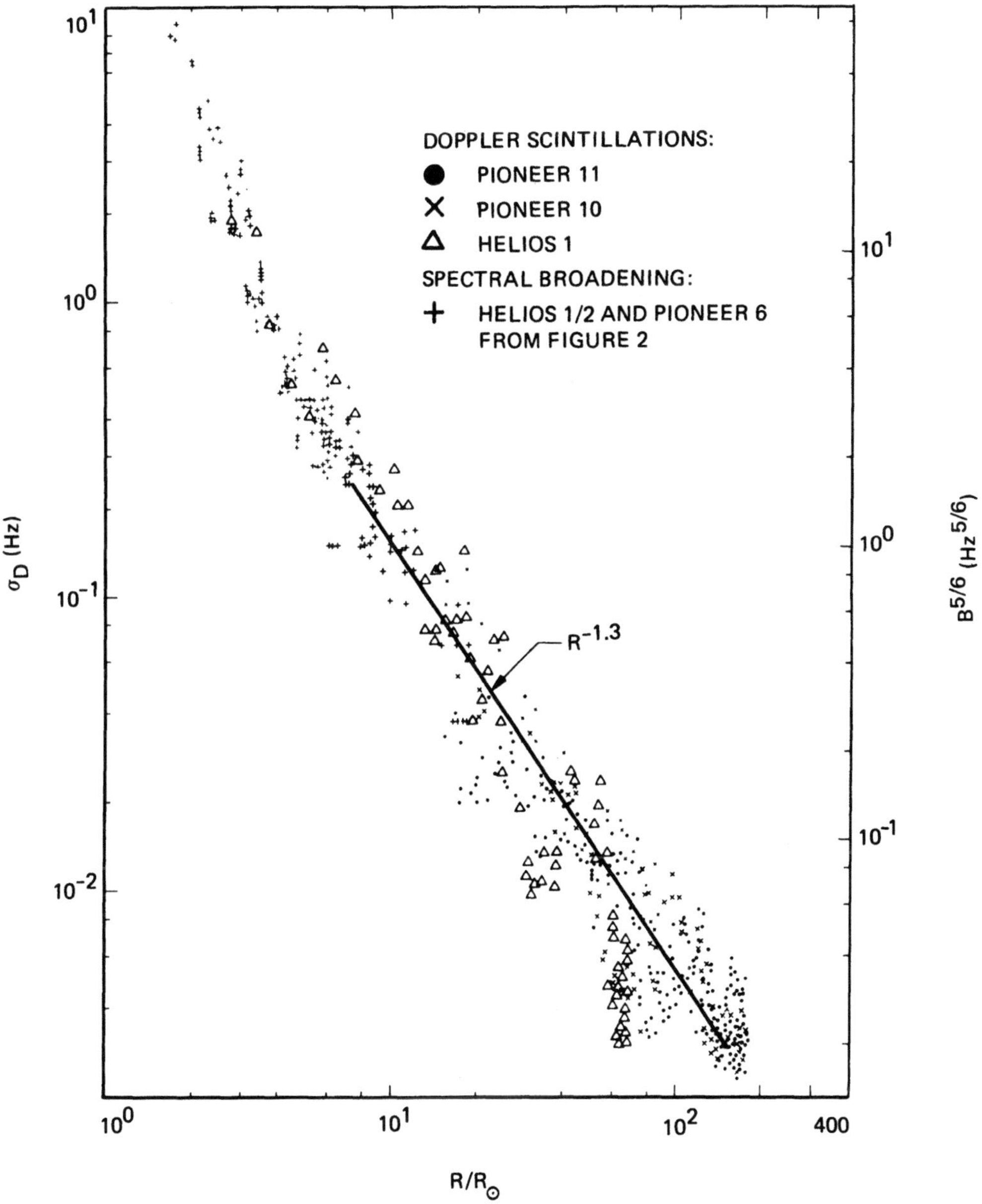

Figure 9. Heliocentric dependence of doppler/phase scintillations and spectral broadening.

approximately proportional to $R^{-1.8}$, suggesting a solar wind that is slightly converging in the equatorial region between approximately 20 to 180 $R_{\odot}$, a result that is not unreasonable (Suess et al. 1977). Intensity scintillations (e.g., Armstrong 1975) show that small-scale density fluctuations have an approximate radial dependence of $R^{-2.05}$. If it is assumed that the small and large-scale sizes are proportional, then, according to equation (6), a radial dependence of $R^{-1.3}$ for σ_D implies that $v \propto R^{0.3}$. Extending the results shown in Figure 6 to 1 AU according to this variation yields a velocity of 509.8 km s^{-1}. An inverse-square law dependence for σ_{ne} leads to a $R^{0.24}$ dependence for v and a velocity of 442 km s^{-1} at 1 AU. These velocities are consistent with spacecraft measured velocities in 1975 (private communication with A. Lazarus).

Results similar to those shown in Figure 9 deduced from DRVID (Callahan 1975) and ranging (Muhleman et al. 1977) measurements over a more limited heliocentric distance range are consistent with those in Figure 9. Clearly, phase is the much preferred data type.

Phase scintillations are far superior to intensity scintillations because they provide information over the full range of scale sizes. Because they respond to large scale irregularities, the correlation distance is correspondingly long. Since the frequency spectrum of the phase scintillations is inverse power-law, there is considerable power in the low-frequency fluctuations which permits phase scintillations to be useful over a greater radial distance range. Phase scintillations are also valuable close to the sun because, like spectral broadening and unlike intensity scintillations, they do not saturate (see Figure 9) when they become strong. These features make phase scintillations very attractive for carrying out multiple-station wind velocity measurements in the acceleration region of the solar wind with a wide range of baseline distances (Woo 1977b), measurements that would add substantially to the results in Figure 6 and complement the extensive multiple-station intensity scintillation measurements farther out (Hewish 1972; Watanabe et al. 1974; Coles et al. 1974).

4. CONCLUDING DISCUSSION

It is obvious that spectral broadening and phase scintillations are two of the most useful radio scattering phenomena that can be observed with monochromatic, coherent, and point-source spacecraft signals. Angular broadening of a spacecraft signal has been measured for the first time using a single receiving antenna, and future interferometric measurements are clearly desirable. Angular broadening is particularly attractive because it responds to density fluctuations only, does not saturate, and can be analyzed with a second-moment theory (Lee and Jokipii 1975; Woo et al. 1977). From the standpoint of probing ability and interpretation, the advantages of the above mentioned phenomena far outweigh those of intensity scintillations. Still, observing intensity scintillations of spacecraft radio signals instead of natural radio sources offers some merit, especially when they are employed to study the density spectrum. This stems from the fact that a spacecraft signal is a point source, and the intensity spectrum is not confused by finite source size effects. Chang (1976) has analyzed the Pioneer 9 intensity scintillations and his results are consistent with those deduced by Lovelace et al. (1970) using natural radio sources (see Figure 7). Scattering by density fluctuations also causes pulse broadening, an effect that has been observed with pulsar measurements (Counselman and Rankin 1973). Although ranging measurements with spacecraft signals have been conducted, useful pulse broadening measurements have not. Woo et al. (1977) have also shown that observations of the scintillations of the total field with multiple stations (mutual coherence function) would provide measurements of the wind velocity near the sun. These measurements are similar to multiple station intensity scintillations in that they require a short baseline, whereas phase scintillations can be observed with short as well as long baselines.

Extensive radio scattering observations using spacecraft signals will clearly complement the vast number of radio astronomical observations that have taken place in the last twenty years and will surely contribute substantially to our understanding of the dynamics of the solar wind especially near the sun.

ACKNOWLEDGMENT

This paper presents the results of one phase of research carried out at the Jet Propulsion Laboratory, California Institute of Technology, under Contract No. NAS 7-100, sponsored by the National Aeronautics and Space Administration.

REFERENCES

Armstrong, J.W.: 1975, Observations of Interplanetary Scintillations at 74 MHz, University of California, San Diego (Ph.D. Thesis).
Blesing, R.G. and Dennison, P.A.: 1972, Proc. Astron. Soc. Australia 2, 84.
Callahan, P.S.: 1975, Astrophys. J. 199, 227.
Cannon, A.R.: 1976, Radio Frequency Probing of the Solar Corona, University of California, Berkeley (Ph.D. Thesis).
Chang, H.: 1976, Analysis of Dual-Frequency Observations of Interplanetary Scintillations Taken by the Pioneer 9 Spacecraft, Stanford University (Ph.D. Thesis).
Coles, W.A., Rickett, B.J. and Rumsey, V.H.: 1974, in C. T. Russell (ed.) Solar Wind III, University of California, Los Angeles, p. 351.
Counselman, C.C., III, and Rankin, J.M.: 1973, Astrophys. J. 185, 357.
Croft, T.: 1973, J. Geophys. Res. 78, 3159.
Cronyn, W.M.: 1972a, Astrophys. J. 171, L101.
Cronyn, W.M.: 1972b, Astrophys. J. 174, 181.
Ekers, R.D. and Little, L.T.: 1971, Astron. Astrophys. 10, 310.
Erickson, W.C.: 1964, Astrophys. J. 139, 1290.
Goldstein, R.M.: 1969, Science 166, 598.
Gurvich, A.S. and Tatarskii, V.I.: 1975, Radio Sci. 10, 3.
Hartle, R.E. and Barnes, A.: 1970, J. Geophys. Res. 75, 6915.

Hewish, A.: 1972, in C.P. Sonett, P.J. Coleman and J.M. Wilcox (ed.) Solar Wind, NASA, Washington, D.C., p. 77.
Hollweg, J.V. and Harrington, J.V.: 1968, J. Geophys. Res. 73, 7221.
Houminer, Z. and Hewish, A.: 1974, Planet Space Sci. 22, 1041.
Hundhausen, A.: 1972, Solar Wind and Coronal Expansion, Springer-Verlag, New York.
Intriligator, D.S. and Wolfe, J.H.: 1970, Astrophys. J. 162, L187.
Ishimaru, A.: 1977, Appl. Opt. (in press).
James, J.C.: 1968, in J.V. Evans and T. Hagfors (ed.), Radar Astronomy, McGraw-Hill, New York, p. 323.
Jokipii, J.R.: 1973, Ann. Rev. Astron. Astrophys. 11, 1.
Jokipii, J.R. and Hollweg, J.V.: 1970, Astrophys. J. 160, 745.
Knight, C.A., Robertson, D.S., Shapiro, I.I., Whitney, A.R., Rogers, A.E.E., Clark, T.A., Marandino, G.E., Vandenberg, N.R. and Goldstein, R.M.: 1971, Bull. Am. Astron. Soc. 3, 447.
Koehler, R.L.: 1968, J. Geophys. Res. 73, 4883.
Landt, J.A.: 1974, J. Geophys. Res. 79, 2761.
Lee, L.C. and Jokipii, J.R.: 1975, Astrophys. J. 196, 695.
Levy, G.S., Sato, T., Seidel, B.L., Stelzried, C.T., Ohlson, J.E. and Rusch, W.V.T.: 1969, Science 166, 596.
Lotova, N.A.: 1975, Soviet Phys. Usp. 18, 292.
Lovelace, R.V.E., Salpeter, E.E., Sharp, L.E. and Harris, D.E.: 1970, Astrophys. J. 159, 1047.
Muhleman, D.O., Esposito, P.B. and Anderson, J.D.: 1977, Astrophys. J. 211, 943.
Mutel, R.L.: 1975, Theory and Observations of Interferometer Visibility Scintillations with Application to the Interplanetary Medium, University of Colorado (Ph.D. Thesis).
Neugebauer, M.: 1975, J. Geophys. Res. 80, 998.
Newkirk, G.: 1967, Ann. Rev. Astron. Astrophys. 5, 213.
Okoye, S.E. and Hewish, A.: 1967, Monthly Notices Roy. Astron. Soc. 137, 287.
Pintér, S.: 1973, Bull. Astron. Inst. Czech. 24, 337.
Rumsey, V.H.: 1975, Radio Sci. 10, 107.
Saito, K.: 1970, Ann. Tokyo Astron. Obs. 12, 53.
Schatten, K.H.: 1970a, Science 168, 395.
Schatten, K.H.: 1970b, Solar Phys. 12, 484.
Sheely, N.R., Harvey, J.W. and Feldman, W.C.: 1976, Solar Phys. 49, 271.
Stelzried, C.T., Levy, G.S., Sato, T., Rusch, W.V.T., Ohlson, J.E., Schatten, K.H., Wilcox, J.M.: 1970, Solar Phys. 14, 440.
Sullivan, J.D., Belcher, J.W., Lazarus, A.J.: 1975, EOS Trans. AGU 56, 440.
Suess, S.T., Richter, A.K., Winge, C.R., Nerney, S.F.: 1977, Astrophys. J. (in press).
Waldmeier, M.: 1977, Nature 265, 611.
Ward, B.D.: 1975, Radio Wave Scattering in the Interplanetary Medium, Adelaide University (Ph.D. Thesis).
Watanabe, T., Shibaski, K. and Kakinuma, T.: 1974, J. Geophys. Res. 79, 3841.

Wolff, C.L., Brandt, J.C. and Southwick, R.G.: 1971, Astrophys. J. 165, 181.
Wilson, D.: 1976, The Three Dimensional Solar Corona--A Coronal Streamer, University of Colorado (Ph.D. Thesis).
Woo, R.: 1975, Astrophys. J. 201, 238.
Woo, R.: 1977a, Astrophys. J. (submitted).
Woo, R.: 1977b, Nature 236, 514.
Woo, R., Yang, F.C. and Ishimaru, A.: 1976a, Astrophys. J. 210, 593.
Woo, R., Yang, F.C. and Ishimaru, A.: 1977, Astrophys. J. (in press).
Woo, R., Yang, F.C., Yip, K.W. and Kendall, W.B.: 1976b, Astrophys. J. 210, 568.
Yakovlev, O.L., Trusov, B.P., Vinogradov, V.A., Efimov, A.I., Kruglov, Yu. M., Matyugov, S.S. and Razmanov, V.M.: 1974, Kosm. Issle. 12, 600.

DISCUSSION

Duffett-Smith: There is evidence of an enhancement in the spectrum of density irregularities at about the proton gyro-radius. Are your measurements sensitive to this? Have you any direct measurements for elongations greater than 20°?

Woo: Our observations measure the density spectrum in the range of the proton gyro-radius wavenumber at the smaller solar elongations. Beyond 20°, we cannot study this at 2.3 GHz because of signal-to-noise limitations.

Houminer: Interplanetary scintillation observations of the power spectrum of the small scale density irregulatities between 0.5-1.0 AU as well as spacecraft observations at 1 AU show that the spectrum of the irregularities cannot be described by a single power-law spectrum. The observed spectrum shows a marked flattening at wavenumbers between 10^{-3} - 10^{-2} km^{-1}. I wonder if your observed power-law spectrum at $\sim 2R_\odot$ can be extrapolated to 1 AU and compared to spectra obtained there. The origin of the small scale density irregularities is still unknown. It is possible that the spectrum of the irregularities at $2R_\odot$ is different from the spectrum at 1 AU.

Woo: The measurements we have made are of the region nearer the sun.

Dryer: Your suggestion concerning the mass flux with a slightly converging flow is very interesting. Do you think that the physical explanation may lie in the chance that higher speed stream flows from the northern and southern hemispheres may be moving the solar wind slightly toward the equatorial plane?

Woo: Yes, that might be one explanation. Another is a meridional temperature gradient as discussed recently by Suess et al. (1977).

Grzedzielski: When converting the phase oscillations into electron density oscillations do you use a thin phase screen theory and if no, do you allow for the lack of uniformity of the medium (n_e gradients, divergence of solar wind flow, etc.)?

Woo: We use an extended medium analysis, and assume that the solar wind is spherically symmetric (R^{-2} dependence for the rms density fluctuations).

Alurkar: What is the main restriction on the scale size of plasma density irregularities, derived from intensity scintillation, that does not limit the scale sizes derivable from phase scintillation?

Woo: The Fresnel filtering effect.

OBSERVATIONS OF INTERPLANETARY SCINTILLATION: SOLAR WIND VELOCITY MEASUREMENTS

T. KAKINUMA
The Research Institute of Atmospherics, Nagoya University
Toyokawa, 442 Japan

ABSTRACT

Recent observations of the solar wind velocity by IPS method are reviewed. Detailed analyses show that the diffraction pattern is anisotropic; thus, we cannot make the assumption of isotropic pattern in the calculation of velocity. A latitudinal distribution of the solar wind velocity is derived from the observations; the solar wind velocity is 800 km/sec for latitudes above 45°. The observations indicate the presence of the corotating stream which extends from this high-speed region to the equator. Comparisons of IPS data with EUV or XUV solar observations show that there is a close relationship between holes and streams, but the association is not one to one. IPS observations show that the most turbulent region of the flare-associated shock wave lies in the piston.

1. INTRODUCTION

The observations of interplanetary scintillation (IPS) of radio sources provide us valuable information on the solar wind over a wide range of heliographic longitude, latitude and radial distance. In 1967, Dennison and Hewish (1967) first showed that the solar wind velocity can be derived from three-station observations of IPS. Similar observations have been made at several observatories since then. The results of these observations have been reviewed by Hewish (1972) in 1972 and by Coles et al. (1974) in 1974. At present the solar wind velocity measurements with spaced receivers are being carried out at 74 MHz by the group of the University of California, San Diego (UCSD) and at 69 MHz by the group of Nagoya University, Toyokawa (TYKW). These observations cover the range of heliographic latitude from 60°S to 80°N and the range of radial distance from 0.3 AU to 1.2 AU. Daily values of the solar wind velocity are regularly published by the UCSD group in *Solar Geophysical Data*.

This paper reviews the recent results of IPS observations and con-

M. A. Shea et al. (eds.), Study of Travelling Interplanetary Phenomena/1977, 101–118.

centrates on the velocity measurements. IPS observations also yield other important parameters for studying the solar wind, such as the IPS index and the IPS spectral index. The observations of these parameters will be reviewed by Houminer in this Proceedings.

The method of analysis of observations, the latitude dependence of the solar wind velocity, the relation between coronal structures and high-speed streams, and the large-scale properties of flare-associated interplanetary disturbances are discussed.

2. ANALYSIS OF IPS OBSERVATIONS

Radiation from a compact radio source is scattered by electron density irregularities in the solar wind; the resultant diffraction pattern drifts at the solar wind velocity across the earth. In order to derive this pattern velocity, successive observations of the whole pattern (a complete picture of the pattern) are required (Briggs, 1968; Little and Ekers, 1971). This is, however, impracticable in our case where only a limited number of receivers is available. Thus we have to make some assumptions to calculate the pattern velocity. Various methods for analyzing three-station observations have been proposed by a number of authors (e.g., Briggs et al., 1950; Little and Ekers, 1971; Armstrong and Coles, 1972; Kakinuma et al., 1973). In these methods, the shape of the pattern (contour of constant correlation) is usually assumed to be elliptical, and then the axial ratio and orientation of the characteristic ellipse and the pattern velocity are derived from cross-correlations and auto-correlations (or power spectra) of intensities observed at three stations. However, most of these methods are very sensitive to the error in normalizing the cross-correlation and can be used only for the data with high signal-to-noise ratio. Therefore, we further make the following assumptions in daily calculations of the pattern velocity: (1) the random changes of the pattern (the pattern rearrangement) are negligible (the frozen assumption), and (2) the direction of velocity is equal to that expected for the radial solar wind (the radial assumption) (Coles and Rickett, 1975). Under these assumptions, we can calculate the pattern speed and its error from three time lags (not sensitive to the normalization error) of cross-correlations of intensities between pairs of sites without determining the shape of the pattern.

Coles and Kaufman (1977) have examined these assumptions with four-site observations. They have found that the average direction of velocity is radial within 4 degrees, in agreement with spacecraft observations. They have also found by the Little and Ekers method that the rms deviation of velocity, by which the pattern rearrangement is characterized, is 20-30% (upper limit) of the mean velocity. According to their estimates, the velocity calculated under the frozen assumption is biased slightly (less than 10%) above the true mean velocity. Lotova et al. (1977) estimated the standard deviation of velocity by dispersion analysis (Lotova and Chashey, 1974) using the UCSD data of 3C144 and 3C48 for four days (this estimate is not sensitive to the normalization error). They found that the deviation of velocity varied from 17% to 37% of the mean and was significantly greater for the high-latitude source, 3C48. The turbulent velocity of density irregularities in the solar wind is considered to be much smaller than the wind velocity, except in the vicinity of the sun; therefore, we may neglect the random changes of the pattern for observations at lower radio frequencies.

The four-site observations of the ecliptic source, 3C144, have also given an upper bound of 1.3 for the pattern axial ratio due to an anisotropic density-irregularity spectrum (Coles and Kaufman, 1977). Kojima (1977) has recently investigated in detail the anisotropy of the pattern by applying the method by Kakinuma et al. (1973) to the TYKW data of the high-latitude source, 3C48, which have high signal-to-noise ratios. He has found that (1) the direction of velocity is radial within a few degrees, (2) the mean axial ratio is 1.7, and (3) the orientation of the ellipse, on the average, coincides with the direction of velocity for the case when the main scattering region on the line of sight is at low latitudes, but deviates by about 6 degrees anticlockwise from the direction of velocity for the case when the main scattering region is at high latitudes, as shown in Figure 1. He has concluded that this anisotropy is

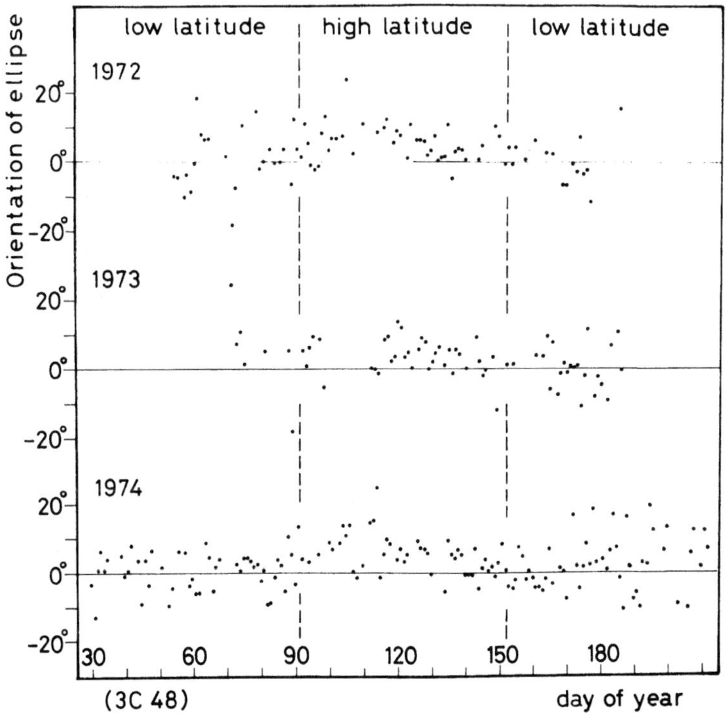

Figure 1. Daily plots of the orientation angles of ellipse with respect to the radial direction obtained by IPS observations of 3C48 during 1972 -1974 (Kojima, 1977).

statistically significant. It is not due to the normalization error, nor is it due to the structure of 3C48, but is due to the anisotropy of electron density irregularities, in agreement with Dennison's conclusion (1969). It may be interesting to examine the relation between this pattern anisotropy and the interplanetary magnetic field.

If the pattern is elongated in different direction from the flow direction (the radial direction), the pattern speed calculated under the assumption of isotropic pattern is biased. At the same time, the speed error (the scatter of speed estimates for three baselines) apparently becomes very large, even though there is no inconsistency among three time lags (the sum of them must be zero). The bias and the apparent error vary depending on the geometry of the observation system. For example, the maximum bias is estimated to be about 15% for an axial ratio of 2 and an orientation angle of 10 degrees, while the apparent error exceeds 20% (Kojima, 1977). The pattern is also elongated for an anisotropic source; Coles and Kaufman (1977) have estimated the maximum bias to be about 15% for 3C161, 3C237, 3C273, and 3C298. Therefore, we can not make the assumption of isotropic pattern in the calculation of velocity. This has also been pointed out by Coles and Kaufman (1977).

In relating the pattern speed to the solar wind velocity we have to take into account the integration effect along the line of sight to the radio source; the pattern speed is the weighted average of the transverse components of the solar wind velocity along the line of sight. The weighting functions along the line of sight have been calculated for a power law model of electron density spectrum under the assumption of a spherically symmetric solar wind and the inverse square dependence of electron density fluctuations on the radial distance from the sun. The calculations for various solar elongations (e.g., Coles and Rickett, 1975) show that the weighting function has a peak at the point of closest approach of the line of sight to the sun except at large elongation angles. Jokipii and Lee (1972, 1973) have calculated the pattern velocity under the same assumptions using the weak scintillation approximation; the pattern velocity is found to be about 10% smaller than the solar wind velocity.

When high and low speed streams with different density fluctuations exist we must consider their distribution along the line of sight. According to Watanabe (1975), if a stream is sufficiently wide, the situation is very similar to that for the spherically symmetric solar wind when the stream center reaches the point of closest approach of the line of sight to the sun; the most effective scattering region on the line of sight is entirely covered by the stream. The average duration of high speed stream observed by a space probe (Intriligator, 1973) is about five days; therefore, in most cases, the pattern speed may be considered to be nearly equal to the solar wind velocity at the point of closest approach. He has pointed out that the comparison between IPS and space probe data (Watanabe and Kakinuma, 1972; Armstrong and Coles, 1972) shows a close agreement of both data when streams with durations of more than about three days are concerned.

Harmon (1975) has studied in detail the turbulent structure of a stream, comparing observed IPS indices and velocities with those expected from IMP 7 observations. He has found that the region of enhanced electron density fluctuations lies one or two days behind density compression but in front of the main body of the stream, and that density fluctuation levels remain relatively high within the density rarefaction. In this study, he directly compared the pattern speeds with the IMP 7 data mapped to the point of closest approach and found a remarkably good agreement. From this result we can conclude that for the quiet solar wind and the corotating stream, the observed pattern velocity is approximately equal to the solar wind velocity at the point of closest approach when the main scattering region is near the ecliptic plane. However, this does not always apply to a shock disturbance; if the turbulent level is extremely high the IPS enhancements due to the shock disturbance will be clearly observed, even though the disturbance does not intersect the line of sight near the point of closest approach.

Interpretation of IPS observations rests on a theory of wave propagation through a turbulent extended medium. For weak scintillations the Born approximation can be used and so far most discussions have made use of this approximation. However, for strong scintillations, we have no adequate theory for the case of an extended medium. Rumsey (1976) has recently discussed intensity fluctuations due to an extended incoherent source for the case of a highly turbulent extended medium (sufficiently strong scintillations). It is important for the interpretation of IPS velocity observations to know the weighting function along the line of sight for strong scintillations.

3. LATITUDE DEPENDENCE OF THE SOLAR WIND VELOCITY

Recent IPS observations show that there is a definite increase of solar wind velocity with heliographic latitude. Coles and Rickett (1976) have reported the variation of the average solar wind velocity with latitude in each year from 1971 to 1975, as shown in Figure 2. The mean gradient is found to be 2.1 km/sec per degree of latitude. In this case, IPS pattern speed is considered to be the solar wind velocity in the vicinity of the point of closest approach. The heliographic longitude, latitude and the radial distance of this point change every day as the sun rotates and the earth orbits the sun; the radial distance decreases with increasing latitude. Thus they have studied the radial-distance dependence using the observations of the four ecliptic radio sources for which the latitude remains within 10 degrees of the equator. As shown in Figure 3, they have found no radial-distance dependence of the solar wind velocity between 0.4 AU and 1.1 AU.

Such an increase in solar wind velocity with increasing latitude was first found by Dennison and Hewish (1967) in 1966, while Hewish and Symonds (1969) and Hewish (1972) reported that there was no evidence for the latitude dependence in 1967 and 1969. (Vitkevitch and Vlasov (1972) reported a decrease in wind velocity with latitude in these years.) Coles

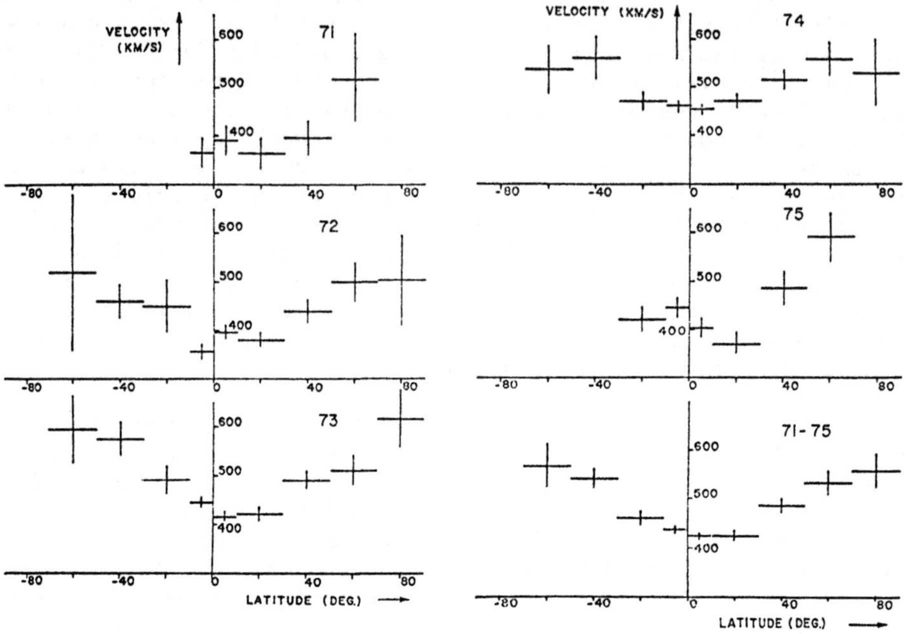

Figure 2. Variation of the solar wind velocity with latitude in each year from 1971 to 1975. The lower right graph is the overall average from the other five graphs. (Figure from Coles and Rickett (1976); reproduced by permission of the American Geophysical Union.)

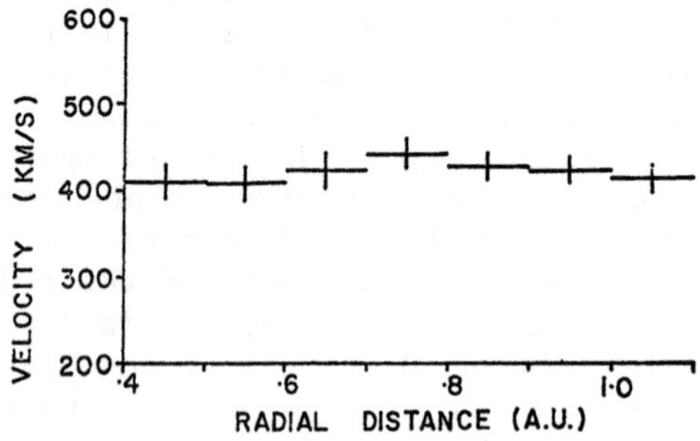

Figure 3. Solar wind velocity versus radial distance during 1971-1975, derived from the observations within 10° of the solar equator. (Figure from Coles and Rickett (1976); reproduced by permission of the American Geophysical Union.)

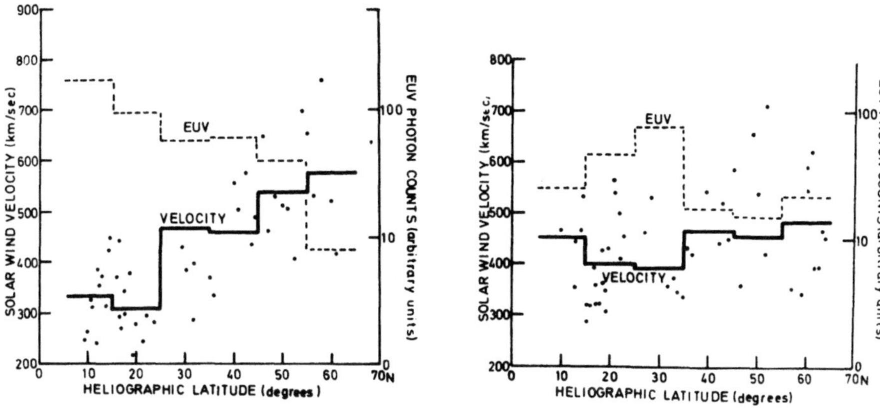

Figure 4. The latitudinal distribution of the solar wind velocity obtained by IPS observations of 3C48 (left) during May-June 1972 and (right) during May-June 1973. Daily values of the solar wind velocity are plotted by dots against the heliographic latitude of the source of observed solar wind. Average solar wind velocity in each 10° of latitude is represented by a solid line, and average EUV photon counts at the source (the intensity of FeXV emission observed by OSO 7) are represented by a broken line. (Figure from Watanabe et al. (1974); reproduced by permission of the American Geophysical Union.)

and Rickett (1976) attributed this discrepancy to the observations of short duration; the average latitude behavior can be masked by streams and slower variations in the solar wind velocity. Watanabe et al. (1974) have studied the relation between latitudinal distributions of the solar wind velocity and of EUV brightness of the corona and found an inverse correlation between them. Their results, as shown in Figure 4, suggest that average EUV brightness at high latitudes might have been approximately the same as that at low latitudes during the periods in 1967 and 1969 when the Cambridge group's measurements were made.

During the declining and the minimum phases of solar activity, the coronal structure remains quite stable and a recurrent pattern is found over periods of several months (Hansen et al., 1976), as shown in Figure 5(a). Spacecraft observations show that there were stable recurrent streams during 1973-1974 (Gosling et al., 1976). The IPS observations of 3C48 during 1974 and 1976, as shown in Figure 5(b), show the existence of stable recurrent streams over periods of several months. We may divide these daily observations of the pattern speed into two parts: the variation due to the quiet solar wind and that due to recurrent high-speed streams. For example, if we take the lower envelope of the observations in 1976, we obtain the variation of pattern speed with latitude of the point of closest approach for an axisymmetrical quiet solar wind, as shown in Figure 6. As described in the previous section, the pattern

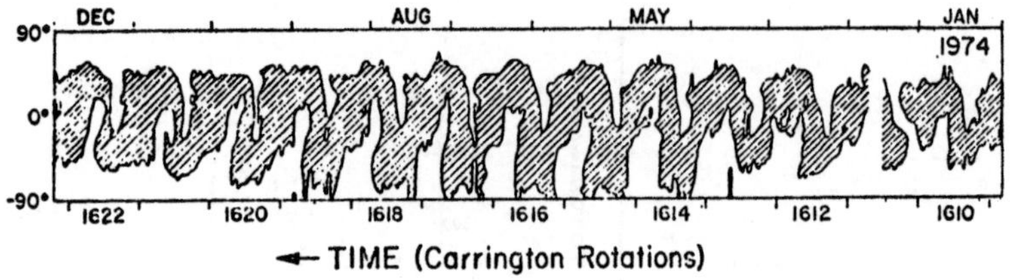

Figure 5(a). Synoptic map of the K corona at 1.5 $R_\odot$ for intensity level 2.0×10^{-8}pB. in 1974. (Figure after Hansen et al. (1976); reproduced by permission of Pergamon Press Ltd.)

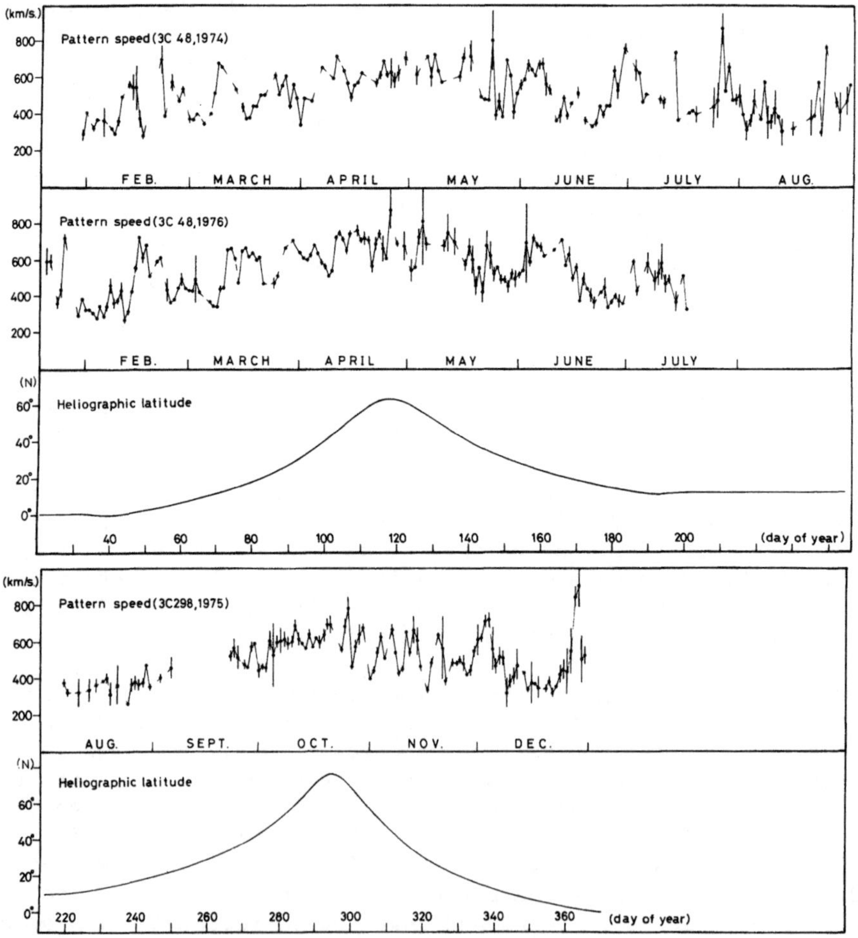

Figure 5(b). IPS velocity observations for 3C48 in 1974 and 1976, and for 3C298 in 1975. The heliographic latitude of the point of closest approach for each source is also shown.

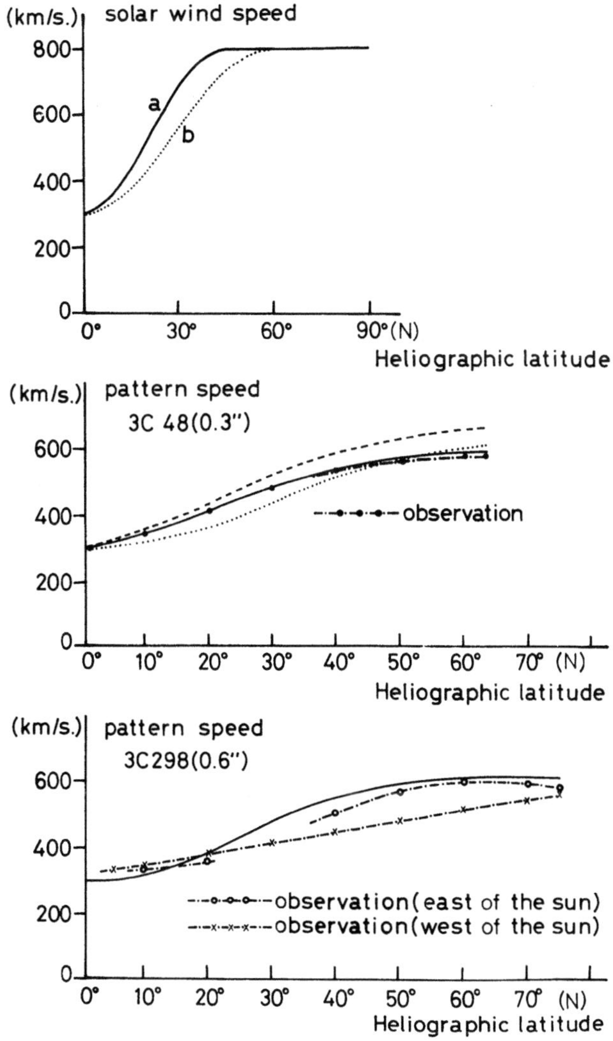

Figure 6. The models (a and b) of latitudinal distribution of the solar wind velocity, and the computed and the observed (the lower envelopes of the observations shown in Figure 5(b)) variations of pattern speed with latitude. A solid line for each source represents the variation of pattern speed for the model a and Δn(equator)=1.8×Δn(pole) (Δn: rms electron density fluctuation). In the middle graph, a broken line represents the variation of pattern speed for the model a and Δn(equator) =Δn(pole); a dotted line represents the variation of pattern speed for the model b and Δn(equator)=Δn(pole).

speed is a weighted average of the solar wind velocity along the line of sight. If there is a latitude dependence of the solar wind, a large variation of the solar wind velocity along the line of sight is expected for high-latitude observations; the lower envelope does not show the actual dependence. We have calculated the pattern speed for various models of the axisymmetrical solar wind with latitude dependence. We first computed the cross-power spectrum from 0.1 Hz to 1 Hz assuming the weak scintillation (Harmon, 1975) and then the intensity cross-correlation by Fourier transform; the pattern speed was calculated from the time lag of the cross-correlation function. We assumed a power law spectrum of electron density fluctuations, the elliptical dependence on latitude and the inverse square dependence on radial distance of electron density fluctuations, and a Gaussian brightness distribution of the source with a diameter of 0.3 arc sec. We took into account the inclination of the solar axis to the ecliptic plane. As shown in Figure 6, we find a latitudinal distribution of the solar wind velocity which can explain the lower envelope of the observations for 3C48 in 1976. In this model, the electron density fluctuation at the equator is 1.8 times as large as that at the pole, and the solar wind velocity is 800 km/sec for latitudes higher than 45 degrees. This high-speed region is considered to correspond to the north polar hole. For a model in which the solar wind velocity increases linearly with latitude, a velocity more than 1000 km/sec is required at the pole.

As seen in the observations in 1974 and 1976, the amplitude of speed enhancement due to the recurrent stream apparently decreases with increasing latitude, and the maximum speed is about 700 km/sec. This may be interpreted to show that the high-speed stream is an equatorward extension of the above high-speed region at high latitudes. The synoptic map of K corona at height 1.5 $R_\odot$ in 1974 (Hansen et al., 1976) also shows that there was a stable region of low K-corona intensity extending as a bay from the north pole to the equator. The coincidence between the development of stable holes which extended from the pole to the equator and the appearance of stable streams has been pointed out by Gosling et al. (1976). IPS observations give evidence for the existence of a wide stream extending from the polar region to the equator. The above model suggests that there is no corotating localized stream at latitudes above about 45 degrees; the enhancements in pattern speed in high-latitude observations are due to corotating streams crossing the line of sight at latitudes below 45 degrees.

We have examined whether the above model can also explain the observations for 3C298 in 1975, which show similar variations of pattern speed. 3C298 is another strong high-latitude source with a diameter of 0.4-0.6 arc sec (Little and Hewish, 1968; Readhead and Hewish, 1974). The calculated pattern speeds are plotted against the latitude of the closest approach in Figure 6. The agreement with the observations is not good in this case; in fact the model b in Figure 6 seems to fit. The observations, however, show that the solar wind was not axisymmetrical during this period; the centroid of the high-speed region seems to be considerably removed from the pole. Furthermore, as the source size is rather large, the observed scintillation index sharply decreases at high latitudes by the effect of strong scintillation, while the calculation (weak scintillation approximation) shows gradual increase of scintillation index. Thus we have to take into account the distribution of the solar wind velocity in longitude as well as in latitude and the effect of strong scintillation for the observations of 3C298 in 1975.

4. RELATION BETWEEN CORONAL STRUCTURES AND THE SOLAR WIND VELOCITY

The IPS observations enable us to study the coronal origin of high-speed solar wind at high latitudes. Watanabe (1975) has compared IPS and space probe observations of the solar wind velocity (May-August 1972 and May-August 1973) with the intensity of EUV emission (FeXV 284Å)

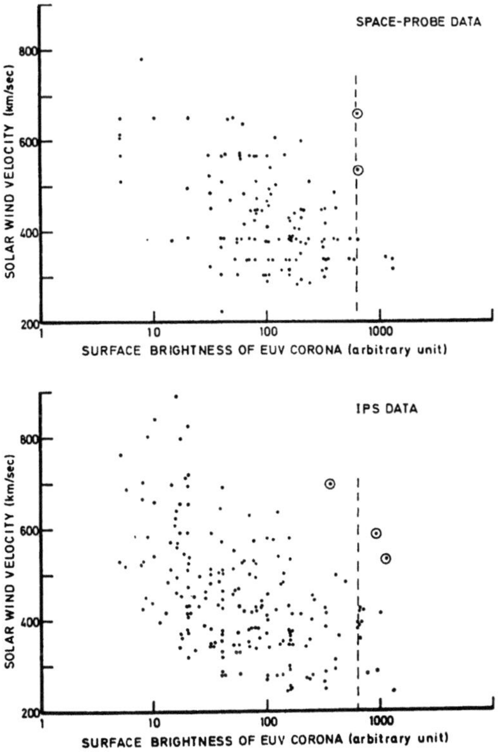

Figure 7. The relationship between the solar wind velocity and the surface brightness of EUV corona at the position of solar wind source. The surface brightness of EUV corona is represented by the count level of EUV photons (in arbitrary units) emitted from 1' square area on the sun. A vertical broken line shows the EUV brightness of 640. The points enclosed with small circles show the extraordinary tendency (Watanabe, 1975).

obtained by OSO 7. He has found an inverse correlation between the observed solar wind velocity and FeXV brightness at the position of the solar wind source, as shown in Figure 7. The heliographic coordinates of the source were calculated under the assumption of ideal spiral model, and the monitoring region of the IPS observation was assumed to be in the vicinity of the point of closest approach. According to his discussion, most of EUV coronal holes are the principal sources of high-speed solar wind faster than 500 km/sec, although he did not identify coronal holes explicitly. The points which correspond to the central regions of EUV coronal holes are distributed in the vicinity of EUV coronal brightness of 5 in Figure 7. As exceptional cases (the points enclosed by small circles), the active regions with multipolar magnetic geometry were found to be the sources of high-speed solar wind in spite of their high EUV brightness. He has also found from Mariner 2 data that the unipolar active regions are another class of sources of high-speed solar wind. His conclusion is that there is a close relationship between the solar wind velocity and the magnetic field geometry (the open field geometry) of the source region. His results also suggest that the association between coronal holes and high-speed streams is not one to one.

Rickett et al. (1976) have made a comparison of coronal holes and the source regions of high-speed streams of heliographic latitudes from +40 degrees to -60 degrees measured at UCSD by the IPS method. Ten coronal holes were identified in wide band XUV solar images taken during the manned portions of Skylab mission (six periods during May 1973 - Feb. 1974). In order to examine the coincidence between holes and stream sources, the IPS observations were summarized in a table (Table 1) in the following manner. First, the source region of the observed solar wind was identified for each IPS observation by tracing back (a constant speed of 400 km/sec was assumed) to the lower corona from the main scattering region about the point of closest approach of the line of sight to the sun; this region (a linear region) was called the subscattering region. Next, "the velocity observations were separated into three classes, 'hole days', 'close approach days' and 'no encounter days', depending on whether the subscattering region fell in a coronal hole, fell within 20 degrees northward or southward of a hole, or missed all holes, respectively." The velocity observations were also divided into two groups, depending on whether a stream was observed or not. In each square of the table, the total number of hole and close approach days for all IPS sources (below the line) and the total number of those days which are also classified as stream days (above the line) are given; therefore, the ratio of these numbers shows the degree of coincidence between streams and holes (the ratio due to chance coincidence was estimated to be 1-3). This table shows that there is a close relationship between streams and holes, but the association is not one to one. Their statistical interpretation of the whole data also shows that the association is not highly significant.

In the above studies, the pattern speed is assumed to be a spatial average of the solar wind velocity about the point of closest approach. However, if the solar wind has a latitude dependence such as that shown

HOLE SEQUENCE \ DATA PERIOD	1	2	3	4	5	6		
1	0/10	x	0/8	1/15	x	x	CH7	N. POLAR
2	14/20	3/3	3/3	x	x	5/11	CH5,2*	S. POLAR
3	5/20	16/30	2/8				CH1	
4	1/8	4/6					CH4	
5	0/4						CH3	
6		0/10	4/13	8/11			CH2	
7				10/15	9/11	x		
8				2/5				
9				6/10			CH2	
10					4/10		CH2*	

Table 1. Summary of observations. The upper part of each box (one box for each hole observed on the central meridian passage in each data period) contains the number of coincidence days for all IPS sources, and the lower part contains the total number of encounter (and close approach) days for all IPS sources and the hole in question. The ratio of these numbers shows the degree of coincidence between holes and streams. A cross means that a hole was identified but did not give rise to an encounter. The coronal hole numbers (CH) adopted by Nolte et al. (1976) are indicated on the right. (Table after Rickett et al. (1976); reproduced by permission of the American Geophysical Union.)

in Figure 6, we must carefully interprete high-latitude observations for which the latitude of the point of closest approach is high. As mentioned in the previous section, an enhancement of pattern speed in high-latitude observations does not always mean the existence of high-speed stream at high latitudes. On the other hand, even though no definable enhancement of speed is observed, the hole over the pole is the source of high-speed solar wind. For example, as pointed out by Rickett et al. (1976), the hole sequence 1 over the north pole for which the velocity series were quite high (500-650 km/sec) is considered to be the source of high-speed solar wind. If this is taken into account, the correlation between holes and streams will be found to be much higher. This shows the importance of the knowledge of the latitude dependence of the solar wind velocity for the statistical study of the origin of high-speed solar wind.

As described by Rickett et al. (1976), the study of counterexamples (the hole 5 and 6 in Table 1, or streams from active regions found by Watanabe) is important for finding the coronal condition generating a fast stream, and studies including longer time intervals will be necessary to obtain more statistically significant results. Hansen et al. (1976) have shown that the K-coronal observations can clearly detect the coronal hole from the ground. It will be interesting to make a comparison between IPS observations and K-coronal maps. (This comparison has recently been made by D. G. Sime (UCSD).)

5. LARGE-SCALE PROPERTIES OF FLARE-ASSOCIATED DISTURBANCES

The large-scale properties of flare-associated disturbances can be investigated by daily observations of a number of scintillating sources in different directions. In early August 1972, the enhancements in solar wind velocity and the IPS index due to three shock waves were detected at four observatories (Armstrong et al., 1973; Rickett, 1975; Watanabe et al., 1973; Houminer, 1973; Ward, 1975a,b). Three-site observations for velocity measurements were made at UCSD and TYKW. The results of these observations have been reviewed by Kakinuma and Watanabe (1976). The positions of the lines of sight to the radio sources on August 7, 1972 are shown in Figure 8.

IPS velocity observations show the anisotropic expansion of the disturbance; the speed of the disturbance is fastest in the direction of the flare normal. The shape of the disturbance from the August 7th flare (No. 4 flare in Figure 8), for which the data coverage was good, was found to be a sphere of radius 0.73 AU, centered at 0.4 AU from the

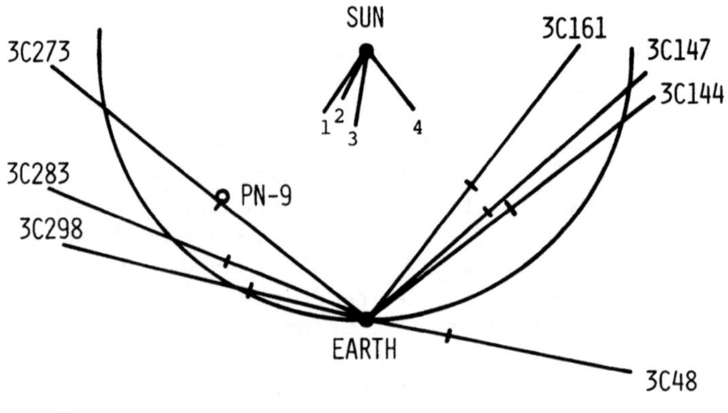

Figure 8. The positions of the lines of sight to the radio sources on August 7, 1972, projected onto the solar equatorial plane (viewed from the north). The tick marks the point of closest approach (or the peak of the weighting function) for each source. The longitude of four flares which occurred in early August 1972 are shown by bars.

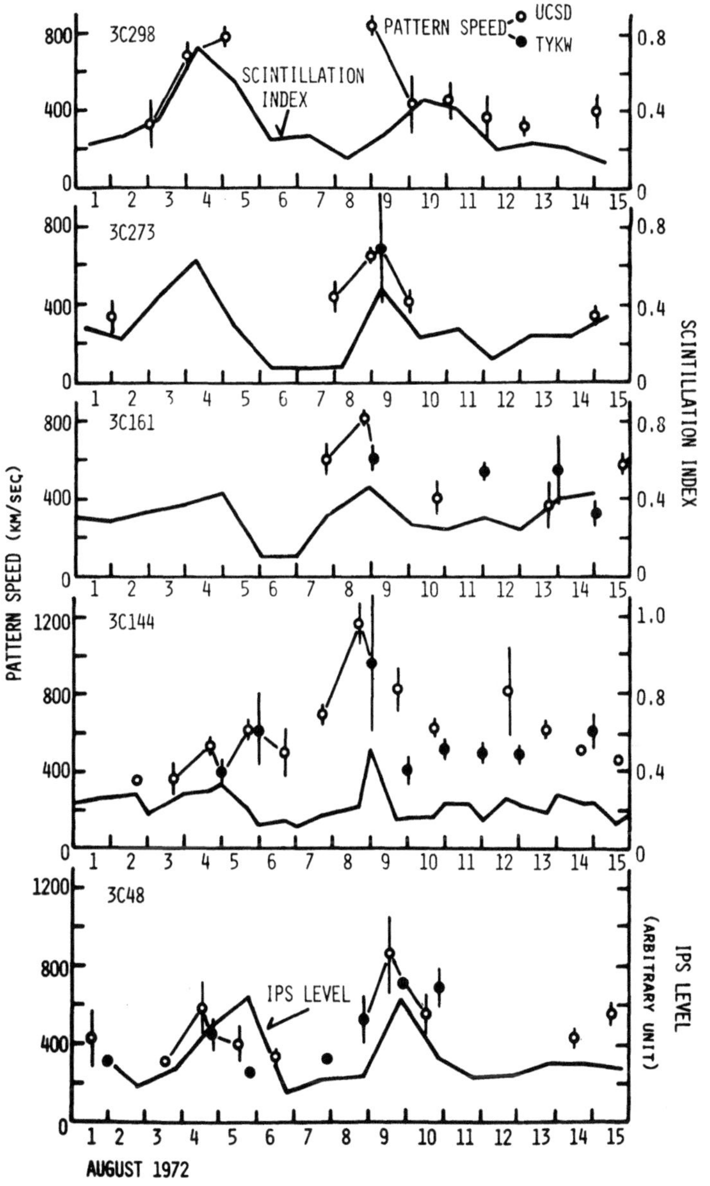

Figure 9. IPS observations in early August 1972. The velocities observed at UCSD and TYKW, the IPS level (rms fluctuations of intensity normalized by the galactic background noise level) for 3C48 obtained at TYKW, the IPS index for 3C144 obtained at UCSD (halved) and Adelaide, and the IPS indices for other sources obtained at Adelaide are shown.

sun along the flare normal (Rickett, 1975). The extent in longitude of this disturbance covered about ±80 degrees from the flare normal at 0.8 AU and about ±60 degrees at 1 AU, and the extent in latitude covered at least ±30 degrees from the flare normal at 0.8 - 1 AU from the sun (Rickett, 1975; Watanabe et al., 1973).

As shown in Figure 9, the observations of 3C144 at UCSD at 16 UT on August 8 measured the peak velocity (1200 km/sec), which is considered to be nearly equal to the velocity of the shock front, but no obvious enhancement in IPS index was observed at that time (Rickett, 1975). However, the observations at TYKW and at Adelaide (Ward, 1975a,b) measured the large index enhancement seven hours later. Similarly, the observations of other sources show that the enhancements in IPS index occurred behind the velocity enhancements. This may be interpreted to show that the most turbulent region of this shock lay at some distance behind the shock front (Kakinuma and Watanabe, 1976). Watanabe (1977) has investigated the deceleration of the shocks from the August 2nd flare and the August 7th flare. He has found that the variations of velocity measured from the observations for 3C48 can be well explained by assuming that the most turbulent region lay in the vicinity of the density maximum behind the shock front. For the shock wave from the August 7th flare, the distance from the shock front is estimated to be about 0.1 AU (about 6 hours). Dryer (1974, 1977) has suggested that the contact region (or piston) predicted by the MHD theory (Steinolfson et al., 1975) is highly turbulent and responsible for the observed enhancements in IPS index. The above results support his idea. Most of flare-associated disturbances detected by the IPS method, though they were rare, are considered to have been of a piston-driven type (Watanabe, 1977).

The disturbance from the flare at 20 UT on August 2 (No. 2 flare in Figure 8) did not appreciably decelerate between Pioneer 9 and Pioneer 10 which were situated at about 20 degrees east of the flare normal (Intriligator, 1976); however, the IPS observations of 3C48 at 30-70 degrees west of the flare normal show that this disturbance was strongly decelerated (deceleration power law index = 1.3) outside the earth's orbit, suggesting the asymmetrical expansion of this disturbance (Watanabe, 1977). This is in agreement with the conclusions of a number of authors as reviewed by Dryer (1975). The observations at Pioneer 9 and 10 (Intriligator, 1976) and the IPS observations (Watanabe, 1977) also show that the disturbance from the August 7th flare was not significantly decelerated in the region eastward of the flare normal.

The IPS method is very useful for studying the interplanetary disturbance, but it is only possible at present to infer the approximate shape and extent of the disturbance because IPS observations are not continuous (each source is only observed for a short period per day at each observatory). In order to study the transient phenomena, it is desirable to improve the time resolution of IPS observations by international cooperation.

ACKNOWLEDGEMENTS

I am grateful to Dr. M. A. Shea and Dr. D. F. Smart for helpful suggestions. I thank Dr. W. A. Coles, Dr. J. J. Kaufman, Dr. N. A. Lotova and Dr. I. V. Chashey for providing their unpublished results of investigations. I am indebted to Dr. T. Watanabe, Dr. H. Washimi and Mr. M. Kojima for valuable discussions and for permission to use some of their unpublished results. I also thank the many authors, the American Geophysical Union and Pergamon Press Ltd. for allowing me to use figures from their publications.

REFERENCES

Armstrong, J. W. and Coles, W. A.: 1972, *J. Geophys. Res.* 77, 4602.
Armstrong, J. W., Coles, W. A., Harmon, J. K., Maagoe, S., Rickett, B. J., and Sime, D. G.: 1973, in H. E. Coffey (ed.), *Collected Data Reports on August 1972 Solar-Terrestrial Events,* World Data Center A for Solar-Terrestrial Physics, Report UAG-28, Part 2, Boulder, Colorado, U.S.A., p. 371.
Briggs, B. H.: 1968, *J. Atmospheric Terrest. Phys.* 30, 1777 and 1789.
Briggs, B. H., Phillips, G. J., and Shinn, D. H.: 1950, *Proc. Phys. Soc. London* B63, 106.
Coles, W. A. and Kaufman, J. J.: 1977, 'Solar Wind Velocity Estimation from Multi-Station IPS', *J. Geophys. Res.* (submitted).
Coles, W. A. and Rickett, B. J.: 1975, *Solar-Geophysical Data,* 366 (supplement), Feb. 1975, US Department of Commerce, Boulder, Colorado, U.S.A., p. 15.
Coles, W. A. and Rickett, B. J.: 1976, *J. Geophys. Res.* 81, 4797.
Coles, W. A., Rickett, B. J., and Rumsey, V. H.: 1974, in C. T. Russell (ed.), *Solar Wind Three,* p. 351, *Proceedings of the Asilomar Solar Wind Conference,* 25-29 March, 1974, Univ. of Calif., Los Angeles, Inst. of Geophys. and Planet. Phys.
Dennison, P. A.: 1969, *Planet. Space Sci.* 17, 189.
Dennison, P. A. and Hewish, A.: 1967, *Nature* 213, 343.
Dryer, M.: 1974, *Space Sci. Rev.* 15, 403.
Dryer, M.: 1975, *Space Sci. Rev.* 17, 277.
Dryer, M.: 1977, private communication.
Gosling, J. T., Asbridge, J. R., Bame, S. J., and Feldman, W. C.: 1976, *J. Geophys. Res.* 81, 5061.
Hansen, R. T., Hansen, S. F., and Sawyer, C.: 1976, *Planet. Space Sci.* 24, 381.
Harmon, J. K.: 1975, 'Scintillation Studies of Density Microstructure in the Solar Wind Plasma', Univ. of Calif., San Diego (Ph.D. Thesis).
Hewish, A.: 1972, in C. P. Sonett, P. J. Coleman, Jr., and J. M. Wilcox (eds.), *Solar Wind,* NASA SP-308, Supt. of Doc., Washington, D. C., U.S.A., p. 477.
Hewish, A. and Symonds, M. D.: 1969, *Planet. Space Sci.* 17, 313.
Houminer, Z.: 1973, in H. E. Coffey (ed.), *Collected Data Reports on August 1972 Solar-Terrestrial Events,* World Data Center A for Solar-Terrestrial Physics, Report UAG-28, Part 2, Boulder, Colorado, U.S.A.,

p. 375.
Intriligator, D. S.: 1973, *High Speed Streams in the Solar Wind,* World Data Center A for Solar-Terrestrial Physics, Report UAG-27 (US Department of Commerce, NOAA Environmental Data Service, Asheville, North Carolina, U.S.A.).
Intriligator, D. S.: 1976, *Space Sci. Rev.* 19, 629.
Jokipii, J. R. and Lee, L. C.: 1972, *Astrophys. J.* 172, 729.
Jokipii, J. R. and Lee, L. C.: 1973, *Astrophys. J.* 182, 317.
Kakinuma, T. and Watanabe, T.: 1976, *Space Sci. Rev.* 19, 611.
Kakinuma, T., Washimi, H., and Kojima, M.: 1973, *Publ. Astron. Soc. Japan* 25, 271.
Kojima, M.: 1977, 'Anisotropic Structure of Solar Wind Plasma Irregularities Observed with Interplanetary Scintillations', *Publ. Astron. Soc. Japan* (submitted).
Little, L. T. and Ekers, R. D.: 1971, *Astron. Astrophys.* 10, 306.
Little, L. T. and Hewish, A.: 1968, *Monthly Notices Roy. Astron. Soc.* 138, 393.
Lotova, N. A. and Chashey, I. V.: 1974, in C. T. Russell (ed.), *Solar Wind Three,* p. 375, *Proceedings of the Asilomar Solar Wind Conference,* 25-29 March, 1974, Univ. of Calif., Los Angeles, Inst. of Geophys. and Planet. Phys.
Lotova, N. A., Chashey, I. V., and Coles, W. A.: 1977, 'Dispersion Analysis of Solar Wind Velocity', Lebedev Physical Institute of the U.S.S.R. Academy of Sciences (preprint).
Nolte, J. T., Krieger, A. S., Timothy, A. F., Viana, G. S., and Zombeck, M. V.: 1976, *Solar Phys.* 46, 291.
Readhead, A. C. S. and Hewish, A.: 1974, *Mem. Roy. Astron. Soc.* 78, 1.
Rickett, B. J.: 1975, *Solar Phys.* 43, 237.
Rickett, B. J., Sime, D. G., Sheeley, N. R., Jr., Crockett, W. R., and Tousey, R.: 1976, *J. Geophys. Res.* 81, 3845.
Rumsey, V. H.: 1976, *Radio Science* 11, 545.
Steinolfson, R. S., Dryer, M., and Nakagawa, Y.: 1975, *J. Geophys. Res.* 80, 1223.
Vitkevitch, V. V. and Vlasov, V. I.: 1972, *Soviet Astron.* 16, 480.
Ward, B. D.: 1975a, 'Radio-wave Scattering in the Interplanetary Medium', Univ. of Adelaide, South Australia (Ph.D. Thesis).
Ward, B. D.: 1975b, *Proc. Astron. Soc. Aust.* 2, 378.
Watanabe, T.: 1975, *Publ. Astron. Soc. Japan* 27, 385.
Watanabe, T.: 1977, 'IPS Observations of Flare-associated Shock Waves in 1972-1974', Res. Inst. Atmos., Nagoya Univ. (preprint).
Watanabe, T. and Kakinuma, T.: 1972, *Publ. Astron. Soc. Japan* 24, 459.
Watanabe, T., Kakinuma, T., Kojima, M., and Shibasaki, K.: 1973, *J. Geophys. Res.* 78, 8364.
Watanabe, T., Shibasaki, K., and Kakinuma, T.: 1974, *J. Geophys. Res.* 79, 3841.

SCINTILLATION OBSERVATIONS OF THE INTERPLANETARY PLASMA

Z. HOUMINER
National Committee for Space Research
Radio Observatory, Haifa, Israel

ABSTRACT

Observations of interplanetary scintillation indicate the presence of corotating scintillation sectors which are aligned along the spiral magnetic field over distances of 0.5 - 1.0 AU. Enhanced scintillation sectors lead high solar wind velocity streams by 2-3 days and are associated with the compressed density regions detected by spacecrafts. Scintillation enhancements tend to disappear at high heliocentric latitudes though high velocity streams exist at all latitudes. This can be explained by the weakening of the stream interaction due to the increase in the mean solar wind velocity at high heliocentric latitudes. Scintillation and velocity enhancements which occur simultaneously for radio sources located on opposite sides of the Sun are due to shock disturbances associated with solar flares. By combining IPS velocity measurements with spacecraft and geomagnetic observations, a good estimate can be obtained of the shape and extent of the shock disturbance in both heliocentric latitudes and longitudes. Recent spacecraft and scintillation observations of the power spectrum of the density irregularities indicate that the spectrum flattens at wavenumbers between 10^{-3} and $10^{-2}km^{-1}$. The flattening of the power spectrum is due to additional population of density irregularities with a scale size of the order of the proton gyroradius.

INTRODUCTION

Radio waves from extragalactic or galactic radio sources are scattered by irregularities of plasma density in the interplanetary medium. If the radio sources are compact so that they illuminate the plasma with sufficient coherence, the interference of the scattered waves produces a diffraction pattern at the Earth. This phenomenon is called interplanetary scintillation (Hewish, et al., 1964).

Studies of the solar wind by interplanetary scintillation are carried out at either spaced sites or at a single station. Studies of

M. A. Shea et al. (eds.), Study of Travelling Interplanetary Phenomena/1977, 119–141.

the motion and correlation of the diffraction pattern at three spaced sites, give information about the solar wind velocity and direction of flow (Dennison and Hewish, 1967; Hewish and Symonds, 1969; Armstrong and Coles, 1972; Watanabe and Kakinuma, 1972; Vitkevich and Vlasov, 1972). The temporal variation of intensity observed at a single site gives information about the power spectrum of the small scale density irregularities (Jokipii and Hollweg, 1970; Hewish, 1971; Cronyn, 1972; Matheson and Little, 1971; Bourgois, 1972; Houminer, 1973a; Rickett, 1973). Day-to-day variation of velocity and scintillation index (defined as the rms intensity fluctuation to the mean intensity of the source) can be used to study the large scale structure of the solar wind, i.e., corotating streams and shock waves associated with solar flares (Dennison and Wiseman, 1968; Sharp and Harris, 1967; Houminer, 1971; Watanabe and Kakinuma, 1972).

This paper reviews current interplanetary scintillation observations of corotating solar wind streams and flare-associated shock waves and discusses their dependence on solar latitude and radial distance from the Sun. Recent scintillation observations of the power spectrum of the small irregularities are reviewed and compared to in-situ measurements. The long term average solar wind velocity dependence on radial distance and latitude and the solar origin of high velocity streams will be discussed in another review paper.

COROTATING STREAMS

Recurrent Solar Wind Streams

Interplanetary scintillation observations of scintillation index and solar wind velocity exhibit large enhancements which last for several days and are mostly related to corotating solar wind streams (Houminer, 1971; Houminer and Hewish, 1972; Watanabe and Kakinuma, 1972; Coles et al., 1974; Erskine et al., 1976). Houminer (1971) showed that the time displacements of the scintillation enhancements for radio sources lying in different directions were consistent with a simple model of corotating sectors of enhanced density irregularities. The model was based on Readhead (1971) who shows, for a spherically symmetrical distribution of irregularities about the Sun, that the scintillation observed at elongations $\varepsilon < 90^{\circ}$ was mainly due to those irregularities located in the region where the line of sight passes nearest to the Sun. For $\varepsilon < 90^{\circ}$ the effect of Fresnel filtering becomes important and the main scattering region is located about 0.2 - 0.3 AU from the Earth. The good agreement between observed and predicted time displacements of scintillation maxima observed at different elongations confirms that the fluctuating density ΔN_e inside enhanced scintillation sectors varies with radial distance from the Sun similarly to the time-averaged plasma irregularities.

Both velocity and scintillation enhancements tend to recur with the solar rotation period (Houminer and Hewish, 1972; Watanabe and

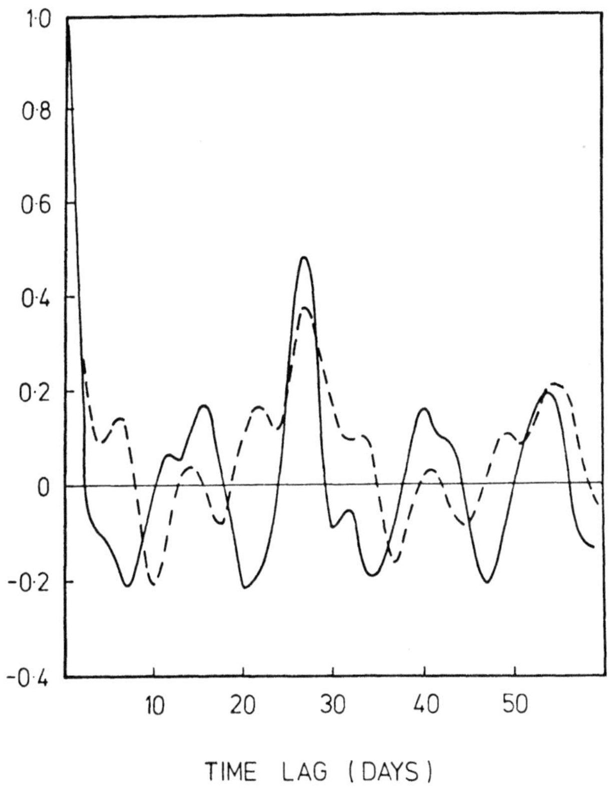

Figure 1. Autocorrelation of scintillation index variations during January-August, 1971 (solid line) and during September 1971-May 1972 (dashed line).

Kakinuma, 1972; Coles et al., 1974). Daily scintillation index results obtained with the 81.5 MHz array at Cambridge were used to determine the rotation period. The results of an autocorrelation analysis for two periods of 9 solar rotations each, during January 1971 - May 1972, are shown in Figure 1.

Only radio sources with lines of sight within $\pm 15^{\circ}$ of the plane of the ecliptic were used in this analysis. The autocorrelation results for the two periods show significant maximum of about 0.5 at a time lag of 27 ± 0.5 days. The average half-life of the recurrent scintillation sectors during the 16 month period was about 1 solar rotation, though some sectors last up to 5 solar rotations.

Autocorrelation analysis of velocity variations for 6 solar rotations during the first half of 1973 was performed by Coles et al.,

(1974) using data of 3 spaced arrays at 74 MHz. Their results show a correlation coefficient of 0.94 at one solar rotation, and 0.7 at two rotations, after correction for the estimation error. The rotation period was 27.1 ± 0.5 days. In contrast, similar analysis for six solar rotations during the second half of 1972 (Coles et al., 1974) show no evidence for periodic features. Autocorrelation analysis was also performed by Erskine et al. (1976) using scintillation index results at 34 MHz during May-December 1974. Their results show a correlation of about 0.4 at a time lag of 28 ± 2 days.

There is a strong correlation between the scintillation enhancements and the velocity structure of the solar wind. Houminer and Hewish (1972) compared scintillation index variations at 81.5 MHz and the solar wind velocity measured by the Pioneer 6 and Vela satellites for January-April 1971. Their cross-correlation results indicate a strong correlation which maximizes at a time shift of 2.5 days. This shows that enhanced scintillation sectors lead streams of high solar wind velocity in the corotating system. Similar analysis, using simultaneous IPS observations of velocity and scintillation index, was carried out by Watanabe et al. (1974) and Coles et al. (1974). Their data verified the Houminer and Hewish (1972) conclusion that the scintillation enhancement occurs at the maximum positive velocity gradient 2-3 days before the velocity maximum.

Houminer and Hewish (1972) suggested that enhanced scintillation is caused by the existence of regions of compressed plasma resulting from the interaction of fast and slow velocity streams. Houminer and Hewish (1974) compared proton density data from Pioneer 6 and 7 and scintillation index results for the period January-April 1971. Both sets of data were corotated to Earth and are shown in Figure 2.

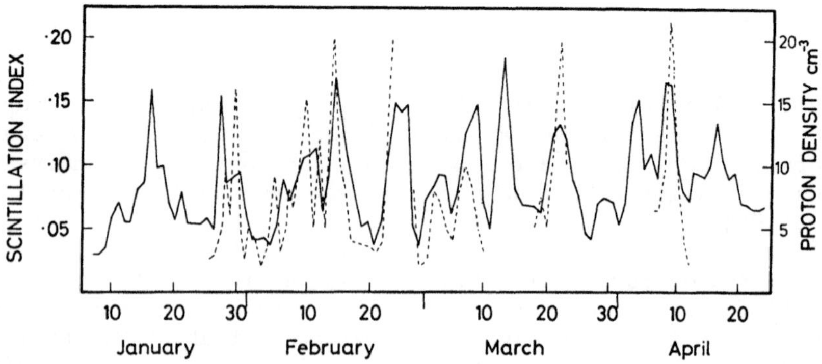

Figure 2. Scintillation index results (solid line) and proton densities as measured by Pioneer 6 (dashed line) for the period January 8-April 25, 1971.

The results show that the variations of plasma density and scintillation index are strongly correlated in direct confirmation of the compressed plasma hypothesis.

This important correlation indicates that scintillation measurements provide a means of monitoring the plasma density of the solar wind. Observations of scintillation enhancements of radio sources at widely separated elongations show that the compressed density region associated with scintillation sectors is aligned along a corotating spiral between 0.5 - 1.3 AU. Figure 7b of Houminer and Hewish (1972) showing scintillation observations of 4 radio sources during the passage of a corotating scintillation sector in February 1971, demonstrates that enhanced irregularity sectors and thus the compressed plasma density can be detected at least 6 days before arriving at the Earth.

Recent observations by Erskine et al. (1976) verify the conclusion that the scintillation index is mainly a measure of interplanetary density structure. Erskine et al. (1976) analyzed scintillation data of more than 80 small diameter radio sources at a frequency of 34 MHz. They showed both by direct comparison and by numerical cross-correlation analysis that what they call "all sky" scintillation enhancements at 34 MHz are caused by corotating compressed density regions sweeping over the Earth. Figure 3 shows solar wind proton density from IMP 7 and 8 spacecrafts for days 330-338 in 1974, together with the scintillation activity index, Q, for several clusters of sources at different sky boxes. The geometry of the sky boxes in a geocentric all-sky projection is also shown in Figure 3. The configuration of the corotating compressed plasma density sector consistent with the "all sky" scintillation enhancement is sketched at the bottom right of Figure 3.

Cross-correlation results derived by Erskine et al. (1976) are shown in Figure 4. The cross-correlation analysis was performed between the scintillation activity index and the solar wind velocity, the proton density and the interplanetary magnetic field magnitude. Their results show that the scintillation activity index correlates better with solar wind density (correlation coefficient of 0.56) than with any of the other correlative parameters examined, and that density and IPS activity are in phase with each other.

Harmon (1975) used model-prediction and inversion techniques in order to determine the distribution of the density irregularities across a solar wind stream. He suggested that the enhanced density irregularities are centered about 2 days after the maximum density compression. This result is in disagreement with the Houminer and Hewish (1974) and Erskine et al. (1976) observations that the scintillation enhancements and the density compression are strongly correlated and are in phase.

Spacecraft observations of bulk density and velocity show clearly that density compressions which last for 1-3 days, are centered, on the average, about 2-3 days in advance of velocity maxima (Figure 3, Belcher

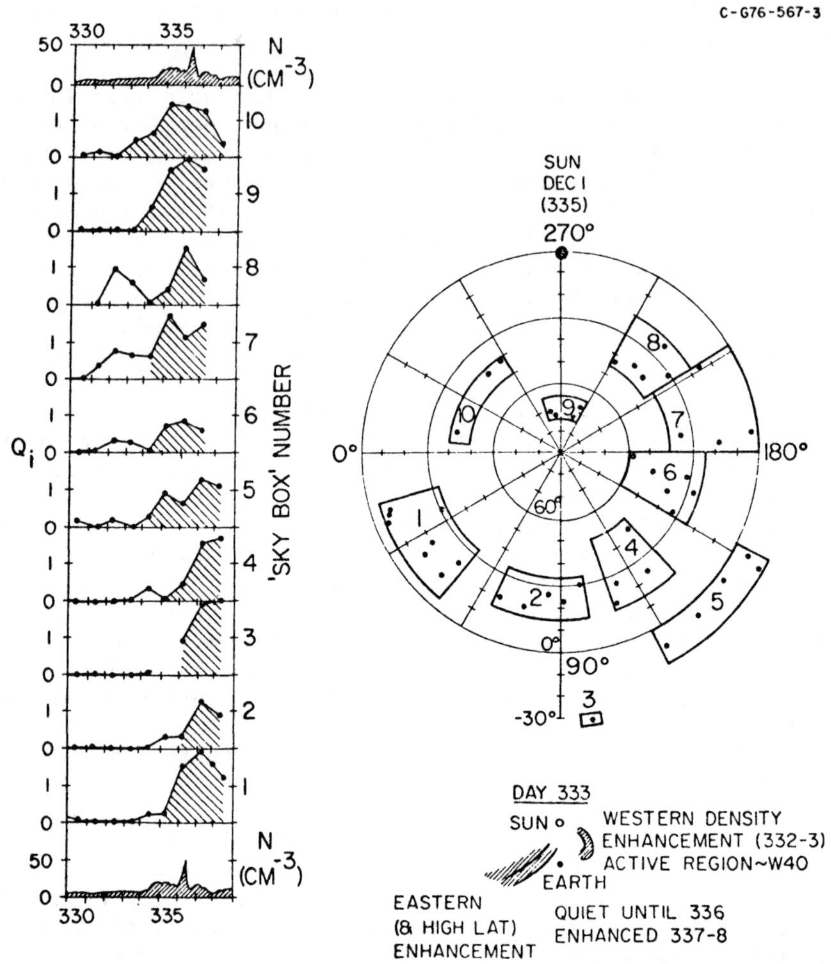

Figure 3. Solar wind proton density from IMP 7 and 8 for days 330-338 in 1974, together with scintillation activity index, Q, for several clusters of sources at different sky boxes. The geometry of the sky boxes in a geocentric all-sky projection is also shown. The configuration of the corotating compressed plasma density sector consistent with the "all sky" scintillation enhancement is sketched at the bottom right. (From Erskine et al., 1976).

and Davis, 1971; Figure 12 and 13, Ness et al. 1971, Hundhausen and Bame, 1971; Figure 8, 9 and 10, Erskine et al., 1976). These observations and the correlation, discussed previously, between scintillation sectors lead streams of high velocity by 2-3 days, tend to support the Houminer and Hewish (1974) and Erskine et al. (1976) conclusion.

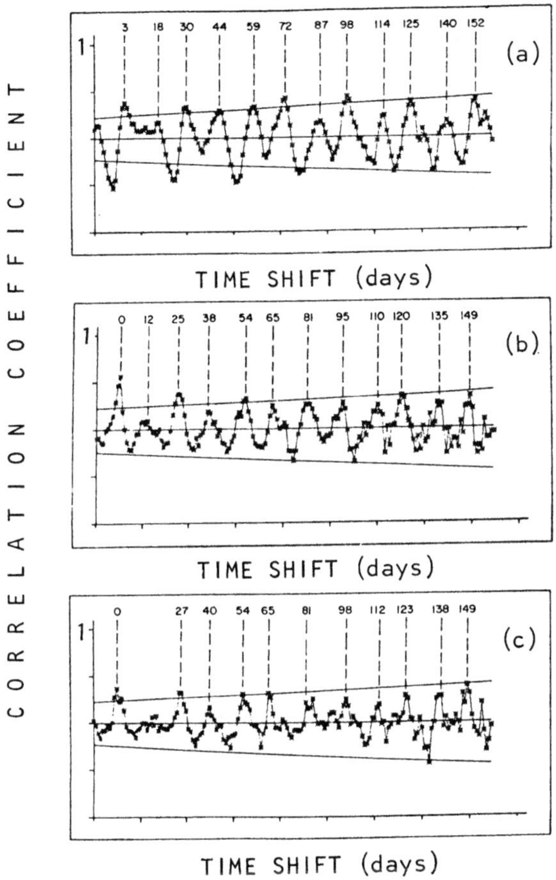

Figure 4. Time series crosscorrelation results for correlation of IPS activity index versus (a) solar wind velocity, (b) solar wind proton density and (c) interplanetary magnetic field magnitude. (From Erskine et al., 1976).

High Helicentric Latitude Observations

Observations of the intensity scintillation of small angular diameter radio sources provide, at present, the only way of studying, day-by-day, the solar wind over a wide range of heliocentric latitudes. Observations of the solar wind velocity outside the plane of the ecliptic have been discussed by many authors (Dennison and Hewish, 1967; Hewish and Symonds, 1969; Coles and Maagoe, 1972; Watanabe et al., 1974; Coles et al., 1974; Coles and Rickett, 1976). These observations will be reviewed elsewhere. In this section scintillation index and wind velocity observations of corotating streams outside the ecliptic and results of their extent and latitude dependence will be presented.

The heliocentric latitude extent of a corotating stream can be determined by noting the scintillation enhancements or velocity variations as a function of time for a grid of sources, during the passage of a single well-defined stream. Alternatively, many observations can be combined in an attempt to study, by statistical means, the scintillation and velocity modulation caused by corotating streams. It is assumed that the mean irregularity distribution is spherical symmetric. This is based on Houminer (1973b) who showed that the time-averaged scintillation results do not show heliocentric latitude dependence.

The latitudinal extent of a single long-lived corotating stream was investigated by observing the scintillation index variations at 81.5 MHz for four radio sources in August 1971. The geometry for sources with lines of sight in and outside the plane of the ecliptic is shown schematically in Figure 5a.

The daily scintillation index variations for the 4 sources during the passage of a scintillation sector are shown in Figure 5b. The elongations and the heliocentric latitude of the point of closest approach to the Sun along the line of sight is given for each source. Shown also in Figure 5b are the theoretical variations assuming a sector which extends uniformly to all latitudes. It is clear that if the scintillation sector extended to all heliocentric latitudes, one would see enhancements for the sources 3C247 and 3C268.4 before August 6. Thus this particular stream does not extend to heliocentric latitudes $\geq 36^{\circ}$.

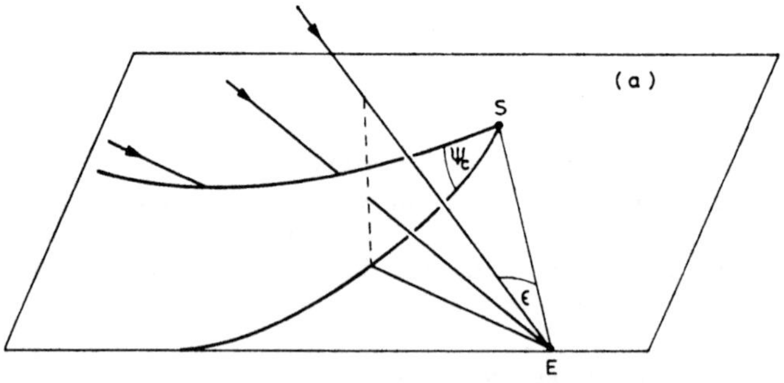

Figure 5a. The geometry for sources with lines of sight in and outside the plane of the ecliptic.

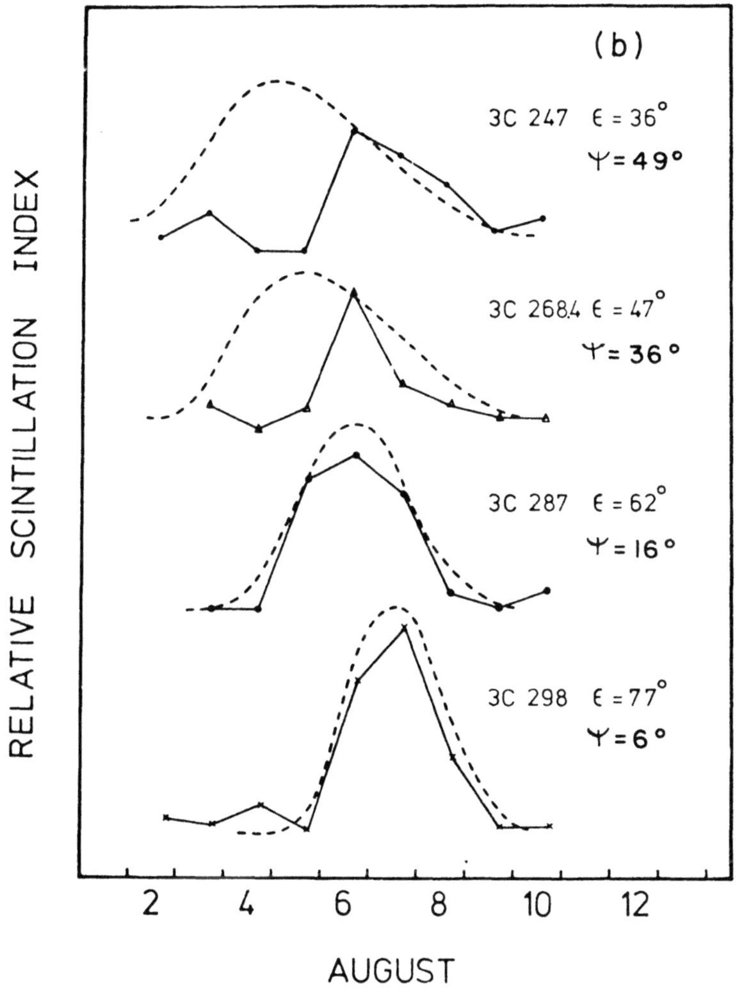

Figure 5b. The scintillation index variations observed for several sources with different elongations ε and heliocentric latitudes ψ, during the passage of a scintillation sector. The theoretical daily variations in scintillation index for the same sources are shown as well (dashed lines).

Coles et al. (1974) investigated the latitude extent of corotating streams by combining velocity and scintillation index measurements for several radio sources. Their velocity results during February-June 1973 indicate that an asymmetric stream observed in the early parts of solar rotations 1598-1601 is confined to heliocentric latitudes south of 10° N. Coles et al. (1974) compared the scintillation data with other solar observations and concluded that the corotating stream is aligned with a southerly hole in the EUV emission. Rickett et al. (1976) tried to identify the sources of the high velocity streams by mapping the streams back to the Sun and comparing their locations with coronal holes identified in wide band XUV solar images. Rickett et al. (1976) found that for more than half of the observed coronal hole passages, high velocity streams were associated with these holes. The observed streams tend to be wider in longitude than the holes and could be detected at latitudes up to about 20° from the holes. Rickett et al. (1976) observations indicate that the width of corotating streams associated with coronal holes will be dependent on the latitudinal width of the holes and should be at least $20-30^o$ wide.

Houminer (1973b) used statistical methods to investigate the latitude dependence of enhanced scintillation sectors. He combined many observations of scintillation index for lines of sight lying in and outside the plane of the ecliptic. The time occurrence of maximum scintillation for sources lying east and west of the Sun provided a method for discriminating between (a) sectors which extend uniformly to all latitudes, (b) sectors sheared by differential rotation and (c) sectors confined to low latitudes. The results of a crosscorrelation analysis showed that the time displacements of scintillation enhancements for sources lying east and west of the Sun are consistent with scintillation sectors confined to a latitude $\leq 35^o$.

The heliocentric latitude dependence of the enhanced scintillation sectors can also be determined by considering the day to day modulation in scintillation caused by regions of enhanced density irregularities (Houminer, 1973b). Scintillation index results for many small diameter radio sources, observed during the period January-December 1971, were grouped according to the heliocentric latitude of the point of closest approach to the Sun along the line of sight. The r.m.s. day-to-day variations in the scintillation index results were then calculated as a function of the heliocentric latitude and the results are shown in Figure 6.

There is a definite reduction in the r.m.s. variation as the heliocentric latitude increases which indicates that the scintillation sectors are confined to a limited latitude range or that the magnitude of the scintillation enhancements are latitude dependent. The expected r.m.s. variations for various ψ_c, where ψ_c is the heliocentric extent of a scintillation sector, are also shown in Figure 6. The results of Figure 6 are consistent with scintillation sectors which do not extend beyond 40° in heliocentric latitude.

Though the average properties of the solar wind velocity outside the plane of the ecliptic were discussed by many authors, the dependence of the velocity modulation, associated with streams, on the solar latitude was not discussed. However, some day-to-day observations of velocity as a function of heliocentric latitude are available (Coles et al., 1974; Watanabe et al., 1974).

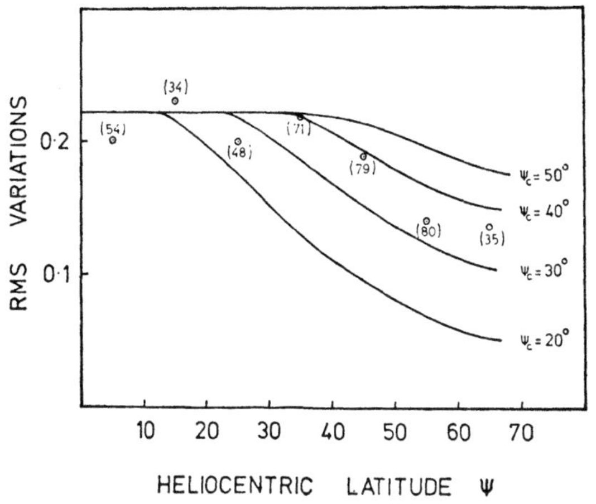

Figure 6. The r.m.s. variation in the scintillation index as function of the heliocentric latitude ψ_c. The numbers in brackets denote the number of days used for each point. The expected variations for various ψ_c are also shown.

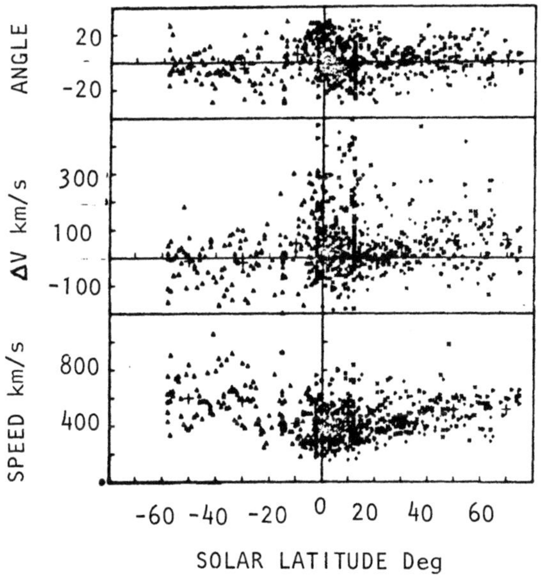

Figure 7. Estimates of the solar wind velocity versus latitude during 1973 from observations of four high-latitude sources, 3C48, 3C147, 3C161 and 3C298. (From Coles et al., 1974).

Figure 7 shows solar wind velocity results versus latitude during 1973 from observations of four high latitude sources, 3C48, 3C147, 3C161 and 3C298 (Figure 6 of Coles et al., 1974). The results show a definite decrease in the velocity modulation at northern heliocentric latitudes but not at southern latitudes. Scintillation observations of 3C48 during May-June 1972 and 1973 (Watanabe et al., 1974) do not show any apparent decrease of the modulation at high latitudes during these periods.

We conclude that corotating high velocity streams exist at all latitudes. There is no definite evidence for a dependence of the velocity modulation on the solar latitude in contrast with scintillation index observations which show a strong heliocentric latitude dependence.

Suess et al. (1975) used a global stream model to explain the absence or weakening of enhanced scintillation at high heliocentric latitudes. They calculated the stream density and velocity profiles as a function of radial distance from the Sun for different initial temperature perturbations at 28 solar radii. They concluded that the solar wind streams generally fall in the regime where a relatively small change in the initial perturbation amplitude can produce a large change in the maximum density associated with the stream interaction near 1 AU but a small change in the maximum velocity in the stream. This result shows that even a small decline in the stream amplitudes at high latitudes can give a large decrease in the strength of the stream interaction. As it was shown that the scintillation enhancements are related to the compressed density at the interaction region, such a small decline could explain the observed high latitude weakening or absence of scintillation enhancements. Alternatively, a stream structure in which the low or ambient solar wind velocity increases with latitude would lead to a similar weakening of the stream interactions and a similar implication regarding scintillations (Suess et al., 1975). Such a model is, in fact, consistent with the recent results of Coles and Rickett (1976) which showed that the average solar wind velocity during 1971-1975, increases away from the solar equator with a mean gradient of 2.1 km sec^{-1} per degree of latitude. Thus the calculations of Suess et al. (1975) explain the observed modulation in velocity and scintillation.

Evidence for differential rotation of the sources of the high velocity streams was presented by Coles et al. (1974). They performed an autocorrelation analysis of velocity data for mid-1973 using 3C161 as a source. During this period the 3C161 subscattering point is at southerly heliocentric latitudes. The results of the autocorrelation indicate a rotation period of 28.6 ± 1.4 days (with 90% confidence) which is about 6% longer than the period observed at the equator.

FLARE-ASSOCIATED DISTURBANCES

Observations of Flare-Associated Disturbances

Observations of the scintillation of a number of radio sources, lying at widely separated positions, show occasional increases in scintillation index and velocity which cannot be ascribed to corotating features since they occur almost simultaneously for radio sources located on opposite sides of the Sun. These increases are short-lived in comparison to corotating streams and typically only last for one or two days. Sharp and Harris (1967) and Dennison and Hewish (1967) first suggested that such enhancements may be related to shock waves associated with solar flares.

Several more authors related scintillation enhancements to flare-associated disturbances propagating over a wide front (Wiseman and Dennison, 1972; Cronyn et al., 1975; Roelof et al., 1976; Houminer, 1976). IPS observations of the complex disturbances of the solar wind in early August 1972 were the most extensively discussed (Armstrong et al., 1973; Watanabe et al., 1973; Houminer, 1973c; Rickett, 1975; Ward, 1975a, 1975b) and were recently reviewed by Kakinuma and Watanabe (1976). During this period IPS velocity measurements were also available (Armstrong et al., 1973; Watanabe et al., 1973).

Houminer (1976) examined day-to-day observations carried out on 30 radio sources over a period of 8 months in 1971 and found only 3 scintillation enhancements due to interplanetary shock waves. He concluded that such enhancements are relatively rare. The periods when flare associated disturbances were observed by interplanetary scintillation are listed in Table 1.

Table 1

Flare-Associated Disturbances Observed by IPS

Period	Type of Observations	Source
March 27-31, 1966	Scintillation index	Sharp and Harris (1967)
June 6-8, 1969	Scintillation index	Wiseman and Dennison (1972)
January 26-27, 1971	Scintillation index	Houminer (1976)
April 14-15, 1971	Scintillation index	Houminer (1976)
May 17-18, 1971	Scintillation index	Houminer (1976)
August 3-5, 1975	Scint. and velocity	Kakinuma and Watanabe (1976)
August 8-9, 1972	Scint. and velocity	Kakinuma and Watanabe (1976)
June 25-26, 1974	Scintillation index	Cronyn et al. (1975)

The Structure of Flare-Associated Disturbances

It would be possible, in principle, to determine the shape and width of a shock disturbance by continuously monitoring the scintillation and velocity of a large number of sources at widely separated elongations and with the main scattering regions along the lines of sight distributed at different radial distances and solar latitudes. However, IPS observations of each source are usually carried out at meridian transit for periods which vary between a few minutes and about an hour. Therefore, it is difficult to disentangle the various unknown characteristics, namely, the velocity of the shock disturbance as a function of radial distance from the Sun, the location of the shock front along the line of sight and finally the shape, width and strength of the disturbance. Thus, the observation of scintillation enhancements during a passage of a flare-associated disturbance can only provide the approximate shape and extent of the disturbance.

Wiseman and Dennison (1972) observed daily a grid of 14 sources during a flare associated disturbance in June 1969. They found that the shock disturbance propagated over a wide front and suggested that the shock structure perpendicular to the ecliptic plane was non isotropic. Ward (1975a) estimated the shape of the disturbance associated with the solar flares on August 2, 1972 by comparing the observed scintillation index variations on five sources with calculated variations assuming different shaped disturbances. He concluded that the shape of this disturbance was consistent with the theoretical shock configuration of De Young and Hundhausen (1971) for plasma ejection into a cone of 60° half-angle. He also estimated the extent in heliocentric latitude to be at least 60°. Houminer (1976) estimated the average shape and velocity of a shock disturbance by calculating the expected variations in scintillation index assuming a nearly spherical disturbance and comparing the theoretical curves with daily scintillation index results at different elongations. He combined scintillation index results during the passage of 3 shocks in the first half of 1971 and found good agreement between the expected and measured results for a nearly spherical disturbance with an average velocity of 500-600 km sec^{-1}.

Rickett (1975) demonstrated that by combining IPS velocity measurements with spacecraft and geomagnetic observations one can get a good estimate of the shape of a shock disturbance and its extent in both heliocentric longitudes and latitudes. Rickett studied the shape of the shock disturbance associated with the large solar flare on August 7, 1972 using velocity results obtained during the passage of the shock from six radio sources together with velocity observations at Pioneer 9. The shape of the shock in longitude and latitude is shown in Figure 8. Rickett concluded that the shape of the shock front can be represented by a circle of radius 0.73 AU centered at 0.4 AU from the Sun along the flare normal. Comparison with the calculations of De Young and Hundhausen (1971) for a blast wave injected into the solar wind over various cone angles show that the injection must have been into a cone of half-angle between 30° and 60°.

In the previous section it was shown that corotating scintillation enhancements are related to the compressed plasma density resulting from the interaction of fast and slow streams. It seems likely that the regions causing flare-associated scintillation may be identified with the regions of high density compressed plasma behind a shock front. The very low scintillation indices observed during the passage of the shock associated with the solar flare on August 4, 1972 (Houminer, 1973c; Ward, 1975a) may be explained by the very low proton number density observed by Pioneer 9 during the passage of this shock (Mihalov et al., 1974).

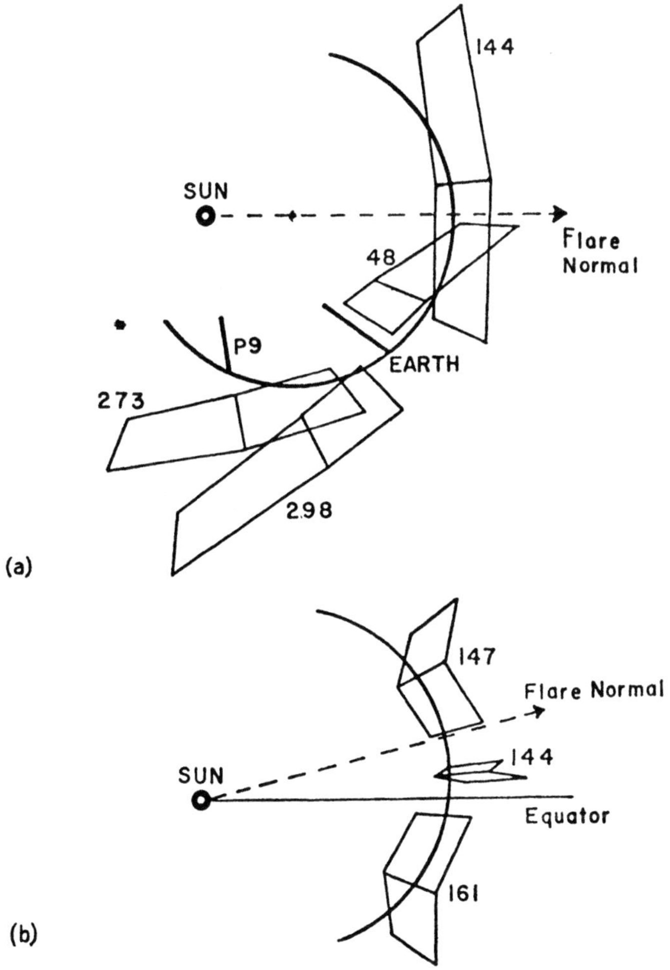

Figure 8. The shape on August 9 at 01 UT of the disturbance associated with the August 7, 1972 solar flare (From Rickett, 1975).

THE SPECTRUM OF THE SMALL SCALE IRREGULARITIES

Observations of interplanetary scintillations indicate that small scale irregularities are always present in the solar wind (Hewish et al., 1964; Hewish and Symonds, 1969), and it has been shown in a previous section that these are most pronounced in compression regions between high and low velocity solar wind streams. To understand the nature and origin of the irregularities it is important to determine their wavenumber power spectrum. Information can be obtained from (a) observations of scintillation index and its dependence on radio wavelength, (b) the correlation of the diffraction pattern measured at three sites, and (c) the temporal power spectrum observed at a single site. In this section some new observations of the power spectrum of the irregularities are presented (Coles and Harmon, 1974; Harmon, 1975) and compared to recent spacecraft measurements of the plasma density variations. These measurements extend now to the high wavenumbers relevant to scintillations (Unti et al., 1973; Neugebauer, 1975, 1976; Unti and Russell, 1976).

Observations of interplanetary scintillations were originally interpreted in terms of a Gaussian power spectrum (Dennison and Hewish, 1967; Cohen et al., 1967; Readhead, 1971). This was questioned by authors who considered the spectrum to be best described by a single power law over the full wavenumber range (Jokipii and Hollweg, 1970; Lovelace et al., 1970; Cronyn, 1972). Hewish (1972) used scintillation and angular broadening observations to show that the measured scale size of the diffraction pattern at the Earth is a characteristic scale of the irregularities and not merely a diffraction limit sometimes referred to as the Fresnel scale. Matheson and Little (1971) and Rickett (1973) suggested that the spectrum must flatten between 10^{-3} and 10^{-2} km^{-1} in order to explain the linear dependence of the scintillation index on the radio wavelength (Hewish, 1971).

Houminer (1973a) presented observations of temporal power spectra from several radio sources at 81.5 MHz. Figure 9 shows the observed spectrum of 3C287 at elongations of 32°, 45° and 60° after correcting for the source diameter effect. Inspection of the spectra in Figure 9 shows that at frequencies $\nu > 0.5$ Hz the slope of the spectra is independent of solar elongation and follows the law $P_I(\nu) \sim \nu^{-1.4}$,

which corresponds to a three dimensional power law exponent $\alpha = 2.4$. For $\nu < 0.5$ Hz there is a marked flattening of the irregularities spectrum which cannot be attributed to Fresnel filtering. Similar results were obtained for other sources at different elongations.

Coles and Harmon (1974) and Harmon (1975) used the Abel transform in order to process a large number of 74 MHz IPS spectra. Their analysis of many spectra obtained during 1972-1973, shows that all have a power law shape at high frequencies. However, about two-thirds of the spectra flatten at the low wavenumbers between 10^{-3} and 10^{-2} km^{-1}. A mean power law exponent was determined by the method of least squares

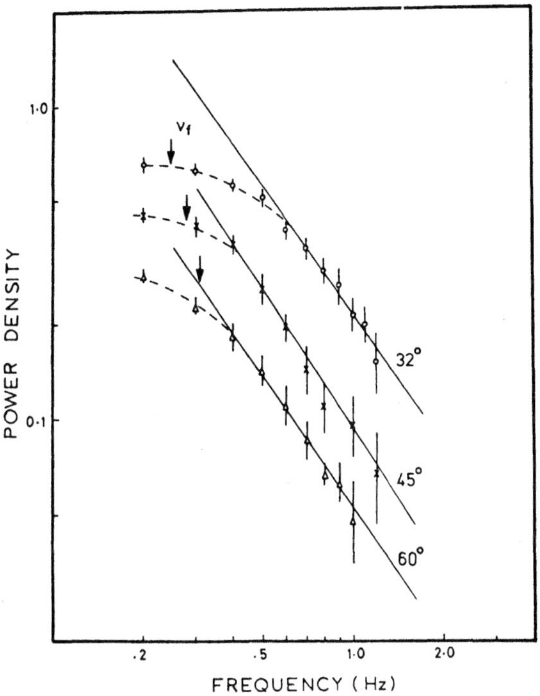

Figure 9. The spectrum of 3C287 at elongations of 32°, 45° and 60° after correcting for the source diameter. The Fresnel frequency ν_f is indicated by an arrow.

to the entire estimated spectrum. A typical value for the mean power law exponent is $\alpha = 2.6$. Examination of day-to-day variation of α, as seen in Figure 10, shows no apparent elongation dependence. However, there are variations which have 28 day periodicities and were found to be associated with corotating structure. Coles and Harmon (1974) and Harmon (1975) concluded that the estimated wavenumber spectra are consistent with the model spectrum proposed by Matheson and Little (1971) and Rickett (1973).

Houminer (1973a) suggested that the flattening of the spectra may be characterized by a width ν_I of the power spectrum which defines a scale length of the small scale irregularities. Houminer's results show that ν_I varies systematically between distances of 0.5 - 1.0 AU according to $\nu_I \alpha_r^{-1.1}$ where r is the radial distance from the Sun. This indicates that the flattening of the wavenumber spectra is related to some characteristic scale of the irregularities. Coles and Harmon (1974) and Harmon (1975) did not discuss the radial dependence of the

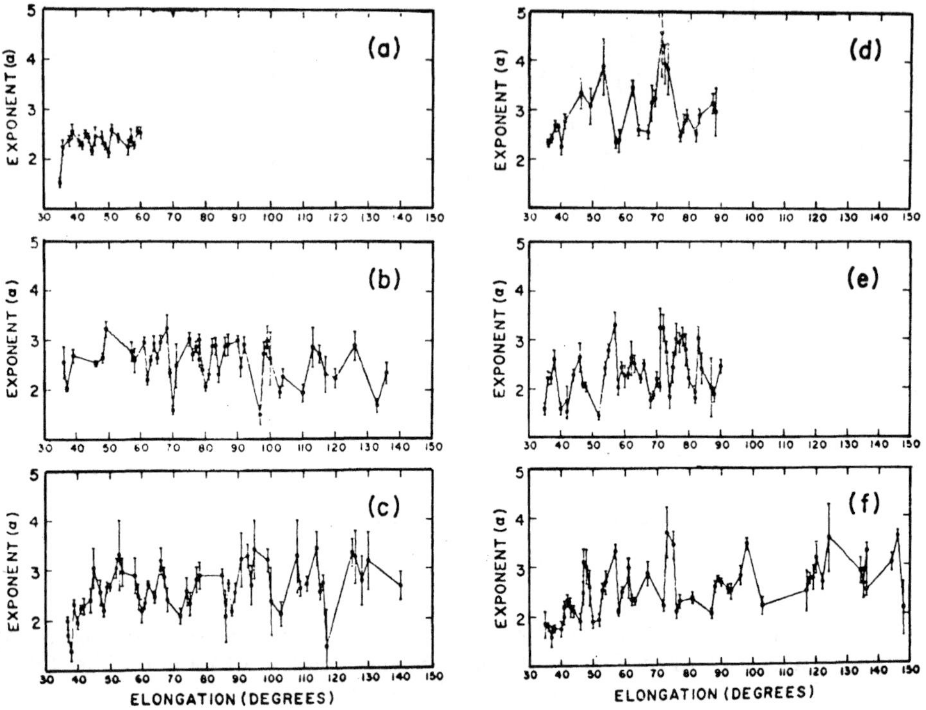

Figure 10. Power law exponent α versus elongation for different periods. (From Coles and Harmon, 1974).

frequency where their spectra flattens. It would be interesting to see whether they find a similar dependence on radial distance from the Sun. Observations of the width of the irregularities spectrum published by Rao et al. (1974) at 327 MHz and by Harris and Hardebeck (1969) at 430 MHz, show significant variation with radial distance from the Sun. They determined the width of the spectrum by the second moment which is mainly dependent on the high frequency portion of the spectrum. It would be better to characterize the width by the first moment or the width to 1/e, which will be more sensitive to the flattening of the spectrum at low frequencies.

Space probe observations of the proton density power spectrum in the range 0.005-13.3 Hz, are now available (Unti et al., 1973; Neugebauer, 1975, 1976; Unti and Russell, 1976). Unti et al. (1973) using a set of 32 spectra from OGO-5, found two-thirds of them to be consistent with a power law with $\alpha = 3.5$. The remainder exhibit a hump between 0.1-1.0 Hz.

Neugebauer (1975) used OGO-5 measurements of positive ion flux to calculate the power spectra of solar wind fluctuations on the frequency range 0.11 - 0.87 Hz. Six hundred and thirty-four individual power spectra judged to be free of contamination by bow shock noise were averaged according to the ratio f/f^*, where f is the observed frequency and $f^* = u/2\pi R$. Here u is the solar wind velocity and R is the proton thermal gyroradius. The resulting sliding average spectra showed significant power enhancements at $f/f^* = 1$. Neugebauer concluded that though, to a first approximation, the spectra of the density irregularities is a power law spectrum, there is, undoubtedly, an additional population of irregularities with scale size approximately equal to the proton gyroradius. This conclusion is in agreement with the suggestions by Hewish (1971), Matheson and Little (1971), Rickett (1973) and the scintillation spectrum results of Houminer (1973a), Coles and Harmon (1974) and Harmon (1975). In another paper, Neugebauer (1976) recalculated the power spectrum of the density irregularities.

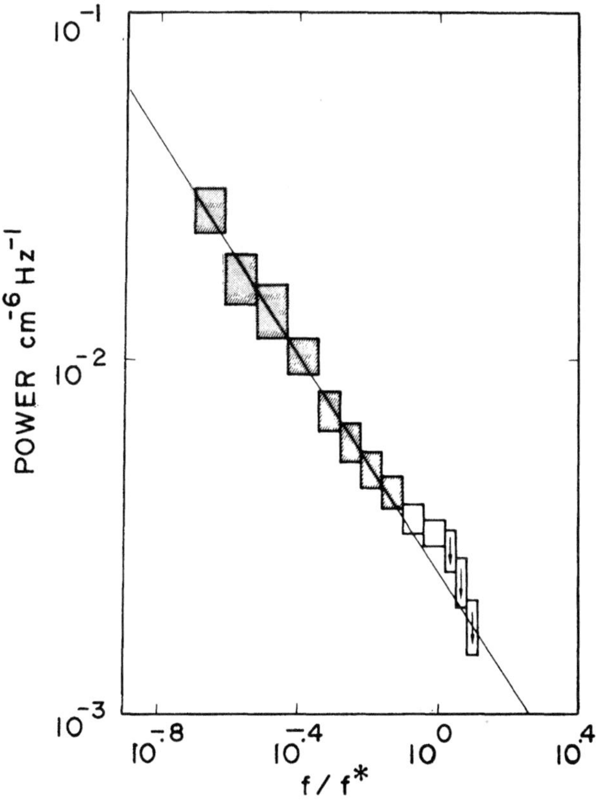

Figure 11. Power spectral density estimates of density fluctuations in the solar wind as function of f/f^*. (From Neugebauer, 1976).

Figure 11 shows the recalculated spectrum using only 104 spectra for which f* is between 0.54 - 0.83 Hz. There is still a small significant enhancement of the spectrum at $f/f^* = 1$. The recalculated spectrum does not alter or change the basic conclusions reached by Neugebauer (1975).

Unti and Russell (1976) also showed that superposition of many density spectra obtained from OGO-5 data, reveals a moderate enhancement at a frequency corresponding to convected irregularities with scale size of a proton gyroradius, consistent with the results of Neugebauer.

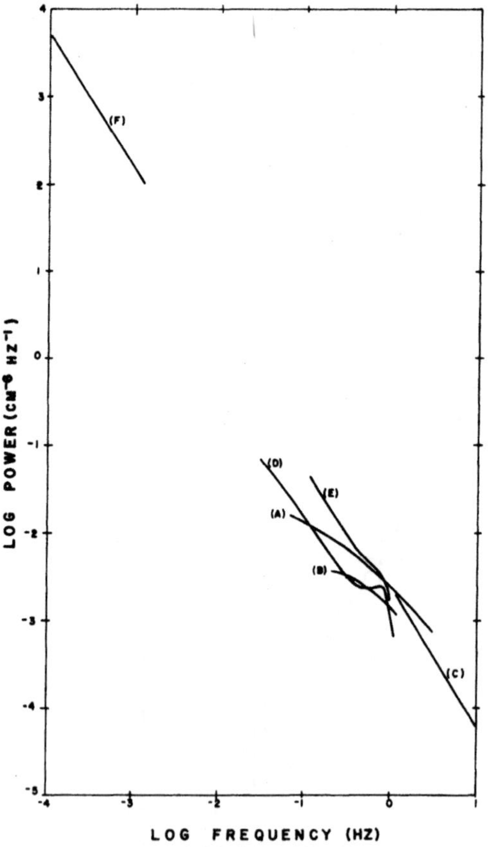

Figure 12. One dimensional density spectra at 1 AU from (a) Coles and Harmon (1974), (b) Houminer (1973a), (c) Cronyn (1972), (d) Neugebauer (1975), (e) Neugebauer (1976) and (f) Intriligator and Wolfe (1970).

Spacecraft and interplanetary scintillation observations of the irregularities density spectrum are combined in Figure 12. The scintillation spectra (Cronyn, 1972; Houminer, 1973a; Coles and Harmon, 1974) were first extrapolated to 1 AU and then converted into a one dimensional spectrum. To convert from the normalized frequency f/f* of the spectra of Neugebauer (1975, 1976), an average value of f* = 0.75 Hz was used. Comparison of the scintillation spectra and the spectra of Neugebauer (1975, 1976) with the low-frequency spacecraft data of Intrilligator and Wolfe (1970) suggests that the overall density spectrum is similar to the model suggested by Matheson and Little (1971). According to Matheson and Little the spectrum which is power law beyond 0.5 Hz, flattens abruptly between 10^{-1}– 10^{-2} Hz and then steepens between 10^{-2}– 10^{-3} Hz in order to meet the low-frequency spacecraft results.

We may conclude that the flattening of the power spectrum of the irregularities is due to additional population of density irregularities with a scale size of the order of the proton gyroradius. As the ion gyroradius is related to the proton temperature T and the magnitude of the interplanetary magnetic field B as $R\alpha(T)^{\frac{1}{2}}/B$, the width of the scintillation spectrum provides a tool for studying the radial dependence of the ratio $B/(T)^{\frac{1}{2}}$. Such an analysis was performed by Houminer (1973a).

ACKNOWLEDGMENTS

I would like to thank Drs. F. T. Erskine, W. M. Cronyn, S. D. Shawhan, E. C. Roelof, B. L. Gotwols, W. A. Coles and J. K. Harmon for permission to use their unpublished data.

REFERENCES

Armstrong, J. W. and Coles, W. A.: 1972, *J. Geophys. Res.* *77*, 4602.
Armstrong, J. W. and Coles, W. A., Harmon, J. K., Maagoe, S., Rickett, B. J. and Sime, D. G.: 1973, in H.E. Coffey (ed.), *Collected Data Reports on August 1972 Solar-Terrestrial Events*, World Data Center A for Solar Terrestrial Physics, Report UAG-28, p. 371.
Belcher, J. W. and Davis, L: 1971, *J. Geophys. Res.* *76*, 3534.
Bourgois, G.: 1972, *Astron. Astrophys.* *21*, 33.
Cohen, M. H., Gundermann, E. J., Hardebeck, H. E., and Sharp, L. E.: 1967, *Astrophys. J.* *147*, 449.
Coles, W. A. and Harmon, J. K.: 1974, '*Interplanetary Scintillation Measurements of the Electron Density Power Spectrum in the Solar Wind.*' University of California, San Diego (preprint).
Coles, W. A. and Maagoe, S.: 1972, *J. Geophys. Res.* *77*, 5622.
Coles, W. A. and Rickett, B. J.: 1976, *J. Geophys. Res.* *81*, 4797.
Coles, W. A., Rickett, B. J. and Rumsey, V. H.: 1974, in C. T. Russell (ed.), *Solar Wind Three*, University of California Press, Los Angeles, California, p. 351.
Cronyn, W. M.: 1972, *Astrophys. J.*, *171*, L101.

Cronyn, W. M., Erskine, F., Shawhan, S. D., Gotwols, B. L. and Roelof, E. C.: 1975, in *Proc. Ionospheric Effects Symp.*, Virginia 1975.
Dennison, P. A. and Hewish, A.: 1967, *Nature 213*, 343.
Dennison, P. A. and Wiseman, M.: 1968, *Proc. Stron. Soc. Aust. 1*, 142.
DeYoung, D. S. and Hundhausen, A. J.: 1971, *J. Geophys. Res. 76*, 2245.
Erskine, F. T., Cronyn, W. M., Shawhan, S. D., Roelof, E. C. and Gotwols, B. L.: 1976, '*Interplanetary Scintillation at Large Elongation Angles: Response to Solar Wind Density Structure.*' (preprint).
Harmon, J. K.: 1975, Ph.D. Dissertation, University of California at San Diego.
Harris, D. E. and Hardebeck, E. G.: 1969, *Astrophys. J. Suppl. 19*, 115.
Hewish, A.: 1971, *Astrophys. J. 163*, 645.
Hewish, A.: 1972, in C. P. Sonnett, P. J. Coleman and J. M. Wilcox (eds.), *Solar Wind*, NASA SP-308, p. 477.
Hewish, A. and Symonds, M. D.: 1969, *Planet. Space Sci. 17*, 313.
Hewish, A., Scott, P. F. and Wills, D.: 1964, *Nature 203*, 1214.
Houminer, Z.: 1971, *Nature Phys. Sci. 231*, 165.
Houminer, Z.: 1973a, *Planet. Space Sci. 21*, 1367.
Houminer, Z.: 1973b, *Planet. Space Sci. 21*, 1617.
Houminer, Z.: 1973c, in H. E. Coffey (ed.), *Collected Data Reports on August 1972 Solar-Terrestrial Events*, World Data Center A for Solar Terrestrial Physics, Report UAG-28, p. 375.
Houminer, Z.: 1976, *Planet. Space Sci. 24*, 951.
Houminer, Z. and Hewish, A.: 1972, *Planet. Space Sci. 20*, 1703.
Houminer, Z. and Hewish, A.: 1974, *Planet. Space Sci. 22*, 1041.
Intriligator, D. S. and Wolfe, J. H.: 1970, *Astrophys. J. 162*, L187.
Jokipii, J. R. and Hollweg, J. V.: 1970, *Astrophys. J. 160*, 745.
Kakinuma, T. and Watanabe, T.: 1976, *Space Sci. Rev. 19*, 611.
Lovelace, R.V.E., Salpeter, E. E., Sharp, L. E. and Harris, D. E.: 1970, *Astrophys. J. 159*, 1047.
Matheson, D. N. and Little, L. T.: 1971, *Planet. Space Sci. 19*, 1615.
Mihalov, J. D., Colburn, D. S., Collard, H. R., Smith, B. F., Sonett, C. P. and Wolfe, J. H.: 1974, in D. E. Page (ed.), *Correlated Interplanetary and Magnetospheric Observations*, D. Reidel Publ. Co. Dordrecht-Holland, p. 545.
Ness, N. F., Hundhausen, A. J. and Bame, S. J.: 1971, *J. Geophys. Res. 76*, 6643.
Neugebauer, M.: 1975, *J. Geophys. Res. 80*, 998.
Neugebauer, M.: 1976, *J. Geophys. Res. 81*, 2447.
Rao, A. P., Bhandari, S. M. and Ananthakrishnan, S.: 1974, *Aust. J. Phys. 27*, 105.
Readhead, A.C.S.: 1971, *Mon. Not. R. Astr. Soc. 155*, 185.
Rickett, B. J.: 1973, *J. Geophys. Res. 78*, 1543.
Rickett, B. J.: 1975, *Solar Phys. 43*, 237.
Rickett, B. J., Sime, D. G., Sheeley, N. R., Crockett, W. R. and Tousey, R.: 1976, *J. Geophys. Res. 81*, 3845.
Roelof, E. C., Krimigis, S. M., Cronyn, W. M., Shawhan, S. D. and McIntosh, P. S.: 1976, *Space Research XVI*, p. 374.
Sharp, L. E. and Harris, D. E.: 1967, *Nature 213*, 377.

Suess, S. T., Hundhausen, A. J. and Pizzo, V.: 1975, *J. Geophys. Res.* *75*, p. 2023.
Unti, T. and Russell, C. T.: 1976, *J. Geophys. Res.* *81*, 469.
Unti, T., Neugebauer, M. and Goldstein, B. E.: 1973, *Astrophys. J.* *180*, 591.
Vitkevitch, V. V. and Vlasov, V. I.: 1972, *Soviet Astron.* *16*, 480.
Ward, B. D.: 1975a, Ph.D. Dissertation, University of Adelaide, Australia.
Ward, B. D.: 1975b, *Proc. Astron. Soc. Aust.* *2*, 378.
Watanabe, T. and Kakinuma, T.: 1972, *Publ. Astron. Soc. Japan* *24*, 459.
Watanabe, T., Kakinuma, T., Kojima, M. and Shibasaki, K.: 1973, *J. Geophys. Res.* *78*, 8364.
Watanabe, T., Shibasaki, K. and Kakinuma, T.: 1974, *J. Geophys. Res.* *79*, 3841.
Wiseman, M. and Dennison, P. A.: 1972, *Proc. Astron. Soc. Aust.* *2*, 79.

DISCUSSION

Duffett-Smith: (Comment) I would like to add something in support of Dr. Houminer's comments about the spectrum of the density irregularities in the region appropriate to IPS. Hewish et al. (1977) have used IPS measurements in four independent tests to discriminate between a power-law model and one with a characteristic scale-size (a gaussian model). In each case the gaussian model fits naturally while the power-law model is inconsistent and does not fit at all in one test. Theoretical curves for the variation of scintillation index with solar elongation, based on the gaussian model, fit the observations very well over a wide range of elongations.
(Hewish, A., Readhead, A.C.S., and Kemp, M.: 1977, submitted to *Mon. Not. R. Astr. Soc.*)

PART III

SOLAR WIND THEORY AND OBSERVATIONS

THE THREE-DIMENSIONAL STRUCTURE OF THE INTERPLANETARY MEDIUM

W.I. AXFORD
Max-Planck-Institut für Aeronomie, Katlenburg-Lindau,
W. Germany

ABSTRACT

The general structure of the interplanetary medium, especially in three-dimensions, is reviewed with special emphasis being given to the solar wind, the interplanetary magnetic field, the modulation of galactic cosmic rays and the propagation of energetic solar particles.

INTRODUCTION

In this paper we will consider the various possibilities and problems which confront the magnetic field, solar wind and energetic particle experiments to be carried on spacecraft launched on out-of-ecliptic (OOE) trajectories over the poles of the Sun. Variations of interplanetary conditions with solar latitude are to be expected for a number of reasons, however there can be no certain predictions as to what will be observed by the OOE spacecraft because we are ignorant of too much of the basic information needed concerning the relationship of the interplanetary medium to features in the solar corona and lower atmosphere. Of course it is the prime purpose of the OOE mission to provide this information, but in the meantime it is necessary to consider what might be happening in order to properly design the experiments and choose a suitable payload.

Our approach will be to consider first the simplest situation in which the Sun itself is assumed to be spherically symmetric and with a very weak magnetic field, so that all latitude variations are due entirely to the spiral form of the interplanetary magnetic field lines. We will observe that this approach leads to an inconsistency when the magnetic field strength is finite and that regions of closed coronal magnetic field must exist as indeed is observed. Our second approach will be to make the simple assumption that the solar wind arises from regions which are magnetically "open" for long periods and which may be identified with coronal holes. This also leads to apparent inconsistencies and we are consequently led to adopt a model in which there is a

M. A. Shea et al. (eds.), Study of Travelling Interplanetary Phenomena/1977, 145–164.

class of solar magnetic field lines which are only transiently open. The implications of such a model, with two rather distinct regions of interplanetary magnetic field, are considered with reference to the behaviour of the solar wind, hydromagnetic waves, energetic solar particles and galactic cosmic rays, as functions of solar latitude.

THE COMPLETELY OPEN MAGNETIC FIELD MODEL

Let us consider first the situation in which the solar magnetic field is very weak and the Sun spherically symmetric. In this case the properties of the interplanetary medium should be more or less as described by the model calculations of Leer and Axford (1972) shown in Figure 1. It is interesting to note that in the shell lying between

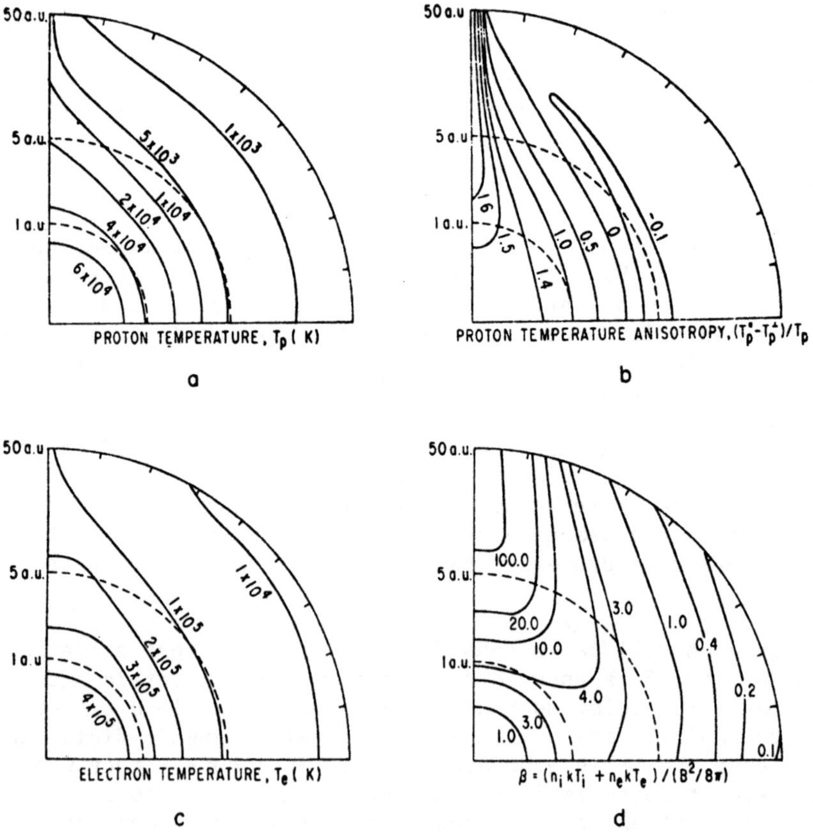

Figure 1. Variation of solar wind plasma bulk parameters according to a model calculation by Leer and Axford (1972). Note that the combination of plasma ß and proton temperature anisotropy is large at high latitudes suggesting that the firehose instability might occur.

heliocentric distances of ∿1 AU and ∿5 AU which contains the proposed OOE trajectory, significant variations in the properties of the solar wind occur only in the region over the poles ($\lambda \gtrsim 75°$). Here the proton temperature anisotropy and the ß of the plasma are both large, suggesting that the firehose instability could be an important process (Parker, 1958). However, one expects the amplitude of Alfvén waves originating in the solar photosphere to be large in this region in any case, so that the temperature anisotropy may not have a chance to develope.

In this model, interplanetary field sector boundaries occur naturally and consist of sheets defined by projecting radially the lines of zero radial magnetic field strength on the surface of the Sun. The neutral sheets thus formed would however be expected to be unstable against reconnection in regions where the total energy of the plasma is inadequate to allow it to draw close loops of magnetic field out from the Sun, i.e. when $[(5/2)p + \rho V^2/2] \lesssim B^2/8\pi$ with p, ρ and V the pressure, density and speed of the solar wind and B the magnetic field strength. Within ∿1 AU the magnetic field is almost radial, so that $B \propto r^{-2}$ and to a rough approximation we may assume that the plasma energy varies also as r^{-2}. The unstable region should be expected on this basis to occur at $r \lesssim (B_1{}^2/4\pi\rho_1 V_1{}^2) \approx 2$ solar radii, where the subscript $_1$ refers to conditions at 1 AU.

This result, although very roughly estimated, suggests that on a scale of a few solar radii the magnetic field is closed off at the roots of sector boundaries in the manner described for example by Svalgaard et al. (1974) and depicted in Figure 2. The most obvious consequence is that one must expect the solar magnetic field to have both closed and open magnetic field lines, as is apparent from coronograph observations, with the former containing stationary coronal plasma and the latter the solar wind.

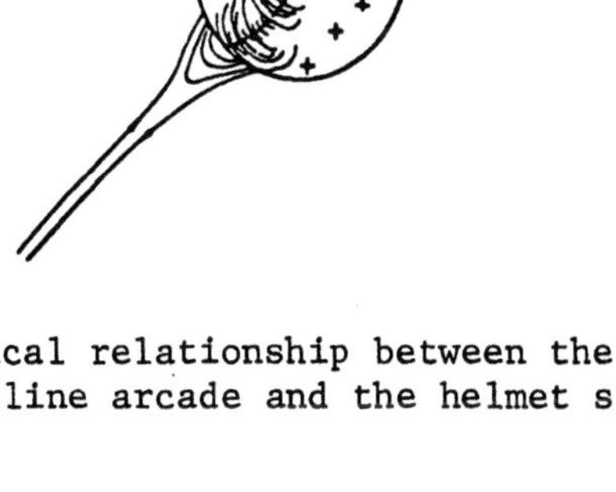

Figure 2. Geometrical relationship between the solar sector boundary, the magnetic field line arcade and the helmet streamers (Svalgaard et al., 1974).

TWO-COMPONENT STATIONARY MODEL WITH OPEN AND CLOSED MAGNETIC FIELD REGIONS

On the basis of the above arguments it is clear that it is desirable to consider a two-component model for the solar corona and solar wind allowing for finite magnetic field strength and the existence of both open and closed field regions. An axisymmetric model of this type has been constructed by Pneuman and Kopp (1970) and is shown in Figure 3 (see also Gold, 1958). Note that only magnetic field lines emanating from latitudes greater than about 45° are open and that the polar caps so defined are the sources of the entire solar wind. The interplanetary magnetic sector boundary is in this axisymmetric case the equatorial plane of the Sun. However this is of course not topologically or conceptually different from the configuration shown in Figure 2 where a slightly more complex magnetic field structure is considered.

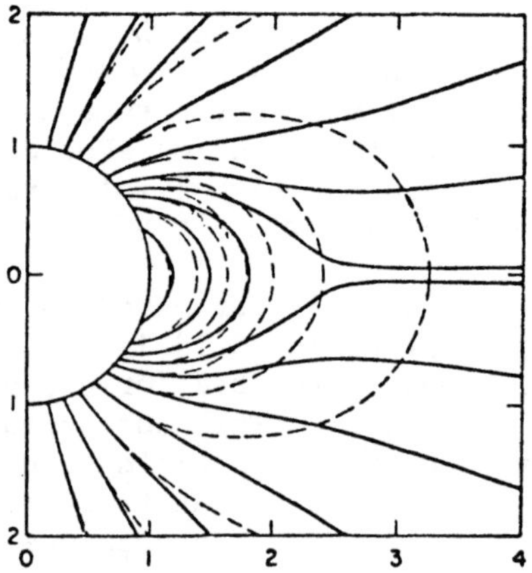

Figure 3. Axisymmetric two-component model for the solar wind and coronal magnetic field calculated by Pneuman and Kopp (1970). Note that the low latitude field lines are closed and hence that the solar wind should originate entirely in the polar caps of the Sun.

What would one expect to observe on the basis of such a model? Wave energy transmitted from the photosphere and dissipated in the closed field regions would maintain high coronal temperatures as observed, and as a consequence of the high thermal conductivity the plasma should be essentially isothermal along field lines. This energy is reradiated as thermal Bremsstrahlung from the high temperature regions and as EUV emissions from the transition region separating the chromosphere and corona. The structure of the transition region in this case would be hydrostatic and determined by a balance between downwards heat conduction and radiative cooling (e.g. Jordan, 1975, 1976). In the open regions, the wave energy and momentum should produce the solar wind and only a small fraction of the energy may be emitted in the form of the above-mentioned radiation. That is, the open regions should connect to coronal holes covering the two polar caps as indicated in both Figures 2 and 3. One cannot assume that the transition region is hydrostatic in these circumstances.

The total energy radiated from the transition layer and corona is $\sim 2.8 \times 10^{28}$ ergs/sec (e.g. Hundhausen, 1972; Axford, 1976; Holzer, 1977), whereas the energy involved in the solar wind is $\sim 3.3 \times 10^{27}$ ergs/sec. This suggests that if the two-component model is valid the fraction of the whole solar surface occupied by open magnetic field lines is $\gtrsim$ 16%. This figure is consistent with observations of the total area of the polar coronal holes (Bohlin, 1977); however, it seems to imply that the average photospheric magnetic field strength in such regions is $\sim$10 gauss, assuming the average radial field at 1 AU is $\sim 3\gamma$. The radial magnetic field strength in the polar coronal holes is not measurable from the earth and it is conceivable that it is indeed sufficient to account for the interplanetary field. However, low latitude coronal holes do not usually have field strengths of this magnitude and we doubt that the polar holes are different.

A further apparent difficulty with this simple concept is that the average solar wind speed at 1 AU near the ecliptic plane is $\sim$400 km/sec whereas at high solar latitudes interplanetary scintillation observations suggest that the average solar wind speed is $\sim$600-800 km/sec (Coles, 1977; Kakinuma, 1977). These high latitude results are however compatible with the observations in "high speed streams" which are clearly identified with coronal holes occurring at relatively low latitudes (Krieger et al., 1973). It is conceivable that some process such as wave reflection at the boundaries of open and closed regions causes the solar wind plasma near the centres of coronal holes to absorb more energy than that coming from the edges (e.g. Piddington, 1975) however in the absence of detailed calculations the argument is not wholly convincing. Alternatively, differences in the divergence of the flow could effect the heat flow in the outer corona in such a manner as to produce larger flow speeds at higher latitudes. One might expect on the other hand that the solar wind is faster near the ecliptic where the effects of solar rotation are greatest.

A more complex version of this model (but without the self consistent treatment followed by Pneuman and Kopp) has been discussed by Schatten et al. (1969) as an extension of earlier work on coronal magnetic field structures by Newkirk et al. (1968). It is assumed here that reconnection does not occur beyond a certain "source surface" so that radial current sheets (corresponding to sector boundaries) exist in this outer region. Within the source surface, reconnection is allowed to take place freely so that the magnetic field relaxes to its lowest energy state (i.e. curl-free). As a result the field beyond the source surface is relatively simple and most of the small features of the photospheric field are absent. An example of a fit between such a model and observed coronal structure is shown in Figure 4. Apart from the possible effects of solar rotation on the solar wind which might be important during periods of low solar activity as pointed out originally by Gold, there are no strong reasons why the current sheet forming the sector boundary should lie in the solar equatorial plane as assumed for convenience by Levy (1975) for example. It should be noted also that the coronal magnetic field is unlikely to be curl-free and indeed the occurrence of flares, surges, coronal transients, and similar phenomena is an indication that this is not the case in general.

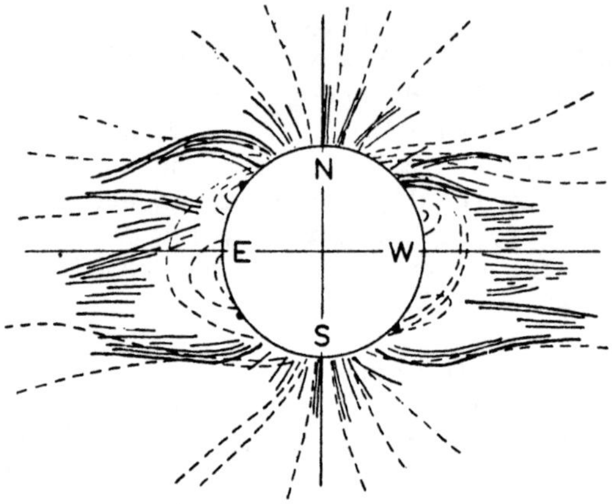

Figure 4. A comparison of the structure of the corona observed during the solar eclipse of October 23, 1976, with a coronal magnetic field model computed by Schatten (Waldmeier, 1977).

Not all interplanetary magnetic field lines in these models are connected to weak field regions in the photosphere which might be expected to correspond to coronal holes. Indeed a significant fraction can also connect to strong field regions. In this respect the models are akin to the three-component model discussed next.

THREE-COMPONENT MODEL OF THE SOLAR MAGNETIC FIELD

There are apparent difficulties with the stationary two-component model described above associated (a) with the large photospheric magnetic fields necessary in coronal holes, and (b) the absence of a good explanation for the difference between the "normal" solar wind observed near the ecliptic plane and that found in high speed streams and over the poles of the Sun. These difficulties can be avoided by the adoption of a model for the coronal magnetic field with three components as follows: (1) long-lasting open regions, associated with coronal holes and high speed solar wind streams; (2) long-lasting closed regions, associated with well-defined coronal magnetic structures; and (3) regions which are open but connected to sources of strong photospheric fields rather than coronal holes and also transiently open and closed regions from which coronal plasma escapes into space intermittently and where the topology of the magnetic field changes continually as a result of reconnection (Figure 5). There is a difference between this three-component model and that suggested by Piddington (1975) in that we do not assume the third component to be essentially free of magnetic field. There is on the other hand a close similarity to the suggestion by Newkirk and Wentzel (1977) concerning the coronal transport of energetic solar flare particles.

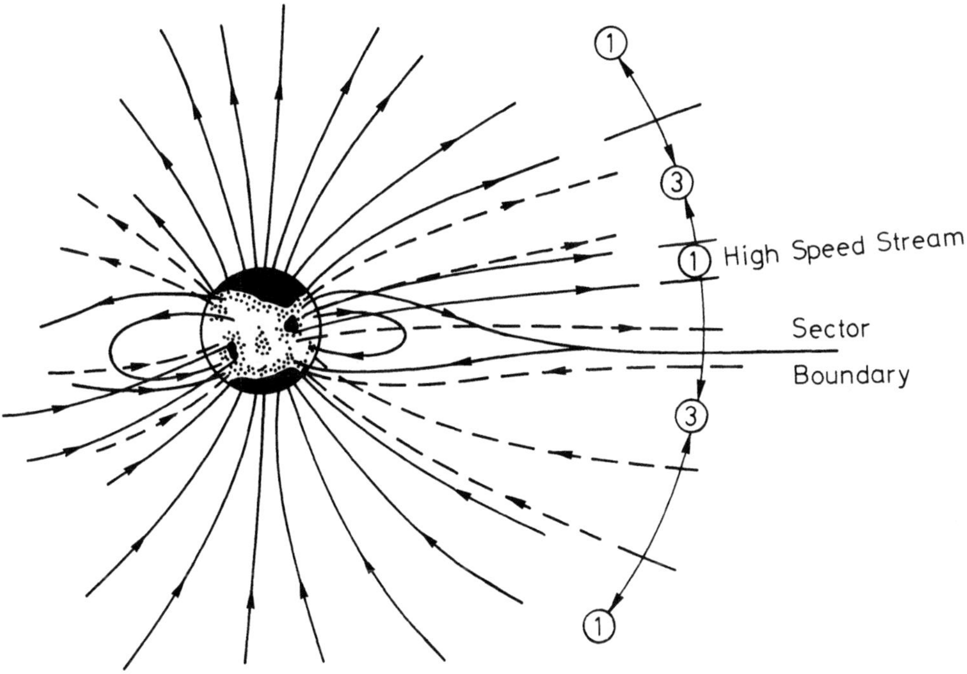

Figure 5. A three-component model for the coronal magnetic field. The shaded regions are intended to indicate coronal holes and the dots are intended to indicate regions where the coronal magnetic field lines are transiently open and closed. Open field lines may also arise in strong field regions. The field lines drawn with a continuous line are either open (region (1)) or closed (region (2)) for long periods. The field lines indicated by dashed lines are transiently open or connect to areas of strong magnetic field in the photosphere (region (3)).

An important distinction between regions (1) and (3) exists in this model which provides a possibility to explain the apparent difference in the solar wind found at high and low latitudes. We suggest that the "transiently open and closed" field lines of region (3) are usually not open long enough for an equilibrium to be achieved with respect to the chromosphere-coronal interface and to the plasma distribution along the field line. As a rough guide one might expect the time scale for equilibrium to be achieved to be of the order of the length of the field line ($\sim 10^{10}-10^{11}$ cm) divided by the speed of sound in the plasma ($\sim$100 km/sec), which is $\sim 10^3-10^4$ sec. There is no reason of course why the field lines should not remain open for shorter periods, but if they are open for much longer one would expect to see corresponding coronal hole features, perhaps on a small scale ("mini holes"). In addition to the transient component of region (3), open field lines which are connected to regions of strong photospheric field may in any case support solar wind flows with atypical characteristics. On the basis of this model the high latitude solar wind and the high speed streams would be associated with region (1) and the slower and more variable "normal" or "quiet" solar wind with region (3). The rather uniform nature of high speed streams (Bame et al., 1977) is understandable since the medium above coronal holes is in equilibrium with the wave field emitted by the underlying solar photosphere which has no particular spatial features in such regions nor any temporal variability.

If we assume that coronal holes (region (1)) cover 15-20% of the surface of the Sun and that region (3) covers say as much as 40-45% of the surface, then the apparent difficulty concerning the total flux of the interplanetary magnetic field does not exist, since the average magnetic field strength in region (1) and (3) would not have to exceed $\sim$2.5 gauss. Alternatively, the area of the photosphere covered by region (3) might be relatively small (e.g. Levine et al., 1977), but the field lines connect with regions of much larger than average magnetic field strength.

An important feature of this model is that the boundaries between the three component regions at the base of the corona are not fixed material boundaries but rather divisions in the character and topology of the coronal magnetic field. Consequently it should be no surprise that the boundaries can evolve in time and even move as a whole relative to the material in the photosphere below. Indeed the boundaries should be determined by the global magnetic field distribution rather than by local properties of the medium and accordingly one can expect the rotation of coronal holes in particular to correspond to that of the dominant magnetic field pattern underlying the solar surface as a whole and not the local material motion of the photospheric surface plasma (Timothy et al., 1975; Bohlin, 1977). Similar remarks apply to the rotation of sector boundaries, which in the three-component model are imbedded in the time varying region (3) but are nevertheless relatively stable since they are defined as the outer boundaries of region (2) which in turn correspond to the coronal magnetic field line arcades described by Svalgaard et al. (1974) and Svalgaard and Wilcox (1975).

THE INTERPLANETARY MAGNETIC FIELD

In general one would expect on the basis of the three-component model that the interplanetary magnetic field is more or less as depicted in Figure 2, although there is no reason, given suitable photospheric fields, why much less simple patterns of ingoing and outgoing field lines should not occur. Indeed, coronograph pictures made during periods of high solar activity suggested that coronal and interplanetary magnetic field structure can become very complicated. The magnetic field strength at high latitudes (i.e. in region (1)) may not be substantially weaker than at low latitudes, although this depends on the high latitude photospheric fields which are not adequately known and also on a degree of divergence of the solar wind relative to purely spherical outflow. The essential difference between regions (1) and (3) is that in the latter region the connection between a given interplanetary magnetic field line (defined by an element of solar wind plasma) and the solar photosphere must be continually changing, perhaps quite rapidly, as a result of reconnection in the corona. This process could provide the most important means of producing random walk of interplanetary magnetic field lines, which is of considerable significance for cosmic ray transport (Jokipii, 1971).

From the point of view of the OOE mission the most important effects to be expected are: (1) a tendency for sector boundaries to disappear with increasing latitude, as observed already by Pioneer 11 (Smith et al., 1977); (2) possible turbulence associated with the firehose instability at high latitudes; and (3) differences in the behaviour of Alfvén waves propagating away from the Sun due to the differing variation of the magnetic field strength with heliocentric distance at high and low latitudes (cf. Völk, 1976).

THE SOLAR WIND

The important effects to be expected in the solar wind according to the three-component model are associated with regions (1) and (3) in the interplanetary medium as indicated in Figure 5. The characteristics of the solar wind in region (1) should be similar to those observed at low latitudes in high speed streams, namely speeds of the order of 600-800 km/sec (e.g. Coles, 1977) and rather uniform properties in terms of density, temperature and composition (Bame et al., 1977). There could be variations with latitude resulting from solar rotation, especially with regard to the behaviour of heavy solar wind ions which can be brought up to the solar wind speed by centrifugal effects if other mechanisms are lacking (McKenzie and Axford, 1977). The divergence of the solar wind may be rapid (relative to purely spherical divergence) in the polar regions (see Figure 5), permitting more rapid acceleration of the solar wind close to the Sun (Kopp and Holzer, 1976). This results in a density decrease which provides a brightness contrast with respect to regions (2) and (3) (e.g. Munro and Jackson, 1976), but does not imply in itself that the solar wind should be faster at large heliocentric distances as the asymptotic speed is determined only by the energy/unit mass given to the flow which is not necessarily related to its divergence.

In region (3) one can expect greater variability in the properties and composition of the solar wind, both as a result of continuous reconnection to closed coronal field loops with varying plasma conditions and also because of the prevalance of stream-stream interactions resulting from longitudinal variations in the underlying solar corona. Since longitudinal variations are probably almost absent at high latitudes (greater than about 60°) there should be no stream-stream interaction with their associated shock waves and 27-day variations. Effects associated with solar rotation should be most pronounced in this region although these may be difficult to detect. The border separating regions (1) and (3) is not likely to be sharply defined, but rather one should expect a gradual transition with increasing latitude to conditions resembling high speed streams. In any case the probability of encountering a high speed stream should increase with increasing solar latitude giving a natural explanation for the observed annual variation of geomagnetic activity (Bohlin, 1977).

If the three components of the coronal magnetic field and plasma represented purely latitudinal structures, there would be no pronounced effects on the solar wind at great distances associated with solar rotation. However longitudinal variations do occur and in particular the polar coronal holes often exhibit excursions to low latitudes which can occasionally even enter the other hemisphere. In such a situation, observations at a fixed solar latitude will indicate a 27-day periodicity and regions of high speed solar wind (i.e. region (1)) will run into and interact with regions of low speed solar wind (region (3)). The interaction region must consist of a contact surface, an outwards-facing and outwards-moving shock and, depending on the strength of the interaction and the distance from the Sun, a backwards-facing outwards-moving shock and a rarefaction wave. The pattern of shocks and rarefaction waves can be further complicated if a number of fast and slow streams are involved.

Model calculations describing the evolution of high speed streams with distance from the Sun which are in good agreement with observations have been carried out by Gosling et al. (1976). In the simple case of a single high speed stream, the flow pattern is analogous to spherically symmetric patterns described by Simon and Axford (1967) and others. The variations of plasma density, temperature and magnetic field strength in a single interaction region are indicated in Figure 6, where can be seen that the essential feature, apart from the forward and reverse shocks, is a region of low temperature and high density and magnetic field strength centered on a contact surface which separates plasma from the fast and slow streams. This region contains the "oldest" plasma in the interaction region, which at the same time is the plasma which has undergone most expansion after first being heated by one shock or the other. The double-peak feature observed in energetic particles in such interaction regions is consistent with the structure in the background plasma (Barnes and Simpson, 1976; Skadron and Axford, 1976). The density gradient associated with the contact surface should be unstable on one side or the other, depending on whether the configuration is accelerating or decelerating, and resulting in density fluctuations which could give rise to intense interplanetary scintillations of radio sources (Dryer, private communication). Similarly it could be expected that heavy ions should diffuse towards the contact surface from one side or the other giving the contact surface a composition signature which could differ from that expected from the differences in composition of the fast and slow streams (Axford, 1970).

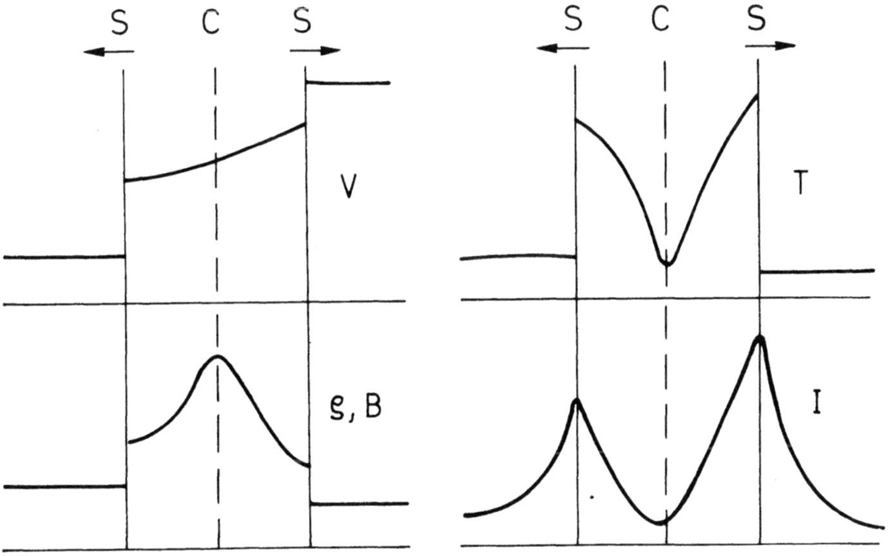

Figure 6. Schematic representation of the variation of the solar wind speed (V), density (ρ), magnetic field strength (B), temperature (T) and the low energy cosmic ray intensity (I) in an interaction region. The symbols S denote forward and reverse shocks and C a constant surface. In equilibrium, the total pressure between the two shocks should be more or less constant and the maxima in ρ and B and the minima in T and I are associated with the strong adiabatic expansion of the medium in the vicinity of the contact surface.

The stream-stream interactions should become weaker and less important with increasing solar latitude and eventually disappear at latitudes where the solar wind plasma observed originates entirely in the polar coronal hole throughout the solar rotation. This provides a clear means of determining whether or not the spacecraft performing the observations is in regions (1) or (3) of Figure 5. Model calculations demonstrating this effect have been carried out by Siscoe (1975) and some of his results are shown in Figure 7. Interplanetary scintillation observations which suggest that scintillation enhancements do not extend beyond 40° in the heliolatitude (Houminer, 1973) are consistent with this picture of stream-stream interactions.

The above remarks refer to the average properties of the solar wind. In the case of disturbances produced by events taking place in the corona, especially solar flares, one should expect the most pronounced effects to be observed at medium latitudes where solar active regions occur most commonly.

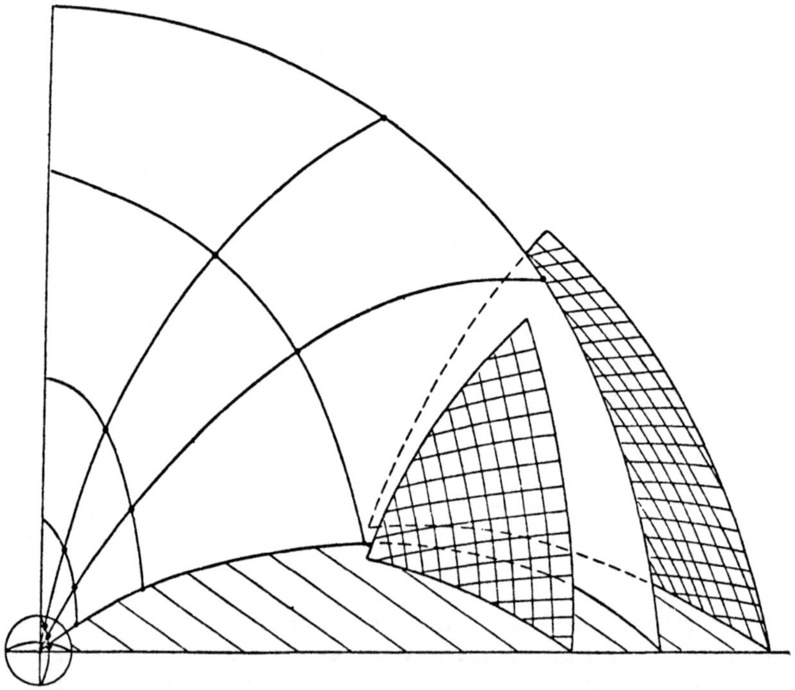

Figure 7. A three-dimensional view of an interaction region showing the forward and reverse shocks (hatched) and the contact surface, all of which are limited to solar latitudes less than about 45° (Siscoe, 1975).

GALACTIC COSMIC RAYS

The energetic particle population in the interplanetary medium contains components which vary on short time scales which we will call "solar" particles and components which are more persistent and vary only slowly with time, which we will call "galactic". In fact, three "galactic" components can be recognized, only one of which appears to represent the true galactic cosmic ray population (see Figure 9), whereas the others may be respectively the result of continuous low level emission of solar particles and the acceleration of interstellar neutral particles which have penetrated deep into the heliosphere before being ionized (Fisk et al., 1974).

Genuine galactic cosmic rays are modulated on penetrating into the interplanetary medium as a result of their interaction with magnetic field irregularities carried by the solar wind. This can be described in terms of a diffusion coefficient κ_r which depends on the power spectrum and distribution of the magnetic fluctuations and on the direction of the average interplanetary magnetic field (Jokipii, 1971). The most detailed model describing the variation of galactic cosmic ray modulation with latitude available at present is that of Völk (1976) whose calculated diffusion coefficients for polar latitudes are shown in Figure 8.

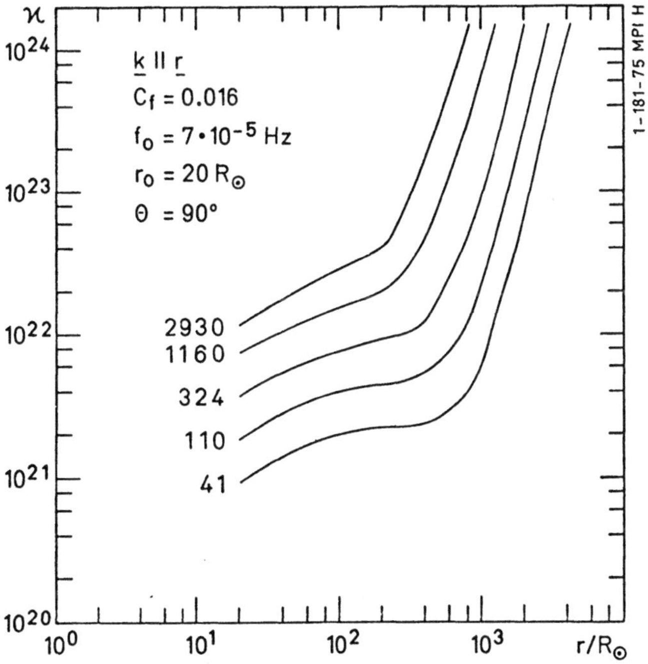

Figure 8. Variation of the radial diffusion coefficient for protons of various energies in MeV as marked. The distance is given in units of solar radii and the values of the diffusion coefficient correspond to a specific set of input parameters (Völk, 1976).

The most important feature of Völk's results is that at a given energy the diffusion coefficient remains small out to large distances in the ecliptic plane, whereas in the polar regions it increases rapidly beyond a distance of the order of ∿1-5 AU, depending on energy. Consequently, the overall modulation, which depends on the quantity $\int_r^\infty (V/\kappa_r)dr$ is very much smaller at high latitudes and in fact may be negligible or sufficiently small to permit a good estimate of its effects to be calculated, even at energies as low as a few tens of MeV per nucleon. This is an extremely important result since it suggests that we should be able to measure, or estimate with confidence, the unmodulated spectra of all species of galactic cosmic rays down to quite low energies. This is impossible at low latitudes near 1 AU for energies below a few hundred MeV per nucleon since the radial diffusion coefficient remains small out to very large distances (e.g. Goldstein et al., 1970; Urch and Gleeson, 1972). Consequently, we should be able, with the help of composition measurements at high solar latitudes, to reach sound conclusions concerning the origin and propagation of galactic cosmic rays as well as understand the modulation process much better than at present.

On the basis of the above arguments one might not expect high energy galactic cosmic rays to exhibit especially interesting behaviour at high solar latitudes since the effects of solar modulation are very small. However, global models of the modulation process must take into account "sinks" of particles which can produce anisotropies even in regions of otherwise small modulation if the particles can move transverse to the average interplanetary magnetic field direction (Cecchini et al, 1975; Fisk, 1975; Moraal and Gleeson, 1975; Levy and Jokipii, 1977). We suggest therefore that the most important measurement involving high energy particles to be made at high solar latitudes concerns their anisotropy rather than the details of their energy spectra which will not be greatly different from those observed near the ecliptic plane.

Cosmic rays of energies less than a few tens of MeV per nucleon are strongly modulated by interplanetary shocks. In particular, interaction regions (see Figure 6) appear to have a very pronounced effect, providing both acceleration in association with the forward and reverse shocks and deceleration in association with the region near the contact surface which has undergone most expansion (Skadron and Axford, 1976). Since these effects are the result of the presence of interaction regions we conclude that they are prevalant only in region (3) and are absent in those parts of region (1) where stream-stream interactions do not occur. It seems possible that the so called "anomalous" component (see Figure 9) is also accelerated by shock waves associated with the interaction regions, so that one might expect this component of the galactic cosmic radiation also to be essentially absent in region (1). It has been suggested that the anomalous component is produced from singly-ionized interstellar neutrals which are accelerated from relatively low energies in the interplanetary medium (Fisk et al., 1974). Whatever the main acceleration mechanism might be, it is evident that the preliminary acceleration to solar wind energies must rely on wave-particle interactions at high latitudes whereas the outwards-moving interplanetary magnetic field provides a very efficient acceleration at low latitudes. We might expect a strong variation in the intensity of the anomalous component with latitude for this reason alone. As long as they are trapped in the high speed streams separating interaction regions these interstellar ions can reach equilibrium in both their energy spectrum and total flux. The ions gain energy on interacting with shock waves and lose energy as a result of adiabatic deceleration in the expanding solar wind elsewhere. The flux is maintained constant by a balance between ionization of interstellar neutrals on the one hand and loss as particles penetrate the shock waves defining the boundaries of interaction regions where only deceleration occurs, on the other.

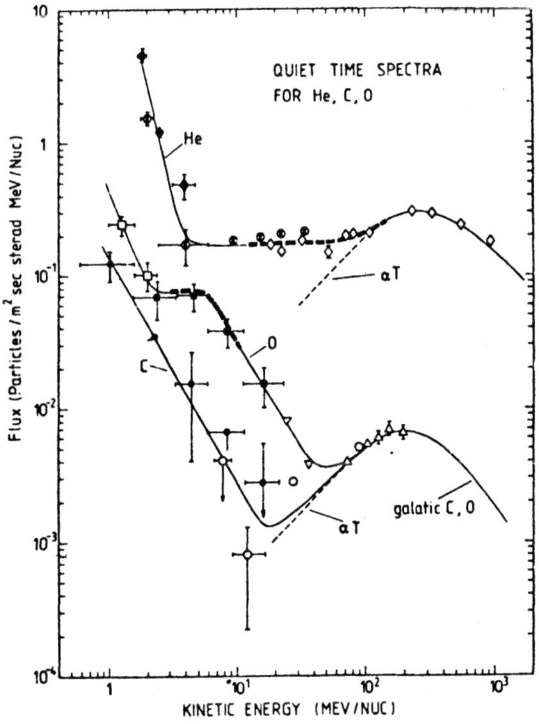

Figure 9. Differential energy spectra for quiet-time helium, oxygen and carbon (Klecker et al., 1977). The anomalous component is apparent as the hump in the oxygen spectrum and the flat region in the spectrum of helium in the kinetic energy range ~3 to ~50 MeV per nucleon. The spectra of all three species at low energies must correspond either to energetic particles emitted by the Sun or to particles of solar origin which have been accelerated in interplanetary space.

Any low level solar contribution to the "galactic" cosmic radiation should be confined mainly to field lines which connect, at least temporarily, with the active regions where the particles are presumed to be accelerated, that is to region (3). However it is not inconceivable that particle acceleration occurs to some extent everywhere in the solar corona, so that such particles could exist also in region (1). Energetic particles produced in planetary magnetospheres (e.g. the Jovian electrons discovered by Chenette et al. (1974) and Teegarden et al. (1974)) should be confined to the vicinity of the ecliptic plane. It is especially important to observe the latitudunal spread of such particles since this provides a clear measurement of the perpendicular diffusion coefficient. The OOE mission, with two spacecraft flying past Jupiter, is very well suited to performing this task.

SOLAR COSMIC RAYS

Energetic solar particles (solar "cosmic rays") are produced in association with active regions or solar flares. They appear in the outer solar system in a wide range of solar longitudes although there is a clear tendency for the most prompt and intense effects to be observed on field lines which connect to the vicinity of the source (Axford, 1970; Wibberenz, 1975; Newkirk, 1975). The spreading has been ascribed either to diffusion in the corona (Reid, 1964; Axford, 1965) or to rapid diffusion across the mean magnetic field direction due to field line wandering (Jokipii, 1971). The possible coronal diffusion process has not been thoroughly discussed apart from the initial concept of a strongly tangled coronal field and the suggestion by Fisk and Schatten (1972) that gradient drifts in current sheets might be important (see Newkirk, 1975). However Newkirk and Wentzel (1977) have pointed out that diffusion associated with continuous field line reconnection in the solar corona could be the most important effect (see also Van Hollebeke et al., 1975). This is consistent with our concept of regions (2) and (3) in the three-component solar magnetic field model and perhaps provides a bridge to the transverse interplanetary diffusion advocated by Jokipii.

In terms of the three-component model, we would expect that solar cosmic rays should be confined largely to region (3) in interplanetary space and perhaps are stored for relatively long periods (i.e., hours) in region (2) of the solar corona. Energetic particles which appear at the edges of region (1), where the field lines remain open for very long periods, should escape immediately from the Sun. Thus we expect that region (1) should be characterized by an absence of solar flare particles unless there is significant transverse diffusion in this region or there are sources of particles in the corona which we have not recognized. In particular, particles originating in solar flares should not be present over the solar poles unless transverse diffusion is very efficient. One must of course allow for the possible existence of weak sources of energetic particles in coronal hole regions and also for in situ acceleration of particles by shocks traversing such open field regions.

CONCLUSIONS

On the basis of the above discussion the features of the interplanetary medium to be expected to be observed as a result of the OOE mission are:

(1) The properties of the solar wind should change from being low speed and variable in region (3) to high speed and uniform in region (1).
(2) High speed streams will be encountered more frequently with increasing latitude until eventually at high latitudes the low speed component disappears entirely.
(3) Stream-stream interaction regions, with their associated pattern of shocks and contact surfaces, will disappear at high latitudes.
(4) Cosmic ray acceleration and deceleration in association with interaction regions in the solar wind must be confined to low latitudes. This might also apply to the anomalous component of the galactic cosmic radiation.
(5) The modulation of galactic cosmic rays should be a minimum at high latitudes and it is expected that even at energies as low as a few tens of MeV per nucleon the degree of modulation is relatively small.
(6) The anisotropy of high energy galactic cosmic rays at high solar latitudes should provide useful information concerning the modulation in a global sense.
(7) Energetic particles produced in planetary magnetospheres, such as the Jovian electrons, can move outside the plane of the ecliptic only as a result of transverse diffusion and will therefore provide a means of determining the effectiveness of this process.
(8) Energetic solar flare particles should be generally absent in region (1) and not normally observable over the solar polar regions unless transverse diffusion is very effective.
(9) Transient disturbances associated with solar flares should have their most pronounced effects at latitudes corresponding to those of solar active regions and in particular the ejecta from solar flares should be encountered most frequently at such latitudes. The disturbances produced by such events should however be detectable everywhere in the hemisphere above the event.

REFERENCES

Axford, W.I.: 1965, *Planet. Space Sci.* 13, 1301.

Axford, W.I.: 1970, *Solar Terrestrial Physics /1970, part II*, D. Reidel Publ. Co., Dordrecht, 110.

Axford, W.I.: 1976, *Physics of Solar Planetary Environments, Proc. Internat. Symp. on Solar-Terrestrial Physics*, 1, Boulder/Colorado, 270.

Bame, S.J., Asbridge, J.R., Feldman, W.C. and Gosling, J.T.: 1977, *J. Geophys. Res.* 82, 1487.

Barnes, C.W. and Simpson, J.A.: 1976, *Astrophys. J.* 210, L91.

Bohlin, J.D.: 1977, *Solar Phys.* 51, 377.

Cecchini, S. and Quenby, J.J.: 1975, *Proc. 14th Internat. Cosmic Ray Conf.* 11, München, 911.

Chenette, D.L., Conlon, T.F. and Simpson, J.A.: 1974, *J. Geophys. Res.* 79, 3551.
Coles, W.A.: 1977, *Space Sci. Rev.*, in press.
Fisk, L.A.: 1975, *Proc. 14th Internat. Cosmic Ray Conf.* 11, München, 905.
Fisk, L.A. and Schatten, K.H.: 1972, *Solar Phys.* 23, 204.
Fisk, L.A., Koslovsky, B. and Ramaty, R.: 1974, *Astrophys. J. Lett.* 190 pt. 2, L35.
Gold, T.: 1958, in B. Lehnert (ed.), *Electromagnetic Phenomena in Cosmic Physics, IAU Symp. No. 6*, Univ. Press, Cambridge, paper 30, 275.
Goldstein, M.L., Fisk, L.A. and Ramaty, R.: 1970, *Phys. Rev. Lett.* 25, 832.
Gosling, J.T., Asbridge, J.R., Bame, S.F. and Feldman, W.C.: 1976, *J. Geophys. Res.* 81, 5061.
Holzer, T.: 1977, to be published.
Houminer, Z.: 1973, *Planet. and Space Sci.* 21, 1367.
Hundhausen, A.J.: 1972, *Coronal Expansion and Solar Wind*, Vol 5 in the series: Physics and Chemistry in Space, Springer-Verlag, Heidelberg.
Jokipii, J.R.: 1971, *Rev. Geophys.* 9, 27.
Jordan, C.: 1975, *Monthly Notices Roy. Astron. Soc.* 170, 429.
Jordan, C.: 1976, *Phil. Trans. Roy. Soc. Lon. A.* 281, 391.
Kakinuma, T.: 1977, in M. A. Shea, D. F. Smart, S. T. Wu (eds.), *Study of Travelling Interplanetary Phenomena/1977*, in the series Astrophys and Space Sci. Library, D. Reidel, Dordrecht, this volume.
Klecker, B., Hovestadt, D., Gloeckler, G. and Fan, C.Y.: 1977, *Astrophys. J.* 212, 290.
Kopp, R.A. and Holzer, T.E.: 1976, *Solar Phys.*, in press.
Krieger, A.S., Timothy, A.F. and Roelof, E.C.: 1973, *Solar Phys.* 29, 505.
Leer, E. and Axford, W.I.: 1972, *Solar Phys.* 23, 238.
Levine, R.H., Altschuler, M.D. and Harvey, J.W.: 1977, *J. Geophys. Res.* 82, 1061.
Levy, E.H.: 1975, *Proc. 14th Internat. Cosmic Ray Conf.* 4, München, 1215.
Levy, E.H. and Jokipii, J.R.: 1975, unpublished.
McKenzie, J.F. and Axford, W.I.: 1977, in preparation.
Moraal, H. and Gleeson, L.J.: 1975, *Proc. 14th Internat. Cosmic Ray Conf.* 3, München, 910.
Munro, R. and Jackson, B.: 1976, *Solar Phys.*, in press.
Newkirk, G.: 1975, *Proc. 14th Internat. Cosmic Ray Conf.* 11, München, 3594.
Newkirk, G., Altschuler, M.D. and Harvey, J.: 1968, in K.O. Kiepenheuer (ed.), *Structure and Development of Solar Active Regions*, Springer-Verlag, New York, 379.
Newkirk, G. and Wentzel, P.: 1977, to be published.
Parker, E.N.: 1958, *Phys. Rev.* 109, 1874.
Piddington, J.H.: 1975, *Astrophys. Space Sci.*, in press.
Pneuman, G.W. and Kopp, R.A.: 1970, *Solar Phys.* 13, 176.
Reid, G.C.: 1964, *J. Geophys. Res.* 69, 2659.
Schatten, K.H., Wilcox, J.M. and Ness, N.F.: 1969, *Solar Phys.* 6, 442.
Simon, M. and Axford, W.I.: 1967, *Astrophys. J.* 150, 105.

Siscoe, G.L.: 1975, *Proc. of the Symposium on the Study of the Sun and Interplanetary Medium in Three Dimensions*, GSFC Greenbelt /Maryland, 166.
Skadron, G. and Axford, W.I.: 1976, AGU Fall Meeting, *EOS Transactions*, SC 22, 980.
Smith, E.J., Davis, L., Coleman, P.J., Colburn, D.S., Dyal, P. and Jones, D.E.: 1977, *J. Geophys. Res.* 82, 1077.
Svalgaard, L., Wilcox, J.M. and Duvall, T.L.: 1974, *Solar Phys.* 37, 157.
Svalgaard, L. and Wilcox, J.M.: 1975, *Solar Phys.* 41, 461.
Teegarden, B.J., McDonald, F.B., Trainor, J.H., Webber, W.R. and Roelof, E.C.: 1974, *J. Geophys. Res.* 79, 3615.
Timothy, A.F., Krieger, A.S. and Vaiana, G.S.: 1975, *Solar Phys.* 42, 135.
Urch, I.H. and Gleeson, L.J.: 1972, *Astrophys. and Space Sci.* 17, 426.
Van Hollebeke, M.A.I., Ma Sung, L.S. and McDonald, F.B.: 1975, *Solar Phys.* 41, 189.
Völk, H.J.: 1976, *J. Geophys. Res.* 81, 5570.
Waldmeier, M.: 1977, *Nature* 265, 611.
Wibberenz, G.: 1975, *J. Geophys.* 40, 667.

DISCUSSION

Schwenn: Coronal holes show nearly no differential rotation, in contrast to sunspots. Why does the constant rotation rate of coronal holes being concentrated mainly around the sun's poles, coincide so well with the rotation rate of equatorial sunspots?

Axford: The boundary of a coronal hole is not a material boundary but is defined apparently by a change in the topology of the magnetic field from open to closed. Consequently it is possible for a coronal hole, or a feature on the boundary of a polar hole, to move with respect to the underlying photosphere, and in particular to rotate at a different rate if the magnetic field is free to undergo reconnection. The rotation rate of the patterns of open field lines should be controlled by the rotation of the magnetic features which determine whether or not a given field line is open, and which may well be related to strong magnetic field regions at low latitudes. (In a somewhat analogous manner, the patterns of open field lines in the earth's magnetosphere do not rotate with the earth).

Wibberenz: What kind of magnetic field structure do you expect over the poles where the polarity changes with time? Will there still be a coronal hole?

Axford: It is not easy to predict what would happen, but in general one would expect coronal holes and open magnetic field lines to always be present in the polar regions. If the polarity of the fields is the source in both polar regions the sector boundary pattern would, of course, be rather complex.

Wibberenz: You mentioned the fire-hose instability which might be produced along interplanetary field lines. What is the spatial scale expected, and would this effect not lead to an increase of scattering of cosmic rays?

Axford: Presumably any Alfvén waves resulting from the fire-hose instability would be limited in amplitude to $\delta B/B \approx 1$. Since this condition is also assumed by Volk in considering the variations in amplitude of solar Alfvén waves one would not expect any great change in the level of hydromagnetic turbulence (if the presumption is correct) and hence the effectiveness of cosmic ray scattering would not change.

Dryer: I'm not sure that the shock/contact surfaces would be restricted to the medium latitudes. We've seen Type II bursts extend over the poles (e.g., Smerd's (Smerd, S. F.: 1970, *Proc. Astron. Soc. Aust.* *1*, 305) observation of the March 1969 event). I would think that even stream-generated shocks (from low latitudes, say) would propagat meridionally up to high latitudes in some cases as well.

Axford: I was referring only to shock/contact surfaces generated by the interaction of high speed streams with the low speed solar wind which can take place only at latitudes below that of the main polar coronal holes. The IPS data of Hewish and Houminer support this argument. The structure of the interaction size must, in any case, be very similar to that of the spherically-symmetric case (e.g., Simon, M., and Axford, W. I.: 1966, *Planetary Space Sci.* *14*, 901) since the flare remains essentially radial and one can relate longitude to time. Of course, there is no reason in which explosive (flare-associated) events should not produce effects in the polar regions as they do in the ecliptic.

THEORETICAL CONTRIBUTIONS TO SOLAR WIND RESEARCH - A REVIEW

S. CUPERMAN
Department of Physics and Astronomy
Tel Aviv University, Ramat Aviv, Israel

ABSTRACT

The theoretical work on the solar wind phenomena done since 1958 can be divided into two main parts: Part I - development and refinement of Parker's initial macroscopic model, the emphasis being placed upon steady state, spherically symmetric flow and the identification of the structure-less background solar wind plasma with the low speed flow. It is in this part that much progress in understanding the solar wind phenomenon has been achieved; Part II - generalization of Parker's initial model such as to include microscopic (kinetic) aspects, temporal variations, deviations from spherically symmetric conditions, complex local magnetic configurations, etc. The last two aspects, in particular, have received considerable attention with the discovery of the coronal holes, their association with high-speed flows and the tentative identification of these flows with the structure-less background solar wind plasma. This review is confined to Part I, as defined above. However, for completeness, several important aspects connected with the subjects enumerated under Part II and which represent the objects of the most recent investigation are also briefly reviewed.

I. INTRODUCTION

Two decades ago, Parker (1958) stimulated by Biermann's (1951) ideas suggesting a steady background stream of plasma from the sun and by the fact that Chapman's (1957) static model leads to non-vanishing pressure at infinity, constructed a hydrodynamic model with the aid of which he demonstrated the existence of a supersonic expansion of the solar corona. The basic predictions of Parker's model were soon confirmed by Mariner II observations (Neugebauer and Snyder, 1966).

The first theoretical description and prediction of the solar wind was given by Parker with the aid of an isothermal one-fluid model in which $n_e = n_p = n$, $v_e = v_p = v$ and $T_e = T_p = T$. Parker (1963) expanded

M. A. Shea et al. (eds.), Study of Travelling Interplanetary Phenomena/1977, 165–194.

the theory by assuming a polytropic equation of state in order to describe the temperature in the solar wind. Noble and Scarf (1963) and Whang and Chang (1965) added an energy equation to the continuity and momentum equations and used numerical methods in order to solve simultaneously all three moment equations. Viscosity (isotropic) was also introduced by Scarf and Noble (1965) and Whang et al. (1966) with no significant change in the results. Weber and Davis (1967) extended the model to include the frozen-in magnetic field, which is pulled out by the expanding solar wind. Their results provide the theoretical basis for the spacecraft and comet-tail observations of a small azimuthal velocity component of several kilometers per second in the corotational direction.

Following the determination of energy transport by electron heat conduction in the solar wind which indicated an anomalously low value (Montgomery et al., 1968; Montgomery, 1972, Ogilvie et al., 1971), Hundhausen (1969), using temperature profiles predicted by existing fluid solutions for the solar wind, estimated the predicted transport energy by electron heat conduction at 1 AU and found it to be larger than the observed magnitude by a factor of 2, at least. Detailed quantitative solutions of the one-fluid equations with anomalous thermal conductivity have been given by Cuperman and Harten (1970a), Gentry and Hundhausen (1969), Durney (1971), Whang (1971), Cuperman et al. (1972), Cuperman and Metzler (1973a, 1973b), etc.

The one-fluid models just mentioned do not differentiate between the electron and the proton temperature and, therefore, cannot predict the observed difference in temperature ($T_{eE}/T_{pE} \simeq 4$). (At 1 AU the subscript E will be used). This is the reason why Sturrock and Hartle (1966) and Hartle and Sturrock (1968) expanded the model equations by using a *collisional* two-fluid model in which electrons and protons were allowed to possess *different temperatures*. The authors justified the use of a *two-fluid* model by the weak collisional coupling between the two species which results from the low densities and high temperatures in the solar wind. They obtained $n_E = 15$ cm^{-3}, $v_E =$ 250 km-sec^{-1}, $T_{eE} = 3.5 \times 10^5$ °K, and $T_{pE} = 4.4 \times 10^3$ °K as a typical set of solutions. The large difference in the two temperatures predicted was considered to be due to the considerably lower thermal conductibity of the protons compared with that of the electrons. The authors tried to obtain smaller T_{eE}/T_{pE} values by varying the coronal base conditions, but this was accomplished by changes in either n_E, v_E, T_{eE}, or T_{pE} in the wrong direction.

Subsequently, it has been shown by Cuperman and Harten (1970b) that the low proton temperature at 1 AU predicted by the two-fluid model can be raised to the desired (observed) value by the use of an "enhanced", essentially noncollisional, coupling between the proton and the electron components in the solar wind. The noncollisional coupling is presumably due to unstable wave particle interactions in the solar wind, which have already been mentioned in connection with the anomalously low thermal conductivity in the one-fluid model. A similar qualitative conclusion has been reached by Nishida (1969) and Toichi (1971). Similar results obtained from the solution of two-fluid equations with noncollisional contributions to the transport coefficients have been published by Wolff et al. (1971). Anomalous thermal conductivities were discussed in Cuperman and Harten (1971a; 1971b; 1972) and Wolff et al. (1971).

A different approach to the "modified" two-fluid model has been considered by Hartle and Barnes (1970) and Barnes et al. (1971). The authors incorporated an ad hoc external energy source assumed to extend between 1 $R_\odot$ and 20 $R_\odot$ which raised the proton temperature and bulk velocity at 1 AU to 3.7×10^4 °K and 320 km-sec^{-1}, respectively. The corresponding particle density, electron temperature, and electron heat flow were 13 cm^{-3}, 3.4×10^5 °K, and 29×10^{-2} erg-cm^{-2}-sec^{-1}, respectively.

In the two-fluid models the solar wind plasma consists solely of electrons and protons. Actually the solar wind contains also other heavier ions and one should consider at least alpha particles whose relative concentration is non-negligible. Theoretical investigations of the three-fluid structure of the solar wind have been carried out by Yeh (1970) for the isothermal, frictionless and very high temperature case, Geiss et al. (1970) for the equal temperatures, polytropic case, Weber (1973) for the critical solution in multi-ion species moment equations, Fahr (1973) for the range between 0.5 and 1 AU, Ryån and Axford (1975) using the complete set of fluid equations with heating functions proportional to the ion mass and neglect of heat conduction, and Cuperman and Metzler (1975) using the complete set of three-fluid equations with inclusion of phenomenological contributions from wave-particle interactions to the transport coefficients.

As a result of these theoretical investigations, a satisfactory agreement between theory and averaged gross features of the solar wind observed at about 1 AU has been achieved. Therefore, it is now possible to use these results for the case of other stars and next to investigate the effect of mass loss by stellar winds on star evolution (see Cuperman and Sternlieb, 1974). This review is devoted essentially to the main aspects of the topics described above and which represent Part I in the solar wind theory, as defined in the Abstract. Thus, in Section II, a general discussion of microscopic and macroscopic description of the multi-component solar wind plasma is given. Special emphasis is placed upon the structure and validity of the classical transport coefficients and their consequences for the solar wind. In Section III, the solutions of macroscopic equations with phenomenologically modified transport coefficients (i.e., with inclusion of non-collisional contributions) are reviewed and compared with satellite observations. Section IV is devoted to a brief description of several important aspects connected with the subjects enumerated under Part II and which represent the object of most current theoretical investigation.

As concerning references, it should be mentioned that despite the relatively large number of references listed, it was not intended to include all the theoretical work published in the last two decades. Many important papers have not been included because of space limitation or because of regrettable omission. Also, the preferential use, for the purpose of illustration, of the results obtained by the Tel Aviv University Plasma Physics Group is due to the author's familiarity with them.

II. MACROSCOPIC VS. MICROSCOPIC DESCRIPTIONS AND STRUCTURE OF TRANSPORT COEFFICIENTS

The solar wind plasma consists primarily of electrons, protons and alpha particles which stream away from the sun. The accurate description of such a plasma system is given by the kinetic equations together with Maxwell's equations. The kinetic equation for the ith particle species is

$$\partial f_{i1}/\partial t + \underline{w}_{i1}\,(\partial f_{i1}/\partial \underline{x}_{i1}) + (\partial/\partial \underline{w}_{i1})\cdot[(\underline{F}_{i1}^{ext}/m_i)f_{i1}] = C_{ii} + C_{ij}$$

$$C_{ii}(f_{ii12}) = -\int(\underline{F}_{ii12}/m_i)\cdot(\partial f_{ii12}/\partial \underline{w}_{i1})\,\underline{dx}_{i2}\,\underline{dw}_{i2} \qquad (1)$$

$$C_{ij}(f_{ij12}) = -\int(\underline{F}_{ij12}/m_i)\cdot(\partial f_{ij12}/\partial \underline{w}_{i1})\,\underline{dx}_{j2}\,\underline{dw}_{j2}$$

where $\underline{f}_{i1}$ $(\underline{x}_{i1}, \underline{w}_{i1}, t)$ is the one-particle distribution function, $\underline{F}_{i1}^{ext}$ - the external force and $C_{ii} + C_{ij}$ - the rate of change of f_{i1} due to the interaction between like particles (ii) and between i and j particles (ij). The C-terms describe both collective (i.e., long range) and individual (i.e., short range, collisional) interactions and depend on $\underline{F}_{ik}$ - the two particle interaction force as well as on f_{ik12} - the two particle distribution function (k = i, j). Actually f_{ik12} obeys an equation similar to eq. (1) which also contains $f_{ik\ell 123}$ - the three-particle distribution function, etc. Thus, the kinetic description requires the simultaneous solution of an infinite number of equations for the reduced functions f (together with Maxwell's equations). If one defines $f_{ik12} = f_{i1}\cdot f_{k2} + \eta_{ik12}$, where $f_{i1}\cdot f_{k2}$ and η_{ik12} represent the collective (correlationless) and the collisional (correlation) contributions to f_{ik12}, respectively, one can write an alternative form for eq. (1), namely

$$\partial f_{i1}/\partial t + \underline{w}_{i1}\cdot(\partial f_{i1}/\partial \underline{x}_{i1}) + (\partial/\partial \underline{w}_{1})\cdot[(\underline{F}_{i1}/m_i)f_{i1}] = B_{ii} + B_{ij} \qquad (2)$$

where

$$\underline{F}_{i1} = \underline{F}_{i1}^{ext} + \underline{F}_{i1}^{collective} \quad \text{and} \quad B_{ik} = B_{ik}(\eta_{ik12}).$$

When collective interactions dominate (i.e., $B_{ik} = 0$) one obtains the Vlasov equation. On the other hand, when the plasma is collision-dominated, one may derive expressions for $B_{ik} \equiv C_{ik}$ in a rather simple way. In the intermediate case, one has to solve consistently the coupled system of kinetic and Maxwell's equations, and this is a formidable task. In spite of the considerable effort invested in the last two decades into this problem in conjunction with Controlled Thermonuclear Research and Astrophysical Research, no general solution has been obtained so far. This is why, in order to obtain quantitative results, a simpler, approximate approach to the problem, namely the macroscopic description, has been successfully used in both fields mentioned above. The macroscopic equations are easily obtained from eq. (1) by multiplying it by 1, $m_i\ (w_i - \langle w_i \rangle)$, $m_i\ (w_i - \langle w_i \rangle)(w_i - \langle w_i \rangle)$, etc. and by integrating successively over velocity space. The left hand side terms of eq. (1) then provide moments of the distribution function, namely

$$n_i(\underline{x},t) \equiv \int f_{i1}(\underline{x}_{i1}, \underline{w}_{i1}, t)\ \underline{d\ w}_{i1}$$

$$\underline{v}_i(\underline{x},t) \equiv n_i^{-1} \int \underline{w}_{i1}\ f_{i1}(\underline{x}_{i1}, \underline{w}_{i1}, t)\ \underline{d\ w}_{i1} \equiv \langle \underline{w}_i \rangle$$

$$\underline{\underline{P}}_i\ (\underline{x},t) \equiv m_i \int (\underline{w} - v_i)\ (\underline{w} - \underline{v}_i)\ f_{i1}\ (\underline{x}_{i1}, \underline{w}_{i1}, t)\ \underline{d\ w}_{i1} \qquad (3)$$

$$\underline{\underline{\zeta}}_i \equiv \underline{\underline{P}}_i - p_i\ \underline{\underline{I}}\ ; \quad p_i \equiv (1/3)\ \mathrm{Tr}\ \underline{\underline{P}}_i$$

$$\underline{q}_i \equiv (\tfrac{1}{2}) m_i \int (\underline{w} - \underline{v}_i)^2\ \underline{v}_i\ f_{i1}\ (\underline{x}_{i1}, \underline{w}_{i1}, t)\ \underline{d\ w}_{i1} \qquad \text{etc....}$$

where n_i, $\underline{v}_i$, $\underline{\underline{P}}_i$, $\underline{\underline{\zeta}}_i$, p_i and $\underline{q}_i$ are the particle number density streaming velocity, pressure tensor, scalar pressure and heat flow vector; $\underline{\underline{I}}$ is the unit dyadic. The right hand side terms, on the other hand, produce transport coefficients as

$$\underline{R}_{ij} \equiv m_i \int (\underline{w} - \underline{v}_i)\ C_{ij}\ \ d\ \underline{w}_i$$

$$Q_{ij} \equiv (\tfrac{1}{2}) m_i \int (\underline{w} - \underline{v}_i)^2\ C_{ij}\ \underline{dw}_i\ , \text{ etc.} \qquad (4)$$

where R_{ij} and Q_{ij} represent, respectively, the mean rate of change in the momentum of the i-type particles, and the random energy added to i-type particles due to the interaction with j-type particles. Symbolically, the resulting equations can be written as continuity equation: $(n_i, \underline{v}_i) = 0$

momentum equation: $(n_i, \underline{v}_i, \underline{\underline{\zeta}}_i) = \underline{R}_i$ (5)

energy equation: $(n_i, \underline{v}_i, \underline{\underline{\zeta}}_i, \underline{q}_i) = Q_i$ etc....

Thus, for each particle species, i, one obtains ℓ equations for $\ell + 1$ moments. Moreover, the equations contain two unknown quantities, $\underline{R}_{ij}$ (C_{ij}) and Q_{ij} (C_{ij}) defined by (4). Thus, as they stand, the macroscopic equations do not remove the difficulties encountered with the kinetic equations. The advantage of the macroscopic description consists of the fact that it suggests simple, approximate ways to solve the problem. A classic way to do this is that of Enskog (1917) and Chapman (1957) which strictly holds for plasmas which are collision-dominated and are characterized by weak space and time variations. That is, in the kinetic equation (eq. (1)) the important terms are assumed to be the collision term and the magnetic term; the other terms which contain space and time derivatives are assumed to be small. Thus, the distribution function f_{i1} is assumed to be approximately a Maxwellian f_i^m with parameters n_i, $\underline{v}_i$ and T_i (which are slowly varying functions of space and time) and is expanded as

$$f_{i1} = f_{i1}^m (n_i, \underline{v}_i, T_i) + f_{i1}^{(1)} + f_{i1}^{(2)} + \ldots.. \qquad (6)$$

Under such simplifying conditions it is possible to specify zero-order collisional terms C_{ik} in eq. (1), obtain and solve an approximate integro-differential equation for $f_{ik}^{(1)}$ and use the solution in order to calculate the moments defined by expressions (3). This has the effect of closing the moment equations indicated by eqs. (5). On the other hand the solution obtained for $f_{ik}^{(1)}$ is used to calculate the collisional transport coefficients $\underline{R}_{ij}^{ik}$, Q_{ij}, etc.

Therefore, the Enskog-Chapman procedure succeeds in closing the system of moment equations together with the specification of the transport coefficients. However the validity of the results is limited to situations such that a) collisions play by far a dominant role and b) the average quantities characterizing the plasma change slowly in space and time. Actually, except for a relatively small region close to the base of the corona, the solar wind is an essentially collisionless plasma. Nevertheless, in spite of the limitations indicated above, the basic understanding of the solar wind phenomenon has come from approximate macroscopic rather than from microscopic descriptions. The reason for this success is that under circumstances similar to those prevailing in the solar wind, collective effects could play a significant collision-like role. Indeed, unstable plasmas may generate random fluctuating fields, which in turn, can produce a strong mixing (randomization) with a significant enhancement of certain transport coefficients above the values expected in a weak collisional plasma. As examples we refer to the heat transfer from electrons to protons or the rate of momentum transfer between protons and α-particles, to be discussed in the next section.

On the other hand, collective effects may contribute to some other transport properties in their own way, thus resulting in transport coefficients which are smaller than expected on pure collisional grounds. This is the case, for example, with the electron heat conduction. In fact, in both Fusion and Astrophysics, measurements clearly indicate occurrence of anomalously low coefficients of thermal conductivity.

The enhancement of some transport coefficients and the inhibition of others by collective interactions are not describable by the macroscopic expressions indicated above. They have to be calculated by solving correctly eq. (1). Since this problem is as yet not completely solved, in the following, the transport coefficients are allowed to include (besides collisional interactions) phenomenological contributions from turbulent wave-particle interactions occurring in the solar wind (i.e., modified transport coefficients).

III. FLUID MODELS AND THEIR SOLUTION

A. Multi-fluid formulation

For simplicity, we consider a multi-fluid model for a steady spherically expanding flow with modified transport coefficients, ignoring effects such as magnetic fields, solar rotation, pressure anisotropy and fluctuations. Thus, following Braginskii (1965) the macroscopic equations are

Continuity: $n_i v_i r^2 = J_i \quad (J_i\text{'s} = \text{constant values});$ (7)-(9)

Momentum: $n_i m_i v_i (dv_i/dr) + (d/dr)(n_i kT_i) + GM_\odot m_i n_i/r^2 = Z_i n_i E + R_{i,mod}$; (10)-(12)

Energy: $(d/dr)(2.5kT_i + 0.5m_i v_i^2 + q_{i,mod}/n_i v_i) = Z_i E - GM_\odot m_i/r^2$

$R_i/n_i + Q_{i,mod}/n_i v_i$ (13)-(15)

[i = electron (e), proton (p), and alpha-particle (α), etc.]. Here r is the radial distance from the Sun; m_i, Z_i, n_i, v_i and T_i are the particle mass, charge, density, radial streaming velocity, and temperature of i-type particles, respectively; k is Boltzmann's constant; G is the gravitational constant and $M_\odot$ is the solar mass; E, the radial electrostatic field which develops as a result of the gravity-induced charge polarization of the solar wind plasma; $R_{i,\,mod} \equiv \Sigma_{j \neq i} R_{ij\,mod}$ with R_{ij} being the dynamical friction (it represents the mean rate of change in the momentum of i-type particles due to collisions with j-type particles); $q_{i,\,mod}$, the heat flux density carried out by type i-particles; $Q_{i,\,mod} \equiv \Sigma_{j \neq i} Q_{ij,mod}$ with Q_{ij} being the heat added to i-type particles due to binary collisions with type j particles. For a three-fluid system (e, p, α), upon elimination of the radial electric field E in equations (10)-(12) and neglect of the electron mass m_e (since it is small compared with the proton and α-particle masses) one easily obtains the following momentum equations for protons and α-particles, respectively:

$$\begin{aligned} & n_p m_p v_p (dv_p/dr) + (n_p/n_e)(d/dr)(n_e kT_e) + (d/dr)(n_p kT_p) + \\ & + GM\, m_p n_p/r^2 = (n_p/n_e)\, R_{e,mod} + R_{p,mod} \end{aligned} \tag{16}$$

and

$$\begin{aligned} & n_\alpha m_\alpha v_\alpha (dv_\alpha/dr) + 2(n_\alpha/n_e)(d/dr)(n_e kT_e) + (d/dr)(n_\alpha kT_\alpha) + \\ & + GM_\Theta m_\alpha n_\alpha/r^2 = 2(n_\alpha/n_e)\, R_{e,mod} + R_{\alpha,mod}\,. \end{aligned} \tag{17}$$

Also, combining equations (13)-(15) with (10)-(12), one obtains the following equivalent energy equations for the three species:

$$\begin{aligned} & 1.5 n_i v_i (d/dr)(kT_i) - v_i kT_i (dn_i/dr) + (1/r^2)(d/dr)(q_{i,mod} r^2) \\ & = Q_{i,mod} \qquad (i = e, p, \text{ and } \alpha). \end{aligned} \tag{18)-(20}$$

The conditions of quasi-neutrality and of absence of electric currents in the solar wind plasma give, respectively,

$$\sum Z_i n_i = 0 \qquad \text{and} \qquad \sum Z_i v_i n_i = 0\,. \tag{21),(22}$$

Now equations (7)-(9), (16), (17), (18) - (20), (21), and (22) are sufficient for the determination of the hydrodynamic quantities n_i, v_i, T_i (i = e, p, and α). However, they do not form a closed set: the quantities R_i and Q_i defined above are integral expressions whose integrands contain the unknown collisional term C_i in the Boltzmann equation. Moreover, since there are more moments than equations, one has to specify (i.e., approximate) the higher moments (as q_i) in terms of the lowest ones. This is discussed in the next sub-section.

B. Conduction, friction and heat transfer

Assuming that binary collisions between the plasma particles are dominant, one may obtain the following expressions for q_i, R_i, and Q_i in terms of the hydrodynamic variables n_i, v_i and T_i (e.g., Spitzer, 1962; Braginskii, 1965; Banks, 1966):

Thermal conduction:

$$q_i = - K_i \partial T_i/\partial r; \quad K_i = \delta_i n_i k^2 T_i/m_i \nu_i \tag{23}$$

where ν_i is the collisional frequency and the δ_i's are constant quantities tabulated by Braginskii (1965). Alternatively, expressions (23) may be written as

$$K_i = K_{i0} T_i^{5/2} \text{ ergs cm}^{-1}\text{ S}^{-1} (^oK)^{-1}. \tag{24}$$

In (24), the K_{i0}'s are almost constant quantities (Spitzer, 1962). However, since the solar wind is essentially collisionless and, in fact, is dominated by noncollisional wave-particle interactions, equations (23) and (24) are not completely adequate. In the absence of a correct, quantitative analytical theory for the thermal conductivity under the conditions prevailing in the solar wind, we use phenomenologically "modified" thermal conductivities, namely,

$$K_{i,mod} = K_{i0,mod}(r) T^{5/2} \text{ ergs cm}^{-1}\text{ S}^{-1} (^oK)^{-1}, \tag{25}$$

where $K_{i0,mod}$ may be substantially different from the ordinary collisional (constant) quantities K_{i0} which are determined mainly by noncollisional interactions. Their possible functional dependence is to be found by the requirement that the theoretical solutions for n_i, v_i, T_i and q_i match the observed quantities. In fact, the modified thermal conductivities (25) simply represent new functions of r, reflecting the existence of mixed collisional and noncollisional interactions in the solar wind. This explains the subscript "mod" appearing in eqs. (16), (17) and (18)-(20) and referring to the quantities q_i, R_i and Q_i.

Dynamical friction:

Following Braginskii (1965) one has

$$R_{ij} = - n_i m_{ij} \nu_{ij} (v_i - v_j), \tag{26}$$

where ν_{ij} is the collisional frequency for momentum transfer and $m_{ij} \equiv m_{ij}/(m_i + m_j)$ is the reduced mass. For two gases with Maxwellian distributions characterized by the temperature T_i and T_j one has (Banks, 1966)

$$R_{ij}=-n_i n_j m_{ij}\nu_{ij}{}^{o}T_i{}^{-3/2}[1+(m_i/m_j)(T_j/T_i)]^{-3/2}(v_i-v_j),$$

$$\nu_{ij}{}^{o}=4(2\pi)^{1/2}Z_i{}^2Z_j{}^2 \ln\Lambda/\ 3m_{ij}{}^2k^{3/2}m_i{}^{-3/2} \tag{27}$$

Analogous to the case of the thermal conductivities q_i discussed above, and for the same reasons, in this paper we consider "modified" friction forces, $R_{ij,mod}$, in which ν_{ij}^{o} may be different from the pure collisional expression (27) and also space dependent: $\nu_{ij}^{o} \rightarrow \nu_{ij\ ,mod}^{o}(r)$.

Heat transfer:

One has:

$$Q_{ij} = -\ 3n_i m_{ij}(m_i+m_j)^{-1}\nu_{ij}k(T_i - T_j) \tag{28}$$

Alternatively,

$$Q_{ij}=-3n_i n_j m_{ij}(m_i+m_j)^{-1}\nu_{ij}{}^{o}T_i{}^{-3/2}[1+(m_i/m_j)(T_j/T_i)]^{-3/2}k(T_i-T_j) \tag{29}$$

with ν_{ij}^{o} defined by (27). Again, we use here "modified" functions $Q_{ij,mod}$, in which ν_{ij}^{o} is replaced by $\nu_{ij\ ,mod}^{o}(r)$. We now define the following space-dependent functions to characterize the ratios of the effective (mixed collisional and noncollisional) to pure collisional interactions in the quiet solar wind:

$$S_i(r) \equiv K_{i,mod}/K_i, \tag{30}$$

$$B_i(r) \equiv R_{ij,mod}/R_{ij}, \tag{31}$$

and

$$C_i(r) \equiv Q_{ij,mod}/Q_{ij} \tag{32}$$

(i,j = e, p, and α). Here the K_i's, R_{ij}'s, and Q_{ij}'s are given by eqs. (24), (27) and (29), respectively.

C. Boundary conditions

The following twelve boundary conditions are used:

I-III. The requirements that the particle fluxes defined by eqs. (7)-(9), namely, $4\pi J_i = 4\pi n_i v_i r^2 = \bar{J}_i$ (constant), to be equal to those observed at 1 AU.

IV-VI. The requirements that the streaming velocities of each ion component pass continuously through a subsonic-supersonic transition; i.e., $v_i(r)$ should pass through the critical point $r_{i,c}$ in the (v_i, r)-plane at which dv_i/dr becomes indeterminate.

VII-IX. The requirements that at some low heliocentric distance r_o the temperatures T_i fit prescribed values.

X-XII. The requirement that the temperature T_i tends to zero at infinity Obviously, assumptions (21) and (22) reduce by two the number of boundary conditons enumerated above.

D. Integration of the fluid equations

The solution of the "modified" three-fluid model equations for the quiet solar wind requires the simultaneous step-by-step integration of the system of equations (7)-(9), (16), (17), (18)-(20), (25), (27) and (29). In the following we first discuss the one-fluid model, next the two-fluid model and finally the three-fluid model. Thus, upon neglecting the α-particle contribution in the three-fluid equations, one obtains a two-fluid system of equations with modified transport coefficients (q_e, q_p and $Q_{ep} = - Q_{pe}$, e.g., eqs. (1)-(4) of Cuperman and Harten (1971a), because $v_e = v_p$ here $R_{ep} = - R_{pe} = 0$. Furthermore, from the two fluid equations one obtains the one fluid model equations by taking $T_e = T_p = T$ and $q = q_e/2$ in the momentum and energy equations. Although the mathematical (numerical) methods used to solve these fluid equations are interesting, because of space limitation, they are not discussed here.

E. One-fluid model

At the present time, there is unaminous agreement on the existence of an anomalously *low* electron thermal conductivity in the solar wind plasma at the Earth's orbit. The anomaly, is explained by the existence of noncollisional wave-particle interactions in the collisionless solar wind, caused by collective electrostatic and electromagnetic instabilities (e.g., Parker, 1964, 1969; Hundhausen, 1969; Cuperman and Harten, 1970a; Forslund, 1970; Eviatar and Schulz, 1970). While a consistent analytical derivation of this transport coefficient obtained from the microscopic plasma structure is as yet an unsolved problem, several works have offered phenomenological substitutes which should lead to satisfactory solutions of the fluid equations that describe the solar-wind propagation. Thus, solving the one-fluid equations with an anomalous (low) coefficient of thermal conductivity, which includes effects coming from both the inhibition of heat transport across the spiraling interplanetary magnetic field and electromagnetic turbulence (due to non-collisional processes,) one obtains the following essential results (see Figure 1).

i) The temperature (T) and heat flux ($H = -\, kdT/dr$) through 1 AU depend mainly on the relative importance of the transport of energy by conduction at *low* heliocentric distances (below 30-40 $R_\odot$), and at larger heliocentric distances (beyond 30-40 $R_\odot$).

ii) If a significant reduction in the transport of energy by conduction occurs only at distances larger than about 30-40 $R_\odot$, then the solutions of the equations predict T and H values at 1 AU which are *larger* than those corresponding to a pure collisional uninhibited thermal conductivity. This is the case, for instance, when only inhibition of heat transport across the spiraling interplanetary magnetic field [a $\cos^2\psi(r)$ reduction factor] is considered.

iii) A sharp reduction of the thermal conduction mainly at low heliocentric distances, presumably due to wave-particle interactions and inhibition across the partially closed, looplike magnetic configurations at the base of the corona, provides satisfactory solutions for T and H (and n,v) through 1 AU. Use of the reduction factor, $\cos^2\psi(r)$, to account for the inhibition across the spiraling interplanetary magnetic field predicts satisfactory T and H solutions *only if exactly the same sharp reduction factor exists at low heliocentric distances as before*. This indicates the essential result, namely, the importance of the reduction of the thermal conductivity as *mainly low heliocentric distances*; this kind of reduction is necessary to reproduce the actual state of the quiet solar wind.

Next, using the anomalous thermal conductivity given by Model II in Figure 1a., the one-fluid equations were integrated in the range 1 AU $\leq r \leq$ 12 AU. As a result, the predicted gross features of the solar wind in the above mentioned range and in particular at distances representing the semimajor axes of the orbits of planets were obtained. The radial dependence of temperature approaches the adiabatic form, $r^{-4/3}$, as a consequence of the zero temperature and thermal conductivity requirement at $r = \infty$ (see Figure 2). Finally, it should be mentioned, these far distant solutions, as well as those inside 1 AU have been confirmed by using a different integration method, namely starting at infinity and progressing toward the sun, rather than starting at the critical point and integrating outwards and then inwards (Cuperman et al., 1975).

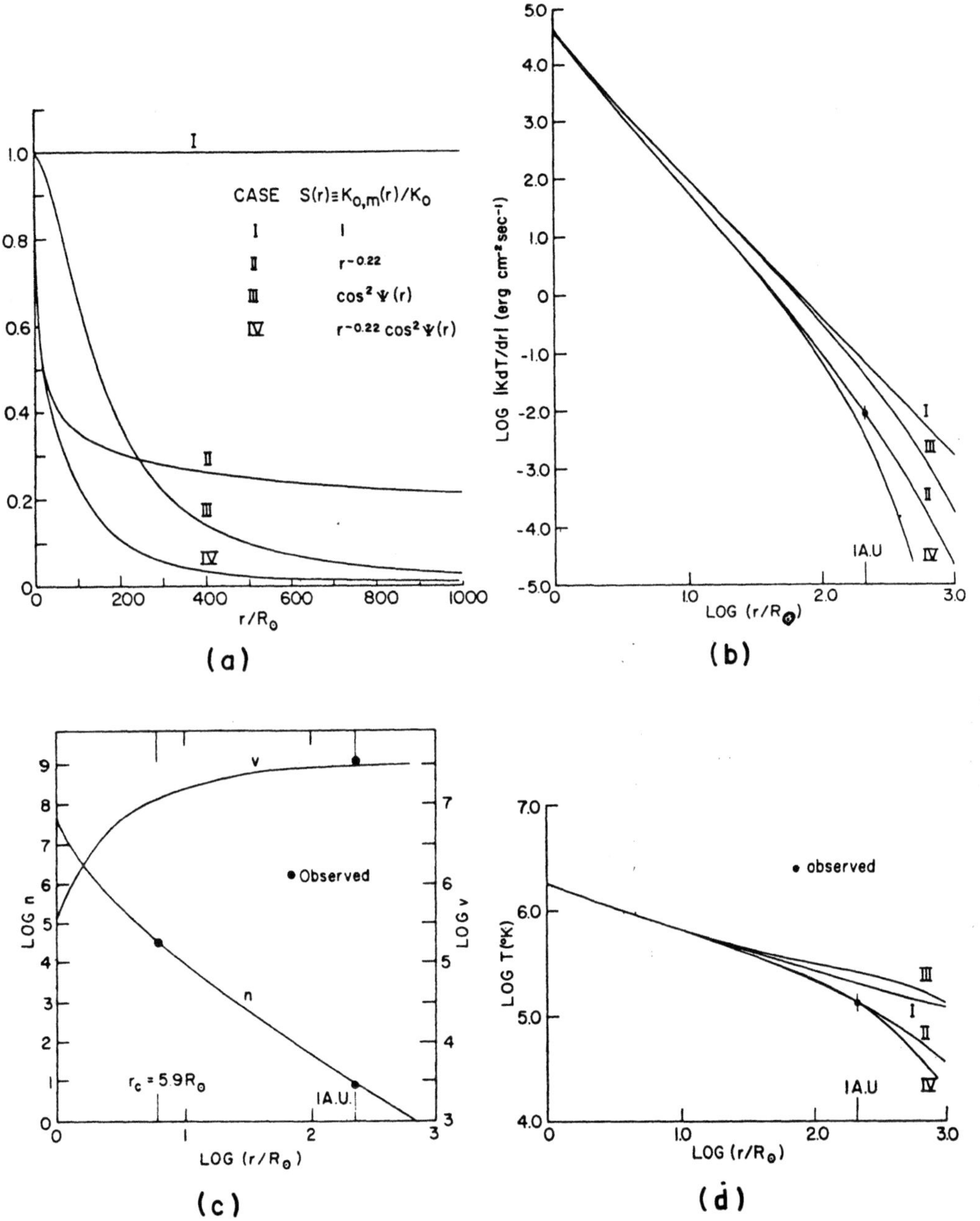

Figure 1. Solutions obtained by solving the "modified" one-fluid equations for the quiet solar wind: (a) "modified" profiles of thermal conductivity (eq. 30); (b) heat flux profiles; (c) particle density and streaming velocity profiles; (d) temperature profiles. (From Cuperman and Metzler, 1973b.)

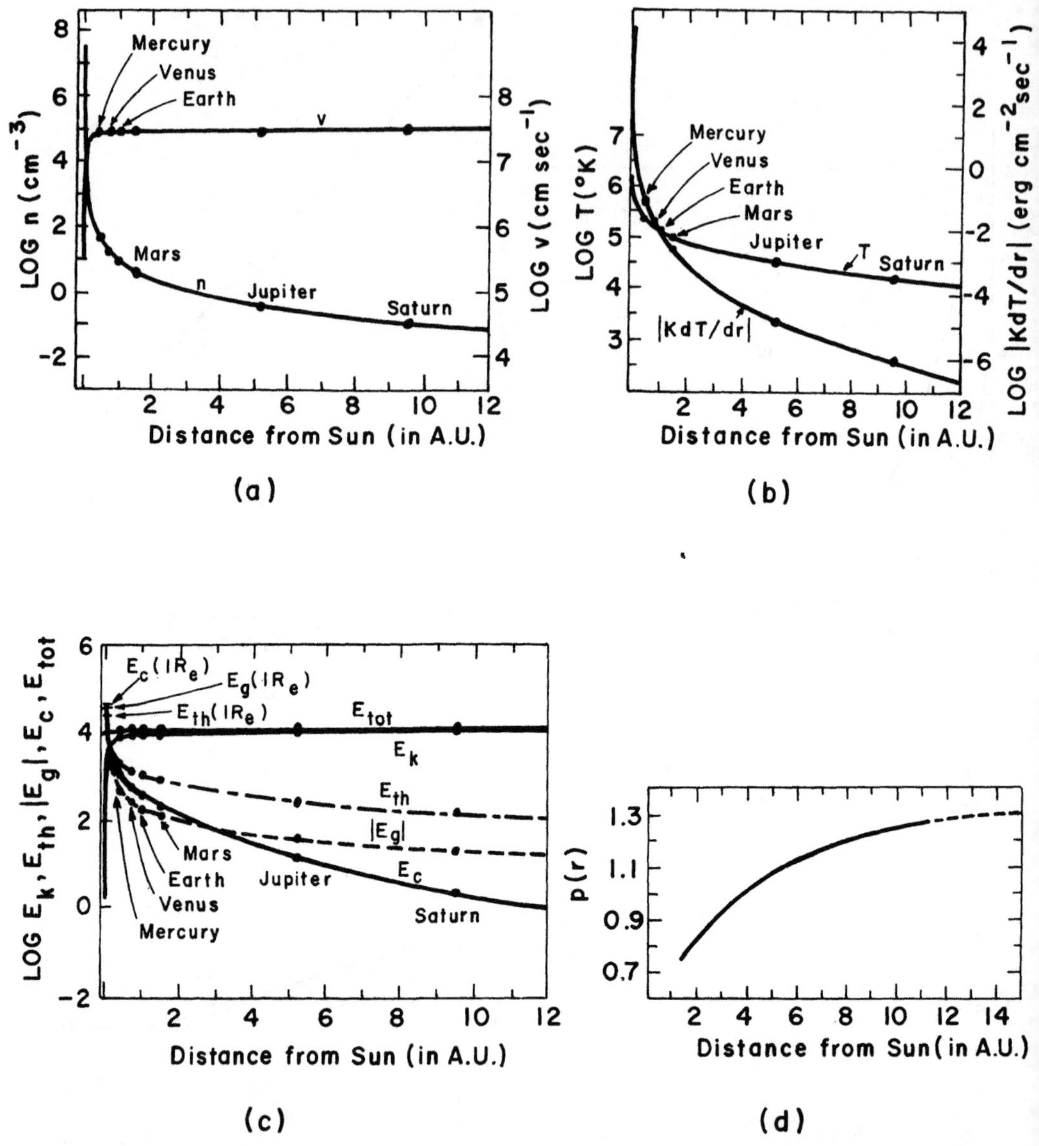

Figure 2. Characteristics of the quiet solar wind beyond the Earth's orbit: (a) particle density and streaming velocity profiles; (b) temperature and heat flux profiles; (c) profiles of the transport of various energies (E_k, kinetic; E_{th}, thermal; E_g, gravitational; E_c, conduction; E_{tot}, total. All energy fluxes are normalized to $4\pi R_\odot^2$, where $r = RR_\odot$); (d) space dependence of the coefficient p, describing the temperature profile $T \propto r^{-p}$. (From Cuperman et al., 1972).

F. Two-fluid model

The one-fluid models do not differentiate between the electron temperature T_e and the proton temperature, T_p while the observations indicate $T_{eE}/T_{pE} \simeq 4$. On the other hand, solution of the two-fluid equations with pure collisional transport coefficients yields electron temperature and particle density larger than observed and proton temperature much lower than measured. This is why the two-fluid equations with anomalous transport coefficients have to be considered. When this is done, it is found that it is possible to obtain solutions for electrons and protons matching the observations at 1 AU in one of the following two ways: (i) by using an "enhanced" energy exchange rate between electrons and protons $\nu_{e,p,mod}$ (cf.eq. 32 in this paper) about 30 times larger than the collisional one (which raises the proton temperature to the observed value but leaves almost unchanged the electron temperature and heat flow) in conjunction with an electron thermal conductivity $K_{e,mod}$, four times lower than the collisional one (which lowers both electron temperature and heat flow to the observed values without further changing the agreement of the proton temperature with the observations); (ii) by using an "enhanced" proton thermal conductivity $K_{p,mod}$, about 2.3 times larger than the collisional one (which like the enhanced energy exchange rate above, raises the proton temperature at 1 AU to the observed value without affecting the electron temperature and heat flow) in conjunction with an electron thermal conductivity about 5.5 times lower than the collisional one (which again, as above, lowers both the electron temperature and heat flow to the observed values). These results are illustrated in Figure 3, and are consistent with recent qualitative models (see Perkins, 1973).

G. Three-fluid model

In subsections IIIe and IIIf it was assumed that the solar wind plasma consists solely of electrons and protons. A general model should also include at least alpha particles, whose relative concentration is non-negligible. Actually, satellite observations indicated three rather unexpected gross features of the α-particles in the quiet solar wind at about 1 AU. (Ogilvie and Wilkerson, 1969; Robbins et al., 1970; Formisano and Moreno, 1971; Bollea et al., 1972; Hirshberg et al., 1974)
i) The ratio of the α-particle to proton streaming velocities is peaked around unity. Actually, because of their smaller charge-to-mass ratio, α-particles are expected to be accelerated less than protons by the interplanetary radial electrostatic field, and because of their greater mass α-particles are also expected to be accelerated less by the pressure gradient.
ii) The long-term average of the α/proton density ratio is 0.037. This value is definitely lower than the lowest value obtained for the solar surface-namely, 0.063 (Lambert, 1967) - and of course lower than the

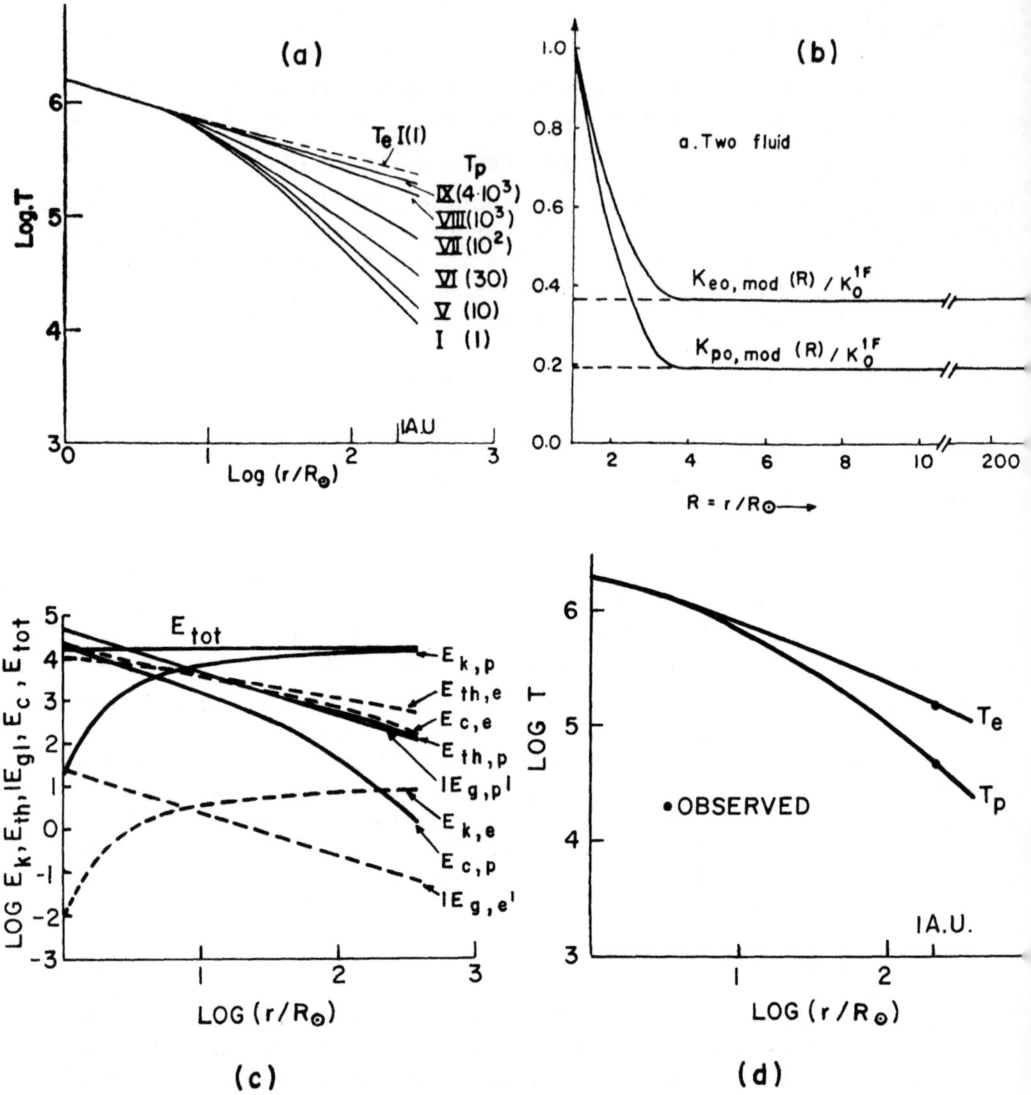

Figure 3. Solutions of "modified" two-fluid equations for the quiet solar wind: (a) effect of relative coupling strengths $\nu_E/\nu_{E,col}$ (indicated in parentheses) between electrons and protons (from Cuperman and Harten, 1970b); (b) profiles of relative coefficients of thermal conductivity, $K_{io,mod}(r)/K_o^{1F}$ used (K_o^{1F} - pure collisional conductivity for one-fluid); (c) profiles of transport of various energies obtained (E_k kinetic; E_{th}, thermal; E_g, gravitational; E_c, conduction; E_{tot}, total. All energy fluxes are normalized to $4\pi R_\Theta^2$, e, electron; p, proton); (d) temperature profiles obtained. (From Cuperman and Metzler, 1975.)

value 0.1 usually accepted for the α/proton density ratio in the lower corona (e.g., Biswas and Fichtel, 1964).
iii) The distribution of the α/proton temperature ratios indicates a mean of approximately 4.0. There is no particular reason for a priori expecting such a situation to exist.

To investigate these aspects, the full set of three-fluid equations as given by eqs. (7)-(9), (16), (17), (18)-(20), (25), (27) and (29) with inclusion of phenomenological contributions from wave-particle interactions to the transport coefficients has been solved. The mathematical solution was achieved by using a perturbation method procedure as follows:
A. Since the α-particle component represents only a relatively small fraction of the total ion content in the quiet solar wind, QSW, one may consider that in a first-order approximation the basic gross features of the electron and protons in the QSW are unaffected by the presence of the α-particles and are satisfactorily described by a two-fluid system of equations. The largest α/p particle density concentration in the QSW, n_α/n_p, is about 10 percent (in the lower corona). Thus one first solves a "two-fluid" problem for the electrons and protons in the way done in the past (e.g., Cuperman and Harten, 1970b). For this one uses equations with the α-terms omitted. The two-fluid equations solved allow for modified (noncollisional) transport coefficients such as increased rate of energy transfer between electrons and protons, anomalously low electron thermal conductivity and enhanced proton thermal conductivity. As a result one obtains the solutions $T_e(r)$, $T_p(r)$, $n_p(r)$, $v_p(r)$ and heat fluxes $-K_e dT_e/dr$ and $-K_p dT_p/dr$ which match satisfactorily the measurements at 1 AU.
B. Next, the two-fluid solutions are used in the equations for the α-particles (9), (17) and (20) which are solved by using a procedure close to that used in the past for the one-fluid problem (e.g., Cuperman and Harten, 1970a). Thus, the solution of the complex three-fluid problem is reduced to that of solving a "modified" one-fluid problem (for the α-particle component) which differs from the ordinary one by the presence in the model equations of additional terms (whose functional dependence is prescribed, as discussed above). This integration procedure represents a generalization of that used in a similar problem by Geiss, Hirt, and Entwyler (1970) who applied it for solving continuity and pure collisional momentum equations only, by assuming that all temperatures are (i) equal and (ii) given by a polytropic law.

An iterative solution procedure providing more accurate solutions is now possible, e.g., using the (first-order) solutions obtained for the α-particles in the two-fluid equations for e and p and solving them to obtain improved two-fluid solutions for e and p, etc.

Table 1 and Figure 4 give the results obtained by the integration of the model equations. For reasons of mathematical (integration)

TABLE 1

SOLUTIONS FOR α-PARTICLES IN THE QUIET SOLAR WIND: PURE COLLISIONAL CASE

Case	$(T_\alpha/T_p)_c$		$r/R_\odot$		n_α/n_p		v_α/v_p		T_α/T_p	
Ia	1.8		215		0.035		1.000		2.55	
IIa	1.9		215		0.035		1.000		3.70	
IIIa	2.0		215		0.035		1.000		5.07	
IIIb		(2.0)*		215		0.155		0.213		0.87
IVa	3.0		215		0.035		1.000		18.10	
IVb		(3.0)		215		0.115		0.305		11.30
Va	6.0		215		0.035		1.000		46.70	
Vb		(6.0)		215		0.041		0.857		48.20
VIa	6.3		215		0.035		1.000		49.60	
VIb		(6.3)		215		0.037		0.946		49.50
Observed	...		215		0.037 ± 0.010		1.000		4.00 ± 2.5	
Ia	1.8		5.50‡		0.089		0.391		1.80	
IIa	1.9		5.52‡		0.088		0.398		1.90	
IIIa	2.0		5.53‡		0.086		0.405		2.00	
IIIb		(2.0)		15.70‡		0.155		0.213		2.00
IVa	3.0		5.45‡		0.073		0.482		3.00	
IVb		(3.0)		10.70‡		0.115		0.305		3.00
Va	6.0		3.99‡		0.039		0.887		6.00	
Vb		(6.0)		4.10‡		0.041		0.852		6.00
VIa	6.3		3.90‡		0.037		0.950		6.30	
VIb		(6.3)		3.92‡		0.037		0.940		6.30
Ia	1.8		4.32§		0.108		0.324		1.86	
IIa	1.9		5.25§		0.102		0.321		1.91	
IIIa	2.0		5.25§		0.100		0.326		2.00	
IIIb		(2.0)		1.00		0.155		0.213		5.63
IVa	3.0		4.65§		0.111		0.295		2.92	
IVb		(3.0)		1.00		0.115		0.305		5.15
Va	6.0		1.40§		0.101		0.345		5.65	
Vb		(6.0)		1.00		0.041		0.852		5.92
VIa	6.3		1.00		0.117		0.299		4.53	
VIb		(6.3)		1.00	0.037		0.940		6.16	
Observed	...		1.00		~0.100		...		...	

* The solutions obtained by neglecting the friction term in the model equations are indicated in parentheses and given in the right-hand side of each column.

† Robbins *et al.* 1970.

‡ Indicates the r-position of the critical point for α-particles.

§ Indicates the distance below which $n_\alpha/n_p \gtrsim 0.1$.

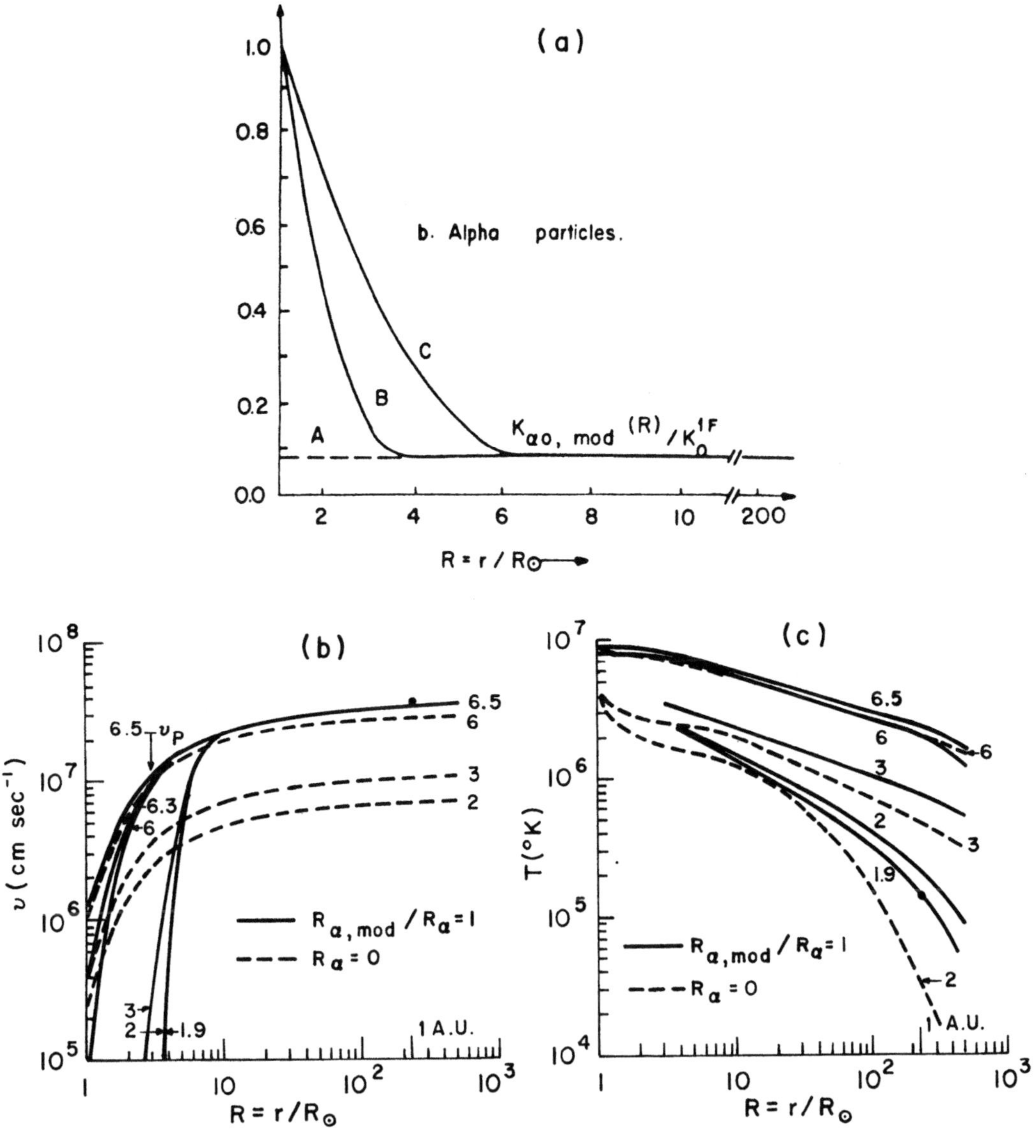

Figure 4. Solutions of three-fluid equations: (a) profiles of relative coefficients of thermal conductivity, $K_{\alpha o,mod}(r)/K_o^{1F}$; (b), (c) profiles of α-particle streaming velocity and temperature respectively (solid line, $R_\alpha \neq 0$; dashed line, $R_\alpha = 0$, i.e., frictionless case. The number attached to each curve indicates the relative value $(T_\alpha/T_p)_c$, used as a convenient integration parameter. In all cases, model C for the thermal conductivity is used (Figure 4a).

convenience, the parameter that was varied in order to obtain the different solution cases shown in Table 1 was $(T_\alpha/T_p)_c$. In fact, this is equivalent to choosing the value of T_α at the base of the corona as a variable parameter. For comparison, the solutions obtained by neglecting the friction term in the model equations are also indicated in parentheses.

Pure collisional transport coefficients. Inspection of Table 1 shows that:

1. The model equations used are able to predict-in all cases with $R_\alpha \neq 0$ (labeled a) in Table 1 - the increase of the relative α-particle streaming velocity v_α/v_p with heliocentric distance to the value 1 and 1 AU. Since the predicted v_p value at 1 AU is 318 kms^{-1}, it follows that also the absolute value of $v_\alpha (=v_p)$ at 1 AU is in good agreement with the observations.
2. The equations also predict the decrease of the relative α-particle abundance with heliocentric distance to the value 0.035 at 1 AU.
3. The predicted values of the ratios T_α/T_p at 1 AU are larger than unity, as observed. In particular, case IIa predicts a value for T_α/T_p at 1 AU which matches very well the measured value.
4. The dynamical friction force, R_α, plays an important role, as illustrated by the comparison of the results for the cases with $R_\alpha \neq 0$ with those for the cases $R_\alpha=0$ (labeled b) in Table 1. Indeed, if $R_\alpha=0$, one finds that the streaming velocities at 1 AU are much smaller than the proton streaming velocity, which contradicts the observations.

Effect of modified transport coefficients. Table 1 indicates that the predicted values for T_α at low heliocentric distances $(r < r_c)$ are rather large; this applies also to case IIa, which provides satisfactory solutions for all measured gross features of the α-particles at 1 AU.

Moreover, unless $(T_\alpha/T_p)_c \gtrsim 6$ (which provides values of T_α that are too large in the interplanetary medium), all the cases in which $R_\alpha \neq 0$ predict streaming velocities which become nonzero only at a distance of several solar radii away from the Sun's surface. This is in contrast to the $R_\alpha=0$ cases which predict nonzero streaming velocities, starting at the very base of the corona.

Modified thermal conductivity. When solving the two-fluid problem for electrons and protons, we learned that in order to obtain equal temperatures T_e and T_p at 1 $R_\odot$ together with energy conservation over all the integration range, we had to assume the two components to be strongly coupled at the very base of the corona (1 R_Θ) (behaving more or less as a "one" fluid plasma) and to "relax" this constraint in the manner indicated in Figure 3b. Analogously we now replace $K_{\alpha o}$ by $K_{\alpha o,mod}(r)$, solve the problem for two different models (B,C), and compare the results with those of the pure collisional model A already discussed in Table 1 (Figure 4b): model B: $K_{\alpha o,mod}(1\ R_\Theta) = K_{eo,mod}(1\ R_\Theta) = K_o^{1F}$, decreases parabolically between 1 R_Θ and $r = r_c^{2F}$ (about 4 R_Θ) where it reaches the value $S_\alpha \equiv K_{\alpha o,mod}/K_{\alpha o} = 1$ and remains constant thereafter; model C: the same as B, except for the fact that the parabolical decrease stops at $r_c^{\alpha}(r_c^{\alpha} \gtrsim r_c^{2F})$.

The results at low heliocentric distances show that the high coronal temperatures predicted by the pure collisional thermal conductivity (model A) are significantly reduced, the other gross features of the solar wind being almost unchanged. Model C best fits the observations.

Modified energy exchange rate. A modest enhancement in the α/p-energy exchange rate may strongly affect the α-temperature at about 1 AU.

Modified friction coefficients. From the results presented so far it follows that the behavior of the α-particle streaming velocity (and consequently particle abundance) at low heliocentric distances is almost unaffected by the consideration of noncollisional contributions to the thermal conductivity and/or energy exchange rate in the model equations. We now consider possible effects due to anomalous dynamical friction forces.

Inspection of the results in Table I shows that neglecting frictional forces in the model equations provides non-zero flow velocity solutions from the very base of the corona but predicts too low streaming velocity and temperature values at 1 AU (e.g., case IIIb). On the other hand, consideration of pure collisional friction coefficients in the model equations provides solutions which match the observations at 1 AU but which at low heliocentric distances predicts nonzero streaming velocities only beyond a distance of few solar radii away from the Sun's surface (e.g., cases IIA and IIIa). These results suggest the strength of the actual dynamical friction force in the quiet solar wind to be intermediate between that corresponding to the frictionless cases, $B_\alpha \equiv R_{\alpha,mod}/R_\alpha = 0$ (label b) and that corresponding to the pure collisional cases, $B_\alpha = 1$ (label a). Since at large heliocentric distances one has (in all $B_\alpha = 1$ cases) $R_\alpha \propto v_p\ (1-v_\alpha/v_p) \simeq 0$, the decrease in B_α below the collisional value of 1 is specifically required at low heliocentric distances where $v_\alpha << v_p$. Essentially, the decrease should be such as to limit the relative importance of the frictional forces in the momentum equation for the α-particles. It should be noted that this situation naturally occurs in the case of large temperatures (e.g., case VIa in Table 1) where, because $v_\alpha/v_p \simeq 1$ starting almost at the base of the corona, one has $R_\alpha \simeq 0$ almost everywhere. This conjecture has been checked by using a profile for the relative friction coefficient, B representing an anomalously low value at low heliocentric distances (especially close to the critical distance) followed by a slow increase to the asymptotic value 1. This indeed provided streaming velocities with non-zero values from the very base of the corona. (Cuperman and Metzler, 1975). Most recently, the possible occurrence of relative streaming velocities $v_\alpha/v_p > 1$ observed at about 1 AU (Ogilvie, 1975; Hirshberg et al., 1974) has been theoretically discussed (Cuperman and Metzler, 1976; Hollweg, 1974).

IV. FURTHER DEVELOPMENTS

A. Kinetic description

It has been shown in Section II that the fluid formulation represents an alternate and equally correct method for describing the solar wind plasma. This formulation's approximate character is introduced by the difficulty met in specifying both the higher moments of the distribution function as well as the transport coefficients. This coefficients' structure in collisionless plasmas should not be necessarily the same as predicted for collision dominated plasma. Thus, qualitatively, one expects the gross features of the solar wind to be adequately described by fluid models. Kinetic (microscopic) theories which consider collective interactions in microscopically unstable plasmas have the role of providing the actual transport coefficients and consequently can modify the fluid solutions in the sense indicated in Section III (in which the required modifications have been phenomenologically considered). A review of waves and instabilities relevant to this subject has been given by Hollweg (1975).

B. Spatial and temporal variations

For methodological purposes one may distinguish between the following large-scale structures: i) those arising from spatially non-uniform conditions that vary little in the time it takes a fluid element to be convected from the Sun to the observer (e.g., corotating high-speed solar wind streams (Carovillano and Siscoe, 1969, etc.), interplanetary magnetic sectors, (Wilcox and Ness, 1968) etc.; ii) transient structures which arise from boundary conditions which vary significantly in a time short in comparison with the sun-observer transient time (eruptive prominences, surges, flares) and iii) combined spatial and temporal varying structures (solar flare-generated shock waves, etc). Reviews on these topics which are not considered here have been written by Hundhausen, 1972 and Gosling, 1975; 1976 (i) and Dryer, 1974; 1975 (ii, iii). In addition, at least some most recent and important work by Dryer et al., 1976; and Wu et al., 1978 should be also mentioned.

C. High-speed solar wind flows

Definition of the background structure-free state of the solar wind. The first extended surveys of observational data (Snyder et al. 1963; Neugebauer and Snyder, 1966) noted the fact that the solar wind flow speed was basically variable and this finding has been confirmed by all subsequent satellite observations. Nevertheless, because of the complexity of the problem, the theoretical approach to the solar wind phenomenon was to extract from the extensive body of solar wind observations that subset which is relevant to the structureless steady state flow. Based on a number of plausible arguments this state was related to conditons of low bulk speed and low ion temperature at 1 AU and was called the quiet solar wind (see Chapter 3 in Hundhausen, 1972). An alternative interpretation of the "quiet" solar wind is that idealized state corresponding to the "average" gross features of the solar wind. This approach resulted in a substantial understanding of the solar wind phenomenon, as reviewed in Sections II and III of this paper.

Recently, an overview of 13 years of solar wind speed data (Bame, et al., 1976 and Gosling et al., 1976) suggested that in fact, not only the high-speed flows play an important role but that it is reasonable that the structureless ambient state of the solar wind may be found at high (~ 750 km^{-1}) flow speeds rather than at low speeds. Similar conclusions have been reached by Feldman et al. (1976) who analyzed the observations at 1 AU from the Los Alamos solar wind instruments on the IMP 6, 7 and 8 satellites for the period March 1971 to July 1974.

Most recently Rosenbauer et al. (1976) presented preliminary results of the Helios Plasma Measurements between 0.3 and 1 AU for the period December 1974 to April 1975. These observations are of major importance for the understanding of the space structure (gradients) of the solar wind. Analysis of these data (Cuperman et al., 1977) show that it is only for the highest-speed flows that the particle flux is conserved, as illustrated in Figure 5. Only for these conditions does the assumption for spherical symmetry, implicit in equations 7-9, hold. This supports the suggestions indicated above on the identification of the background solar wind with high-speed solar wind flows.

Association with high latitude coronal holes. Coronal holes are regions of greatly reduced X-ray, EUV and metric radio emission (e.g., Munro and Withbroe, 1972; Neupert and Pizzo, 1974; Dulk and Sheridan, 1974; Timothy et al., 1975). These structures are regions of low density and/or temperature, and apparently occur in weak, open, diverging, unipolar magnetic field regions (Altschuler et al., 1972; Krieger et al., 1974; Vaiana et al., 1973; Timothy et al., 1975). Moreover, 20% of the 1973 Skylab-period sun (the declining phase of the solar

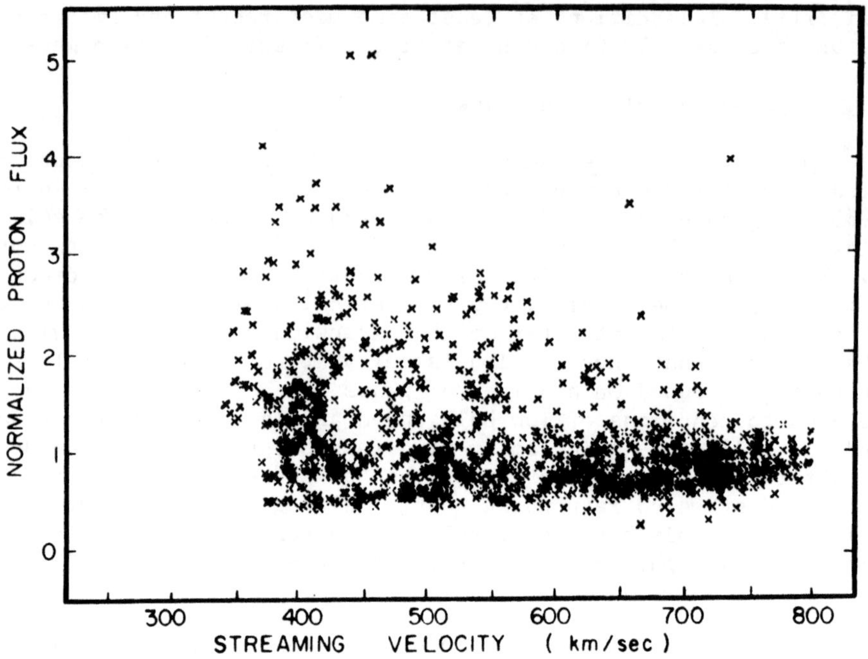

Figure 5. Normalized particle flow, $(nvr^2)/(nvr^2)_E$ vs. streaming velocity for the heliocentric distances $0.3 \leq r \leq 1$ (from Cuperman et al., 1977).

cycle) was covered by coronal holes, of which an essentially constant 15% was contained in the polar holes. All the other lower latitude holes (<60°) provided from 2 to 5% of the total hole areas (Bohlin, 1976).

On the other hand coronal holes have been found to be associated with recurrent high-speed solar wind streams (Krieger et al., 1973; Neupert and Pizzo, 1974; Krieger et al., 1974; Nolte et al., 1976). These investigations indicate the coronal hole sources to be located at high latitude coronal holes. This finding was recently confirmed by a long term correlation investigation between the latitude dependent 5303Å coronal line intensity and the solar wind plasma streams measured at 1 AU heliocentric distance, during the period 1962-1970 (Cuperman and Dryer, 1977; see also Cuperman and Sternlieb, 1972). A significant anticorrelation is found at high latitudes. (See Figure 6.)

The last two discoveries, namely the identification of the most probable state with one of high-flow speeds and the association of high-flow speeds with high latitude coronal holes poses new problems to the solar wind theorists whose effort was directed in the past towards the

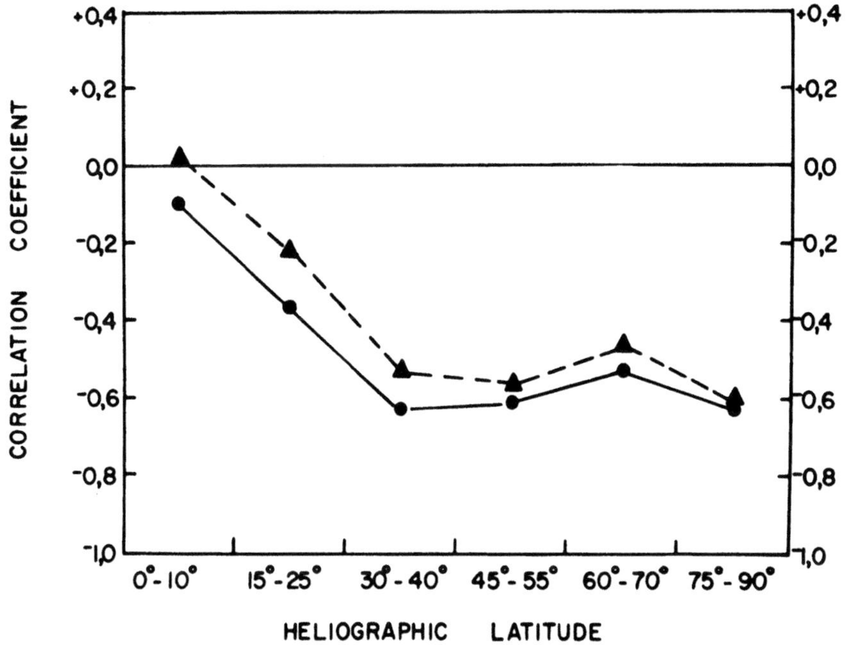

Figure 6. Long term correlation coefficients between the latitude dependent solar activity (green coronal line) and interplanetary plasma streams (solid line-streams with amplitudes greater than 350 km/sec.; dashed line-streams with maximum speeds greater than 700 km/sec). (From Cuperman and Dryer, 1977.)

investigation of steady state, spherically symmetric low speeds at 1 AU models. There are already a number of papers devoted to these problems. For example, Kopp and Holzer (1976) have explored the consequences of rapidly diverging flows in coronal holes, between $r = R_\odot$ and $r = 2R_\odot$. Using a polytrope model the authors found that the expansion speed in the corona is increased well above and the density is decreased well below the values expected from the standard solar wind model with spherical symmetry. This is due to the fact that a rapidly diverging flow geometry tends to decrease the density scale height of the expanding corona and thus to increase the outward-directed pressure gradient force which enhances the acceleration of the flow. An alternative way to obtain high-speed flows is by direct momentum addition (associated with mechanical wave flux (Holzer, 1976, 1977)). On the other hand three dimensional solar wind models whose importance is evident from above are now under investigation (e.g., Suess, 1974).

V. CONCLUDING REMARKS

The main achievements of the theoretical investigations on topics belonging to Part I (defined in Abstract) in the theoretical solar wind research are: i) a basic understanding of the solar wind phenomenon; ii) satisfactory agreement between theory and observed gross features; iii) prediction of the characteristics of the solar wind at heliocentric distances far beyond 1 AU; iv) phenomenological modulation of the transport coefficients of interest - as a basis for further kinetic (microscopic) research; and v) valuable information on the role stellar winds may play in the evolution of stars.

The list of achievements should be completed with those referring to topics belonging to Part II - as discussed in Section IV. Some of these topics represent the object of most current and rewarding investigations. The rather high degree of mathematical sophistication required for their treatment represents a challenge for solar wind theorists.

REFERENCES

Altschuler, M. D., Trotter, D. E. and Orrall, F. Q.: 1972, *Solar Phys.*, *26*, 354.
Bame, S. J., Asbridge, J. R., Feldman, W. C. and Gosling, J. T.: 1976, *Astrophys. J.*, *207*, 977.
Banks, P. M.: 1966, *Planetary and Space Sci.*, *14*, 1108.
Barnes, A., Hartle, R. E. and Bredekamp, J. H.: 1971, *Astrophys. Lett.*, *166*, 153.
Biermann, L.: 1951, Zeit. *Astrophys.*, *29*, 274.
Biswas, S., and Fichtel, C. E.: 1964, *Astrophys. J.*, *139*, 941.
Bohlin, J. D.: 1976, in D. J. Williams (Ed.) *Physics of Solar Planetary Environments*, American Geophysical Union, Washington, D.C., p. 114.
Bollea, D., Formisano, V., Hedgecock, P. C. Moreno, G., and Palmiotto, F.: 1972, in C. P. Sonett, P. J. Coleman, Jr. and J. M. Wilcox (eds.) *Solar Wind*, Asilomar Conf., NASA SP-308, p. 588.
Braginskii, S. I.: 1965, *Reviews of Plasma Physics*, Consultant Bureau, New York.
Carovillano, R. L. and Siscoe, G. R.: 1969, *Solar Phys.*, *8*, 401.
Chapman, S.: 1957, *Smithsonian Contrib. Astrophys.*, *2*, 1.
Cuperman, S. and Dryer, M.: 1977, (submitted to *Astrophys. J.*).
Cuperman, S., and Harten, A.: 1970a, *Cosmic Electrodyn.*, *1*, 205.
Cuperman, S., and Harten, A.: 1970b, *Astrophys. J.*, *162*, 315.
Cuperman, S., and Harten, A.: 1971a, *Astrophys. J.*, *163*, *383*.
Cuperman, S., and Harten, A.: 1971b, *Astrophys. J.*, *169*, 165.
Cuperman, S., and Harten, A.: 1972, in C. P. Sonett, P. J. Coleman, Jr., and J. M. Wilcox (eds.) *Solar Wind*, Asilomar Conf., NASA SP-308, p. 244.
Cuperman, S., Harten, A. and Dryer, M.: 1972, *Astrophys. J.*, *177*, 555.

Cuperman, S., Levush, B., Dryer, M., Rosenbauer, H. and Schwenn, R.: 1977 (in preparation).
Cuperman, S. and Metzler, N.: 1973a, *J. Geophys. Res.*, *78*, 3187.
Cuperman, S. and Metzler, N.: 1973b, *Astrophys. J.*, *182*, 961.
Cuperman, S. and Metzler, N.: 1975, *Astrophys. J.*, *196*, 205.
Cuperman, S. and Metzler, N.: 1976, *Astrophys. and Space Sci.*, *45*, 411.
Cuperman, S., Metzler, N. and Spiegelglass, M.: 1975, *Astrophys. J.* *198*, 755.
Cuperman, S. and Sternlieb, A.: 1972, *Solar Phys.* *25*, 493.
Cuperman, S. and Sternlieb, A.: 1974, *Monthly Notices Roy. Astron. Soc.* *167*, 183.
Dryer, M.: 1974, *Space Sci. Rev.* *15*, 403.
Dryer, M.: 1975, *Space Sci. Rev.* *17*, 277.
Dryer, M., Steinolfson, R. S. and Wu, S. T.: 1976, in M. T. Rycroft (Ed.) *Space Research XVI*, Akademie-Verlag, Berlin, p. 685.
Dulk, G. A. and Sheridan, K. V.: 1974, *Solar Phys.*, *36*, 191.
Durney, B. R.: 1971, *J. Geophys. Res.*, *77*, 4042.
Enskog, D.: 1917, Kinetische Theorie der Vorgänge in Mässig Verdünten Gasen, Thesis, Uppsala.
Eviatar, A. and Schulz, M.: 1970, *Planet. Space Sci.*, *18*, 321.
Fahr, H. J.: 1973, *Solar Phys.*, *30*, 193.
Feldman, W. C., Asbridge, J. R., Bame, S. J. and Gosling, J. T.: 1976, *J. Geophys. Res.*, *81*, 5054.
Formisano, V., and Moreno, G.: 1971, *Nuovo Cimento*, *1*, 365.
Forslund, D. W.: 1970, *J. Geophys. Res.*, *75*, 17.
Geiss, J., Hirt, P., and Entwyler, H.: 1970, *Solar Phys.* *12*, 458.
Gentry, R. A., and Hundhausen, A. J.: 1969, *Trans. AGU* *50*, 302 (abstract).
Gosling, J. T.: 1975, *Rev. Geophys. Space Phys.*, *13*, 1053.
Gosling, J. T.: 1976 in D. J. Williams (ed.) *Physics of Solar-Planetary Environments*, American Geophysical Union, Washington, D. C., p. 286.
Gosling, J. T., Asbridge, J. R., Bame, S. J. and Feldman, W. C.: 1976, *J. Geophys. Res.*, *81*, 5061.
Hartle, R. E. and Barnes, A.: 1970, *J. Geophys. Res.*, *75*, 6915.
Hartle, R. E. and Sturrock, P. A.: 1968, *Astrophys. J.*, *151*, 1155.
Hirshberg, J., Asbridge, J. R. and Robbins, D. E.: 1974, *J. Geophys. Res.*, *79*, 934.
Hollweg, J. V.: 1974, *J. Geophys. Res.*, *79*, 3845.
Hollweg, J. V.: 1975, *Rev. Geophys. Space Phys.*, *12*, 263.
Holzer, T. E.: 1976, in D. J. Williams (ed.) *Physics of Solar-Planetary Environments*, American Geophysical Union, Washington, D. C., p. 366.
Holzer, T. E.: 1977, *J. Geophys. Res.*, *82*, 23.
Hundhausen, A. J.: 1969, *J. Geophys. Res.*, *74*, 5810.
Hundhausen, A. J.: 1972, *Coronal Expansion and Solar Wind*, Springer-Verlag, New York.
Kopp, R. A. and Holzer, T. E.: 1976, *Solar Phys.*, *49*, 43.
Krieger, A. S., Timothy, A. F. and Roelof, E. C.: 1973, *Solar Phys.* *29*, 505.
Krieger, A. S., Timothy, A. F., Vaiana, G. S., Lazarus, A. J., and Sullivan, J. D.: 1974, in C. T. Russell (ed.), *Solar Wind Three*, University of California, Los Angeles, CA., p. 132.

Lambert, D. J.: 1967, *Nature*, *215*, 43.
Montgomery, M. D., Bame, S. J. and Hundhausen, A. J.: 1968, *J. Geophys. Res.*, *73*, 4999.
Montgomery, M. D.: 1972, in C. P. Sonett, P. J. Coleman, Jr., and J. M. Wilcox (eds.) *Solar Wind*, Asilomar Conf., NASA SP-308, p. 208.
Munro, R. H. and Withbore, G. L.: 1972, *Astrophys. J.*, *176*, 511.
Neugebauer, M. and Snyder, C. W.: 1966, *J. Geophys. Res.* *71*, 4469.
Neupert, W. M. and Pizzo, V.: 1974, *J. Geophys. Res.*, *79*, 3701.
Nishida, A.: 1969, *J. Geophys. Res.*, *74*, 5155.
Noble, L. M. and Scarf, F. L.: 1963, *Astrophys. J.*, *138*, 1169.
Nolte, J. T., Krieger, A. S., Timothy, A. F., Vaiana, G. S. and Zombeck, M. V.: 1976, *Solar Phys.*, *46*, 291.
Ogilvie, K. W., and Wilkerson, T. D.: 1969, *Solar Phys.*, *8*, 435.
Ogilvie, K. W., Scudder, J. D., and Sugiura, M.: 1971, *J. Geophys. Res.*, *76*, 3574.
Ogilvie, K. W.: 1975, *J. Geophys. Res.* *80*, 1335.
Parker, E. N.: 1958, *Astrophys. J.*, *128*, 664.
Parker, E. N.: 1963, *Interplanetary Dynamical Processes*, Interscience, New York.
Parker, E. N.: 1964, *Astrophys. J.*, *139*, 93.
Parker, E. N.: 1969, *Space Sci. Rev.*, *3*, 325.
Perkins, F.: 1973, *Astrophys. J.*, *179*, 637.
Robbins, D. E., Hundhausen, A. J., and Bame, S. J.: 1970, *J. Geophys. Res.*, *75*, 1178.
Rosenbauer, H., Miggenrieder, H., Montgomery, M., and Schwenn, R.: 1976 in D. J. Williams (ed.) *Physics of Solar Planetary Environments*, American Geophysical Union, Washington, D. C., p. 319.
Ryan, J. M. and Axford, W. I.: 1975, *J. Geophys. 41*, 221.
Scarf, F. L. and Noble, L. M.: 1965, *Astrophys. J.*, *141*, 1479.
Snyder, C. W., Neugebauer, M., and Rao, U. R.: 1963, *J. Geophys. Res.*, *68*, 6361.
Spitzer, L., Jr.: 1962, *Physics of Fully Ionized Gases* (2nd ed.), (2nd ed.), Interscience, New York.
Sturrock, P. A. and Hartle, R. E.: 1966, *Phys. Rev. Lett.*, *16*, 628.
Suess, S. T.: 1974, in C. T. Russell (ed.), *Solar Wind Three*, University of California, Los Angeles, p. 311.
Timothy, A. F., Krieger, A. S. and Vaiana, G. S.: 1975, *Solar Phys.*, *42*, 135.
Toichi, T.: 1971, *Solar Phys.*, *18*, 150.
Vaiana, G. S., Krieger, A. S. and Timothy, A. F.: 1973, *Solar Phys.*, *32*, 81.
Weber, E. J.: 1973, *Astrophys. and Space Sci.*, *20*, 391.
Weber, E. J., and Davis, L., Jr.: 1967, *Astrophys. J.*, *148*, 217.
Whang, Y. C.: 1971, *Astrophys. J.*, *169*, 369.
Whang, Y. C., and Chang, C. C.: 1965, *J. Geophys. Res.*, *70*, 4175.
Whang, Y. C., Liu, C. K., and Chang, C. C.: 1966, *Astrophys. J.*, *145*, 255.

Wilcox, J. M. and Ness, N. F.: 1968, *Space Sci. Rev.*, *8*, 258.
Wolff, C. L., Brandt, J. C., and Southwick, R. G.: 1971, *Astrophys. J.*, *165*, 181.
Wu, S. T., Dryer, M., Nakagawa, Y., and Han, S. M.: 1978, *Astrophys. J.*, (in press).
Yeh, T.: 1970, *Planetary and Space Sci.*, *18*, 199.

DISCUSSION

Armstrong: What is known of the stability of the time-independent multi-fluid solutions against perturbations? You would not expect to see any unstable equilibrium in nature. Why should an equilibrium solution ever apply to solar wind that is never time-independent?

Cuperman: The steady state, spherically symmetric multi-fluid models describe the average gross features of the solar wind. This relatively simple concept of a structureless equilibrium state, which corresponds to a steady and uniform corpuscular emission from the sun's surface (analogous to light emission) is able to explain the basic physics underlying the coronal expansion, including the subsonic-supersonic transition. Spatial and temporal variations of the conditions at the sun's surface produce significant perturbations of the equilibrium state defined above and generate many additional phenomena which, however, develop on the background of the quiet solar wind plasma.

Smith: You have suggested that the "structureless" solar wind, which is most likely to represent spherically symmetric, time-independent solar wind flow, is associated with high velocities (the region of decreasing solar wind speed). Since the solar wind parameters in this region are probably very different from their average values, shouldn't they be used in making comparisons with the theoretical models?

Cuperman: The existing theoretical models for steady state, spherically symmetric flow solar wind (implying $J \equiv nvr^2$ = const.) are able to describe only subsonic-supersonic transitions leading to relatively low streaming speeds at 1 AU ($\lesssim$ 350 km sec^{-1}). Therefore, despite the puzzling discovery that it is only for high-speed streams that the conservation of particle flux $J \equiv nvr^2$ holds, these models cannot explain the production of the observed high speeds. However, the models should be able to explain the relationship between the measured quantities at all heliocentric distances above 10 $R_\odot$. This work is now in progress.

Jockers: Is it really necessary to assume that T = 0 at large distances? Isn't it sufficient if the heat transport remains finite? If I recall my own work on collision-free two-component solar wind correctly, there were solutions where the proton temperature remained finite at large distances and which must be correct because they are exact solutions of the Vlasov equations.

Cuperman: The requirement $T \to 0$ for $r \to \infty$ is usually used as a boundary condition for the integration of the temperature equation in fluid models for the quiet solar wind. This condition implies $p \equiv nkT \to 0$ and $H = -K_0 T^{5/2}\, dT/dr \to 0$ for $r \to \infty$ (some authors used directly the requirement $p \to 0$ for $r \to \infty$). Actually, a more appropriate boundary condition would be the continuity of solar wind and the (relatively very small) interstellar pressures. At any case there should be no significant difference in the solutions within 10-15 AU.

Bird: If the "structureless solar wind" is associated with the high speed regions ($v \approx 600$ km s^{-1}), will this pose a more difficult problem for magnetohydrodynamic models that are usually hard pressed to achieve even 400 km s^{-1} at 1 AU?

Cuperman: The answer is yes. One way to obtain high-speed streams is to consider rapidly diverging flows in coronal holes (cf. Kopp and Holzer, 1976); the rapidly diverging flow geometry tends to decrease the density scale height of the expanding corona and thus to increase the outward-directed pressure gradient force which enhances the acceleration of the flow. Another way to obtain high-speed flows is by direct momentum addition (Holzer, 1976).

Wallis: Concerning the energy input into the solar wind from dissipation of high-speed streams, as described by Ed Smith; is it significant in order of magnitude, and could the arbitrary assumption of enhanced coupling between electrons and protons be avoided if the energy input went primarily to the protons?

Cuperman: The solar wind plasma (including the high-speed streams) at distances larger than several solar radii is essentially collisionless. Any dissipation of streaming energy and heating of protons occurs via wave-particle interaction. Microfluctuations are enhanced to a significant level of wave energy (by the conversion of relative streaming energy) which in turn heats low energy particles. This is exactly what we call non-collisional (or enhanced) electron-proton coupling.

THE LARGE SCALE AND LONG TERM EVOLUTION OF THE SOLAR WIND SPEED DISTRIBUTION AND HIGH SPEED STREAMS

DEVRIE S. INTRILIGATOR
University of Southern California
Los Angeles, California 90007, U.S.A.

ABSTRACT

The spatial and temporal evolution of the solar wind speed distribution and of high speed streams in the solar wind are examined. Comparisons of the solar wind streaming speeds measured at earth, Pioneer 11, and Pioneer 10 indicate that between 1 AU and 6.4 AU the solar wind speed distributions are narrower (i.e., the 95% value minus the 5% value of the solar wind streaming speed is less) at extended heliocentric distances. These observations are consistent with the exchange of momentum in the solar wind between high speed streams and low speed streams as they propagate outward from the sun. Analyses of solar wind observations at 1 AU from mid 1964 through 1973 confirm the earlier results reported by Intriligator (1974) that there are statistically significant variations in the solar wind in 1968 and 1969, years of solar maximum. High speed stream parameters show that the number of high speed streams in the solar wind in 1968 and 1969 is considerably more than the predicted yearly average, and in 1965 and 1972, less. Histograms of solar wind speed from 1964 through 1973 indicate that in 1968 there was the highest percentage of elevated solar wind speeds and in 1965 and 1972 the lowest. Studies by others also confirm these results although the respective authors did not indicate this fact. The duration of the streams and the histograms for 1973 imply a shifting in the primary stream source.

I. INTRODUCTION

During the last few years we have had the unique opportunity of obtaining simultaneous data from a number of spacecraft as close to the sun as 0.3 AU and as far away as ~10 AU. This has made it feasible to study travelling interplanetary phenomena over distances as large as ~10 AU. In Section II of this paper we focus on the radial evolution from 1 AU to 6.4 AU of solar wind speed distributions and high speed streams. In Section III of this paper we will discuss the long term temporal evolution of high speed streams in the solar wind

M. A. Shea et al. (eds.), Study of Travelling Interplanetary Phenomena/1977, 195–225.

based on the available solar wind data in the vicinity of the earth from 1964 through 1973.

II. RADIAL EVOLUTION

The radial evolution of the solar wind is of considerable interest since it can help us to ascertain the relative significance of the many possible physical mechanisms and phenomena that have been suggested as playing a major role in the solar wind propagation to large heliocentric distances. Parker (1958) in his original paper on the solar wind predicted some aspects of its variation over heliocentric distance. In the intervening years numerous papers have been written concerning the role of such phenomena as waves, turbulence, interaction regions, high speed streams, longitudinal and latitudinal asymmetries, and the interstellar medium in affecting the radial evolution of the solar wind over large heliocentric distances. With the successful launch of Pioneer 10 on March 3, 1972 and Pioneer 11 on April 4, 1973 direct observations became available of the solar wind at extended heliocentric distances. Collard and Wolfe (1974) comparing Pioneer 10 and Pioneer 11 daily samples over a period of 200 days found that in the solar wind the same major features, such as high speed streams, were observed by the two spacecraft out to distances as great as 5 AU. Intriligator (1975a) reported that there is no apparent change out to distances of at least 3 AU in the shape of the power spectrum associated with fluctuations in the solar wind streaming speed. This power-law frequency dependence in the $\sim 10^{-4}$ Hz to $\sim 10^{-3}$ Hz frequency range implied that large scale turbulence plays an important role in the solar system out to heliocentric distances of at least 3 AU. Smith and Wolfe (1976) using simultaneous Pioneer 10 and 11 plasma and magnetic field data found that there were corotating interaction regions in the solar wind at extended heliocentric distances. These regions are characterized by a simultaneous increase in the solar wind speed and the magnetic field magnitude which result in the formation of a forward shock. Smith and Wolfe (1976) reported that in many cases these forward shocks are followed by reverse shocks. The following paper in this volume is by Smith and Wolfe (1977) and this paper extends their previous analyses and contains a detailed discussion of these interaction regions.

In this section of the present paper we extend the analyses of Collard and Wolfe (1974) and Intriligator (1975b) by studying the radial evolution of the solar wind streaming speed and the radial evolution of high speed streams in the solar wind.

1. Observations

The Pioneer 10 and Pioneer 11 solar wind plasma data employed in this paper were obtained by the Ames Research Center Plasma Analyzer experiments. Each of these analyzers consists of two concentric quadrispherical electrostatic analyzers with multiple collectors (Wolfe et al., 1974a, b; Intriligator and Wolfe, 1974, 1976). The measurements of the

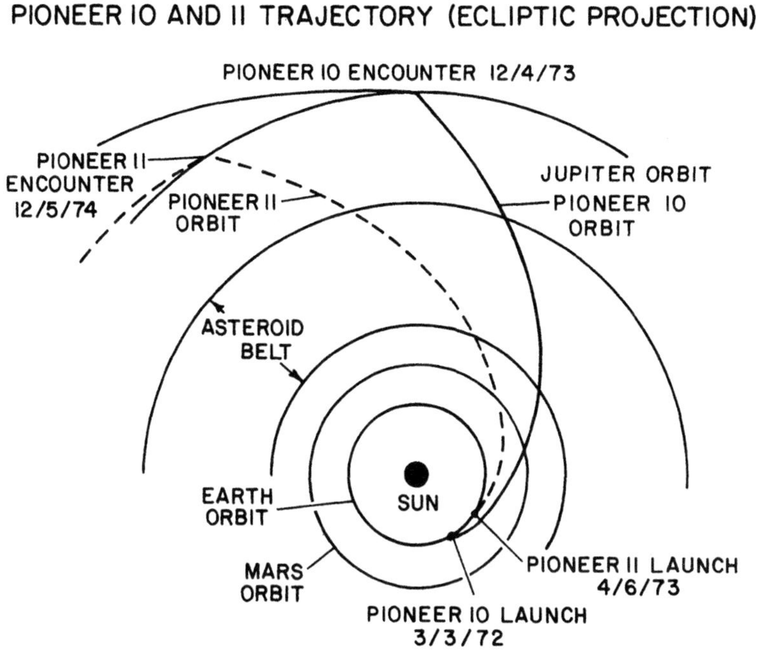

Figure 1. Pioneer 10 and Pioneer 11 trajectories (ecliptic projection).

solar wind proton data are employed in this study. These observations were analyzed using the calibrated instrument transmission function and a least squares iteration technique to fit the flight data to an isotropic Maxwellian distribution. Three hourly averages were used for this study where each three hourly average is the average of the available least squares samples that were analyzed at intervals of approximately one hour.

Figure 1 shows the Pioneer 10 and 11 trajectories. Figure 2 shows the earth-sun-probe (spacecraft) angle and the radial position of Pioneer 10 and Pioneer 11 for the time intervals employed in this study. The solar rotation numbers refer to the solar rotation times as seen at earth.

Comparisons of the solar wind streaming speeds measured at earth, Pioneer 11, and Pioneer 10 indicate that the solar wind speed distributions are narrower (i.e., the 95% value minus the 5% value of the solar wind streaming speed is less) at extended heliocentric distances. These comparisons also indicate, however, that there are significant temporal variations so that the magnitude of this effect can vary from one solar rotation to the next.

Figure 3 shows for solar rotations 1912, 1913, and 1914 at each spacecraft the relative frequency of occurrence of the three hourly

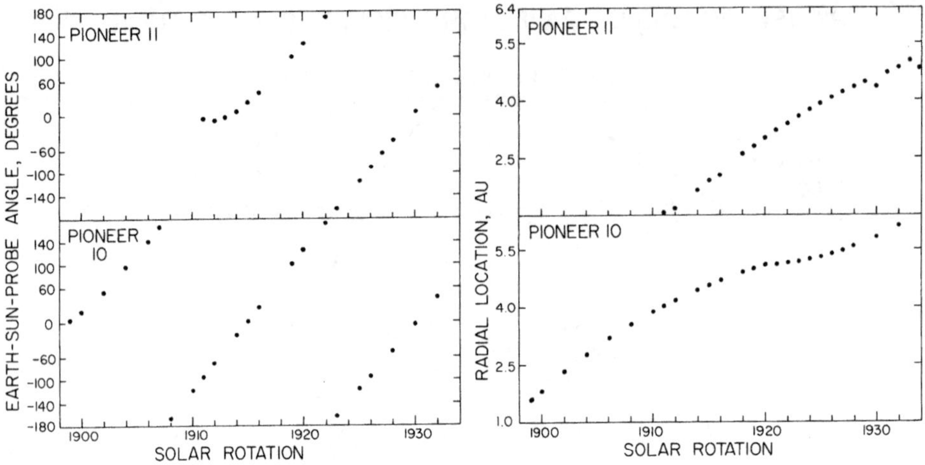

Figure 2. Pioneer 10 and 11 locations for the solar rotations examined in this paper. The earth-sun-probe (spacecraft) angle (in degrees) denotes the longitude of the Pioneer spacecraft with respect to the earth-sun line. The radial location (in AU) of the Pioneer spacecraft is also shown.

averages of the solar wind streaming speed in intervals of 25 km/sec. All the histograms in this figure and subsequent figures have been normalized so that there are equal areas under the histograms. It is clear from Figure 3 that the speed distribution is much narrower at Pioneer 10 than at Pioneer 11 or IMP. Some additional information concerning these histograms is listed in Table I. The data in Table I indicate that the width (the 95% value minus the 5% value) of the histograms became more narrow at extended heliocentric distances. That is, the width of the histogram at 1 AU (at earth) was 393 km/sec, at Pioneer 11 (which travelled between 1.16 AU and 1.88 AU during this time) it was 377 km/sec, and at Pioneer 10 (which travelled between 4.16 AU and 4.54 AU during this interval) it was only 224 km/sec. The corotational delay (Intriligator and Neugebauer, 1975) between IMP and Pioneer 11 varies during these three solar rotations from less than one day to ~2 days, and between IMP and Pioneer 10 the corotational delay varies between ~16 to ~19 days. By combining all of the solar wind streaming speed data for three solar rotations together, a high percentage of overlap between the data sets is achieved.

Figure 4 is similar in format to Figure 3 and shows for solar rotations 1916, 1917, and 1918 the solar wind speed histograms at IMP, Pioneer 11, and Pioneer 10. In Figure 4 it is even more evident than in Figure 3 that the width of the solar wind speed histograms become narrower at increasing heliocentric distances. The data in Table I indicate that the width of the histograms for solar rotations 1916, 1917, and 1918 at IMP, Pioneer 11, and Pioneer 10 are, respectively, 300 km/sec, 198 km/sec, and 186 km/sec.

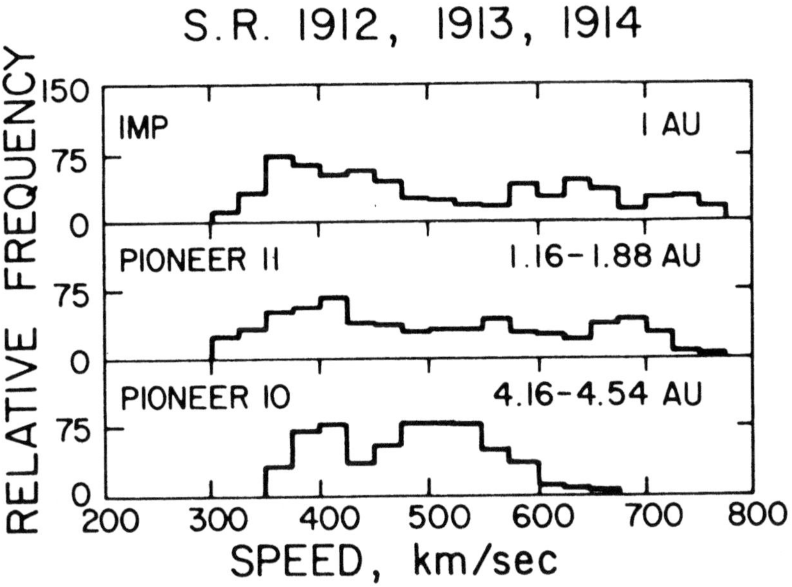

Figure 3. Solar wind speed histograms for solar rotations 1912, 1913, and 1914. The relative frequency of occurrence of the three hour averages of the solar wind streaming speed in intervals of 25 km/sec The histograms have been normalized so that there is the same area under each histogram. The normalization values, the mean solar wind speeds, and the width of the histograms are given in Table I.

There is a near radial alignment of Pioneer 10 with Pioneer 11 during solar rotations 1920 through 1926. This is indicated in Figure 2 where for solar rotations 1920 through 1926 the earth-sun-probe (spacecraft) angle for Pioneer 10 and Pioneer 11 are extremely close (within ~4° of each other). The histograms of the solar wind speed distributions at Pioneer 10 and Pioneer 11 during this period of near radial

TABLE I

Solar Rotation	Parameter	IMP	Pioneer 11	Pioneer 10
1912, 1913, 1914	Mean speed, km/sec	510	503	481
	95% minus 5%, km/sec	393	377	224
	Normalization	1.20	1.04	1.02
	Radial location, AU	1	1.16 – 1.88	4.16 – 4.54
1916, 1917, 1918	Mean speed, km/sec	436	392	417
	95% minus 5%, km/sec	300	198	186
	Normalization	1.45	1.01	1.01
	Radial location, AU	1	2.13 – 2.81	4.67 – 4.99

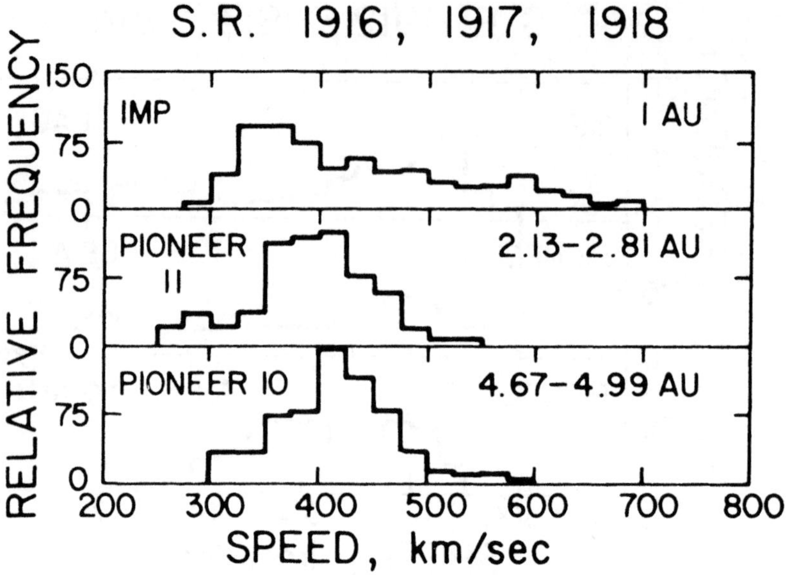

Figure 4. Solar wind speed histograms for solar rotations 1916, 1917, and 1918. See Figure 3 caption and Table I.

alignment are shown in Figure 5. It is evident from this figure that the solar wind speed distributions are considerably narrower at Pioneer 10 (in the vicinity of 5.06 AU to 5.46 AU) than at Pioneer 11 (in the vicinity of 3.02 AU to 4.21 AU). The widths of the histograms during this period of radial alignment are summarized in Table II.

During solar rotation 1920 through 1926 there were large temporal variations in the solar wind. This is indicated in Figures 6 and 7 where the histograms of the solar wind speed for solar rotations 1920, 1921, and 1922 are contrasted with those for solar rotations 1923, 1924, and 1925. The IMP data at 1 AU are also shown in these figures although there is not a radial alignment between IMP and the two Pioneer spacecraft during this time. Additional information concerning these histograms is presented in Table II.

The temporal variations between solar rotations 1920, 1921, and 1922 as compared with solar rotations 1923, 1924, and 1925 are further emphasized in Figure 8 where the data from Figure 6 and 7 are superimposed. It is clear from this figure that the solar wind speed histograms for solar rotations 1923, 1924, and 1925 are skewed more toward the higher speeds than the histograms for the three earlier solar rotations. This period at the end of 1973 and at the beginning of 1974 was, in fact, notable for the broad high speed streams observed at 1 AU which had relatively long durations and reached peak speeds in excess of 600 km/sec.

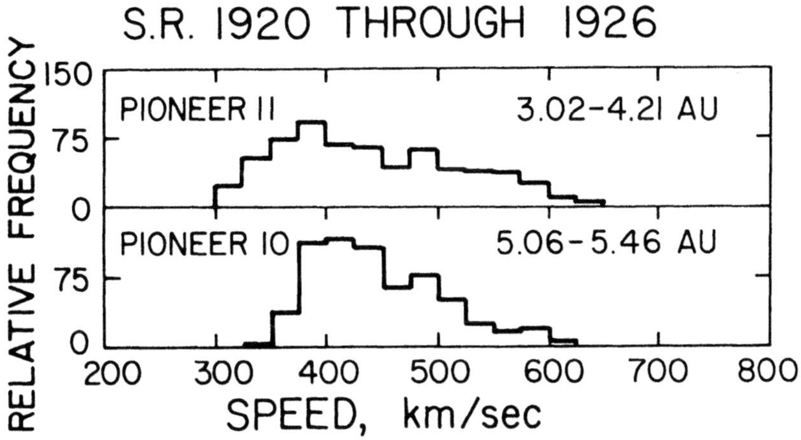

Figure 5. Solar wind speed histograms for the near radial alignment of Pioneer 10 and 11 during solar rotations 1920 through 1926.

There is also a great contrast in the type of solar wind streams in the vicinity of earth following the launch of Pioneer 10 in 1972 as compared with the solar wind streams in the vicinity of earth following the launch of Pioneer 11 in 1973. An indication of these temporal variations in the solar wind is given by contrasting the data in Figures 9 and 10. Figure 9 shows the solar wind speed distributions at IMP and Pioneer 10 for solar rotation 1898 (in 1972). Figure 10 shows the solar wind speed distributions at IMP and Pioneer 11 for solar

TABLE II

PIONEER 10/11 RADIAL ALIGNMENT

Solar Rotation	Parameter	IMP	Pioneer 11	Pioneer 10
1920 through 1926	Mean speed, km/sec		444	447
	95% minus 5%, km/sec		260	197
	Normalization		0.52	0.57
	Radial location, AU		3.02–4.21	5.06–5.46
1920, 1921, 1922	Mean speed, km/sec	502	423	442
	95% minus 5%, km/sec	346	252	142
	Normalization	1.09	1.31	1.64
	Radial location, AU	1	3.02–3.58	5.06–5.15
1923, 1924, 1925	Mean speed, km/sec	531	453	457
	95% minus 5%, km/sec	351	243	225
	Normalization	1.02	1.18	1.16
	Radial location, AU	1	3.58–4.06	5.15–5.36

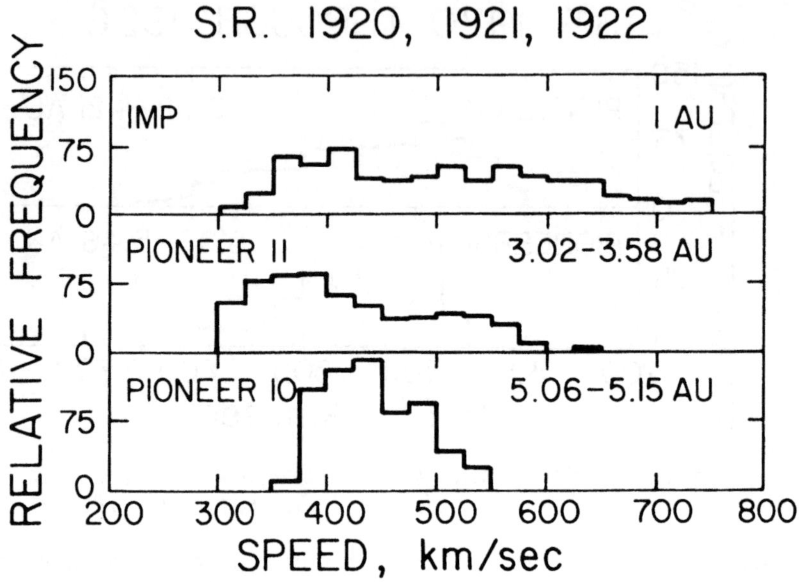

Figure 6. Solar wind speed histograms 1920, 1921, 1922. See Figure 3 and Table II.

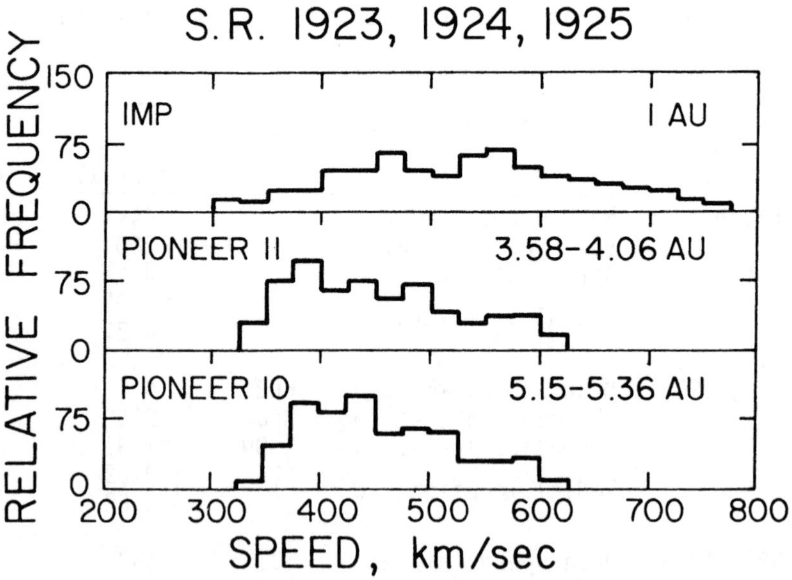

Figure 7. Solar wind speed histograms for solar rotations 1923, 1924, and 1925. See Figure 3 and Table II.

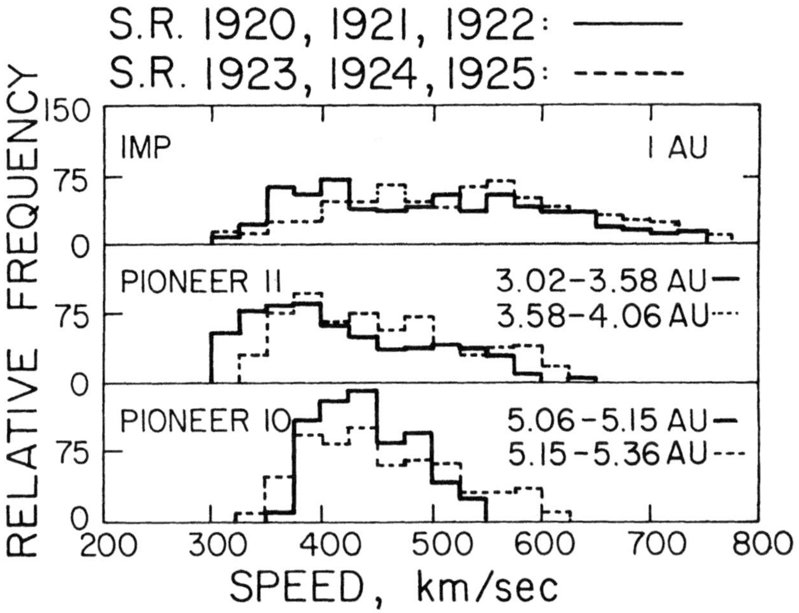

Figure 8. Superposition of histograms for solar rotations 1920, 1921, and 1922 with solar rotations 1923, 1924, and 1925. The difference between the two sets of histograms is indicative of significant temporal variations in the solar wind.

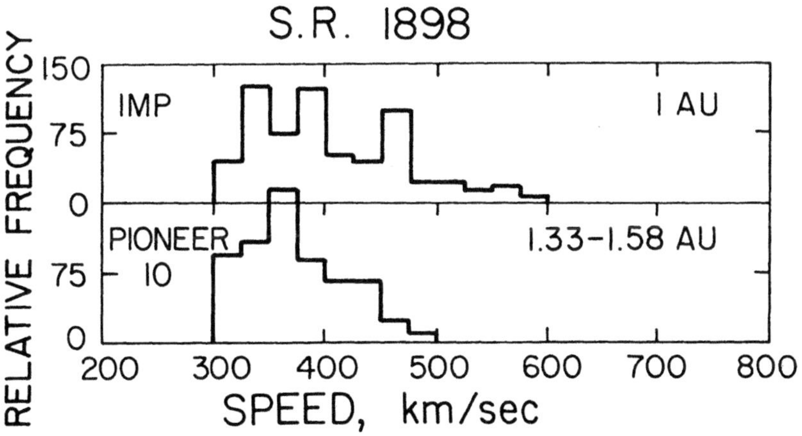

Figure 9. Solar wind speed histograms for solar rotation 1898 for the near radial alignment of Pioneer 10 and IMP. See Table III.

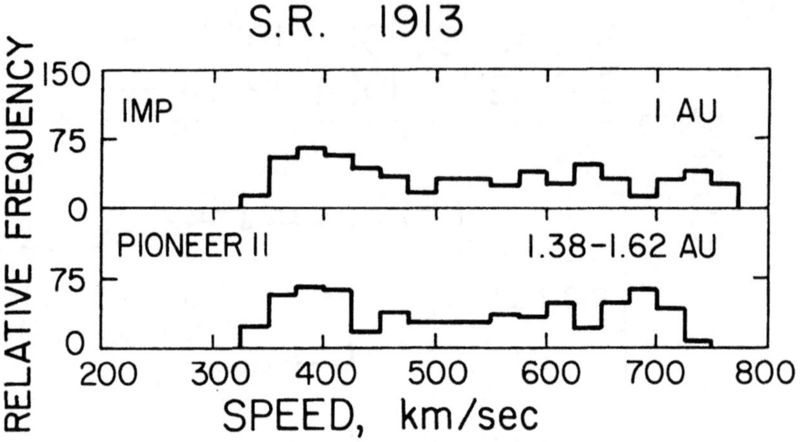

Figure 10. Solar wind speed histograms for solar rotation 1913 for the near radial alignment of Pioneer 11 and IMP. See Table III.

rotation 1913 (in 1973). The corotational delay between IMP and Pioneer 10 in solar rotation 1898 and between IMP and Pioneer 11 in solar rotation 1913 is on the order of 1 to 2 days. The mean solar wind speed and the widths of the histograms for Figures 9 and 10 are given in Table III. The data in Figures 9 and 10 and Table III indicate that while in both solar rotations the width of the solar wind speed distributions narrow between IMP and the Pioneer spacecraft, there are basically higher mean speeds and broader speed distributions in 1973 as compared with 1972. These variations in the solar wind speed as observed near 1 AU may be indicative of a shifting in the primary source at the sun of high speed streams in the solar wind (as suggested

TABLE III

PIONEER/IMP RADIAL ALIGNMENT

Solar Rotation	Parameter	IMP	Pioneer 10
1898	Mean speed, km/sec	406	337
	95% minus 5%, km/sec	221	155
	Normalization	3.93	3.79
	Radial location, AU	1	1.33–1.58
		IMP	Pioneer 11
1913	Mean speed, km/sec	531	526
	95% minus 5%, km/sec	390	356
	Normalization	3.48	3.01
	Radial location, AU	1	1.38–1.62

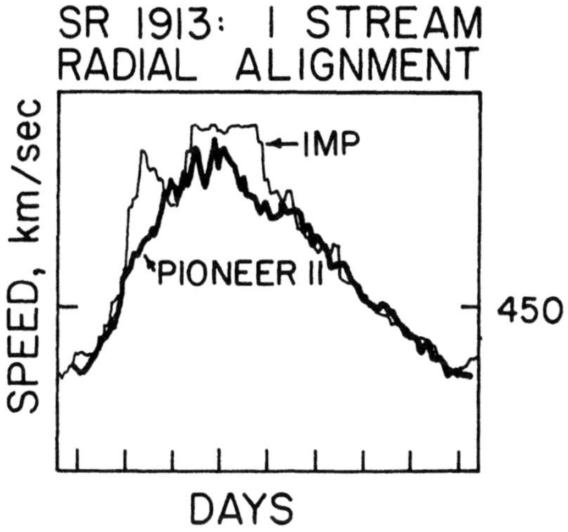

Figure 11. Superposition of solar wind speeds associated with a high speed stream at Pioneer 11 and at IMP.

in Intriligator, 1977b). This temporal variation in the solar wind and other long term variations in the solar wind are discussed in Part III.

The radial evolution of a single high speed stream during this interval of extremely broad streams in 1973 is examined in Figures 11 and 12 and Table IV where a single high speed stream is followed from IMP to Pioneer 11. As indicated in Figure 2, during this interval there is a near radial alignment between the two spacecraft. The data in Figure 11 indicate that as the stream propagated from 1 AU to 1.5 AU

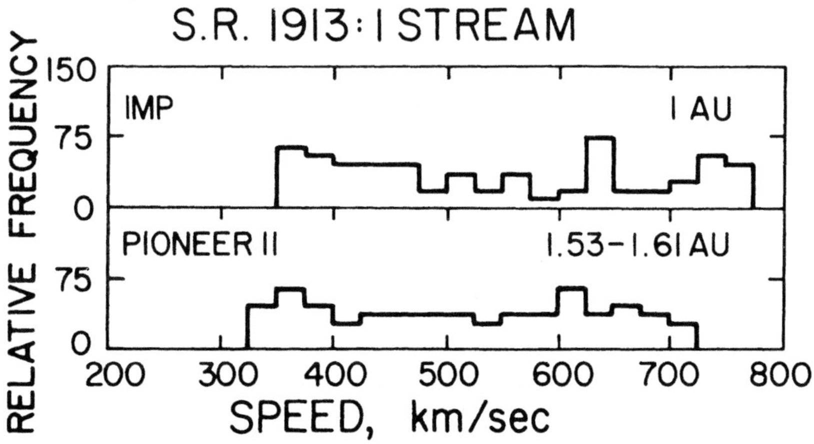

Figure 12. Solar wind speed histograms for the high speed stream shown in Figure 11. See Table IV.

TABLE IV

RADIAL EVOLUTION OF STREAM PROFILES

Solar Rotation	Parameter	IMP	Pioneer 11
1913 - 1 stream	Mean speed, km/sec	548	517
	95% minus 5%, km/sec	395	354
	Normalization	9.39	9.26
	Radial location, AU	1	1.53 – 1.61

there was a decrease in the mean solar wind speed and a significant narrowing of the width of the solar wind speed histogram for this stream.

During solar rotation 1924 when Pioneer 11 is at ~3.7 AU there is again a near radial alignment of Pioneer 11 and IMP at earth (see Figure 2). The radial evolution of high speed streams during this interval are shown in Figures 13 and 14 and Table V. Once again the data indicate that the width of the streams were much less at Pioneer 11 and the mean solar wind speed associated with the streams was less at Pioneer 11.

To date the largest radial separation between two sites for which solar wind data are available occurred during solar rotations 1932 and 1933 (in 1974). During this interval there are IMP data obtained at earth and Pioneer 10 data obtained between 6.09 AU and 6.40 AU. As indicated in Figure 2, the earth-sun-probe (spacecraft) angle for Pioneer 10 at this time is quite large indicating that there is no near

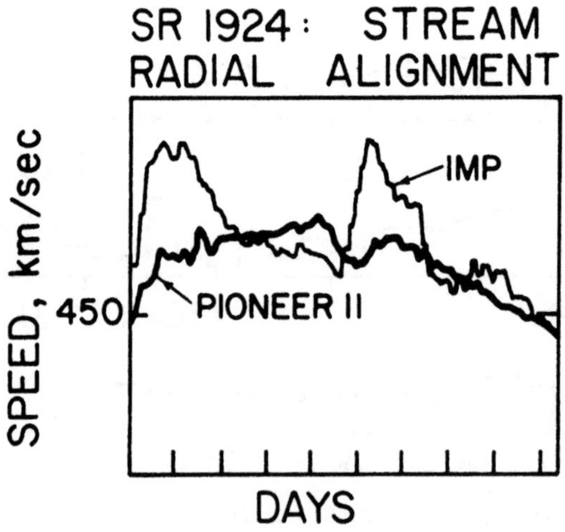

Figure 13. Superposition of solar wind speeds associated with the high speed streams at Pioneer 11 and at IMP.

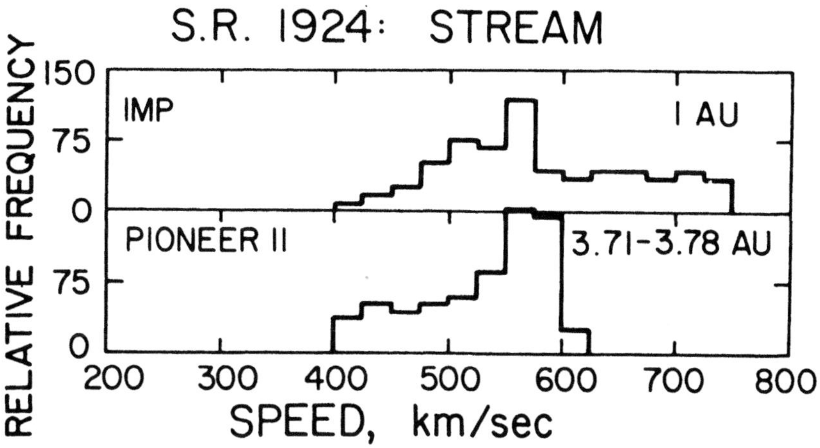

Figure 14. Solar wind speed histograms for the high speed streams shown in Figure 13. See Table V.

TABLE V

RADIAL EVOLUTION OF STREAM PROFILES

Solar Rotation	Parameter	IMP	Pioneer 11
1924 - stream	Mean speed, km/sec	580	530
	95% minus 5%, km/sec	269	175
	Normalization	8.64	8.53
	Radial location, AU	1	3.71–3.78

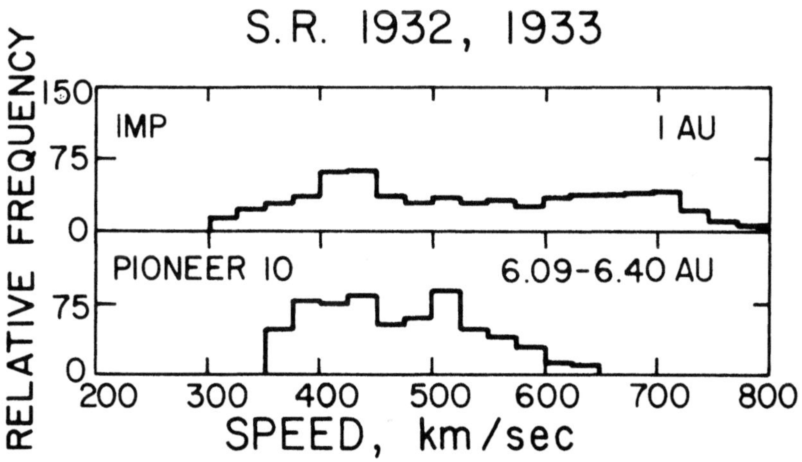

Figure 15. Solar wind speed histograms for solar rotations 1932 and 1933. These histograms show data for the largest radial separation reported to date. See Table VI.

TABLE VI

LARGEST RADIAL SEPARATION			
Solar Rotation	Parameter	IMP	Pioneer 10
1932, 1933	Mean speed, km/sec	532	471
	95% minus 5%, km/sec	383	228
	Normalization	1.67	2.20
	Radial location, AU	1	6.09–6.40

radial alignment between these two sites. The corotational delay time between earth and Pioneer 10 is on the order of 20 days during this interval so that by presenting two solar rotations of data a high percentage of overlap is achieved. The solar wind speed histograms at IMP and Pioneer 10 for solar rotations 1932 and 1933 are shown in Figure 15. As indicated in this figure and in Table VI the mean solar wind speeds and the width of the Pioneer 10 histogram obtained beyond 6 AU are considerably less than those obtained by IMP at 1 AU.

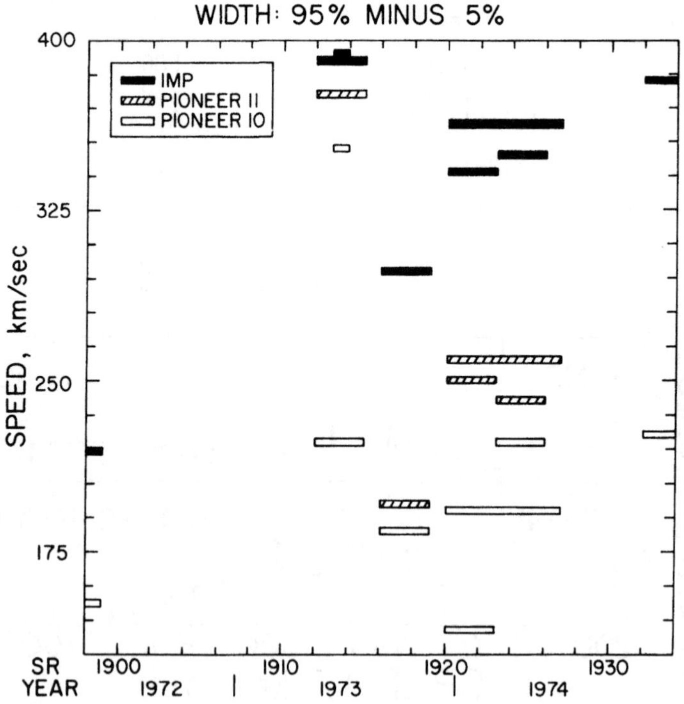

Figure 16. Summary of the width (95% value minus 5% value) of the solar wind speed distributions at IMP, Pioneer 11, and Pioneer 10 for solar rotations examined in the previous figures and tables.

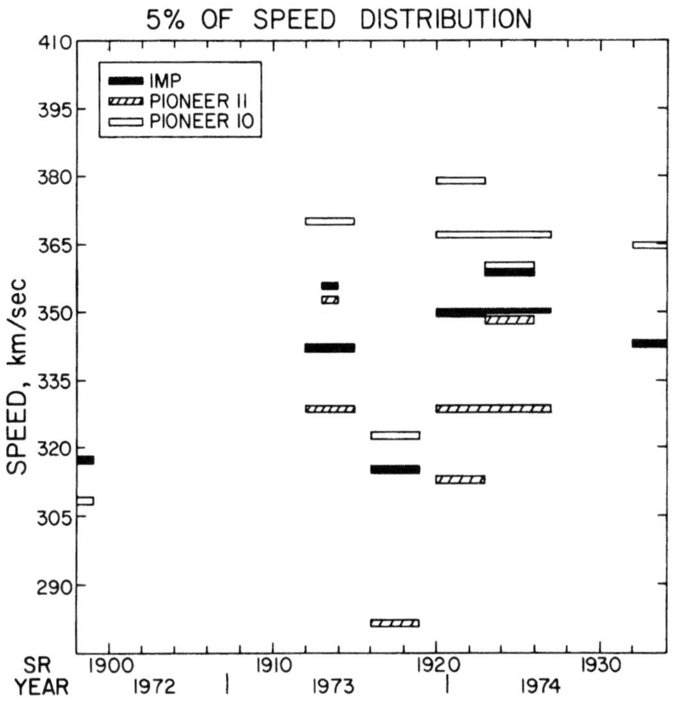

Figure 17. Summary of the 5% values of the solar wind speed distributions at IMP, Pioneer 11, and Pioneer 10 for the same intervals as those shown in Figure 16.

Figure 16 summarizes the width of the histograms for each of the intervals discussed above. The data in Figure 16 indicate that in each epoch discussed above the width of the solar wind speed distribution narrows with increasing heliocentric distance. In other words, in each interval in which there are IMP data, Pioneer 11 data, and Pioneer 10 data the width of the histograms at IMP (the dark bars) are always the highest value and the width of the histograms at Pioneer 10 (the open bars) are always the lowest value, with the width of the histograms at Pioneer 11 (the dashed bars) lying in between.

In Figure 17 there is a summary of the 5% values of the speed distributions at each of these locations for the same time intervals shown in Figure 16. These 5% values of the speed distribution are exceedingly interesting since they indicate that at extended heliocentric distances the 5% values of the Pioneer 10 solar wind speed distributions are higher than the 5% values of the speed distributions closer to the sun.

The increasing of the 5% value of the histograms at extended heliocentric distances and the narrowing of the width of the histograms at these extended distances are consistent with each other and may imply the physical significance of the exchange of momentum between high speed

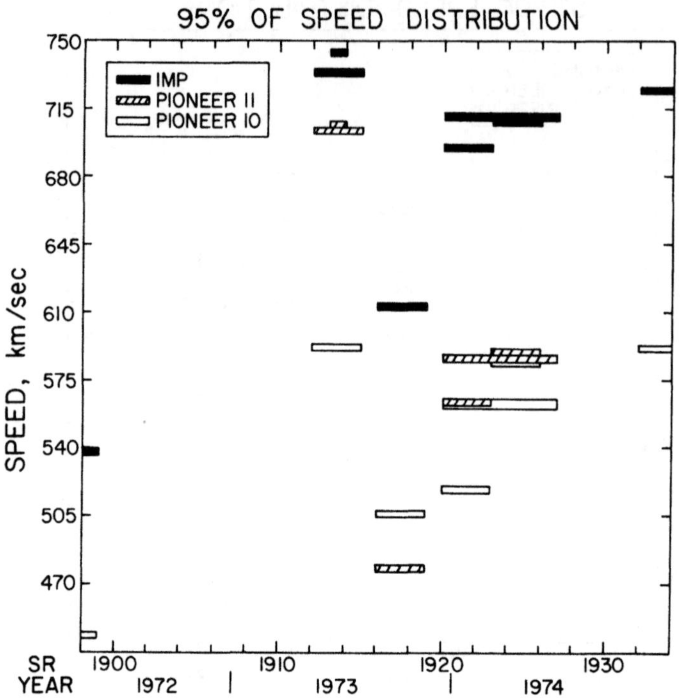

Figure 18. Summary of the 95% values of the solar wind speed distributions at IMP, Pioneer 11, and Pioneer 10 for the same intervals as those shown in Figures 16 and 17.

TABLE VII

Solar Rotation	Year	Spacecraft	Radial Range, AU	Radial Distance For 1/e Decay, AU	Distance At Which Width Of Histogram ≃20 km/sec AU
1898	1972	IMP Pioneer 10	1 1.33-1.58	1.28	2.63
1913	1973	IMP Pioneer 11	1 1.38-1.63	5.54	15.9
1920-1926	1973/ 1974	Pioneer 11 Pioneer 10	3.02-4.21 5.06-5.46	5.93	13.6
1920,1921,1922	1973/ 1974	Pioneer 11 Pioneer 10	3.02-3.58 5.06-5.15	3.15	6.17
1923,1924,1925	1974	Pioneer 11 Pioneer 10	3.58-4.06 5.15-5.36	18.6	45.1

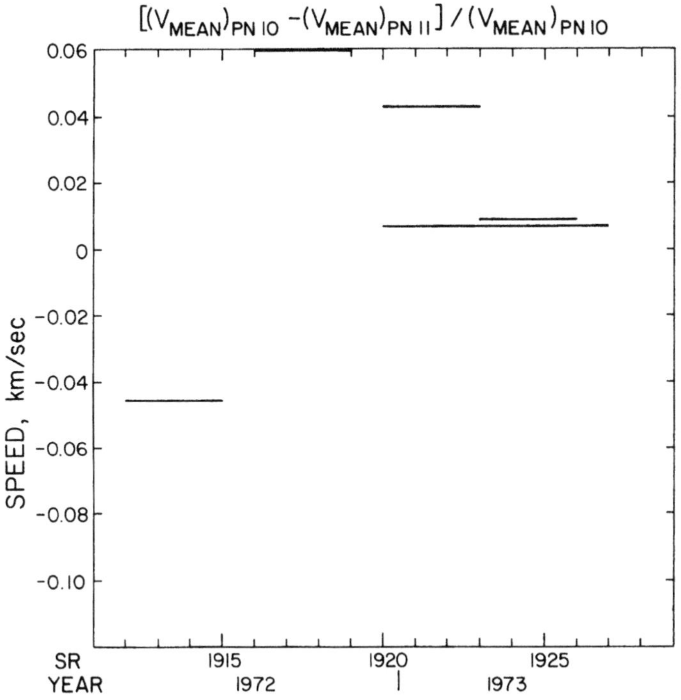

Figure 19. Summary of $\frac{(V_{mean})_{Pioneer\ 10} - (V_{mean})_{Pioneer\ 11}}{(V_{mean})_{Pioneer\ 10}}$. *The width of the histograms (see Figure 16) and the 5% values of the histograms (see Figure 17) appear to have more physical significance than the mean speed associated with the histograms.*

streams and low speed streams as the solar wind propagates outward from the sun. For completeness, in Figure 18 there is a summary of the 95% values of the solar wind speed distributions at each of the spacecraft for the epochs discussed above. As expected, the highest 95% values are obtained at 1 AU. However, during some intervals the lowest values are obtained at Pioneer 11 rather than at Pioneer 10. The fact that the highest 95% values are obtained at 1 AU is again consistent with the exchange of momentum between high speed streams and low speed streams which results in the eventual narrowing of the width of the speed distribution. In Figure 19 there is a summary of $[(V_{mean})_{Pioneer\ 10} - (V_{mean})_{Pioneer\ 11}]/[(V_{mean})_{Pioneer\ 10}]$ for the time intervals of the data in Figure 3 (S.R. 1912, 1913, 1914), Figure 4 (S.R. 1916, 1917, 1918), and Table I when there was no radial alignment between Pioneer 10 and 11 and for the time intervals of the data in Figure 5 (S.R. 1920 through 1926), Figures 6-8 and Table II when there was a radial alignment between Pioneer 10 and 11. Negative values in Figure 19 indicate a deceleration in the mean speed of the solar wind as it propagates outward. These data (e.g., S.R. 1920 through 1922 as compared with S.R. 1923 through S.R. 1925) again emphasize the temporal variations in the solar wind. Furthermore these data may imply that it is more meaningful to emphasize the change in the width of the solar wind speed distributions than the simple mean of the distribution. As an indication of how the width of the solar wind speed distributions is narrowing with increasing heliocentric distance we can compute the exponential rate of decay of the width of the histograms. Table VII presents this information for all the intervals presented above during which there was a near radial alignment of the two sites of observation. Also presented in Table VII is the distance at which a width of ≈20 km/sec for the solar wind speed distribution would be obtained assuming the exponential decay. The results in Table VII indicate that these values range from on the order of a few AU to tens of AU and that they appear to be very time dependent.

It is also important to study the propagation of interplanetary shocks over large heliocentric distances. In the next paper, Smith and Wolfe (1977), there is a discussion of the propagation over several AU of the forward and reverse shocks which form beyond 1 AU and which are associated with the corotating interaction regions. These shocks are relatively common beyond 1 AU. At times there are interplanetary shocks at extended heliocentric distances associated with solar flares. The interplanetary shocks associated with the August 1972 solar flares have recently been studied (Intriligator, 1976, 1977a; Smith, 1976; Smith et al., 1977). During this time there was a radial alignment between Pioneer 9 at 0.8 AU and Pioneer 10 at 2.2 AU. Using simultaneous plasma and magnetic field observations from Pioneer 9 and Pioneer 10, Intriligator

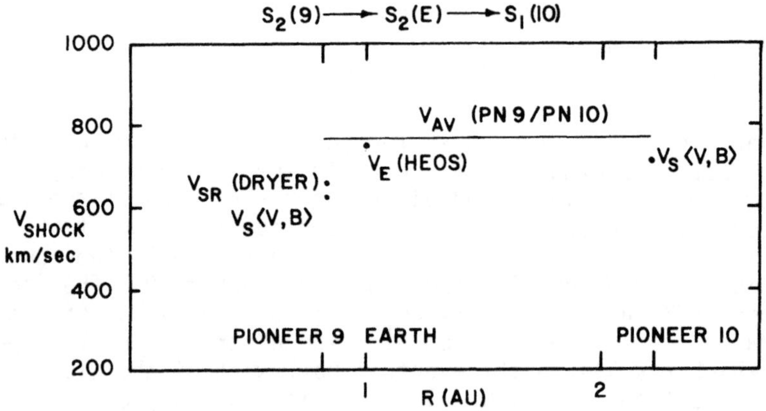

Figure 20. Summary of local shock speeds obtained for the forward shocks $S_2(9)$, $S_2(E)$, and $S_1(10)$ and the average shock speed V_{AV}(Pn 9/Pn 10) implied by associating $S_2(9)$ with $S_1(10)$. The shock speed V_E(HEOS) was calculated near earth. This figure emphasizes the similarity of each of these speeds and implies that the fast forward shock $S_2(9)$ propagated between 0.8 and 2.2 AU without strongly decelerating.

(1977a) reexamined the times of passage of the shocks and calculated the local shock speeds at each spacecraft. The analyses indicate that two interplanetary shocks associated with the August 1972 solar events propagated at approximately constant speed between Pioneer 9 at 0.8 AU and Pioneer 10 at 2.2 AU. This result is contrary to some theoretical expectations and earlier reports for these events of a strong deceleration of these shocks with increasing heliocentric distance. Figure 20 is from Intriligator (1977a). This figure indicates the similarity of the local shock speeds (700 km/sec ± 10%) calculated at Pioneer 9, earth (HEOS), and Pioneer 10. The line in Figure 20 indicates the average speed of propagation of the shock, assuming that the Pioneer 9 shock $S_2(9)$ propagated radially outward to Pioneer 10 where it was identified as $S_1(10)$. The good agreement between this average shock speed and the local shock speeds leads to the important conclusion that there was no strong deceleration of the shock as it propagated outward from Pioneer 9

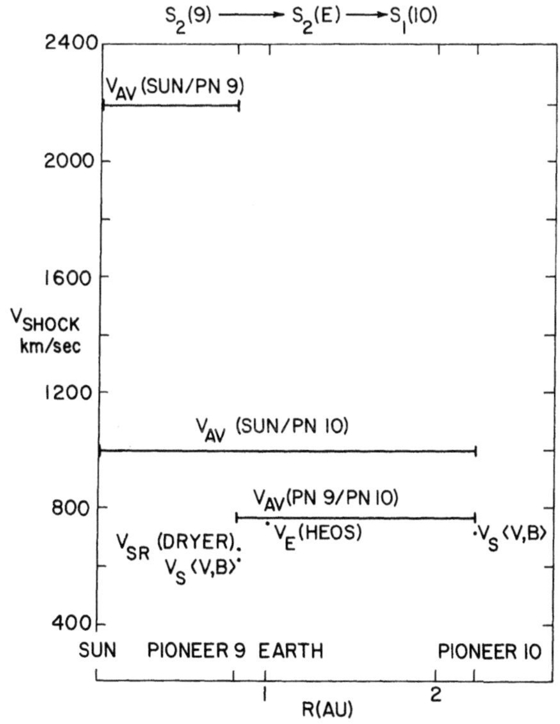

Figure 21. Summary of the average shock speeds and local shock speeds for the forward shocks $S_2(9)$, $S_2(E)$, and $S_1(10)$. The high average speeds of propagation from the sun to Pioneer 9, V_{AV}(sun/Pn 9), and from the sun to Pioneer 10, V_{AV}(sun/Pn 10), are shown. These high average shock speeds emphasize the extremely strong deceleration of the shock closer to the sun (within 0.8 AU). The local shock speeds and V_{AV}(Pn 9/Pn 10) imply that between 0.8 and 2.2 AU there was no strong deceleration of the shock.

at 0.8 AU to Pioneer 10 at 2.2 AU. This evidence for the approximately constant speed of propagation between 0.8 AU and 2.2 AU also implies that these interplanetary shocks previously underwent extremely strong deceleration between the sun and 0.8 AU. The average speed of propagation of the shock from the sun to Pioneer 9 is ~2200 km/sec and the average speed of propagation from the sun to Pioneer 10 is ~1000 km/sec. These high average shock speeds are shown in Figure 21 which indicates that strong deceleration of the interplanetary shocks took place between the sun and 0.8 AU and that there was no significant deceleration between 0.8 AU and 2.2 AU.

2. Summary of Radial Evolution

The Pioneer 10 and 11 missions have provided our first opportunity to study travelling interplanetary phenomena over large heliocentric distances. The analyses presented above have extended our knowledge of the basic evolution of the solar wind out to distances as large as 6.4 AU The observations of the solar wind presented above indicate that there continues to be significant narrowing of the width (the 95% value minus the 5% value) of the solar wind speed distributions between 1 AU and 6.4 AU. The narrowing of the width of the distributions with increasing heliocentric distance (see Figure 16) and the increase in the 5% values of the histograms (see Figure 17) at the most extended heliocentric distances are consistent with each other and may imply the significance of the exchange of momentum between high speed streams and low speed streams as the solar wind propagates outward from the sun.

There are significant temporal variations from one solar rotation to the next in the solar wind at 1 AU. The observations presented above indicate that the temporal variations in the solar wind at 1 AU propagated outward to large heliocentric distances and affect the large scale properties of the interplanetary medium. In comparing solar wind speed distributions which were associated with broad solar wind speed histograms at 1 AU with solar wind speed distributions which were associated with narrow solar wind speed histograms at 1 AU we found that all of the speed distributions became more narrow as they propagated outward. However, some of the histograms at 1 AU gave rise to significantly broader solar wind speed histograms at extended heliocentric distances. When we computed the radial distance for an exponential (1/e) rate of decay of the solar wind speed histograms we found a much larger radial distance for 1/e decay associated with some of the histograms (e.g., the histogram for S.R. 1923, 1924, and 1925). Similarly, for these histograms we calculated (assuming exponential decay) that they attain an approximately constant speed of expansion at considerably further distances (e.g., for S.R. 1923, 1924, and 1925 we calculated a distance of ~45 AU, whereas for S.R. 1920, 1921, and 1922 we calculated a distance of ~6 AU). The interplanetary shocks observed at Pioneer 10 at 2.2 AU associated with August events at the sun indicated that shorter term temporal variations can also propagate to extended heliocentric distances. In the next section of this paper there is a discussion of long term temporal variations in the solar wind as observed by spacecraft in the vicinity of 1 AU.

III. LONG TERM VARIATIONS

The activity of the solar wind varies in phase with sunspot activity from 1964 through 1972. The purpose of this section is to confirm and extend the first report of this result in Intriligator (1974), and to point out that other researchers' results confirm this observation (although the authors did not indicate this fact), and to suggest that the rise in stream activity in 1973 implies a shift in the primary source for streams near solar minimum. In this study the original data set employed by Intriligator (1974) has been substantially augmented as discussed below.

Intriligator (1974) emphasized the importance of high speed streams in the solar wind and initiated the use of a number of parameters to characterize the high speed streams and their duration. Previously, other investigators (Gosling et al., 1971; Wolfe, 1972; Diodato et al., 1974) had always computed averages of various solar wind parameters rather than focusing specifically on high speed streams as indicators of solar wind activity. Intriligator (1974) initiated the use of three parameters as indicators of solar wind activity: the number of high speed streams, the duration (in days) of these high speed streams, and the product of the yearly average solar wind proton speed times the duration. A high speed stream is defined as one having a rapidly rising increase in solar wind speed, V, and a peak speed greater than or equal to 450 km/sec. The duration of a high speed stream is defined from the start of the increase in solar wind speed through the peak ($V \geq 450$ km/sec) to the time that the speed declines to 400 km/sec (or to the start of the next high speed stream if the speed at the beginning of the following stream exceeds 400 km/sec). The value of the duration in days is rounded to the nearest whole day. The third parameter is the product of the yearly average solar wind speed (based on the available daily speeds for that year) times the duration in days. It should be emphasized that the same criteria for selecting high speed streams have been employed in both the present paper and the original paper (Intriligator 1974).

1. Data Set

In the present study we have substantially supplemented the original data in Intriligator (1974) with the additional data that have become available as summarized in Table VIII. Many days for which there were only quick-look data (Intriligator, 1974) have been replaced by three hourly averaged data and some previous data gaps have been eliminated. In addition, the coverage of the data set has been extended so that it now includes solar wind data from mid 1964 through 1973. Column (2) in Table VIII shows the yearly totals of the number of days for which data (>0 points) are available. An indication of the quality of the data coverage is given in Column (3) which shows the percentage of the days listed in Column (2) for which four or more three hourly averages are available. Column (4) summarizes the numerous spacecraft data sets used. Data recorded near earth were used when available.

TABLE VIII

YEAR (1)	DAYS (> 0 PTS) (2)	≥ 4 PTS / > 0 PTS % (3)	SPACECRAFT (4)	STREAMS OBSERVED (5)	%≥450 (6)
1964	122	31	a	22	34
1965	284	52	a,b,c	33	23
1966	322	53	b,c,d,e,f,g	33	32
1967	353	81	a,b,e,f,h,i	47	33
1968	352	71	a,d,e,f,i,j,k,l	59	54
1969	336	75	a,d,e,k,l,m	57	34
1970	351	62	a,d,l,m	51	34
1971	344	62	a,d,j,l,n	44	39
1972	329	69	a,j,n,o	39	26
1973	329	90	a,n,o,p	47	53
'64-'73	3122	67	all	432	38

a=VELA/IMP, b=VELA 3, c=PIONEER 6 (AMES), d=PIONEER 6 (MIT), e=EXPLORER 33, f=PIONEER 7 (MIT), g=PIONEER 7 (AMES), h=EXPLORER 34, i=EXPLORER 35, j=PIONEER 6 QUICK LOOK, k=PIONEER 9, l=OGO 5, m=HEOS, n=IMP, o=PIONEER 9 QUICK LOOK, p=PIONEER 8 QUICK LOOK

Data gaps were sometimes filled in with data obtained elsewhere, corotated (Intriligator and Neugebauer, 1975) to the earth. However, if the general trend of the corotated data was not very similar to the available near earth data, then it was not used to fill gaps. This has resulted in a few decreases in the total number of days of coverage in a year as compared with the coverage in Intriligator (1974) but the present data set has a high percentage of days with high quality data coverage (as indicated in Column (3) in Table VIII), and represents the most complete, consistent coverage available near the earth.

2. Observations

Using this new data set and the same criteria as those employed in Intriligator (1974), tables were compiled (Intriligator, 1973, 1974) of high speed streams in the solar wind. (The differences in the tabulated stream parameters, particularly the durations, reported below compared to those in Intriligator (1974) are due to the replacing of quick-look data by more complete data coverage and the replacing of corotated deep space data by data obtained in the vicinity of the earth.) Figure 22 summarizes some of the results obtained with the aid of these tables. The shaded histograms show the total number of high speed streams, with an increase of solar wind speed (ΔV) $\gtrsim$ 150 km/sec, observed in each year from mid 1964 through 1973. If one were to assume that there were no solar cycle variation and that there was a constant rate of high speed streams during this time it can be approximated by

$$\sum_{(\text{mid})\,1964}^{1973} (\text{number of streams}) \Big/ \sum_{(\text{mid})\,1964}^{1973} (\text{days } (> 0 \text{ points})) \simeq .14 \text{ streams/day}$$

This daily rate can then be multiplied by 365 days to obtain the (constant) yearly number of streams. The line at ~50.5 indicates this value. Using the daily rate for streams obtained above we can predict the number of streams that should have been seen based on the days of observations. These predicted values, adjusted for days of observations, are indicated by the stars and were obtained by multiplying the daily rate (from the equation) by the days (>0 pts) for each year as listed in Column (2) of Table VIII. The dashed lines denote the yearly sunspot numbers. The dotted/dashed lines denote the rough estimate of the normalized (extrapolated) yearly number of streams. These were obtained by simply assuming a constant rate of streams for that year based on the

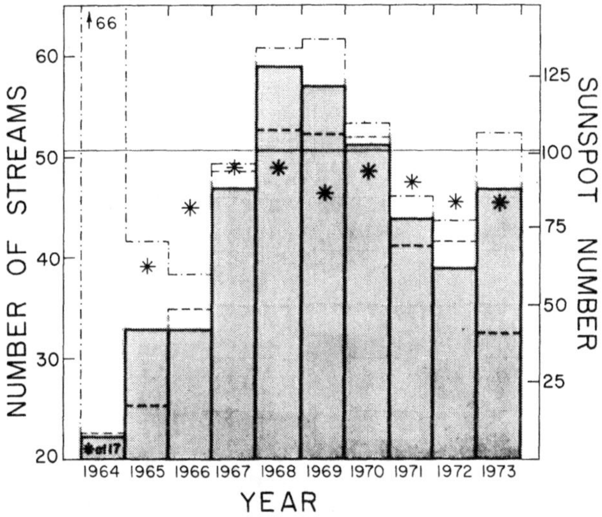

Figure 22. The shaded histograms show the number of high speed streams (with $\Delta V \gtrsim 150$ km/sec) in the solar wind that were observed based on the days of observations (Table VIII, Column (2)). The horizontal line at ~50.5 indicates the (constant) yearly number of high speed streams obtained by assuming that there is no solar cycle variation (see text). The stars indicate the predicted numbers of high speed streams obtained by multiplying the daily rate (.14) by the days (>0 pts) of observation listed in Table VIII, Column (2). The dashed line indicates the yearly sunspot number. The dotted/dashed line indicates the normalized (extrapolated) yearly values based on assuming a constant rate for each year (see text). In 1968 and 1969 the observed (shaded histograms) number of high speed streams exceed by ~1 standard deviation (σ) the predicted (constant) yearly value (the horizontal line). In 1965, 1966, and 1972 the observed number of high speed streams are less than the predicted values (the stars). These features are additional evidence for a solar cycle variation in the solar wind.

measured streams that year and the corresponding days of observations. (The poor data coverage for 1964 makes this (extrapolated) value very dubious).

In 1968 and 1969 the observed number of high speed streams exceed by ~1 standard deviation ($\sigma = \sqrt{N}$ for counting phenomena) the predicted (constant) yearly value. In addition, in 1965, 1966, and 1972 the observed number of high speed streams are less than the predicted values. Both of these features are additional evidence for a solar cycle variation in the solar wind.

The total duration (in days) of the high speed streams, with $\Delta V \geq 150$ km/sec, observed in each year from mid 1964 through 1973 are shown by the shaded histograms in Figure 23. If we estimate a daily rate by assuming that there is no solar cycle variation, then using the same approach as in the equation above, we obtain ~ .40 duration/day. As in Figure 22, the line represents the (constant) yearly value; the stars, the predicted values adjusted for days of observation; the dashed lines, the yearly sunspot numbers; and the dotted/dashed lines, the extrapolated yearly normalizations of the observations.

In 1968 and 1973 the observed durations exceed the predicted (constant) yearly value. Also, in 1965, 1966, and 1972 the observed durations are less than the predicted values adjusted for the number of days of observation. The departures (above in 1968, and below in 1965, 1966,

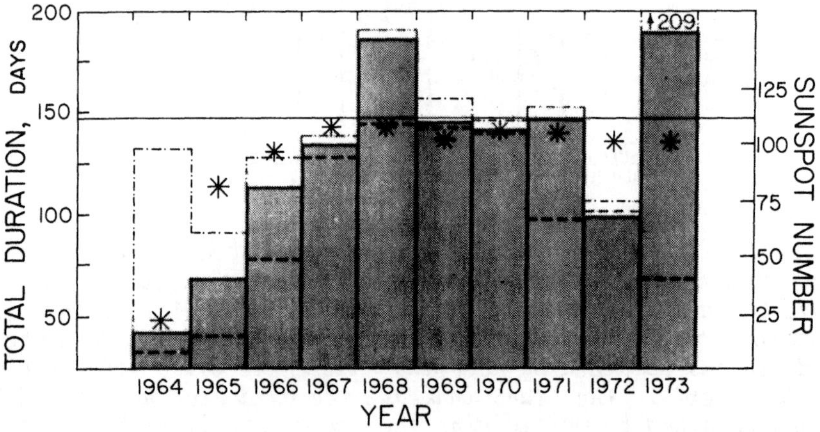

Figure 23. The shaded histograms show the total duration (in days) of high speed streams (with $\Delta V \geq 150$ km/sec) in the solar wind that were observed based on the days of observation. The horizontal line is the (constant) yearly value obtained from assuming that there is no solar cycle variation (see text and Figure 20 caption). As in Figure 20, the stars indicate the predicted durations adjusted for days of observations. See text and Figure 20 caption for explanation of other symbols.

and 1972) from predicted values provide additional evidence for a solar cycle variation in the solar wind. The increased duration in 1973 may be consistent with a shifting at solar minimum of the primary source of streams observed in the ecliptic plane as discussed below.

Similar results are obtained with the parameter the yearly average solar wind speed (in km/sec) for each year multiplied by the corresponding observed total duration (in days) of high speed streams (with $\Delta V \geq 150$ km/sec) for that year. Again assuming that there is no solar cycle variation, we estimate a daily rate of $\sim .18 \times 10^3$ (km sec^{-1} days) per day, thus yielding a (constant) yearly value of $\sim 6.6 \times 10^4$ km sec^{-1} days. In 1968 and 1973 the observed quantities exceed the predicted (constant) yearly value, and in 1965, 1966, 1967, and 1972 the observed quantities are less than those predicted. These trends are similar to those discussed above for the duration of high speed streams, as observed in the ecliptic plane, and provide additional evidence for a solar cycle variation in the solar wind and may imply a shift in 1973 of the primary stream source. It should be noted that in Vasyliunas (1975) there are also significant peaks in 1968 in both the "total duration" and "speed x duration."

In Bame et al. (1976) in the graph of the "average half-width (days)," if one considers the data from 1962 through 1971, there is a definite peak centered on 1968, a year of solar maximum. There are a number of other peaks in 1968 in the figures in their paper (e.g., "average maximum speed," "streams with $V_{MAX} > 700$ km/sec") that indicate an increased solar wind activity in 1968 as compared with the years 1965 through 1971. These authors, however, did not note these peaks and focused only on the high values obtained in years of solar minimum. The figure of the number of 'all streams' in Bame et al. (1976) fails to show a peak in 1968, which at first glance is difficult to reconcile with the obvious increase in high speed streams in 1968 found in all of our data. They, however, omitted from their data set the available (Intriligator and Neugebauer, 1975) solar wind data in late 1968 and early 1969, at the time of highest mean monthly sunspot count (Dodson, et al., 1974). If they had included this data in their study then their figure 'all streams' would have shown a definite peak in both 1968 and 1969. One should also note that their "normalization" of their data for 1968 and 1969 drastically underestimates the total number of streams they would have observed if they had included the data from late 1968 and early 1969. This argues against the reliance on the normalization techniques employed by Vasyliunas (1975) and Bame et al. (1976) and indicates the necessity of including all the available data and basing all substantive conclusions on the observations themselves as emphasized in Intriligator (1974).

The dark lines in Figure 24 indicate the relative frequency of occurrence of the solar wind streaming speed in intervals of 25 km/sec for each year in our data set from 1964 through 1973. The frequency histogram for the entire (parent) population composed of data from mid 1964 through 1973 is indicated by the lighter line. All the histograms

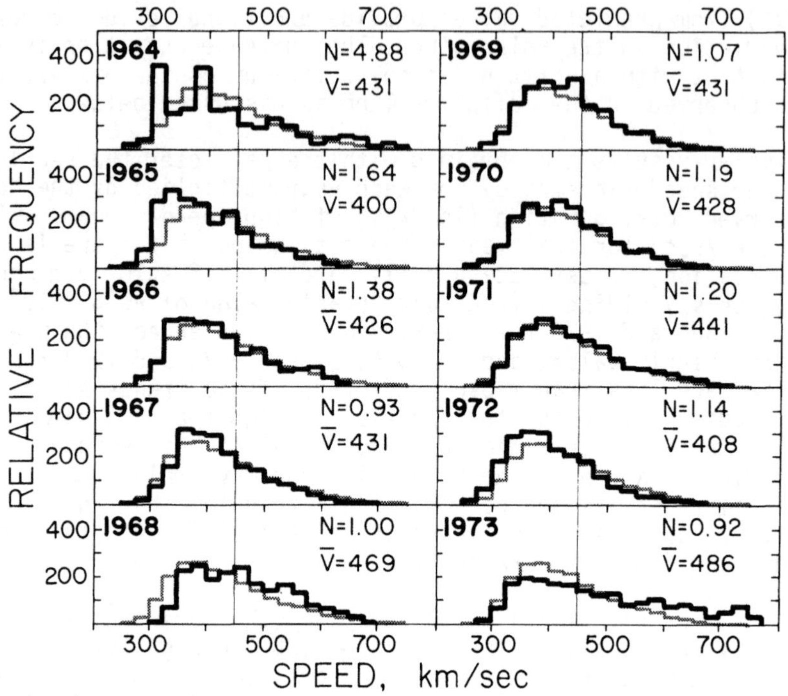

Figure 24. The dark lines indicate relative frequency of occurrence of the solar wind streaming speed in intervals of 25 km/sec for each year from 1964 through 1973. The light line indicates the frequency histogram of the speed for the entire (parent) population composed of all the data from mid 1964 through 1973. The histograms have been normalized (there is the same area under each histogram) as indicated by the N's in the figure. The yearly average solar wind speeds are shown by the $\bar{V}$'s. The histograms indicate that there is an increased number of high solar wind speeds (≥450 km/sec) in 1968. Table VIII, Column (6) shows that the highest percentage (54%) of elevated speeds is observed in 1968. This trend is present in the histograms of Gosling et al. (1971), however, they and Bame et al. (1976) did not note it.

have been normalized, as indicated by the N's in the figure, so that there are equal areas under the histograms. The histogram for 1968 is skewed toward high speeds and the histograms for 1965 and 1972 are skewed toward low speeds. As listed in Table VIII Column 6, the highest percentage (54%) of time during which the solar wind speed was ≥ 450 km/sec occurred in 1968 and the lowest percentages (23% and 26%) occurred in 1965 and 1972. This evidence for a solar cycle variation in the solar wind is also shown in the histograms for 1965 through 1970 in Gosling et al. (1971) which indicate a definite increase in speed in 1968. These authors, Bame et al. (1976), and Bridge (1976), however, did not note this trend in their data and erroneously concluded that there was no discernible upward shift in the bulk speed distributions with rising solar activity. The skewing of the 1973 histogram toward high speeds is consistent with the 1973 data in Figure 23, and may imply the shifting of the primary stream source in 1973 as discussed below. The probability is $\geq 95\%$ that in 1965, 1968, 1972, and 1973 the deviation of these yearly histograms from the parent is not merely a random fluctutation.

3. Summary of Long Term Variations

The observations above provide a great deal of additional evidence that confirm the earlier results reported by Intriligator (1974) that tabulations of high speed streams yield many statistically significant variations in the solar wind that are in phase with the solar cycle. There is likewise an evident variation in the yearly histograms of the solar wind speed. In addition, these data may imply a shift in the primary source of high speed streams in 1973.

The tabulations of the high speed stream parameters for the solar wind data available from mid 1964 through 1973 indicate that during this period there is a statistically significant peak in 1968 for each of these parameters. These peaks in 1968, the year of solar maximum, provide additional evidence for a variation in the solar wind that is in phase with the solar cycle. This evidence for a peak in 1968 is also indicated in several figures in Vasyliunas (1975), Bame et al. (1976), and Bridge (1976), however, these authors did not note these peaks in their data. Furthermore, if Bame et al. (1976) had included in their data set the available solar wind data for late 1968 and early 1969 (the time of highest mean sunspot count) then their figure 'all streams' would also have shown a definite peak in both 1968 and 1969. Their normalization of this data drastically underestimates the total number of streams present and argues against the reliance (as in Vasyliunas, 1975 and Bame et al., 1976) on their normalization techniques and indicates the necessity of including all available data and basing all substantive conclusions on the observations themselves as emphasized in Intriligator (1974) and the present paper. Histograms of the solar wind speed are presented that indicate that the highest percentage (54%) of elevated speeds ($V \geq 450$ km/sec) is observed in 1968. This is also indicated in the histograms in Gosling et al. (1971), however, these authors and also Bame et al. (1976) and Bridge (1976) did not note this trend. The data for 1973 in Figures 23 and 24 are consistent and may imply the shifting of the primary stream source in 1973.

Two possible sources of the increase in high speed streams, observed in the ecliptic plane in the vicinity of earth, in 1968 and 1969 are 1) sunspots or other solar equatorial phenomena and/or 2) coronal holes or some other solar feature mainly at higher latitudes, but for which there is some focusing effect or latitude effect that results in this manifestation in the ecliptic plane. The peak in sunspot number in 1968 and 1969 of solar cycle 20 is not as high as that associated with the peaks in cycles 18 and 19. If at solar maximum many high speed streams originate in sunspots, there would have been more if the sun had been more active; but if non-equatorial features caused these streams a more active sunspot cycle might have masked this effect. The rises observed in 1968 and 1973 suggest that equatorial sources predominate at the peak of the cycle, while off-equatorial sources produce streams observed near the ecliptic plane during this sunspot minimum (e.g., in 1973 coronal holes cover more latitude extending closer to the equator and the plasma and the field lines from coronal holes may extend further toward the ecliptic plane without interacting with the particles and fields associated with sunspots and other equatorial sources).

One can further speculate on the possible relation between the sunspots and high speed streams near solar maximum by calculating the correlation coefficient between the high speed stream parameters and the yearly sunspot number. For the years 1965 through 1972 extremely high correlation coefficients are found between each of the yearly normalized values of the high speed stream parameters (i.e., the dotted/dashed lines in Figures 22 and 23) and the yearly sunspot numbers. The correlation coefficients are, respectively, >.8 (more than 98% confidence), and ~.8 (more than 98% confidence). The times of highest solar wind stream activity in late 1968 and early 1969 coincide with the time of highest monthly mean sunspot count (Dodson et al., 1974). A similar calculation of the correlation between the high speed stream parameters and the inverse (in %) of the yearly average of the Deep River cosmic ray neutron monitor data for 1965 through 1972 yields, respectively, correlation coefficients of >.8, and ~.8 (with significances, respectively, above 99.5% and 98%). This may suggest a relationship between the high speed streams and the modulation of cosmic rays (Jokipii, 1971).

IV. Synopsis

In Section II of this paper we discussed the radial evolution of the solar wind proton streaming speed out to heliocentric distances as large as 6.4 AU. Individual high speed stream profiles were also examined as they propagated radially outward in the solar wind. We found that the evolution of the solar wind speed distributions and of individual high speed streams are consistent with the exchange of momentum in the solar wind between high speed streams and low speed streams as they propagate outward from the sun. Corotating interaction regions (Smith and Wolfe, 1976, 1977) are formed at extended heliocentric distances in association with the high speed streams. In Section III of this paper we discussed the long term evolution of the solar wind and found that there were significantly more high speed streams in 1968 and in 1969 than would have been predicted if there were a constant rate of high speed streams from mid 1964 through 1973. The long term high speed stream data and the long term solar wind speed histograms are consistent and may imply the shifting of the primary high speed stream source in 1973. The correlation coefficients between the high speed stream parameters and the inverse of the cosmic ray neutron monitor data may suggest a relationship between the high speed streams and the modulation of cosmic rays. This is consistent with the formation of the corotating interaction regions in association with the high speed streams at extended heliocentric distances. Some preliminary studies indicate that Forbush decreases have been seen on Pioneer 10 and 11 in association with the passage of corotating interaction regions. It is anticipated that additional Pioneer 10 and 11 data and further analyses of the data at extended heliocentric distances as well as other spacecraft data in the vicinity of earth will substantially add to our understanding of the radial variations and long term variations of the solar wind in the ecliptic plane and their relationship to other phenomena.

ACKNOWLEDGMENTS

I am indebted to J. H. Wolfe for use of the Pioneer data and to W. D. Miller for his participation in the data analysis. Some of the data used for this study were supplied by the NSSDC. This work was funded by NASA under Contract NAS2-7969 and Grant NGR-05-018-181, and by the University of Southern California.

REFERENCES

Bame, S. J., Asbridge, J. R., Feldman, W. C., and Gosling, J. T.: 1976, Astrophys. J. 207, 977.
Bridge, H. S.: 1976, in D. J. Williams (ed.), Physics of Solar Planetary Environments, Proceedings of the International Symposium of Solar-Terrestrial Physics, Boulder, Colorado, June, 1976, Vol. 1, American Geophysical Union, p. 47.
Collard, H. R. and Wolfe, J. H.: 1974, in C. T. Russell (ed.), Solar Wind Three, University of California, Los Angeles.
Diodato, L., Moreno, G., Signorini, C., and Ogilvie, K. W.: 1974, J. Geophys. Res. 79, 5095.
Dodson, H. W., Hedeman, E. R., and Mohler, O. C.: 1974, Rev. Geophys. and Space Physics 12, 329.
Gosling, J. T., Hansen, R. T., and Bame, S. J.: 1971, J. Geophys. Res. 76, 1811.
Intriligator, D. S.: 1973, Solar Geophysical Data, Report UAG-27.
Intriligator, D. S.: 1974, Astrophys. J. Letters 188, L23.
Intriligator, D. S.: 1975a, Astrophys. J. Letters 196, L87.
Intriligator, D. S.: 1975b, Conference Papers 14th International Cosmic Ray Conference, Paper P-10, Vol. 4, Max-Planck-Institut für Extraterrestrische Physik, Müchen, Germany, p. 1534.
Intriligator, D. S.: 1976, Space Sci. Rev. 19, 629.
Intriligator, D. S.: 1977a, J. Geophys. Res. 82, 603.
Intriligator, D. S.: 1977b, (preprint).
Intriligator, D. S. and Neugebauer, M.: 1975, J. Geophys. Res. 80, 1332.
Intriligator, D. S. and Wolfe, J. H.: 1974, Geophys. Res. Letters 1, 281.
Intriligator, D. S. and Wolfe, J. H.: 1976, in T. Gehrels (ed.) Jupiter, University of Arizona, p. 848.
Jokipii, J. R.: 1971, Rev. Geophys. and Space Phys. 9, 27.

Parker, E. N.: 1958, Astrophys. J. 128, 664.
Smith, E. J.: 1976, Space Sci. Rev. 19, 661.
Smith, E. J., Davis, L., Jr., Coleman, P. J., Jr., Colburn, D. S., Dyal, P., and Jones, D. E.: 1977, J. Geophys. Res. 82, 1077.
Smith, E. J. and Wolfe, J. H.: 1976, Geophys. Res. Letters 3, 137.
Smith, E. J. and Wolfe, J. H.: 1977, in M. A. Shea, D. F. Smart, S. T. Wu (eds.), Study of Travelling Interplanetary Phenomena/1977, in the series Astrophys. and Space Sci. Library, D. Reidel, Dordrecht, this volume.
Vasyliunas, V. M: 1975, Astrophys. J. Letters 202, L1.
Wolfe, J. H.: 1972, The Solar Wind (NASA SP-308), 170.
Wolfe, J. H., Collard, H. R., Mihalov, J. D., and Intriligator, D. S.: 1974a, Science 183, 303.
Wolfe, J. H., Mihalov, J. D., Collard, H. R., McKibben, D. D., Frank, L. A., and Intriligator, D. S.: 1974b, J. Geophys. Res. 79, 3489.

DISCUSSION

Jockers: Is there a correlation between the exponential decay length of the width of the velocity distribution and the number of streams?
Intriligator: The exponential decay length is correlated with the width of the speed distribution itself since the broader the speed distribution the larger the exponential decay length. We are currently gathering statistics on the number of high speed streams in various solar rotations and the width of the speed distributions for these same rotations.

Iucci: Do you have information on the magnetic field polarity along the streams?
Intriligator: The observations of the magnetic field associated with these streams will be discussed in the next paper by Smith and Wolfe.

Iucci: You have suggested the existence of two different origins for fast streams - Do you think there is any connection between the changes in the decay-length in the stream velocities and their origin?
Intriligator: In terms of the long term variations in the solar wind, we are suggesting a different origin for the streams in 1973 from those in the years near solar maximum. On the other hand, in terms of the large scale variations in the solar wind speed distributions and the implied decay-length of these variations in the examples that I showed, there were large differences between successive solar rotations in 1973 and 1974. So then the question would be whether there is a difference in origin between one large broad high speed stream in 1973 compared with other large broad high speed streams in 1973 (or 1974).

Gall: Has any study been done on correlation between the fast streams and coronal holes?

Intriligator: We have not done such a study. We agree that it is important for this type of study to be performed and we have, in fact, tried to carry out one but we could not obtain the coronal hole data. If any of you have access to the coronal hole data we would be very interested in conducting a joint study and we will contribute our solar wind data.

Dryer: It has previously been reported by Gosling and his colleagues that an opposite effect (that is, that the number of solar wind streams during solar maximum was smaller, not larger) was found. Was their result based on their examination of a smaller set of spacecraft data?

Intriligator: The histograms in Gosling et al. (1971) actually indicate that there is a definite increase in the solar wind speed in 1968; however, these authors and also Bame et al. (1976) and Bridge (1976) did not note this trend in their data and erroneously concluded that there was no discernible upward shift in the bulk speed distributions with rising solar activity.

In terms of the high speed stream parameters, there are peaks in both the "total duration" and "speed times duration" in Vasyliunas (1975). In Bame et al. (1976) in the graphs of "average half-width (days)" if one considers the 1962 through 1971 data there is a definite peak centered in 1968. There are a number of other peaks in 1968 in their figures such as "average maximum speed" and "streams with V_{Max} > 700 Km/sec." These authors and also Bridge (1976) did not note these peaks and focused only on the high values obtained in 1973. In Bame et al. (1976), the figure of the number of "all streams" fails to show a peak in 1968 since these authors chose to exclude from their data set the Pioneer 9 and OGO 5 data (Intriligator and Neugebauer, 1975) obtained at the height of solar maximum. If they had included this data in their data set then they would have had a definite peak in both 1968 and 1969.

PIONEER 10, 11 OBSERVATIONS OF EVOLVING SOLAR WIND STREAMS AND SHOCKS BEYOND 1 AU

E.J. SMITH
Jet Propulsion Laboratory
Pasadena, California 91103, U.S.A.

J.H. WOLFE
Ames Research Center
Moffett Field, California 94035, U.S.A.

ABSTRACT

Observations between 1 and 5 AU by Pioneers 10 and 11 have led to the identification of large numbers of interplanetary shocks. Both forward and reverse shocks, which begin to develop beyond 1.5 AU and which frequently appear as shock pairs, are found to accompany solar wind streams. The number of forward shocks continues to increase out to at least 5 AU. Reverse shocks are seen less often than forward shocks and, in some instances, disappear at larger distances. There is evidence that the shocks are corotating in the solar frame, as anticipated theoretically. The evolution of solar wind streams beyond 1 AU is profoundly affected by the shocks. A thick interaction region, with large enhancements in density, temperature, field strength and fluctuation level, forms in the region originally characterized by a positive velocity gradient. The solar wind and magnetic field properties adjacent to, and within, the interaction regions have been studied to determine their qualitative behavior and characteristic changes with distance. Several interplanetary shocks generated by solar flares have also been identified and analyzed.

INTRODUCTION

Probably the most characteristic feature of the solar wind near 1 AU is the existence of high speed streams. It is possible to organize most of the interplanetary parameters, such as the plasma density, N, temperature, T, and magnetic field strength, B, in terms of the stream structure represented by variations in velocity, V.

M. A. Shea et al. (eds.), Study of Travelling Interplanetary Phenomena/1977, 227–257.

Thus, it is known that in general N anticorrelates with V, T is well correlated with V, and increases and decreases in B are also related to the peak velocity associated with the stream (Neugebauer and Snyder, 1966; Belcher and Davis, 1971; Ness et al., 1971).

Although it has been customary to view the high velocity solar wind stream as an entity, it is probably more meaningful physically to distinguish between regions on the basis of the gradient in V. The positive gradient at the leading part of a stream tends to be dominated by physical processes associated with compression, whereas, in the trailing part, rarefaction effects are important. In the past, the positive gradient region, in which so called stream-stream interaction effects are occurring, has received the most attention.

The Pioneer mission has successfully fulfilled expectations that it would provide answers to basic questions about the evolution with distance of solar wind streams and thereby reveal much about the internal dynamics of the solar wind. One of the issues which has been settled concerned the possible development of hydromagnetic shocks as the fast solar wind pushed its way through the slower moving wind preceding it (Hundhausen, 1973). Theoretical speculation that stresses propagating away from the compression region would steepen into shocks beyond 1 AU has been strikingly confirmed by the Pioneer field and plasma observations (Smith and Wolfe, 1976; Hundhausen and Gosling, 1976).

The observations have shown that beyond a distance of about 1.5 AU a pair of shocks develop at the inner and outer edges of a relatively thick region, called the interaction region, within which the field and plasma are greatly compressed. One of the shocks is a forward shock which propagates outward from the sun. The other half of the shock pair is a reverse shock, so-called because it travels backward toward the sun in the solar wind frame or, in other words, advances into the trailing solar wind (Razdan et al., 1965; Colburn and Sonett, 1966). Because the velocity of the reverse shock is substantially less than the speed of the solar wind through which it propagates, the reverse shock is actually convected outward in the inertial frame.

It is the purpose of this article to review the observations of interplanetary shocks and high speed streams obtained with Pioneers 10 and 11. Simultaneous plasma and field data are available from Pioneer 10 between 1 and 8.5 AU and from Pioneer 11 between 1 and 5 AU (see Figure 1). The data include several years of observations made simultaneously at both spacecraft. Fortunately, the telemetry coverage of both spacecraft over these distances has been excellent and essentially continuous data are available for intervals of several years.

The observations have revealed very clearly the evolution of solar wind streams over this distance range. They have revealed also the profound effect this evolution has on the structure of the distant

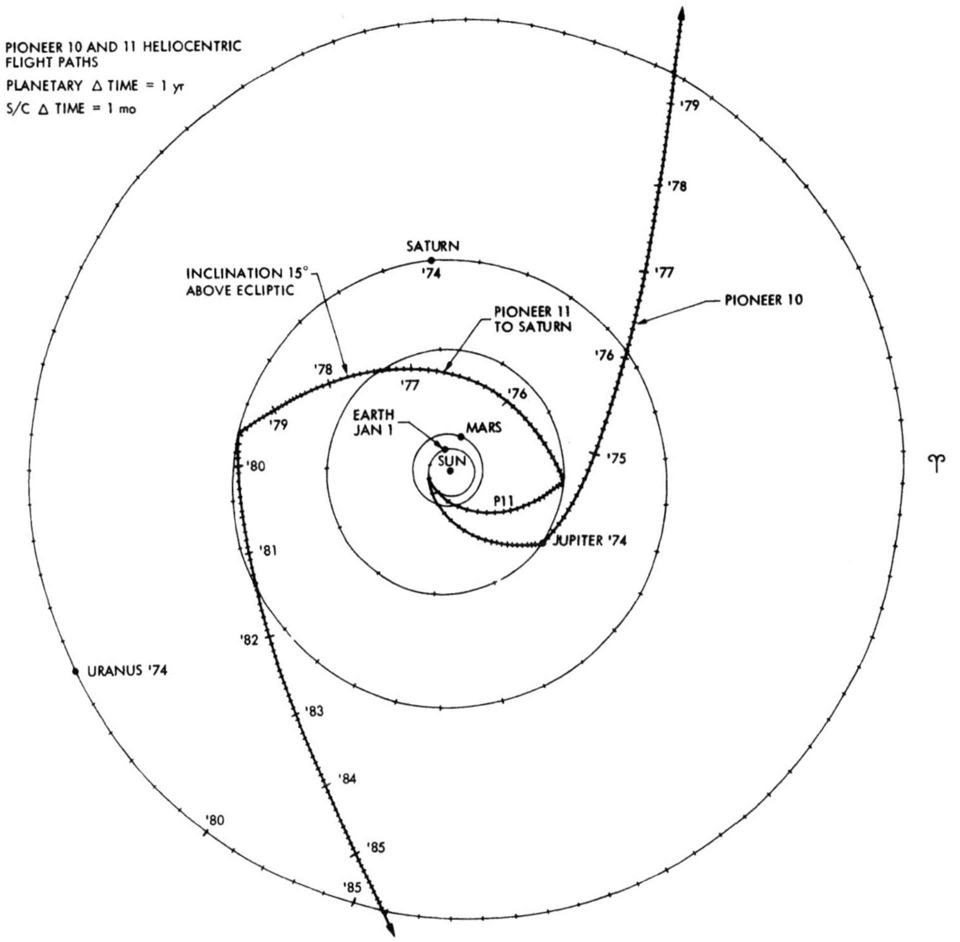

Figure 1. Heliocentric Flight Paths of Pioneers 10 and 11. This ecliptic view, with T representing the first point in Aries, shows the encounters with Jupiter and the planned encounter of Pioneer 11 with Saturn. The post-Saturn-encounter trajectory which is shown as going in the direction of Uranus is only one of several possibilities.

interplanetary medium. The observations have established that the shocks, and the changes which they effect in the medium, play an important role in accelerating energetic particles in interplanetary space and strongly influence the propagation of galactic cosmic rays, solar flare particles and Jovian electrons. Finally, there have been several observations of flare-generated shocks at large distances and these will also be discussed briefly.

STRUCTURE OF THE INTERPLANETARY MEDIUM BEYOND 1 AU

The medium through which fast solar wind streams propagate appears to be reasonably consistent with a spherically symmetric solar wind model (Parker, 1963). The most probable values of the field components and spiral angle over a solar rotation, which correspond in general with intervals of magnetic quiet, follow the theoretical predictions reasonably well (Smith, 1974). Long term averages of the interplanetary field have also been shown to be consistent with the predicted r^{-2} decrease of the radial component and with a direction corresponding to an Archimedes spiral (Parker and Jokipii, 1976; Rosenberg et al., 1977). On the average, the solar wind speed is relatively constant while the ion flux, NV, exhibits an r^{-2} decrease as expected.

However, the evolution of the solar wind streams as they propagate outward actually leads to a basic change in the structure of the medium. Two types of regions can be distinguished which are shown schematically in Figure 2. Interaction regions develop rapidly beyond 1 AU and are observed as clearly defined enhancements in B, N, T and the general level of fluctuation (Smith and Wolfe, 1976). They replace solar wind regions nearer the sun within which a positive velocity gradient occurs. Use of the term, interaction region, implies that within such regions fast and slow moving plasmas are interacting, principally by means of a transfer of momentum and energy involving large amplitude waves including shocks. Between the interaction regions are quiet regions in which B, N, T and their fluctuation levels have relatively low values. In quiet regions, the solar wind velocity tends to decrease monotonically.

Since the streams which give rise to these regions are usually long-lived, and persist from one solar rotation to another, the interaction regions can be considered to be corotating with the sun. Accordingly, they have come to be called, Corotating Interaction Regions or CIRs. During the interval from 1973-1974, the solar wind stream structure was unusually stable, a condition which may be characteristic of the minimum in the sunspot cycle (Wolfe, 1972; Bame et al., 1976). Typically, two large and well defined streams were seen each solar rotation and were observed for many solar rotations. The corotating nature of the interaction regions and quiet regions is clearly demonstrated by the observation of specific regions, first at Pioneer 11 and then, after the appropriate time delay, at Pioneer 10.

Adjacent interaction regions and quiet regions are usually separated by a corotating shock. Interaction regions usually begin at a forward shock and end at a reverse shock (Figure 3; Smith and Wolfe, 1976). Although not all interaction regions begin or end with a shock, their onsets and terminations are, nevertheless, easy to identify

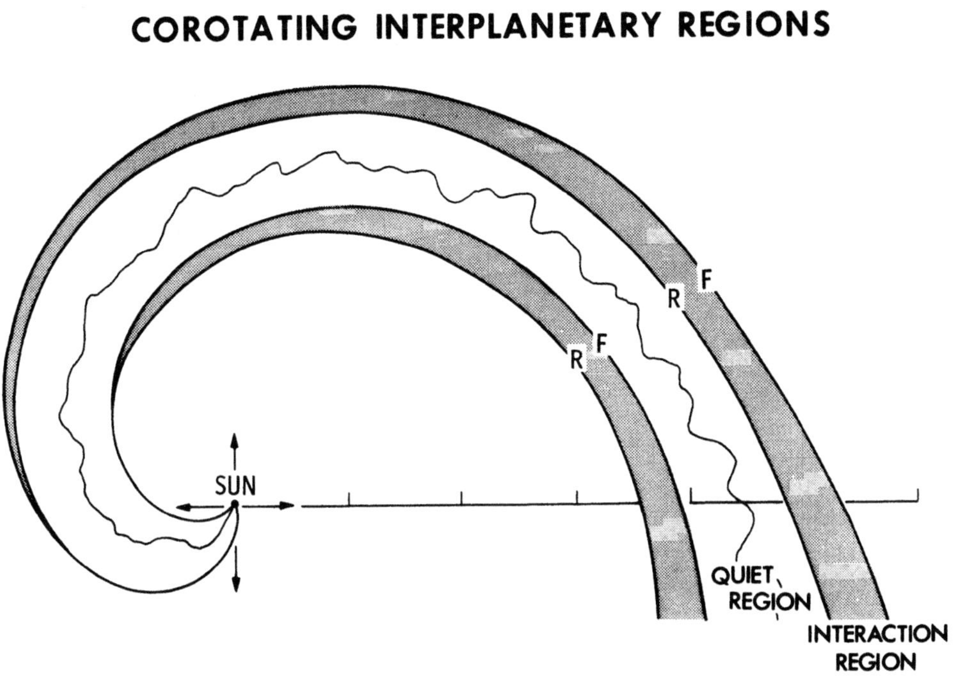

Figure 2. Corotating Interplanetary Regions. Two interaction regions and an intervening quiet region are shown. F and R indicate forward and reverse shock, respectively. An interplanetary field line is shown within the quiet region.

because the field and plasma parameters almost always change abruptly on a time scale of approximately one hour. Presumably, these boundary surfaces are in the process of steepening into shocks which will be fully developed at larger heliocentric distances. There are also other possibilities; e.g., at large distances, the shocks may have already developed nearer the sun and may be decaying back into large amplitude waves. As shown in Figure 2, the leading and trailing edges of the regions (indicated as F, forward shock and R, reverse shock) take the form of spirals, much like the Archimedes spirals traced out by the interplanetary magnetic field.

The characteristic evolution of high speed streams from 1 to 5 AU and beyond is clearly revealed by the Pioneer observations (see Figure 4). The following is an oversimplified but convenient description of this evolution which represents the majority of cases. Once the shocks form, they tend to obliterate positive gradients in the solar wind

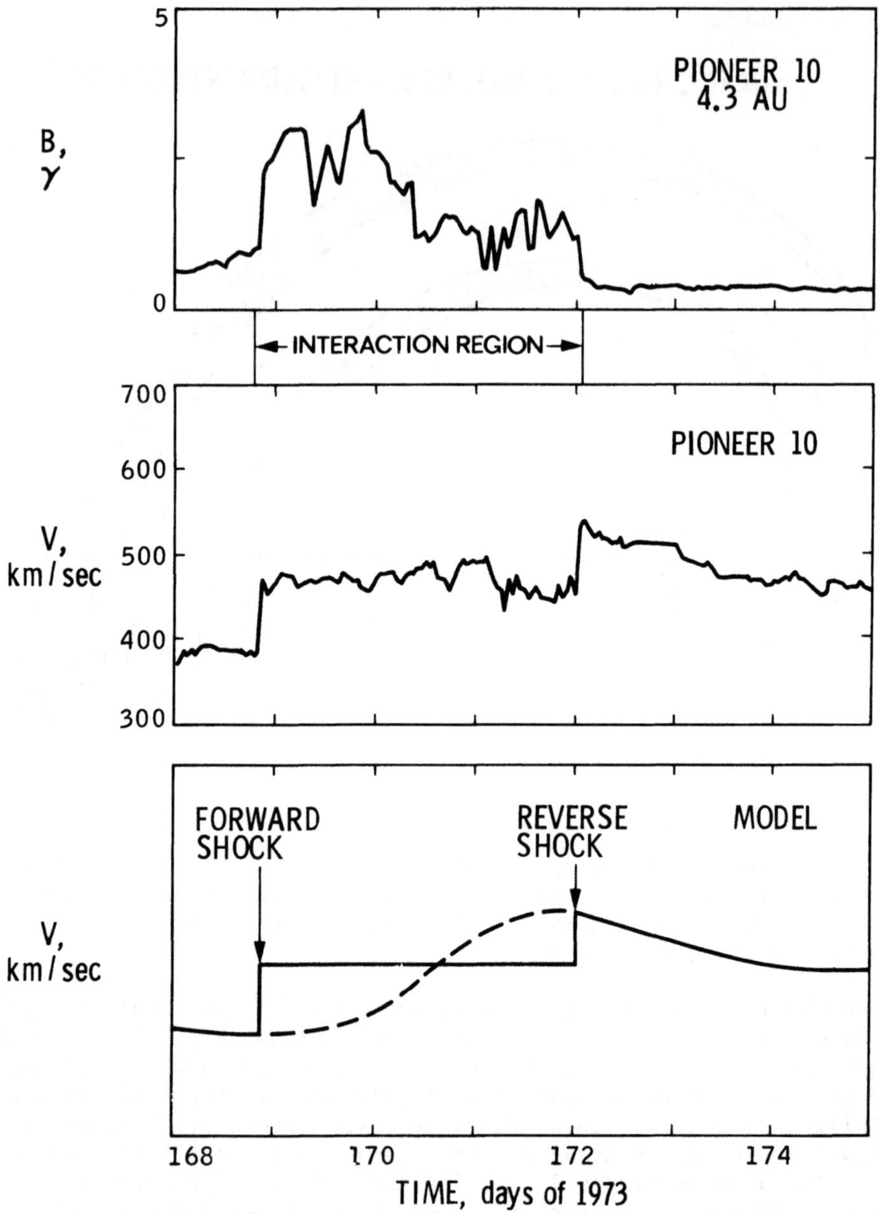

Figure 3. Example of an Interaction Region. The magnetic field and velocity changes in the upper two panels may be compared with the model in the bottom panel in which the dashed curve represents V near 1 AU.

speed and, eventually, to eliminate the large velocity differences originally associated with the stream. The shock at the leading edge of the interaction region continues to propagate forward to accelerate

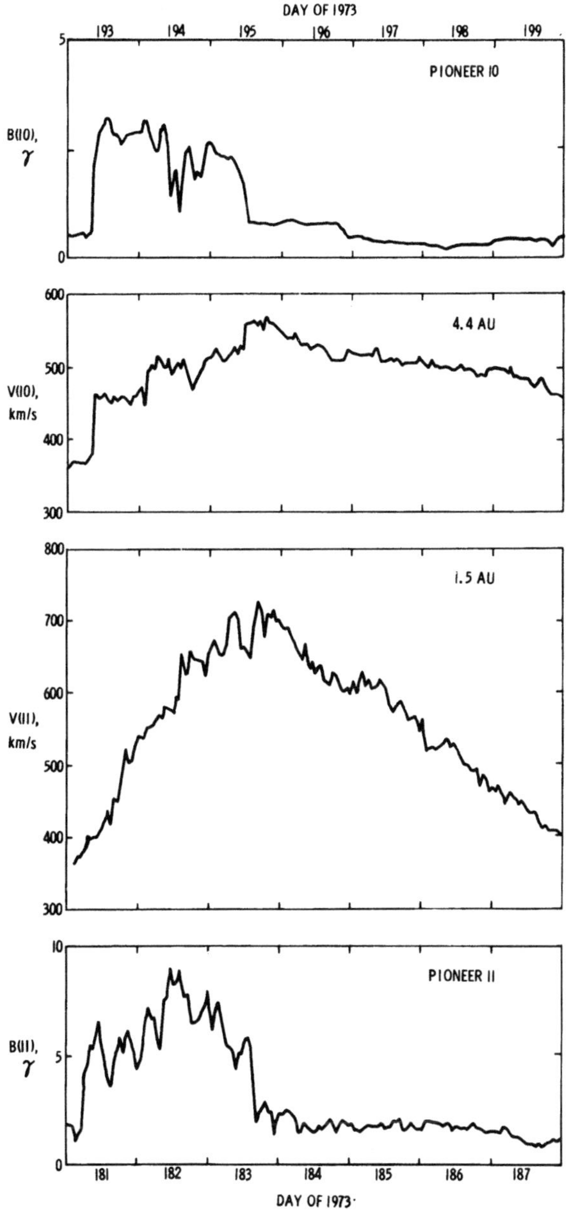

Figure 4. Evolution of a High Speed Stream Between Pioneer 11 and Pioneer 10. The two upper panels show the field and velocity at Pioneer 10, B(10) and V(10), and the two lower panels show V(11) and B(11) corresponding to the same stream at Pioneer 11.

the slower moving solar wind while the reverse shock gradually erodes the peak of the stream and then proceeds into the region of negative gradient or decreasing solar wind speed. Of course, other processes, such as the generation of waves and discontinuities, may also contribute to the elimination of velocity gradients. These effects apparently take place without significantly effecting the average velocity of the solar wind which is approximately the same at large distances as near 1 AU (Collard and Wolfe, 1974).

As the shocks recede from one another, the interaction regions continually widen at the expense of the quiet regions. This feature is also shown schematically in Figure 2. At a radial distance of several AU, the interaction regions are already approximately 1 AU wide and, thus, are comparable in width to the quiet regions. Eventually, at large distances, estimated to be in the vicinity of 10-20 AU, depending on the number and nature of the high speed streams, the interaction regions will have obliterated the quiet region and will begin to interact with one another. The interplanetary medium will then presumably undergo a change to another regime, perhaps characterized by the formation of large numbers of discontinuities as a result of shock-shock interactions and by the presence of widespread irregularities throughout the plasma.

OBSERVATIONS OF COROTATING INTERPLANETARY SHOCKS

As the heliocentric distance increases, shocks gradually begin to appear in the Pioneer plasma and field data. The first shocks are seen near the orbit of Mars (1.4 AU) and their number gradually increases. The first shocks to be observed are forward shocks but at larger distances reverse shocks begin to appear in increasing numbers. From 3 to 5 AU, the typical structure consists of a shock pair.

Several examples of shock pairs as observed in the velocity data are shown in Figure 5. The data consist of one hour samples of the solar wind speed. As can be seen, the usual positive velocity gradient at the front of the streams has been replaced by two abrupt jumps associated with the forward and reverse shocks. Based on inspection of a large number of shock pairs, the average solar wind speed between the forward and reverse shocks appears typically to be constant.

The identification of these abrupt changes in the plasma parameters as shocks has been confirmed by examining high time resolution magnetic field and plasma data. Examples of the field data are shown in Figure 6 which contains several forward shocks. Time resolution is typically a fraction of one second. Each shock involves a large jump in field magnitude. The general similarity to the Earth's bow shock may be noted, including the presence of large amplitude irregularities, which are frequently quasi-periodic and are presumably waves. Waves are also often seen upstream of the interplanetary shocks.

Reverse shocks are also seen clearly in the magnetic field data. Figure 7 shows several examples at high time resolution. Again abrupt changes (in this case decreases) are observed in B. As expected, these decreases in B are accompanied by simultaneous decreases in N

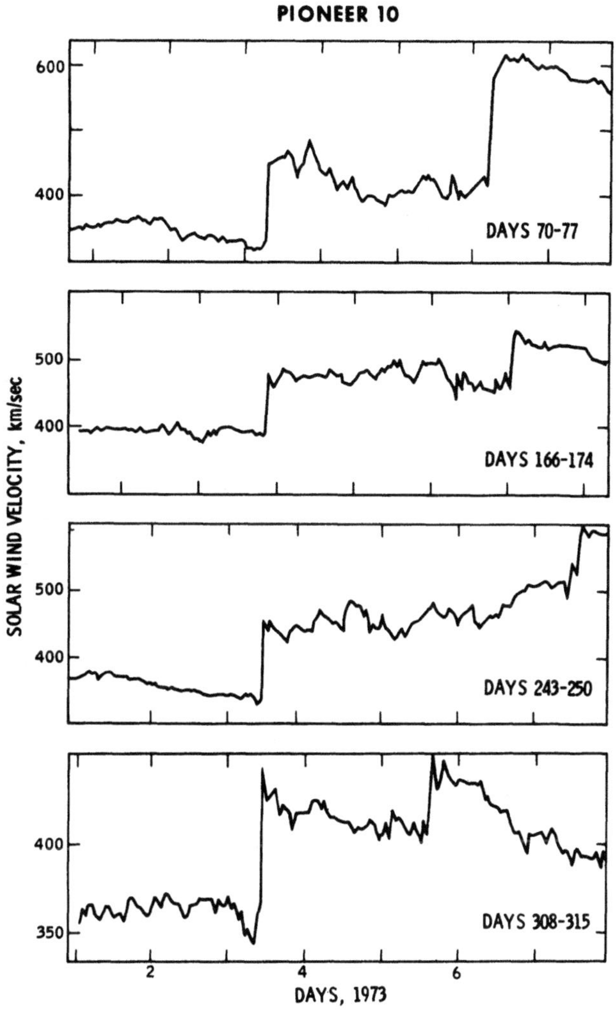

Figure 5. Examples of Shock Pairs.

and T. The reverse shocks are also noisy; however, the largest fluctuations appear before the shock arrives at Pioneer, i.e., they occur principally on the downstream side of the shock as expected.

Both the forward and reverse shocks have the appearance of being perpendicular or quasi-perpendicular shocks, i.e., they involve large changes in field magnitude with little or no change in direction. These are the characteristics of shocks which have their normal perpendicular to the average field direction. Perpendicular shocks

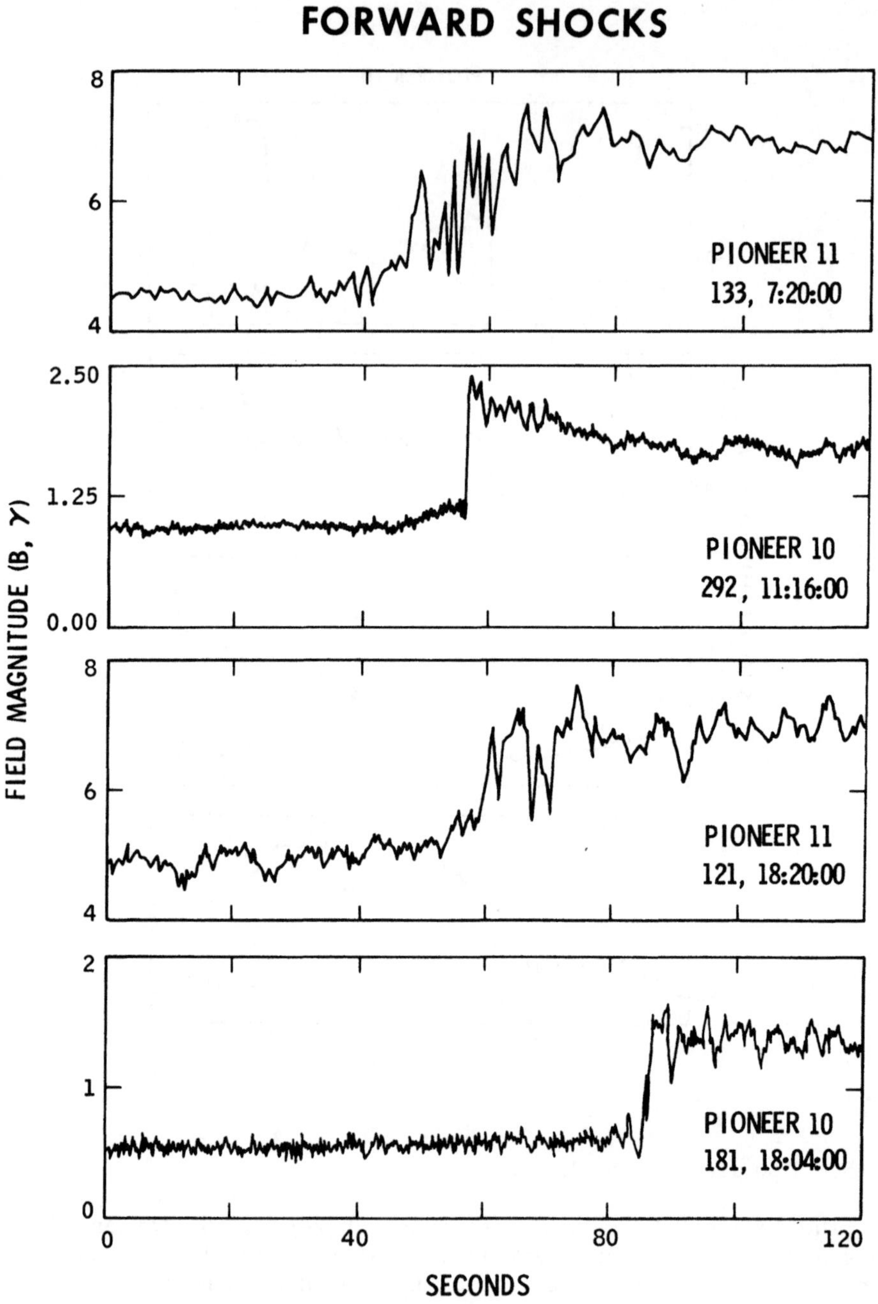

Figure 6. Examples of Forward Shocks.

travel across the field and are, therefore, to be expected in the Pioneer data, since the field at large distances tends to be princi-

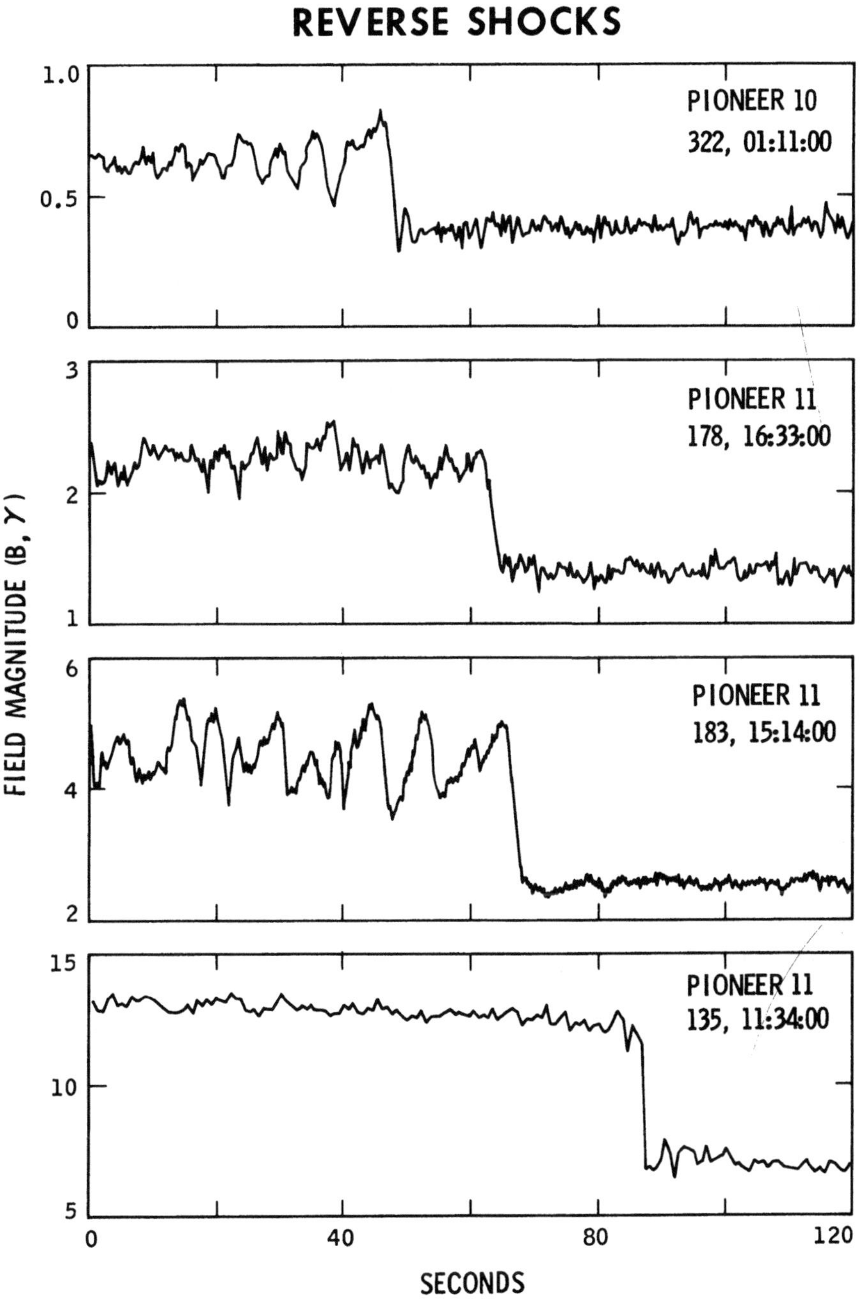

Figure 7. Examples of Reverse Shocks.

pally azimuthal while the shocks are expected to be propagating in essentially the radial direction.

The high resolution field data can be used to infer shock thicknesses at large heliocentric distances. As Figures 6 and 7 show, the typical transit time for the shocks to pass by Pioneer is approximately 2 seconds. Using an average velocity of propagation of 453 km/sec, which is derived below, it can be inferred that the thickness is typically 10^3 km.

The observation of a large number of shocks beyond 1 AU has permitted statistical studies to be carried out and the simultaneous observations of the same shocks at the two spacecraft has made it possible to study their propagation at large heliocentric distances. Whereas interplanetary shocks are observed relatively infrequently near 1 AU, principally in conjunction with solar flares, a very large number of shocks have been identified in the Pioneer data. Approximately 50 shocks have been identified in the Pioneer data from 1973-1974. Furthermore, it has been possible to identify corresponding shocks at both locations, a task that was facilitated by the nearly radial alignment of the spacecraft for a relatively long interval at the end of 1973 and beginning of 1974.

A statistical study has been carried out of the rate of occurrence of shocks and its dependence on heliocentric distance. The rate at which forward and reverse shocks develop as a function of distance is shown in Table I and Figure 8. Each shock tabulated is associated with an interaction region. Since the interaction regions are easy to identify, with or without shocks, they serve as a convenient normalization constant in this analysis. The rate of occurrence has been calculated based on the number of possible shocks corresponding to the number of interaction regions observed.

There is clear evidence in the table and Figure 8 that forward shocks continue to develop between 1 and 5 AU. Near 5 AU, approximately 95% of all possible forward shocks have actually formed. Reverse shocks develop more slowly and form less often than forward shocks. For example, at 5 AU only 70% of all possible reverse shocks are seen.

Furthermore, unlike forward shocks, some reverse shocks seen at Pioneer 11 disappear before reaching Pioneer 10, approximately 2 AU further away. This result has been checked by examining in detail several cases where a reverse shock appears at Pioneer 11 but is not seen at Pioneer 10. Presumably reverse shocks decay back into large amplitude waves. One explanation of this phenomenon is that the reverse shock propagates into a region of gradually decreasing solar wind speed until, finally, it is unnecessary for the inflowing plasma to undergo a large change in V.

Observations of these same shocks at both Pioneers have been used to study their transit time, T, and to derive average shock velocities. Figure 9 is a plot of the transit times as a function of the radial distance of Pioneer 11. Transit times corresponding to 24 shocks and

STATISTICS OF SHOCK OBSERVATIONS

SHOCK TYPE \ SPACECRAFT	P(11), 1-3 AU	P(10), 4.1-5.1 AU	P(11), 3.1-4.9 AU	P(10), 5.1-6.3 AU
FORWARD	23/26=0.88	25/26=0.96	22/23=0.96	18/20=0.90
REVERSE	18/30=0.60	16/23=0.70	14/19=0.74	10/18=0.56

TABLE I

representing a range in heliocentric distance from 1 to 3 AU are shown. There is a characteristic monotonic decrease in the transit times with increasing time or, equivalently, with increasing Pioneer 11 distance. The transit times approach a value of 6 days corresponding to the interval in which the two spacecraft were nearly radially aligned.

The transit times show clear evidence that the shocks are corotating. When the longitude separation of the two Pioneers is large, there is a correspondingly large contribution to the transit time associated with a rotation of the solar source region from the longitude of Pioneer 11 to the longitude of Pioneer 10. During the early part of the mission, the longitude separation of the two spacecraft implied a corotation delay, $T_\phi = \Delta\phi/\Omega_s$, of 8 days. The radial separation during the same interval was $\simeq$ 2.5 AU and was decreasing much more gradually than the longitude separation. By the end of 1973, the two spacecraft were nearly radially aligned and the corotation delay was essentially zero. This reduction in T_ϕ basically accounts for the monotonic decrease in T observed in Figure 9.

The influence of the corotation delay on the transit times has been demonstrated by computing the transit times corresponding to two different values of radial velocity. The transit times were computed by adding the radial delay, $T_r = \Delta R/V$, to T_ϕ. These total delays are shown in Figure 9 as the two curves which bracket the observations. It is seen that the observed transit times lie within these two bounds and, thus, are consistent with radial propagation velocities between 360 and 550 km/sec.

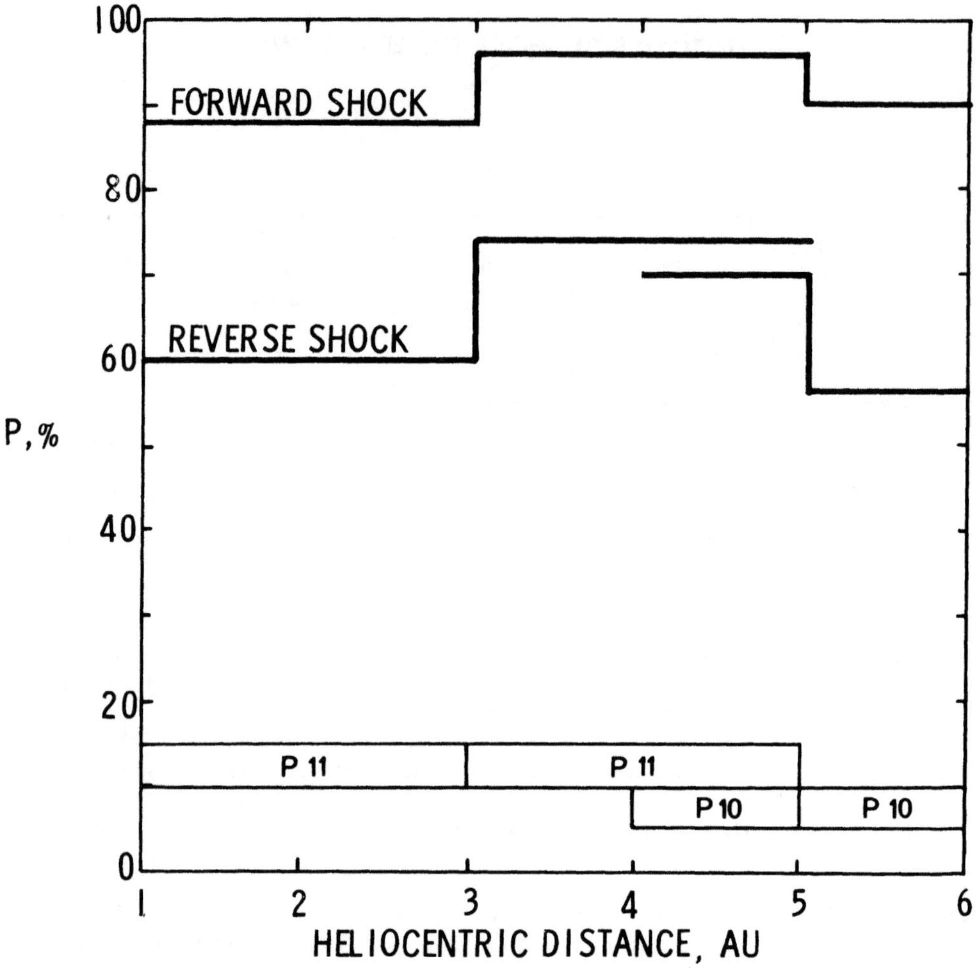

Figure 8. Rates of Occurrence of Shocks as a Function of Heliocentric Distance. The bars (solid for forward shocks and for reverse shocks) correspond to the percentages given in Table I and each bar represents one year of observations at each spacecraft.

The radial velocities of propagation of the shocks between the two Pioneers have also been derived. The contributions due to corotation have been subtracted from the shock transit times to obtain the radial propagation velocities: $V_R = (R_{10}-R_{11})/(T-T_\phi)$. The results are shown in Figure 10. In this, and the preceding figure, the forward and reverse shocks have been treated separately. The average velocities are about the same for both. Furthermore, there is no obvious change of the velocity with heliocentric distance for either

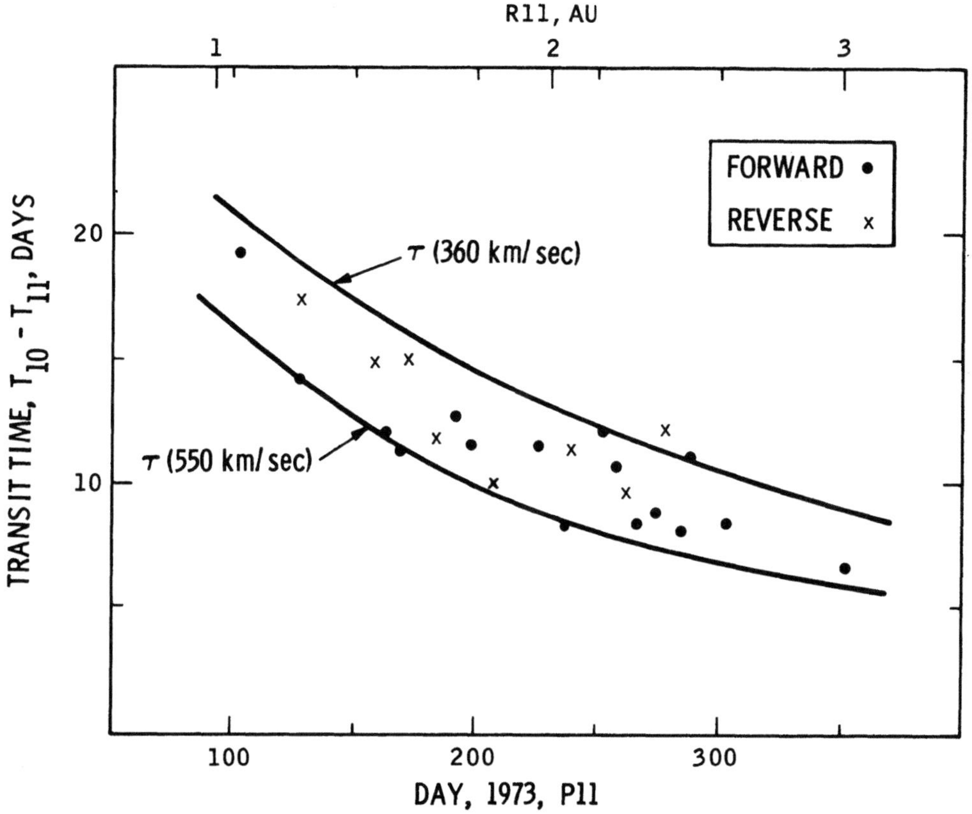

Figure 9. Transit Times of Shocks Between Pioneer 11 and Pioneer 10. The abscissa is the time of arrival of the shocks at Pioneer 11 with the corresponding heliocentric distance being given at the top.

type of shock. The average velocity and standard deviation of all the shocks have been computed with the resulting values shown in the figure: 453 $\pm$ 64 km/sec. It should be noted that these are inertial velocities corresponding to the sum of the shock velocity in the solar wind and the inertial velocity of the solar wind.

The approximate equality of the speeds of the forward and reverse shocks probably has a simple explanation. The velocity of the forward shock is added to the speed of the solar wind through which it propagates, but the latter tends to be the trailing edge of a high speed stream and is, accordingly, relatively slow. The velocity of a reverse shock, on the other hand, is subtracted from the solar wind speed. However, since the reverse shock propagates into high velocity

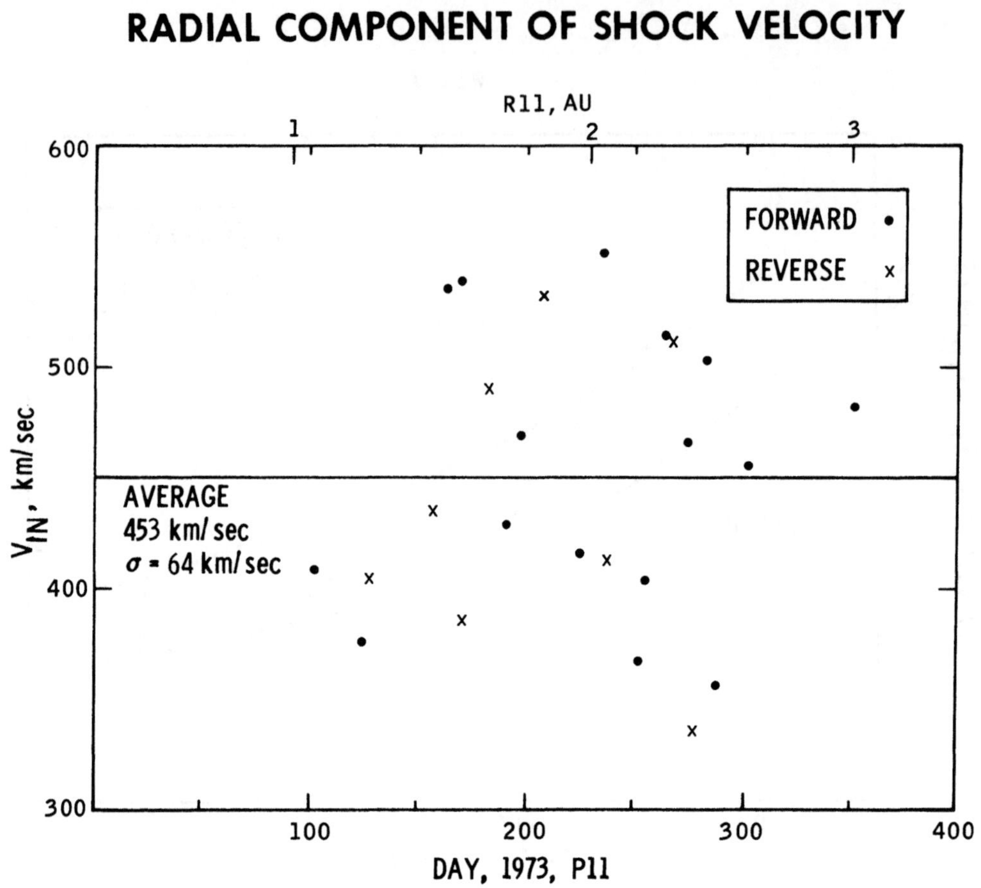

Figure 10. Radial Velocities of Propagation of Shocks. The velocities are plotted at the day of 1973 that they were observed at Pioneer 11. Distance is shown at the top.

solar wind, its inertial speed remains relatively high. Another reason for this result, as discussed below, is that both forward and reverse shocks propagate at relatively slow speeds in the solar wind so that their inertial speeds do not deviate greatly from the solar wind speed.

If allowance is made for the solar wind speed, the shock velocities in the solar wind frame can be inferred. Figure 5 shows typical solar wind speeds of 350 km/sec upstream and 550 km/sec downstream of the interaction regions. Thus, the average inertial speeds of the forward and reverse shocks imply velocities in the solar wind frame of approximately 150 km/sec for both types of shock.

These velocities may be compared with typical values of Alfvén and magnetosonic phase velocities at these large heliocentric distances and an estimate obtained of the corresponding Mach numbers. Thus, for

$B = 3\gamma$, $N = 2\ cm^{-3}$ and $T = 4 \times 10^4 K$, these speeds are 50 to 60 km/sec. It is seen that the shocks have relatively low Mach numbers between 2 and 3.

OBSERVATIONS OF COROTATING INTERACTION REGIONS

Interaction regions can be identified using any of a large number of solar wind parameters or preferably a combination of parameters. Our identification has been based principally on B which exhibits relatively abrupt increases and decreases at the onset and termination of the interaction region. Ambiguities sometimes arise in connection with small interaction regions which involve only modest increases in B and the other parameters. Furthermore, two interaction regions will occasionally merge before reaching one or the other of the spacecraft. It is then difficult to determine when the first interaction region ended, although it is still easy to determine when the other begins.

The recurrent nature of the interaction regions demonstrates that they are corotating (Smith and Wolfe, 1976). The same interaction region associated with a recurrent high speed stream is seen for many solar rotations. One interaction region, for example, has been seen for at least 12 successive solar rotations. With our simultaneous Pioneer 10, Pioneer 11 observations, interaction regions can be followed from the inner to the outer spacecraft. As in the case of corotating shocks, during the early part of the mission, a significant part of the transit time delay is contributed by the corotation delay associated with the separation in longitude of the two Pioneers.

Evidence of the gradual widening of the CIRs is provided by Figure 11 which shows a pair of histograms of the widths of CIRs observed at Pioneer 11 and Pioneer 10, respectively. During this interval, Pioneer 11 traveled over a radial distance range of 1.0 to 3.3 AU while Pioneer 10 was traveling from 3.0 to 5.1 AU. The most probable value of the width near 2 AU is seen to be 1.5 days, whereas near 4 AU it has increased to 2.5 days. Several histograms of this kind, covering different distance ranges, have been derived and the most probable values are plotted as a function of distance in Figure 12. As can be seen, the CIRs widen at a rate of about 1 day/AU.

The widening of the CIRs is to be expected whether or not shocks form. Stresses that develop within the CIRs propagate both inward and outward as large amplitude hydromagnetic waves which gradually steepen into shocks. The leading edge travels faster than the solar wind plasma ahead of, and within, the interaction region while the trailing edge travels more slowly than the plasma within the interaction region.

In spite of this gradual widening, a substantial compression of the solar wind plasma occurs within the CIRs. This compression compensates for the decrease in plasma parameters that would be expected in the case of unimpeded solar wind flow. For example, the peak value in B has been found in numerous cases to be the same at 4 to 5 AU

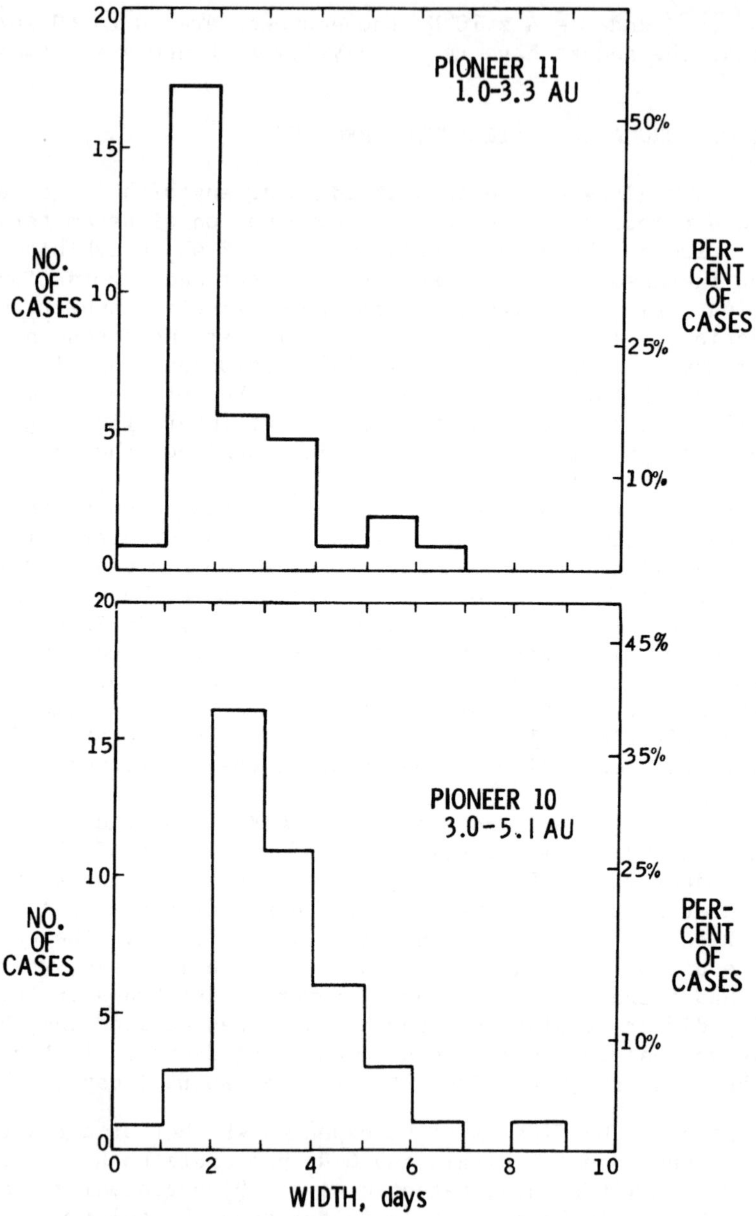

Figure 11. Widths of Corotating Interaction Regions Over Two Different Distance Ranges.

as it is nearer the sun at 2 to 3 AU. This effect produces substantial deviations of B, N, T, etc. from the radial gradients anticipated on the basis of spherically symmetric solar wind flow, e.g., Parker's model.

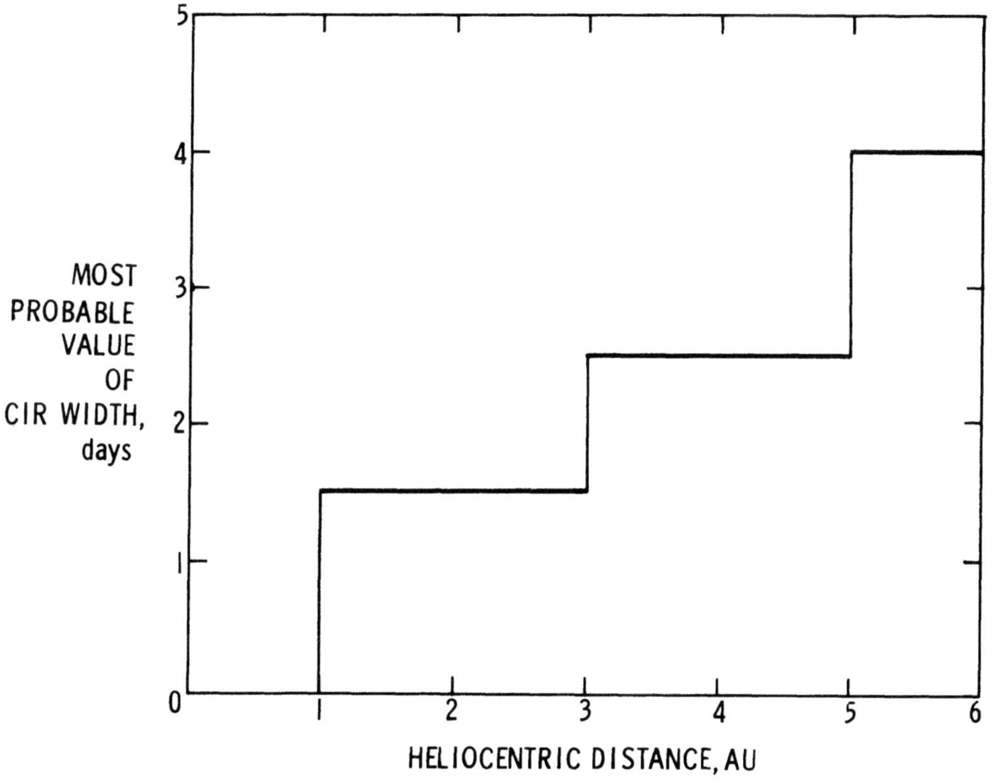

Figure 12. Variation of CIR Widths with Distance. Each bar shows the most probable value derived from histograms like those shown in Figure 11.

Variations in the plasma parameters within the CIRs have been investigated with a view to identifying their characteristic changes. These changes provide important clues concerning the physical processes associated with the interaction. There have been several theoretical attempts to model the stream-stream interaction at large distances (Hundhausen, 1973; Steinolfson et al., 1975; Gosling et al., 1976; Goldstein and Jokipii, 1977), making it possible to compare the theoretical predictions with observations.

As an aid in organizing the various plasma parameters within the CIRs an attempt has been made to identify the interface between plasma which was originally slowly moving and has been accelerated and plasma that was rapidly moving and has been decelerated. The concept is illustrated in Figure 13 which has been adapted from Siscoe (1976).

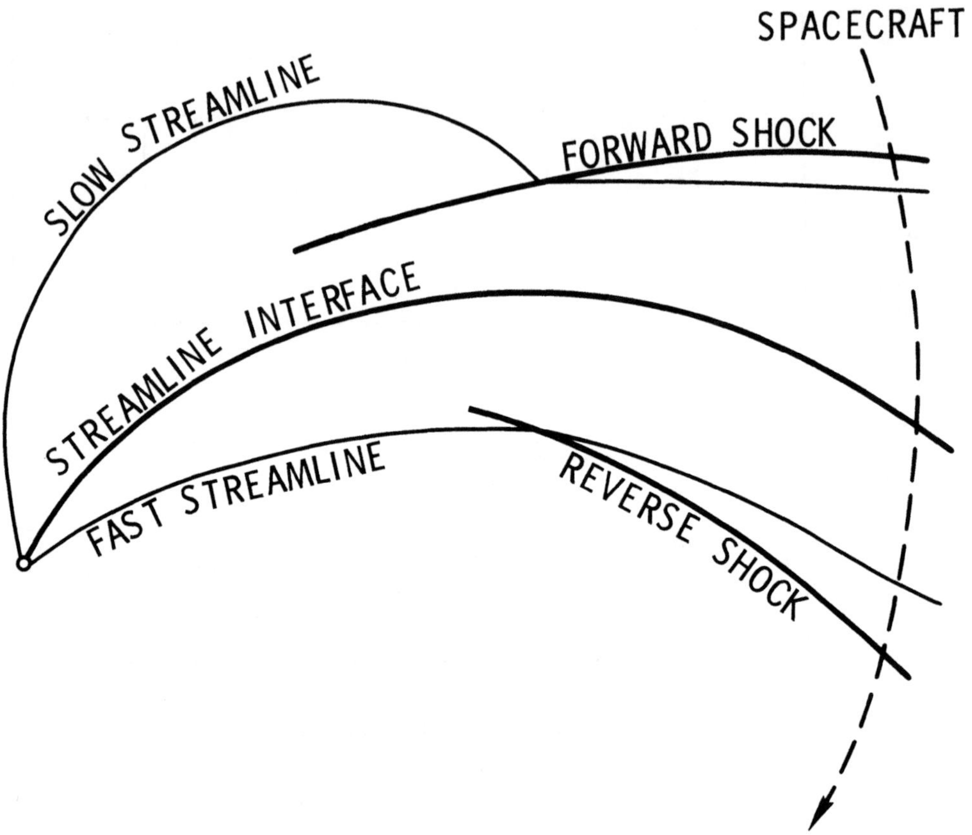

Figure 13. Solar Wind Streamlines and Shock Surfaces. The solar wind streamlines are shown in a reference system corotating with the sun (so that spacecraft appear to rotate clockwise as shown). The streamlines also correspond to lines of force of the magnetic field.

The figure shows three streamlines along which solar wind plasma is flowing in a reference frame which is rotating with the sun. The streamlines also coincide with the lines of force of the interplanetary magnetic field. They may be thought of as the paths that might be followed by solar energetic particles released at the sun (ignoring convection, diffusion, gradient drifts, etc.). The surfaces corresponding to the forward and reverse shocks are also shown and can be seen to diverge or widen as discussed above. The slow streamline intersects the forward shock and the fast streamline crosses the reverse shock. There is also a streamline lying in the middle of the CIRs which identifies the interface, I, between accelerated and decelerated plasma.

There have been several studies of the interface within stream-stream interaction regions near 1 AU. The interface has been identified in these studies as a region of finite thickness across which there are characteristic changes in the plasma parameters: a negative gradient in N and simultaneous positive gradients in T and B (Belcher and Davis, 1971; Siscoe, 1972; Burlaga, 1974). Another useful identification is provided by the change in azimuthal velocity which is well known from these observations to show a deflection of the solar wind such that the slow plasma appears to come from east of the sun and the fast plasma from west of the sun. A parameter which also helps identify I is the total pressure, $P = NkT + B^2/8\pi$. It has been anticipated theoretically and demonstrated observationally by Siscoe (1972) that P peaks near I so that the gradient on the two sides gives rise to stresses that are exerted outward (acceleration) ahead of I and inward (deceleration) behind I.

We have attempted to identify the interface within the Pioneer CIRs. An example of a CIR observed just inside 5 AU and its designated interface are shown in Figure 14. The velocity and field magnitude appear in the bottom two panels and serve to identify the interaction region. The interface has been chosen to coincide with the peak total pressure. At this point, there is also a change in the gradient of N from positive to negative and T begins to increase to its maximum value.

A substantial number of CIRs have been studied to determine the characteristic behavior of the solar wind parameters with respect to I and from the front to the back of the interaction region. The results of this study are summarized qualitatively in Figure 15 which contains sketches of T, N, V, and B. The characteristic velocity profile associated with the CIRs can be seen in the bottom panel. At the top of the figure, fiducial marks are shown to identify F, the forward shock, I, the interface, and R, the reverse shock.

The field strength and plasma density tend to increase from just behind F to maxima in the vicinity of I. They then decrease to a relatively constant value throughout the remainder of the CIRs and undergo an abrupt decrease at R. The tendency for N to be less behind the CIR than preceding it, as shown in the figure, is an observational result. It is thought to be a consequence of the rarefraction that occurs in the region of negative solar wind gradient. The tendency for N and B to peak in the vicinity of I makes their profiles similar to the profile of P.

The temperature is the most variable parameter as indicated in the second panel from the top. The most common behavior is shown by the

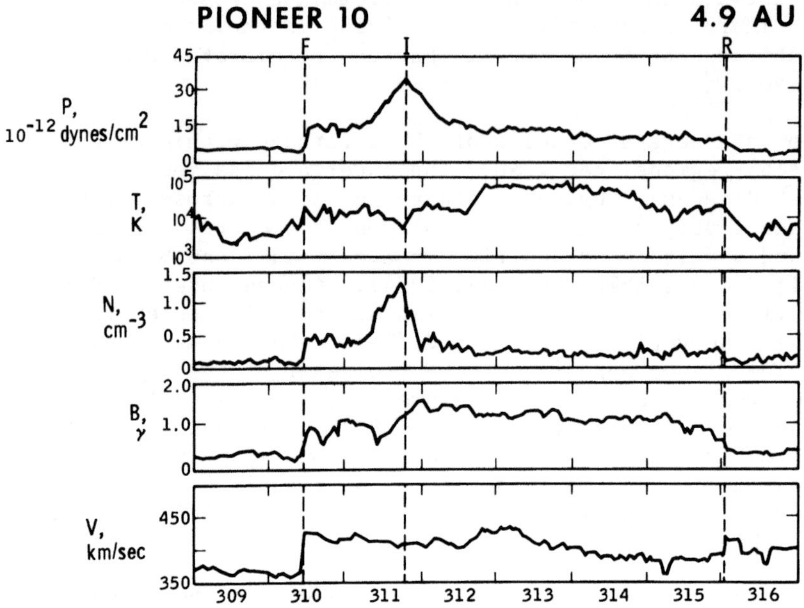

Figure 14. Example of an Interaction Region Observed Near 5 AU. The dashed vertical lines identify the forward shock (F), the interface (I) and the reverse shock (R).

solid line, i.e., an increase from F to a location beyond I, followed by a steady decrease to near R. Other types of variation in T are seen, however, and these are shown by the dashed profiles. Sometimes the temperature will remain constant throughout the CIRs. Occasionally, T will temporarily decrease near I and then increase behind it. The variability in T is not surprising; the temperature variations are basically a manifestation of the regions in which waves are dissipated and this process can occur anywhere within the CIR.

The results shown in this figure are similar to those obtained with numerical simulations of stream-stream interactions (Steinolfson et al., 1975; Hundhausen and Burlaga, 1975). They should be useful in testing, and ultimately, in improving these models.

OBSERVATIONS OF FLARE-GENERATED SHOCKS

Flare generated shocks occur much less frequently in the Pioneer data than corotating shocks associated with high speed streams. However, a few flare shocks have been observed and there may be more in the data which will ultimately be identified.

The most carefully documented and best studied flare shocks are those associated with the events of August 1972. Review articles have recently appeared which include the Pioneer plasma and field observa-

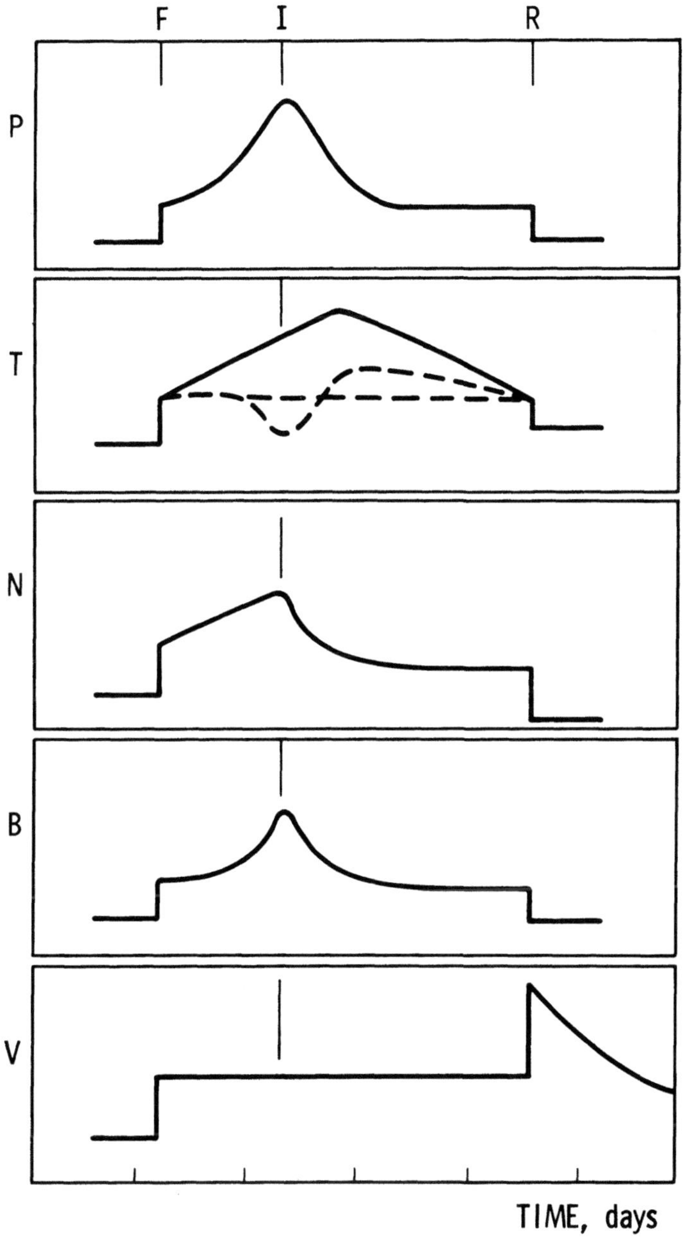

Figure 15. Characteristic Changes in Plasma Parameters Within Interaction Regions.

tions in the vicinity of 2.2 AU (Intriligator, 1976; Smith, 1976). In addition, articles based on the plasma and field data have also been published recently (Intriligator, 1977; Smith et al., 1977).

The principal observational findings regarding the August 1972 shock structure were as follows. In general, it was possible to associate interplanetary shocks observed at Pioneer 10 at 2.2 AU with specific solar flares and SSCs seen at Earth. However, two of the four forward shocks observed at Pioneer 9 near 0.8 AU and at Earth were not subsequently seen at Pioneer 10. In addition, a reverse shock was seen at Pioneer 10 that was not observed at either Pioneer 9 or at Earth. The local velocities of the two forward shocks at Pioneer 10 were derived from V and B and independently from V and N.

The chief conclusion that has been drawn concerns the deceleration of the shocks with heliocentric distance. The shocks decelerated substantially in the vicinity of the sun but were propagating at essentially constant speed between 0.8 and 2.2 AU. This inference is based on a comparison of the local shock speeds with the average speeds between the sun and Pioneer 9, the sun and Pioneer 10 and between Pioneer 9 and Pioneer 10. The results of these comparisons are summarized in Figure 16 (Smith et al., 1977).

The disappearance of the two forward shocks between Pioneer 9 and Pioneer 10 is the subject of continuing study. Several alternative explanations have been advanced, e.g., that the slower moving shocks were overtaken by faster moving shocks with which they merged. Another possibility is that, as the trailing shocks propagated into the wake of the preceding shocks, the heavy disturbed nature of the medium caused them to decay into large amplitude waves. One of the most promising approaches to understanding the dynamic interaction of these multiple events is to use numerical simulations, especially those based on using the Pioneer 9 data at 0.8 AU to predict the Pioneer 10 observations at 2.2 AU (Dryer et al., 1977).

INTERACTION OF SHOCKS AND CIRs WITH ENERGETIC PARTICLES

The extensive changes that occur in the structure of the interplanetary medium beyond 1 AU are bound to have significant consequences for the energetic particles of all types that propagate through this medium. Galactic cosmic rays, solar flare particles and recurrent streams of energetic protons have all been observed at large distances by the Pioneer particle investigators. Furthermore, one of the most important Pioneer 10, 11 results has been the observation of energetic electrons escaping from Jupiter's magnetosphere and propagating into both the inner and outer solar system (Chenette et al., 1974; Teegarden et al., 1974). All these particles will be subjected to large scale drifts associated with solar wind convection, field gradient and curvature drifts and adiabatic acceleration and deceleration. In addition, the plasma turbulence and other irregularities in the interplanetary field will govern the diffusion of the energetic particles and, in conjunction with the shocks, may lead to energy-dependent processes, such as acceleration.

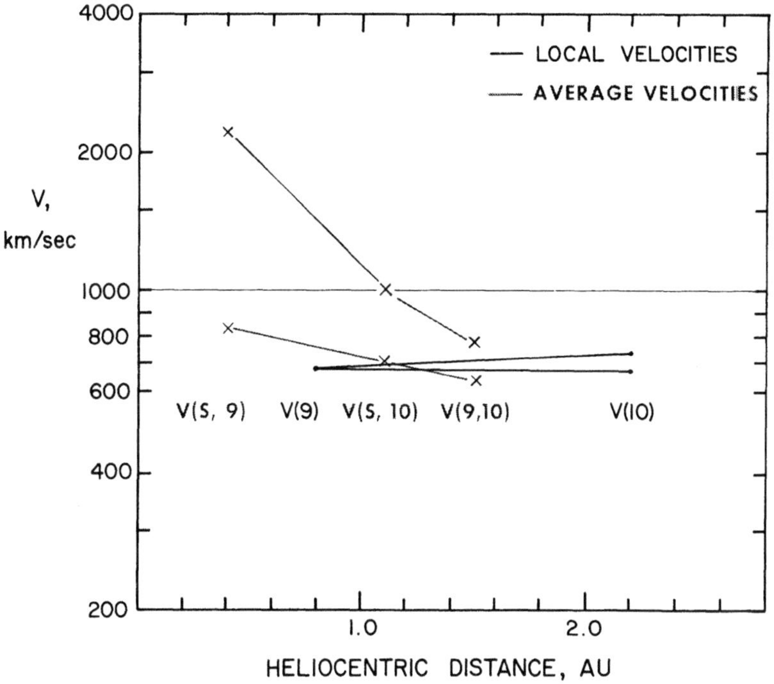

Figure 16. Propagation Velocities of Shocks Associated with the Solar Events of August 1972. The average and local shock velocities are shown as a function of distance. The average velocity from the sun to Pioneer 9 is designated V(S,9), the local velocity at Pioneer 9 is V(9), etc.

One of the topics that has received considerable attention has been the possible acceleration of the recurrent energetic protons ($E_p \sim 1$ mev) in the vicinity of corotating shocks. McDonald et al. (1976) compared the intensities of corotating proton streams near Earth and at Pioneer. They observed that their intensities did not decrease with distance, as anticipated on the basis of an adiabatic deceleration of the particles in the diverging interplanetary magnetic field, but were observed either to be constant or to increase significantly. McDonald et al. (1976) inferred that the most plausible explanation is some process of interplanetary acceleration

Barnes and Simpson (1976) also provided additional evidence for the interplanetary acceleration of these protons, as well as of helium. They found that beyond 2.5 AU the proton streams developed a structure consisting of a double peak. The two peaks were shown to be correlated with the leading and trailing edges of simultaneously present CIRs (Figure 17). The double peak structure and its correlation with the boundaries of the CIRs have also been observed by Pesses et al. (1977). Many aspects of the correlation between shocks, CIRs and the energetic protons, including the physical processes involved, are poorly understood at present and are the subject of intense study.

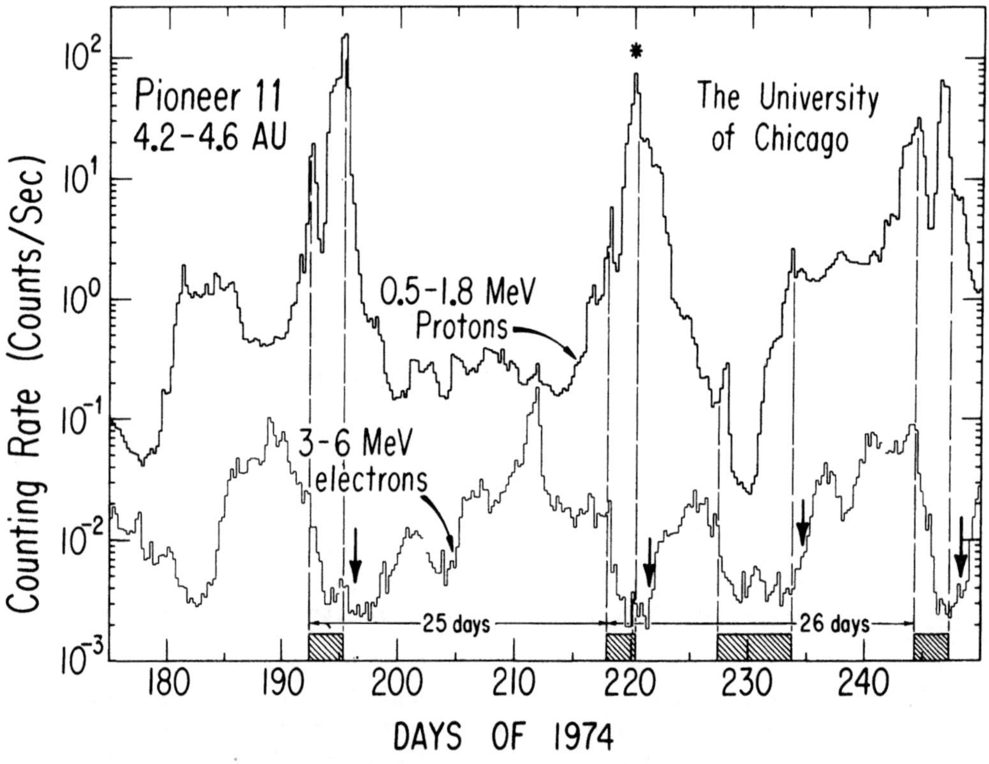

Figure 17. Mev Protons and Electrons in Interplanetary Space and Their Association with CIRs. Times of passage of CIRs by Pioneer 11 are shown as dashed boxes. The CIRs are associated with marked increases in the proton flux and with decreases in the electrons.

Other particles which are known to interact strongly with CIRs are relativistic Jovian electrons ($E_e \simeq 3$ mev). The intensity of these electrons is observed to undergo large variations in intervals corresponding to a solar rotation. These changes have been found to be related to CIRs (also shown in Figure 17). Conlon and Simpson (1977) have developed a model that is able to explain the observations.

The model assumes essentially continuous emission of electrons by Jupiter with propagation both along and across the interplanetary field, the latter as a result of cross-field diffusion. The CIRs appear to act as impenetrable barriers so that the Jovian electrons are basically confined to the quiet region between adjacent CIRs. Jovian electrons are observed at Pioneer and at Earth (Chenette et al., 1977) when both Jupiter and the point of observation occupy the same quiet region. As a CIR corotates past the spacecraft, it establishes a barrier between the point of observation and Jupiter and the electron flux is drastically reduced. The opportunity to study the propagation of energetic electrons from a known source, and with a known energy spectrum, is expected to provide a significant test of current theories regarding the effect of diffusion.

Although the effect of shocks and CIRs on solar particles is a less well developed subject at this time, clear evidence of such effects have been identified for a few specific events. Of particular interest are the observations of solar energetic particles associated with the solar terrestrial events of August 1972, which included one of the largest solar proton events ever recorded. Figure 18 is adapted from Roelof et al. (1974) and is essentially a plot of the count rate at Pioneer 10 during the interval that energetic protons were being detected. We have added arrows showing the times when the two forward shocks and the reverse shock were observed at Pioneer 10. It is evident that, in each case, the arrival of these shocks at Pioneer caused significant reductions in the intensity of the flare particles.

Finally, the effect of CIRs in causing a temporary reduction in the intensity of galactic cosmic rays has also been noted. An earlier study has shown a correlation with recurrent decreases in cosmic radiation, as observed by superneutron monitors on Earth ($E \gtrsim 1$ Gev), and the presence of interaction regions in the vicinity of Earth (Barouch and Burlaga, 1975). A similar correspondence with the cosmic ray decreases has been found in the Pioneer data by computing the times at which CIRs observed at Pioneer 11 would have passed by the Earth. Figure 19 is a plot over many solar rotations of the recurrent CIRs and cosmic ray decreases. There is an obvious correspondence between the two phenomena. The relation has been pursued further by comparing cosmic ray measurements at Pioneer with the arrival at the spacecraft of CIRs. Again a close correspondence has been found between the presence of a CIR and the decrease in the cosmic ray intensity. As with the other correlations mentioned briefly above, further studies will be necessary to identify and understand the physical processes responsible for this relationship.

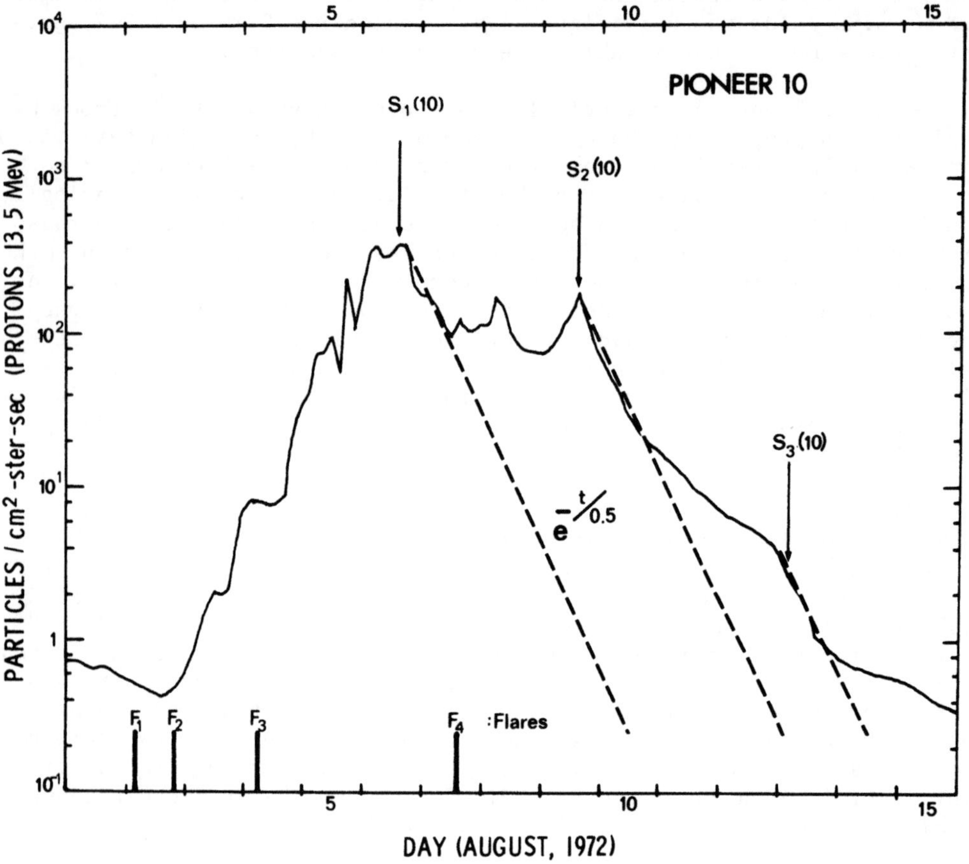

Figure 18. Effect of Flare Shocks on Solar Protons (E_p > 13.5 mev). The proton intensities are a superposition of several impulsive events associated with the four flares (F_1,..F_4). Intensity decreases follow the arrival of each of the shocks (S_1, S_2, S_3) with a characteristic decay of about 0.5 day.

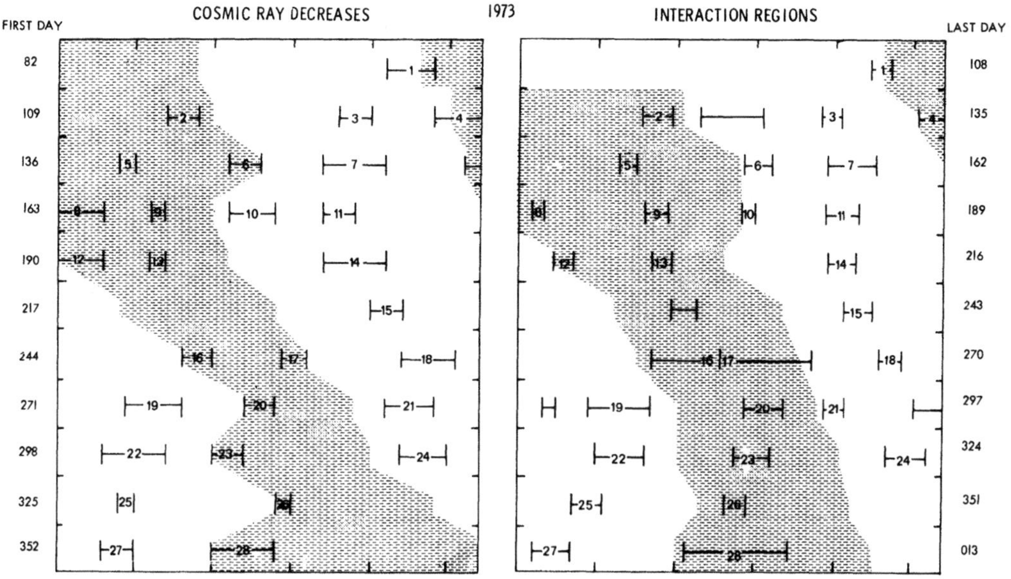

Figure 19. Cosmic Ray Decreases and CIRs. The left hand panel shows ten solar rotations at Earth containing information about the sector structure (crosshatched for negative polarity and open areas for positive polarity) and recurrent cosmic ray decreases (bars showing the duration of the flux decrease at Thule). The sector structure was obtained from Hedgecock (1975) and oversimplified somewhat to help identify corresponding features at Pioneer. The right panel shows the simplified sector structure and duration of CIRs (bars) at Pioneer 11. Corresponding cosmic ray decreases and CIRs are identified by numbers (1, 2, etc.).

ACKNOWLEDGEMENTS

The magnetometer data were reduced at JPL by J. Hull who was assisted by C. Stanley, J. Wolf, L. Shaw, and W. Kulas. E. Parker (JPL) assisted in the analysis of the field and plasma data with help from E. Karspeck. J. Mihalov directed the plasma analyzer data reduction at ARC. Helpful comments were provided by B. Tsurutani. The work at JPL represents the results of one phase of research carried out for NASA under Contract NAS7-100.

REFERENCES

Bame, S. J., Asbridge, J. R., Feldman, W. C., and Gosling, J. T.: 1976, Astrophys. J. 207, 977.
Barouch, E. and Burlaga, L. F.: 1975, J. Geophys. Res. 80, 449.
Barnes, C. W. and Simpson, J. A.: 1976, Astrophys. J. 210, L91.
Belcher, J. W. and Davis, L., Jr.: 1971, J. Geophys. Res. 76, 3534.
Burlaga, L. F.: 1974, J. Geophys. Res. 79, 3717.
Chenette, D. L., Conlon, T. F., and Simpson, J. A.: 1974, J. Geophys. Res. 79, 3551.
Colburn, D. S. and Sonett, C. P.: 1966, Space Sci. Rev. 5, 439.
Collard, H. R. and Wolfe, J. H.: 1974, in C. T. Russell (ed.), Solar Wind Three, Geophys. and Planet. Phys., U. of California, Los Angeles, p.281.
Conlon, T. F. and Simpson, J. A.: 1977, Astrophys. J. 211, L45.
Dryer, M., Candelaria, C., Smith, Z, K., Steinolfson, R. S., Smith, E. J., Wolfe, J. H., Mihalov, J. H., and Rosenau, P.: 1977, Dynamic MHD Modeling of the Solar Wind Disturbances During the August 1972 Events, J. Geophys. Res. (submitted).
Goldstein, B. E. and Jokipii, J. R.: 1977, J. Geophys. Res. 82, 1095.
Gosling, J. T., Hundhausen, A. J., and Bame, S. J.: 1976, J. Geophys. Res. 81, 2111.
Hedgecock, P. C.: 1975, Solar Phys. 44, 205.
Hundhausen, A. J.: 1973, J. Geophys. Res. 78, 2035.
Hundhausen, A. J. and Burlaga, L. F.: 1975, J. Geophys. Res. 80, 1845.
Hundhausen, A. J. and Gosling, J. T.: 1976, J. Geophys. Res. 81, 1436.
Intriligator, D. S.: 1976, Space Sci. Rev. 19, 629.
Intriligator, D. S.: 1977, J. Geophys. Res. 82, 603.
McDonald, F. B., Teegarden, B. J., Trainor, J. H., Von Rosenwinge, T. T., and Webber, W. R.: 1976, Astrophys. J. 203, L149.
Ness, N. F., Hundhausen, A. J., and Bame, S. J.: 1971, J. Geophys. Res. 76, 6643.
Neugebauer, M. and Snyder, C. W.: 1966, J. Geophys. Res. 71, 4469.
Parker, E. N.: 1963, Interplanetary Dynamical Processes, Interscience, New York.
Parker, G. D. and Jokipii, J.R.: 1976, Geophys. Res. Lett. 3, 561.
Pesses, M. E., Van Allen, J. A., and Goertz, C. K.: 1977, J. Geophys. Res. (submitted).
Razdan, H., Colburn, D. S., and Sonett, C. P.: 1965, Planet. Sp. Sci. 13, 1111.
Roelof, E. C., Lezniak, J. A., Webber, W. R., McDonald, F. B., Teegarden, B. J., and Trainor, J. H.: 1974, in D. E. Page (ed.), Correlated Interplanetary and Magnetospheric Observations, D. Reidel Pub., Dordrecht, p. 563.
Rosenberg, R. L., Kivelson, M. G., Coleman, P. J., Jr., and Smith, E. J.: 1977, J. Geophys. Res. (in press).
Siscoe, G. L.: 1972, J. Geophys. Res. 77, 27.
Siscoe, G. L.: 1976, J. Geophys. Res. 81, 6235.

Smith, E. J.: 1974, in C. T. Russell (ed.), Solar Wind Three, Geophys. and Planet. Phys., U. of California, Los Angeles, p.257.
Smith, E. J. and Wolfe, J. H.: 1976, Geophys. Res. Lett. 3, 137.
Smith, E. J.: 1976, Sp. Sci. Rev. 19, 661.
Smith, E. J., Davis, L., Jr., Coleman, P. J., Jr., Colburn, D. S, Dyal, P., and Jones, D. E.: 1977, J. Geophys. Res. 82, 1077.
Steinolfson, R. S., Dryer, M., and Nakagawa, Y.: 1975, J. Geophys. Res. 80, 1223.
Teegarden, B. J., McDonald, F. B., Trainor, J. H., Webber, W. R., and Roelof, E. C.: 1974, J. Geophys. Res. 79, 3615.
Wolfe, J. H.: 1972, in C. P. Sonett et al. (ed.), Solar Wind, NASA SP-308, p.170.

DISCUSSION

Eyni: Did you try to compare the measured width of the active region with some calculations?

Smith: No. However, the rate at which the interaction regions widen is consistent with relatively low Mach number shocks. The latter have been predicted by theorists to be what would be expected for these corotating shocks.

Gall: You have shown the "box" type magnetic field intensity increase in the shock wave ensemble. Do you have data on field components between the reverse and forward shocks; in particular, say, at 1.5 AU has the azimuthal component been especially enhanced?

Smith: Yes, we have vector field measurements so we know all three components as well as the field magnitude. At these large radial distances, the field is principally azimuthal. Since most of the shocks are perpendicular or quasi-perpendicular (with their normals "perpendicular" to the field), the principle change occurs in the azimuthal component.

WAVE-PARTICLE INTERACTION PHENOMENA ASSOCIATED WITH SHOCKS IN THE SOLAR WIND

F. L. Scarf
TRW Defense and Space Systems Group
Redondo Beach, California 90278, U.S.A.

ABSTRACT

Microscopic wave-particle interaction phenomena must generally affect the evolution of a traveling interplanetary discontinuity such as a collisionless shock, and solar wind plasma instabilities should also be associated with interplanetary acceleration, diffusion, and dissipation. Recent local measurements from diagnostics on widely separated spacecraft illustrate some examples of these interaction phenomena, and two bounding cases are considered in detail here. It is shown that the interplanetary shock of September 14, 1974 (detected on IMP-7,8, Hawkeye-1, and Pioneer II) was associated with intense local electrostatic noise and ion acceleration, while the shock of January 6, 1975 (detected on IMP-7 and Helios-1) had no detectable electrostatic turbulence.

INTRODUCTION

Collisionless shocks commonly develop in the solar wind plasma, and microscopic studies of shock characteristics from space provide basic information on dynamical phenomena, such as the dissipation and acceleration processes associated with plasma instabilities and wave-particle interactions. Most of the detailed information presently available in this area comes from analysis of observations made in the region of the earth's bow shock. This natural standing shock forms in the solar wind plasma, and Greenstadt and Fredricks (1977) recently summarized the status of knowledge about the earth's bow shock system. They noted that the dependence of the bow shock macrostructure on the flow Mach number, the plasma β-value, and the upstream field orientation is fairly well understood, that the "foreshock" region populated with reflected particles is being analyzed, and that the localized electrostatic current-driven plasma instabilities associated with laminar thin shocks have been identified. Although they noted that much remains to be done here, Greenstadt and Fredricks (1977) showed conclusively that the development of bow shock data analysis is proceeding well, and that related theoretical concepts can be tested using the observations.

M. A. Shea et al. (eds.), Study of Travelling Interplanetary Phenomena/1977, 259–275.

Traveling interplanetary shock phenomena are also of great interest from a number of points of view. It is well known that interplanetary shocks are associated with significant geomagnetic perturbations (sudden commencements, magnetic storms, etc.). Interplanetary observations have also been used to show generally how stream-stream interactions evolve into shocks (Smith and Wolfe, 1976), in accordance with recent theoretical predictions (Hundhausen and Gosling, 1976; Gosling et al., 1976). Interplanetary shock events are also of interest because they are frequently associated with localized acceleration mechanisms (see, for instance, Sarris and Van Allen, 1974). Finally, we note that interplanetary shocks are usually associated with a much lower range of Mach numbers than the bow shock, and from a general plasma physics point of view, studies of interplanetary shocks are needed to extend and complement the insights derived from analysis of the standing shocks.

Although the need to analyze microstructure at an interplanetary discontinuity is evident, the task is a very difficult one because these discontinuities are generally thin, while the solar wind flows rapidly past the spacecraft, so that extremely high temporal resolution is needed simply to detect and measure the variation in the parameters. The analysis of wave-particle interaction effects is also complicated by the rapid plasma motion because many interplanetary wave modes have rest-frame phase speeds small in comparison with the nominal 400 km/sec wind speed past the spacecraft. Thus, Doppler shifts play a large role, and it is not at all a simple task to identify the rest frame wave mode in many cases, especially since the phase speeds depend strongly on the wave propagation direction and the rest frame wave frequency. The characteristics of the wave spectrum in the spacecraft frame can also be strongly affected by instrumental details, such as rise and decay times for amplifiers.

Recently Gurnett and Anderson (1977) showed how the comprehensive spectral analysis capability of the Helios electric field investigation and its good temporal resolution can be used to analyze Doppler shift effects, and Neubauer et al. (1977) demonstrated that the "shock alarm" from the Helios magnetic field investigation (a similar fast sampling device is present in the E-field instrument) provides high resolution data on interplanetary shock characteristics. With instrumental approaches along these lines, we can expect that in the future, solar wind plasma physics phenomena will be studied in detail. However, the necessary instrumental progress must not be restricted to the area of measurements of DC and wave fields, since many important interplanetary phenomena also involve simultaneous changes in the characteristics of the plasma and energetic particles.

In this report, we consider some unusual cases involving wave, field, plasma and energetic particle variations in the solar wind, in order to point out a few areas where important dynamical questions are still quite open. These open questions provide the framework for an assessment of the needs for future instrumental improvements.

OVERVIEW OF THE SEPTEMBER 15, 1974 AND JANUARY 6, 1975 INTERPLANETARY SHOCKS

The relatively unusual shock events of interest here were associated with SSC-events (storm sudden commencements) at 1343 UT on September 15, 1974 and at 1957 UT, January 6, 1975. On September 15, 1974, IMP-7 was 34 R_e upstream from the earth, almost precisely on the earth-sun line, and on January 6, 1975, IMP-7 was slightly over (less than 20°) toward the dawn side of the earth-sun line as the discontinuity swept by. Thus, in each case IMP-7 was about as far upstream from the earth's bow shock as possible, and in an excellent position to make measurements of interplanetary phenomena, with minimum magnetospheric clutter.

Both shocks were clearly detected by a number of instruments on IMP-7. The September 1974 event was also detected on IMP-8 (in the magnetosheath at $X_{SM} = -25.6\ R_e$, $Y_{SM} = -23.0\ R_e$, $Z_{SM} = 1.4\ R_e$) and on Hawkeye-1, which was in the magnetosheath over the pole until 1343:30-1345 UT, when the advancing interplanetary shock moved the spacecraft out into the solar wind (J. Van Allen, D. Gurnett, private communication). The September 15 event has also been traced to a specific 2B solar flare (at 1510 UT, September 13, 1974), and the same traveling shock was apparently detected on Pioneer 11 at 1115 UT, September 29, 1974, when the spacecraft was at 4.7 AU (E. Smith, private communication).

The January 6 event was apparently also detected on a number of other spacecraft; Helios A, located at 0.924 AU, with spacecraft-sun-earth angle of -7.4°, detected the shock at 2044 UT, and IMP-8, which was in the distant magnetosheath [at X(GSE) = $-32.6\ R_e$, Y(GSE) = $-10.8\ R_e$, Z(GSE) = $-25.8\ R_e$] may have detected the event immediately before 1950 UT (this qualification is necessary because IMP-8 had a data gap just after 1950).

Sarris et al. (1976) recently discussed the September 15, 1974 event, and they noted that the shock was extremely unusual because the related acceleration "spike" involved protons, alpha particles, and $Z \geq 6$ nuclei extending in energy to > 25 MeV, in contrast with more general cases where intensity enhancements do not extend beyond 5 MeV. Some aspects of this unusual result are illustrated in the top panels of Figure 1, where the September 15 flux profiles for 0.4,3 MeV protons, 3.5 MeV alphas, and 2 MeV/nucleon mediums are compared with the corresponding measurements from January 6, 1975. It is quite evident that the very distinct September 15 2 MeV/N ion spike (labeled "mediums") had no counterpart on January 6, although the pre-shock flux levels in this channel were comparable for both events. The bottom panels of Figure 1 show some other unusual features of the two shocks. On September 15, 1974 no electromagnetic turbulence was detected in the low frequency channel which is sensitive down to 5-10 Hz, and on January 6, 1975 no low frequency electrostatic turbulence could be detected.

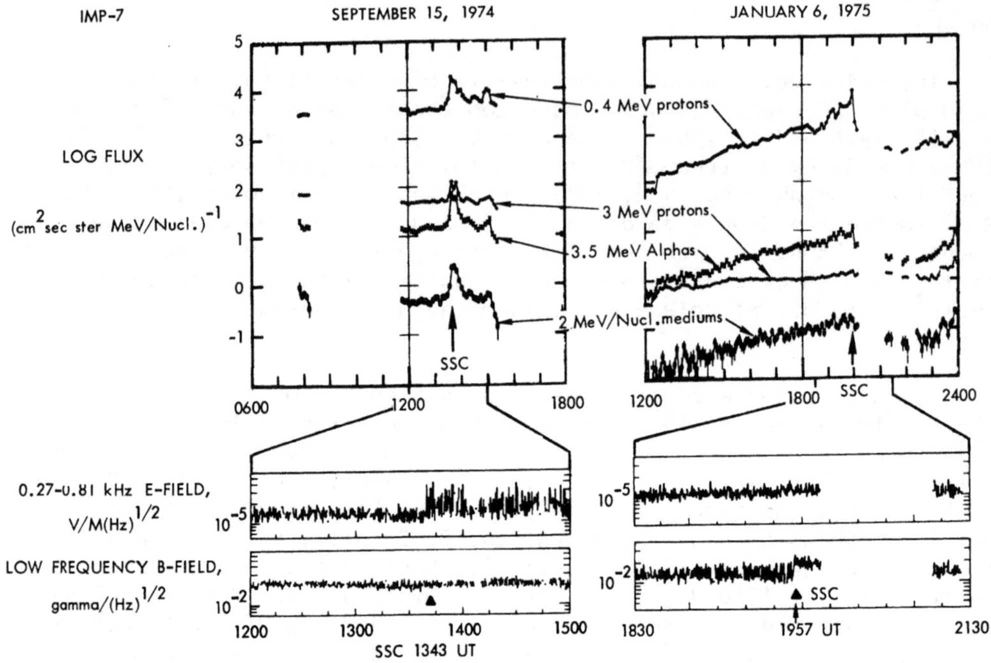

Figure 1. The top panels give 5.5-minute average flux values from the Johns Hopkins/APL charged particle instrument on IMP-7, and the bottom panels contain low frequency E and B wave level data from the TRW plasma wave instrument.

The IMP-7 plasma wave instrument customarily detects enhanced E and B turbulence at an interplanetary shock, so both of the events of Figure 1 represent unusual cases. The conventional responses in the low frequency B-field channel and in the 270-810 Hz E-field channel were previously indicated in Figures 9 and 10 of Scarf et al. (1974); in these figures the IMP-7 B- and E-wave observations for the period 1630 to 1715 UT, October 31, 1972 were displayed, and it was shown that highly variable enhanced turbulence levels developed for an extended period after passage of an interplanetary shock at 1653 UT. The detailed E-field spectral densities (corrected for sheath effects) during a three-minute period centered around 1653 UT on October 31 are shown in the left side of Figure 2, and it can be seen that wave level enhancements with E up to 10^{-4} volts/meter $(Hz)^{1/2}$ were detected for $f \lesssim 600$ Hz. These peak wave levels are somewhat lower than typical bow shock E-field wave level peaks [Rodriguez and Gurnett (1975) show characteristic shock spectra with $E \simeq 10^{-3} - 10^{-5}$ volts/meter $(Hz)^{1/2}$ for $f < 1$ kHz], but, in fact, the October 31, 1972 event was an unusually strong perpendicular shock, and Ipavich and Lepping (1975) described many of the associated unusual phenomena.

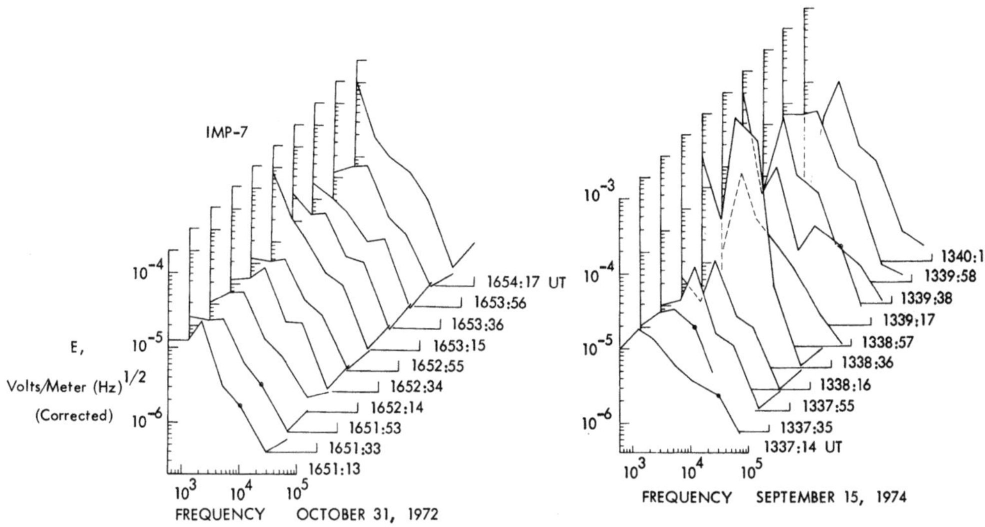

Figure 2. Comparison of electric field wave level spectra for the two interplanetary shocks associated with the sudden commencements at 1654 UT, October 31, 1972 and 1343 UT, September 15, 1974.

With this background in mind, it is somewhat surprising to compare the left- and right-hand sides of Figure 2 because the drawing shows that the corrected E-field amplitudes on September 15, 1974 were as much as an order of magnitude more intense than the peak levels detected after the strong October 31, 1972 event. The plots of Figure 2 show that the E-field observations of October 31, 1972 and September 15, 1974 were also very different in terms of the detection of higher frequency electric field turbulence. The September 15 observations of elevated 30 kHz wave levels in association with passage of the shock are extremely unusual and at first this was regarded as evidence that upstream electrons from the earth's bow shock were producing magnetospheric disturbances at $f \simeq f_p^e$ [see Scarf et al., 1971; Rodriguez and Gurnett, 1975]. However, at 1343 UT on September 15, 1974, the Hawkeye-1 plasma wave instrument did detect upstream noise at $f \simeq 56$ kHz (D. Gurnett, private communication) corresponding to $N \simeq 40$ electrons/cm^3. Both the 30 kHz and 70 kHz channels on IMP-7 have comparable response near 56 kHz (approximately 20 db less sensitive than for a signal at the respective center frequency), and the full set of observations indicates that it is highly unlikely that the 30 kHz channel output represented the f_p^e-mode at this time; we therefore attribute the response shown here to phenomena associated with the September 15 shock itself, rather than with proximity to earth.

THE SEPTEMBER 15 SHOCK OBSERVATIONS

In their paper on the high energy ion shock spike of September 15, 1974, Sarris, Krimigis and Armstrong (1976) showed 18 detailed count rate and anisotropy profiles for protons, electrons, alpha particles, and medium nuclei, based on use of 5.5-minute averages similar to those displayed in the top panels of Figure 1. The central panels of Figure 3 contain a few of these plots for the entire interval 1200-1500 UT, on the same time scale as the low frequency wave profiles of Figure 1, which are shown again on the bottom. Figure 3 also shows the simultaneous variation in the solar wind density and velocity, as determined by the Los Alamos (LASL) plasma probe on IMP-7, and the panel at the very top gives an expanded view of the magnetic field observations from the GSFC magnetometer on IMP-8 (the IMP-7 magnetometer stopped operating in April of 1973).

This figure illustrates some interesting aspects of the complex event. It can be seen that the shock was again a perpendicular one at IMP-8 (i.e., no change in field direction) and that the enhancement in count rate for most energetic protons was detected just before the enhancements for the lower energy particles. It is also interesting to see that the onset of the electrostatic noise enhancement was well correlated with the detection of spikes in the lower energy protons, the alphas and the mediums [this detailed correlation is still strikingly apparent when the unaveraged observations are examined]. Finally, it is worth noting that the University of Iowa LEPEDEA also detected enhanced fluxes of low energy ($\sim$ 100-900-volt) protons in the sector perpendicular to the sun line for about an hour after passage of the shock.

For this shock, the IMP-7 LASL plasma probe data can be put together with the IMP-8 B-field observations in order to analyze the shock orientation and shock speed, and in order to relate the APL particle observations to conventional theories of acceleration at almost perpendicular shocks. Sarris et al. (1976) noted that a one-step, DC-type acceleration process is positively excluded by their data, and they argued that the high energy shock spike can be "reasonably" accounted for by conventional concepts involving acceleration at almost perpendicular shocks. This agreement is reassuring, but the close correlation between the onset of the intense electrostatic waves and the detection of a proton flux increase also suggests that it is important to identify the waves in order to be able to consider possible wave-particle interaction effects.

The LASL density curve verifies very well Gurnett's Hawkeye-1 identification of $f_p^e = 56$ kHz for the post-shock plasma, and this implies that in the rest frame of the wind the ion plasma frequency (f_p^+) was 1.3 kHz, with a Buneman mode frequency [$f \simeq (M_e/M_i)^{1/3} \times f_p^e$] of 4.5 kHz; the IMP-8 F-value plot suggests that the electron gyrofrequency, f_c^e, was somewhat higher than 1 kHz just after 1340 UT, and this means that in the frame of the post-shock wind, the characteristic frequencies were either less than 5 kHz, or just at 56 kHz. Since the spectra displayed

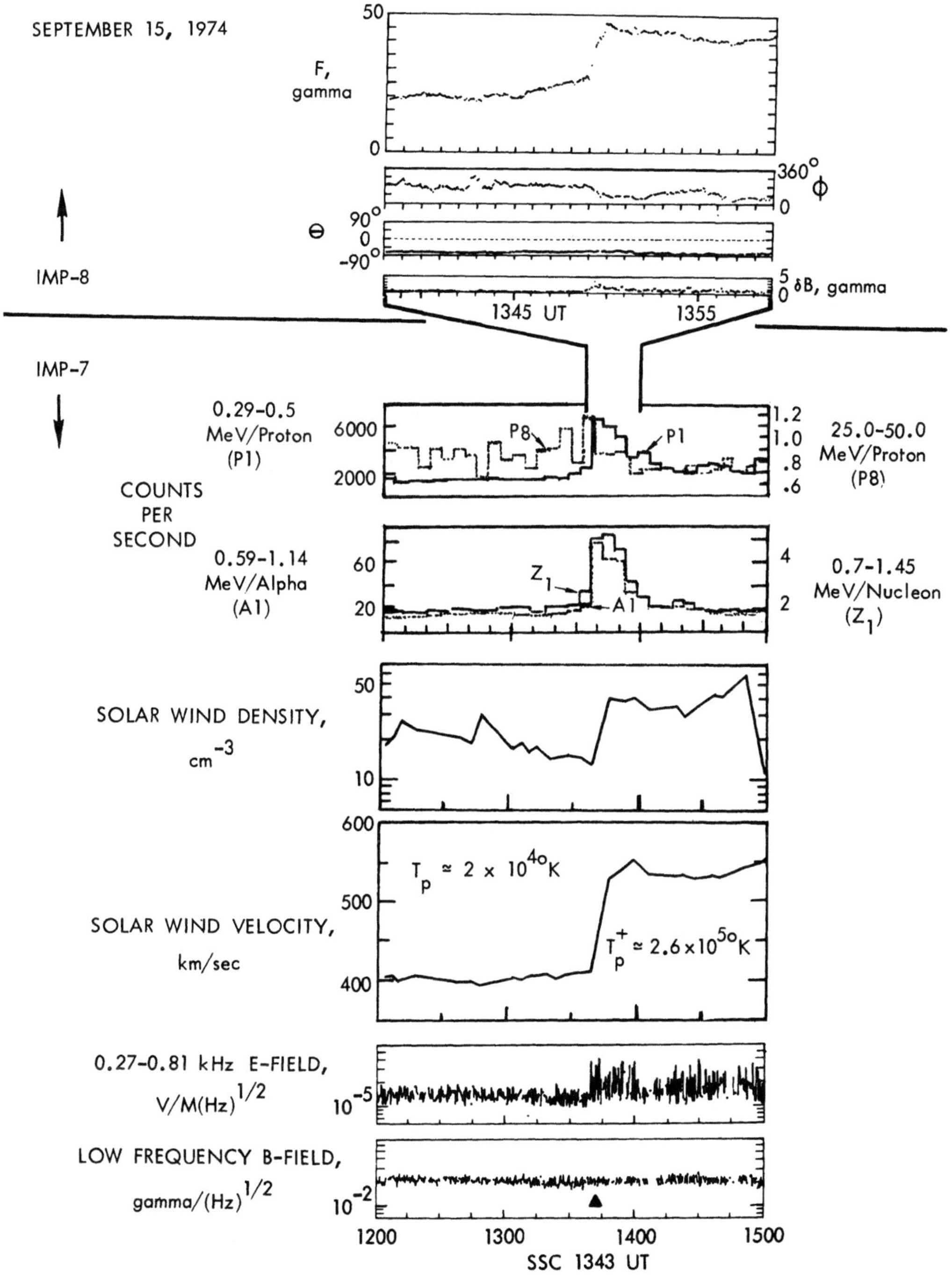

Figure 3. Details of the September 15 shock observations.

in Figure 2 show considerable activity in all E-field channels extending up through 30 kHz (see also Figure 6, below), it appears likely that Doppler shifts must be invoked in order to explain these wave observations in a simple manner. However, since the Buneman and the electron plasma oscillation modes have phase speeds comparable to the electron thermal speed, which is large compared to the post-shock wind flow speed, the resultant Doppler shift is negligible in these cases. The most likely candidate for Doppler shifting would seem to involve a slow mode such as the ion acoustic mode.

In the plasma rest frame, a simplified form of the ion acoustic wave dispersion relation is (Stix, 1962)

$$\frac{1}{\omega^2} = \frac{M_i}{\kappa T_e}\left(\frac{1}{k^2\cos^2\phi} + \frac{1}{K_D^2}\right) \tag{1}$$

where the Debye wave number, K_D, is given by $K_D^2 = (\omega_p^+)^2 \; M_i/\kappa T_e$, with $\omega_p^+ = 2\pi f_p^+$. Here $\omega = 2\pi f$, $k = 2\pi/\lambda$, and ϕ is the angle between $\underset{\sim}{k}$ and the B-field direction. When $\phi = 0$ and $k \simeq K_D$ (for the minimum wavelength signals), we find $\omega \simeq \omega_p^+/\sqrt{2}$; Equation (1) also has a continuum of solutions with longer wavelengths and lower frequencies, and for $\lambda >> 2\pi/K_D$,

$$V(\text{phase}) = \frac{\omega}{k} \simeq \left(\frac{\kappa T_e}{M_i}\right)^{1/2} \cos\phi + \text{----}\,, \tag{2}$$

Since the phase speeds for ion sound waves vary over a restricted range of values, all of which have $(\omega/k) << V(\text{solar wind})$, Doppler shifts can be sizeable, and in the spacecraft frame we have

$$\omega' = \omega + \underset{\sim}{k}\cdot\underset{\sim}{V}(\text{wind}) \tag{3}$$

$$f' = f + [V(\text{wind})/V(\text{phase})]\cos\alpha \tag{4}$$

where $\underset{\sim}{k}\cdot\underset{\sim}{V}(\text{wind}) = kV(\text{wind})\cos\alpha$. Thus, in order to assess the various magnitudes for the Doppler shifts, it is clearly necessary to determine the ranges for ϕ and α, as well as the ranges for V(phase) and f.

In fact, ion acoustic waves can propagate with a wide range of ϕ-angles, and this fact is evident in many of the discussions of Stix (1962). For instance, the simplified ion acoustic wave dispersion relations on page 42 and on page 214 have k_x and k_z, and in the summary of the Spitzer theory of enhanced diffusion (page 222), Stix notes that ion acoustic waves have perpendicular E-fields that yield finite E x B drifts. The top panel in Figure 4 illustrates this Spitzer concept, in which ion sound waves propagating "along" adjacent field lines with random phases necessarily generate transverse E-fields that are effectively associated with a distribution of non-parallel k-vectors (i.e., $\phi \neq 0$). Many other authors have noted that ion sound waves generally have a finite and broad range of ϕ-values, and this point is clearly evident

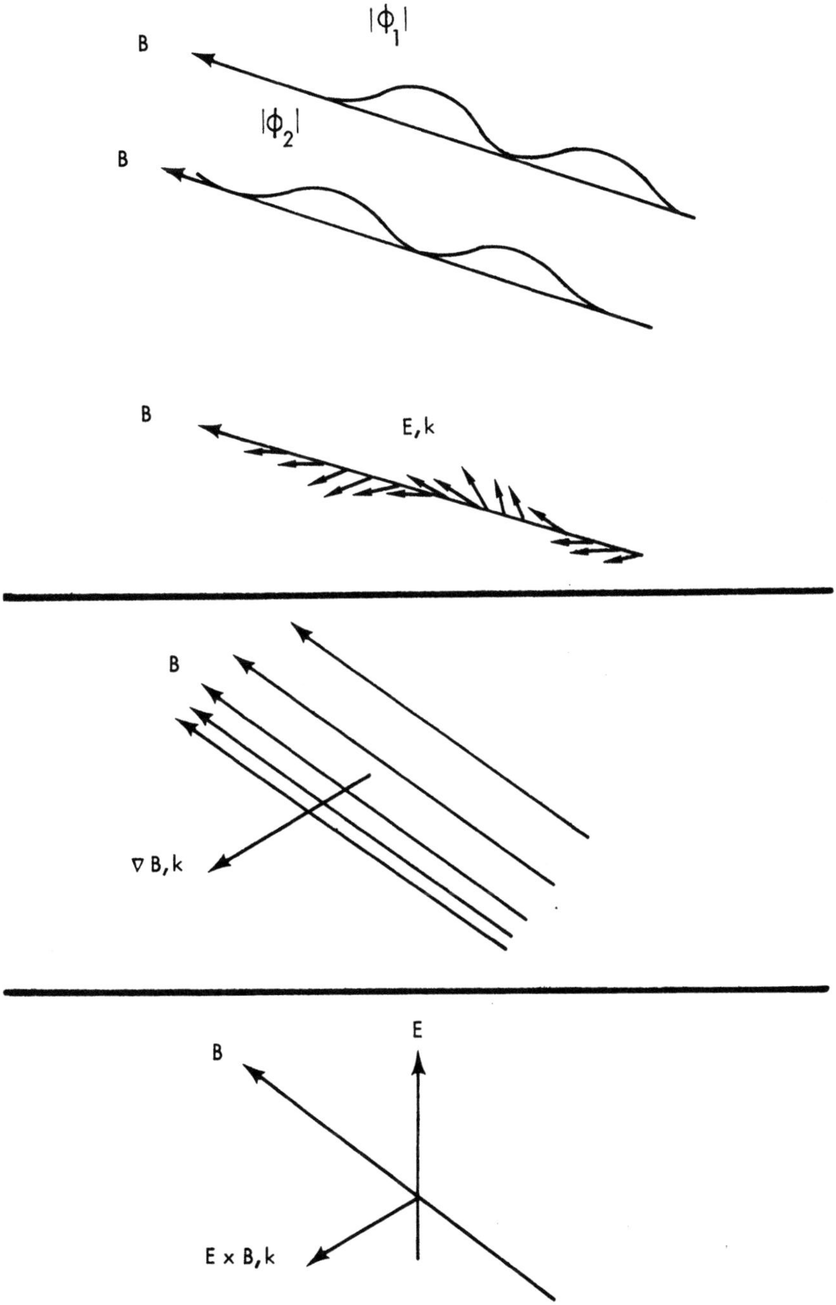

Figure 4. Plausible mechanisms that lead to electrostatic wave vectors at finite angles to the magnetic field. Top: randomly-phased potential ripples lead to fluctuating E-direction; Middle: gradient B-drift gives $\underset{\sim}{k}$ perpendicular to $\underset{\sim}{B}$. Bottom: $\underset{\sim}{E}$ x B drift gives $\underset{\sim}{k}$ perpendicular to $\underset{\sim}{B}$.

in a series of papers by Gary and Sanderson (1970) and by Gary (1970, 1971, 1972) concerning electrostatic waves in perpendicular collisionless shocks. In general, the waves which grow most readily have $\underset{\sim}{k}$ parallel to the electron-proton drift speed; this may or may not be parallel to $\underset{\sim}{B}$, and for gradient B drifts, $\underset{\sim}{E} \times \underset{\sim}{B}$ drifts (see the lower panels in Figure 4), it is clear that $\underset{\sim}{k}$ is not along B.

In order to proceed with the specific Doppler shift analysis for the observations of September 15, 1974, we select sequences of ω-k or f-λ pairings that represent solutions of Equation (1), and we evaluate $f' = f'[f,\lambda,V(\text{wind}),\alpha]$ as a function of α for each sequence; we then inspect other data to try to determine the range of α-values. For instance, let us consider the initial post-shock region and examine the shortest wavelength ion acoustic wave with $k = K_D$. The LASL measurements give N = 40, $f_p^+/\sqrt{2}$ = 939 Hz, V = 540 km/sec and $2\pi\lambda_D = \lambda(\text{minimum})$ = 34 meters, and $f' = [0.939 + 15.88 \cos\alpha]$ kHz; this curve is plotted in the lower part of Figure 5, and it can be seen that Doppler shifts could lead to apparent frequencies for the shortest wavelength signal (just after 1343 UT) that range anywhere between zero and 16.8 kHz.

The actual Doppler shift will, of course, depend on α, the angle between $\underset{\sim}{k}$ and $\underset{\sim}{V}$(wind). Since $\underset{\sim}{k}$ should be primarily oriented along the electron-proton drift direction, it is plausible that $\underset{\sim}{k}$ will also be parallel to the heat flux direction, since the heat flux vector is a direct measure of the drift of suprathermal electrons. Fortunately the LASL plasma probe is capable of determining the heat flux direction, and Dr W. Feldman generously provided the information plotted in the upper part of Figure 5. This conclusively shows that the heat flux was essentially perpendicular to the solar wind direction after passage of the shock. Thus, the analysis can readily account for the fact that large amplitude waves with $0 < f' \lesssim 2(f_p^+/\sqrt{2})$ were detected, in terms of ion acoustic waves propagating more or less perpendicular to $\underset{\sim}{V}$(wind). It is also not difficult to understand the varying E-field enhancements in the 1.3, 2.3, 5.4 and 10.5 kHz channels (see Figure 2 and Figure 6, below) in terms of the simple dispersion relation of Equation (1), along with the Doppler shift analysis of Figure 5; since the (f'/f) function climbs so steeply as α deviates from 90° or 270°, small fluctuations in this angle can very easily lead to large increases in f'. In fact, the upper panel in Figure 5 shows that ψ did fluctuate over a substantial range after passage of the shock, and we should therefore expect to detect a fairly wide range of Doppler shifts.

It is somewhat more difficult to interpret the 30 kHz post-shock wave level enhancement in terms of Equations (1) & (4), but it is certainly not impossible. For instance, although it is customarily stated that the minimum wavelength for an electrostatic wave has $k(\text{min}) \simeq K_D$ so that $\lambda(\text{min}) \simeq 2\pi\lambda_D = 2\pi/K_D$, this lower bound on the wavelength is only roughly defined by considerations involving Landau damping, and the number of plasma particles in a Debye sphere. In a real situation with non-Maxwellian plasma distributions, the Landau damping may well be modified or even absent, and we certainly cannot exclude waves with $\lambda < 2\pi/K_D$.

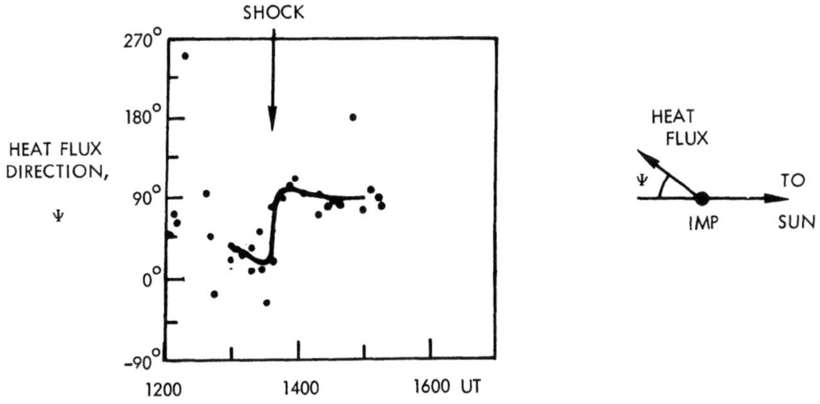

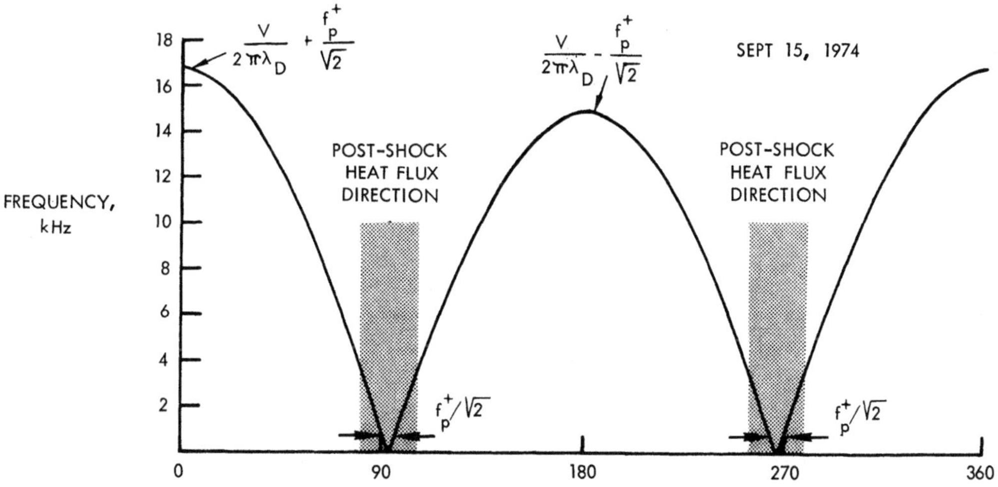

Figure 5. Doppler shift analysis for the September 15, 1974 shock. Top: Heat flux directions deduced from the IMP-7 LASL plasma probe data. After passage of the shock, the heat flux was essentially perpendicular to the wind velocity vector. Bottom: apparent wave frequency as a function of angle between $\vec{k}$ and $\vec{V}$(wind). The shaded areas, with numerically small **Doppler** shift, are appropriate for $\vec{k}$ parallel to the heat flux direction.

Thus, for the post-shock conditions of September 15, 1974, an oscillation with wavelength = 18 meters ($\simeq \pi\lambda_D$) could be Doppler shifted to appear in the 30 kHz channel. Other likely explanations for this observation might involve more complex dispersion relations than the ion sound wave relation of Equation (1).

THE JANUARY 6 SHOCK OBSERVATIONS

The September 15 observations provide promising material for studies of shock phenomenology, local particle acceleration and wave-particle interaction effects, but it is easy to find examples where the available data do not seem to fit together at all. The January 6, 1975 interplanetary shock is in this category.

The discontinuity was detected on Helios-1 at 2044 UT, and Neubauer et al. (1977) carried out a detailed analysis of the shock transition. They showed that the shock was (again) approximately perpendicular, with $F_2 = 12.96\gamma$ and $F_1 = 7.26\gamma$, where $F_{1,2}$ refer to the field magnitude just before and after the jump. Since $N_2/N_1 = F_2/F_1$ in a perpendicular shock (Tidman and Krall, 1971), we would expect to measure a shock density jump with a ratio of approximately 1.8. However, if the shock detected on Helios-1 at 2044 UT, January 6, 1975 is indeed associated with the SSC of 1957 UT, this prediction of a density jump by a factor of 1.8 is not consistent with any of the near-earth observations.

The relevant IMP-7 LASL plasma probe measurements near the January 6, 1975 shock, together with summaries of the E-B wave measurements are shown in Figure 6, and these results are contrasted with the corresponding set of observations for the September 15, 1974 event. The problem is readily apparent in the data because the IMP-7 density jump is $(N_2/N_1) \simeq (20/5) \simeq 4$. This directly-measured ratio does not at all agree with the inference drawn from the Helios-1 observations (i.e., $N_2/N_1 < 2$, using F_2, F_1 and the near perpendicularity of the event). Moreover, since $N_2/N_1 \simeq 4$ for a shock involves an exceedingly strong (and rare) event, one would expect intense electrostatic noise, but as shown in Figure 6, no such plasma turbulence was detected in any of the E-field channels; instead the IMP-7 plasma wave instrument detected a noticeable increase in the low frequency magnetic wave level at 1957 UT, corresponding to development of enhanced electromagnetic turbulence in the post-shock plasma.

The MIT plasma probe on IMP-7 was also measuring solar wind parameters at this time, and Dr. A. Lazarus kindly furnished corresponding data for the present study. The MIT data plots indicate that the Faraday cup measured a density range quite comparable to the one derived from the LASL electrostatic analyzer, and the responses differed primarily in the times and frequencies of sampling. That is, for the 1957 UT discontinuity, the high resolution Faraday cup data also indicated that very large amplitude density fluctuations were present and that these were very rapidly convected past the spacecraft as the wind speed jumped up past 700 km/sec. Thus, since large amplitude fluctuations in N were present during this event, it is likely that none of the IMP-7 experimenters obtained data with high enough time resolution to investigate with accuracy the shock jump relations.

The energetic particle and plasma wave measurements from IMP-7 are not inconsistent with this explanation. For instance, the highest reso-

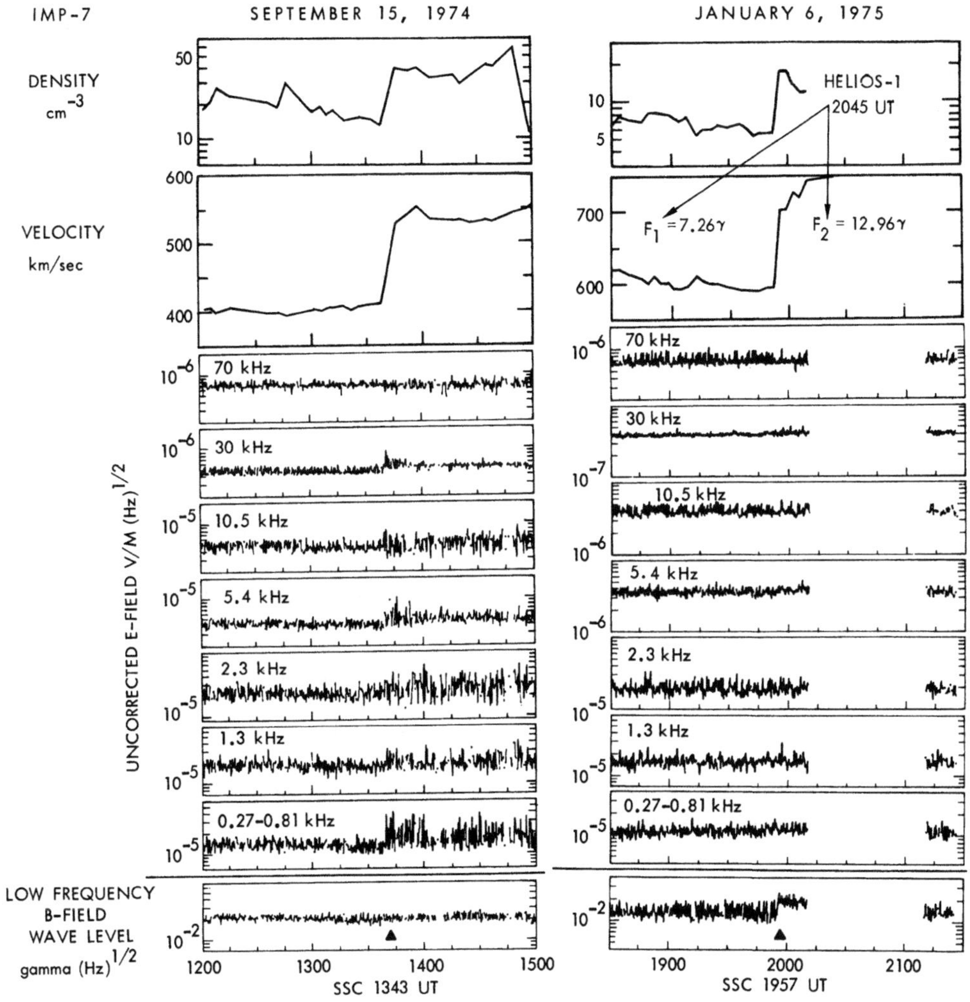

Figure 6. Comparison of eight-channel wave observations and plasma parameters near the two shocks of interest.

lution, spin-averaged measurements in the P-1 channel (0.29-0.5 MeV/proton) of the APL instrument showed a shock-associated proton spike that was extremely narrow (duration much less a minute), and since the E-field plasma wave scans are repeated only once per 20.5 seconds, it is quite possible that an intense short-duration electrostatic noise spike was not detected simply because of the slow instrumental scan sequence. It is also possible that the IMP-7 plasma wave system, with its short, relatively insensitive E-field antenna, and with its complicated response to spacecraft sheath effects (Scarf et al., 1974) simply had an in-flight threshold too high to detect the E-field turbulence that developed at this particular shock.

Of course, we cannot discount the possibility that more complicated processes were underway here. For instance, it is conceivable that the Helios-1 and IMP-7 instruments measured different aspects of a stream-stream interaction with significant spatial and temporal structure, so that it was not appropriate to compare the IMP and Helios data in detail.

DISCUSSION

The case studies discussed above suggest that many important questions about interplanetary shocks have not yet been answered conclusively or unambiguously. For instance, while some theories of proton acceleration at nearly perpendicular interplanetary shocks (Sarris and Van Allen, 1974; Sarris et al., 1976) appear to have considerable merit, we cannot now dismiss the possibility that local wave-particle interactions also contribute to the acceleration. This possibility is of special interest when we consider the population of lower energy ions such as the $E \lesssim 10$ keV protons which were found to be streaming along the heat flux direction after passage of the September 15, 1974 shock. Indeed, Bame et al. (1975) presented evidence for local heating of even lower energy ions in high speed solar wind streams, and they discussed these observations in terms of a number of local plasma instabilities.

Other large scale questions that are still open concern the ways in which these microscopic processes control or affect the stability and evolution of the macroscopic discontinuities. It is noteworthy that virtually all of the interplanetary shock events discussed here or in the cited references are perpendicular shocks. Do parallel shocks decay rapidly, or does the large scale geometry favor the perpendicular shock evolution?

In order to make progress in these areas, suitable instrumentation must be flown to acquire high resolution deep space data that can be used to analyze the events from a plasma physics point of view. The Helios-1,2 wave investigations, with their shock alarm circuits, clearly provide an advance over the slow IMP-7 data-taking sequence discussed here. The pending ISEE-C plasma wave investigation will also provide a significant improvement, since one spectrum analyzer will give a full 16-channel spectral scan twice each second. However, it is not enough to have high time resolution for the DC magnetometer and the wave instrument, since the wave-particle interaction phenomena involve the energetic particles and the full distribution of solar wind plasma particles, extending up to suprathermal energies. For all instruments, very high time resolution is needed if we are to obtain unambiguous answers to the many interplanetary plasma physics questions which are presently open.

ACKNOWLEDGMENTS

I am especially grateful to Dr. W. Feldman and to Dr. T. Krimigis for furnishing me with the previously unpublished IMP-7 data contained in Figures 1, 3, 5, 6 and for allowing me to use these observations here.

I am also very grateful to Drs. L. A. Frank, D. A. Gurnett, A. Lazarus, R. P. Lepping, F. M. Neubauer, E. J. Smith, J. A. Van Allen, and D. Williams for providing me with information about their associated observations of these events from instruments on IMP-7, IMP-8, Hawkeye-1, Helios-1, and Pioneer 11. This research was sponsored by the National Aeronautics and Space Administration under Goddard Space Flight Center Contract NAS 5-23150.

REFERENCES

Bame, S. J., Asbridge, J. R., Feldman, W. C., Gary, S. P., and Montgomery, M. D.: 1975, Geophys. Res. Lett. 2, 373.
Gary, S. P. and Sanderson, J. J.: 1970, J. Plasma Phys. 4, 739.
Gary, S. P.: 1970, J. Plasma Phys. 4, 753.
Gary, S. P.: 1971, J. Plasma Phys. 6, 561.
Gary, S. P.: 1972, J. Plasma Phys. 7, 417.
Gosling, J. T., Hundhausen, A. J., and Bame, S. J.: 1976, J. Geophys. Res. 81, 2111.
Greenstadt, E. W. and Fredricks, R. W.: 1977, Shock Systems in Collisionless Plasmas, Solar Systems Plasma Physics: A Twentieth Anniversary Overview, North-Holland Publishing Company (to be published).
Gurnett, D. A. and Anderson, R. R.: 1977, J. Geophys. Res. 82, 632
Hundhausen, A. J. and Gosling, J. T.: 1976, J. Geophys. Res. 81, 1436.
Ipavich, F. M. and Lepping, R. P.: 1975, Conference Papers of the 14th International Cosmic Ray Conference, Munich, Fed. Rep. of Germany, Vol. 5, p. 1829.
Neubauer, F. M., Beinroth, H. J., Barnstorf, H., and Dehmel, G.: 1977, J. Geophys. Res. 82 (in press).
Rodriguez, P. and Gurnett, D. A.: 1975, J. Geophys. Res. 80, 19.
Sarris, E. T. and Van Allen, J. A.: 1974, J. Geophys. Res. 79, 4157.
Sarris, E. T., Krimigis, S. M., and Armstrong, T. P.: 1976, Geophys. Res. Lett. 3, 133.
Scarf, F. L., Fredricks, R. W., Frank, L. A., and Neugebauer, M.: 1971, J. Geophys. Res. 76, 5162.
Scarf, F. L., Fredricks, R. W., Green, I. M., and Crook, G. M.: 1974, J. Geophys. Res. 79, 73.
Smith, E. J. and Wolfe, J. H.: 1976, Geophys. Res. Lett. 3, 137.
Stix, T. H.: 1962, The Theory of Plasma Waves, McGraw-Hill, New York, New York.
Tidman, D. A. and Krall, N. A.: 1971, Shock Waves in Collisionless Plasmas, John Wiley-Interscience, New York.

DISCUSSION

Dryer: I have a question and a comment. Has the January 1975 shock at Helios been determined on the basis of Rankine-Hougoniot conditions to be a perpendicular shock?

Concerning the change of this shock's characteristic from Helios (where it is perpendicular) to the Earth's position (one hour later where you indicated that it might be oblique), this point might not be surprising since the field orientation can change very quickly-thereby changing the shock's character. This was suggested some years ago by Greenstadt et al. (1974) in the HAO Symposium (Greenstadt, E. W., Dryer, M., and Smith, Z. K.: 1974, in A. J. Hundhausen and G. Newkirk, Jr. (eds.), *Flare-Produced Shock Waves in the Corona and in the Interplanetary Space*, NCAR, Boulder, p. 245.)

Scarf: The Helios investigators have definitely established that this event was a shock, using data from the plasma probe and the magnetometer. The open question concerns the IMP observations, which do not have the benefit of measurements from a DC magnetometer. I noted that the high resolution MIT plasma probe observations show rapid, large amplitude fluctuations in N, V, and for a standing base shock one would tend to think of this as being associated with a parallel shock rather than a perpendicular one; however, for a rapidly moving interplanetary event, such fluctuations may be commonplace events when the shock is perpendicular. Of course, we cannot rule out the type of field orientation changes that you describe either. The lesson here is that we really need comprehensive high resolution measurements for all important parameters.

Schwenn (additional comment to question by Dryer): We did analyze the HELIOS plasma data: The shock of Jan. 6, 1975 as seen from HELIOS was a perpendicular shock.

Armstrong: You mentioned the possibility of wave-particle interactions accelerating particles. Do you have in mind resonant or non-resonant interactions?

Scarf: I believe that non-resonant acceleration is much more likely because the turbulence is broadbanded and impulsive.

Armstrong: Could the electrostatic noise be produced by an ion-streaming instability thus resulting from rather than causing the particle acceleration?

Scarf: It is difficult to distinguish between different cause-and-effect chains here, but the suprathermal ions detected by the LEPEDEA had a very low density, and it seems unlikely that they could have driven the intense waves observed on September 15, 1974.

Galeev: I want to point out a possibility that could explain the dissipation within the shock front in the case where you did not detect the electric field noise. As many laboratory experiments indicate now, the ion thermalization within the high Mach number perpendicular shock takes place due to lower hybrid turbulence which is out of your instrumentation frequency range.

Scarf: This certainly is a possibility for the Jan. 6 shock event, if it was indeed a perpendicular shock at IMP-7. As you note, the IMP-7 wave instrument could not analyze electrostatic lower hybrid noise in the post shock region [$f(LHR) \simeq 8$ Hz for $F_2 = 12\ \gamma$] although the low frequency B-field channel (which did detect activity) is sensitive in this frequency range.

Wibberenz: The Jan. 6, 1975, shock occurred within a fast solar wind stream, and it had the appearance of a driven shock. Have you looked into possible differences of the ambient medium for the two shocks?

Scarf: No I haven't, but I might mention that both Helios and IMP-7 also detected shocks about 24 hours later. Specifically, an SSC at 2322 UT, Jan. 7, 1975 was associated with an intense low frequency wave level enhancement on IMP-7, and this event was detected by the Helios-1 fluxgate and search coil at 0023 UT, Jan. 8, 1975. Thus, it appears that this period early in January was a relatively interesting one, from the point of view of traveling interplanetary disturbances.

Scarf (question to E. Smith): The September 1974 shock was seen in the Pioneer magnetometer data beyond 4 AU. Was the shock perpendicular at Pioneer?

Smith: It appeared to be at least quasiperpendicular although the shock has not been analyzed in detail. In regard to Wibberenz's question, the September 1974 shock occurred within an interaction region at Pioneer and hence, was indeed "associated" with a high speed stream at that distance.

PART IV

COMETARY TOPICS

COMETS IN THE STIP CONTEXT

Max K. WALLIS
Mathematical Institute,
24 St. Giles, Oxford, U.K.

ABSTRACT

Fluid descriptions of plasma motion through a cometary coma are briefly sketched, distinguishing the bow shock and ionizing flow region mainly within it, the tail region and ray structure, and the 'ionosphere' coupled closely to the expanding cometary gas. Whether there is a contact discontinuity or continuous transition between the incoming flow and the ionosphere depends on solar fluxes rather than comet size. A discontinuity as observed requires much faster ionization in the inner coma or severe collisional cooling of incoming plasma.

Changes in structure and brightness may reflect solar UV and solar plasma variations, but may also be evidence of intrinsic instabilities of hydrodynamic/MHD or chemically-reactive flow. Molecular ionization and dissociation processes strongly influence the stagnation region ahead of the comet, and make it particularly susceptible to flow instabilities. Solar UV variations are energetically dominant within the ionosphere, changing the evaporation and chemical reaction rates, and probably stimulate dust halos. Shocked changes in the solar wind propagating through the head can trigger structural and intensity fluctuations in the plasma, notably disruptions of the plasma tail.

INTRODUCTION

A general difficulty in theoretical modelling of comets has for long been the lack of knowledge of the basic gaseous and solid constituents. The huge brightness and structural variations have been ascribed to special internal properties, such as chemically-explosive constituents or fragile eruptive structures.

Two facts revealed over the past few years have, however, clarified the situation and compelled consideration of basic simple processes, that are stimulated by the solar radiation and solar wind. The first fact is that H_2O is a major and perhaps the dominant

M. A. Shea et al. (eds.), Study of Travelling Interplanetary Phenomena/1977, 279–289.

constituent; the second that gas densities are generally so high that collisional processes dominate both fluid and chemical behaviour within a few hundred or thousand kilometres radius.

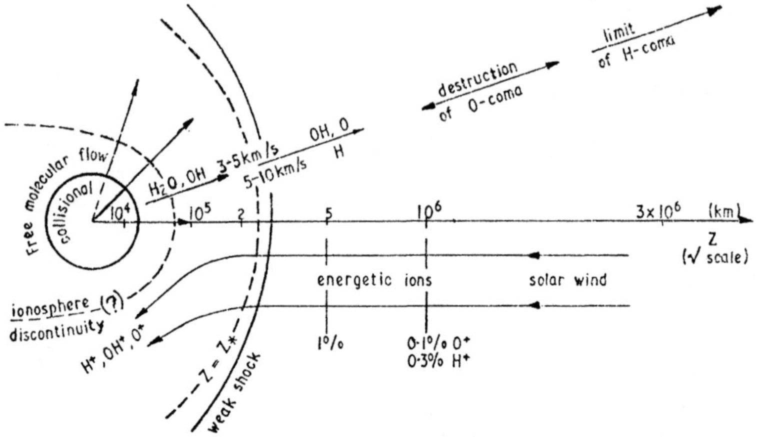

Fig.1. Coma of the H_2O comet, with the scales of its interaction with the solar wind (Wallis 1973a). Note that the upstream distance is measured on a square root scale. The solar plasma begins to decelerate at 5×10^5 km, but is hardly deviated in the supersonic regime to 3×10^5 km.

SOLAR PLASMA PENETRATION OF THE COMA

The appropriate starting point for models is a comet-source of pure H_2O, expanding into interplanetary space. Figure 1 shows (on a non-linear scale) the various regions in a sizeable H_2O comet, typified by comet Bennett (10^{29} molecules ster^{-1} s^{-1}) around ½-1 AU from the sun in a steady, average solar wind (Wallis 1973a). Far out in the coma, to a few times 10^6 km, new ions created by photoionization and charge exchange processes are accelerated by E and B fields and join the mean flow. They have high energies, with thermal speeds around solar wind speed; still, the gyration radius for eg. the O^+ ions is some 5000 - 2000 km and far less than the hydrodynamic scale. How far plasma turbulence effects thermalization of the energetic ions is **uncertain. early claims of rapid thermalization (Wu and Davidson 1972)** have been retracted (Wu and Hartle 1974), while satellite data from planetary neighbourhoods regularly show persistent weak beams of non-thermal particles (Bogdanov et al. 1975, Vaĭsberg et al. 1976). Though the degree of thermalization influences specific rates of ionization and cooling processes, it is unimportant for most purposes and the solar wind plus cometary plasma can be treated as a single fluid.

The quasi-one-dimensional form for the conservation equations modified by source terms is the most useful for qualitative discussion:

$$A^{-1}\frac{\mathrm{d}}{\mathrm{d}s}\rho u A = \mathscr{M} \tag{1}$$

$$A^{-1}\frac{\mathrm{d}}{\mathrm{d}s}\rho u^2 A + \frac{\mathrm{d}p}{\mathrm{d}s} = \mathscr{S} \tag{2}$$

$$A^{-1}\frac{\mathrm{d}}{\mathrm{d}s} uA\left(\tfrac{1}{2}\rho u^2 + \frac{\gamma p}{\gamma - 1}\right) = \mathscr{E}. \tag{3}$$

where A is the stream-tube cross-section. On introducing the Mach number of the flow $M = \sqrt{(\rho u^2/\gamma p)}$ equations (1 - 3) can be reduced to

$$(1 - M^{-2})\frac{\mathrm{d}u}{u} = M^{-2}\frac{\mathrm{d}A}{A} - \{\tfrac{1}{2}(\gamma + 1)\mathscr{M} - \gamma\mathscr{S}/u + (\gamma - 1)\mathscr{E}/u^2\}\frac{\mathrm{d}s}{\rho u} \tag{4}$$

$$\mathrm{d}M^{-2} + (\gamma + M^{-2})\frac{\mathrm{d}u}{u} - M^{-2}\frac{\mathrm{d}A}{A} = -\{(\gamma + M^{-2})\mathscr{M} - \gamma\mathscr{S}/u\}\frac{\mathrm{d}s}{\rho u}\,. \tag{5}$$

Equation (4) has the form of the standard "throat" equation of gas dynamics (Shapiro 1953). It shows how the source terms contribute to a generalized drag, as given qualitatively in Table 1, and is useful for examining possible transitions from supersonic solar wind to subsonic flow near the comet.

	Supersonic	Subsonic
Mass increase	−	+
momentum loss	−	+
Energy loss	+	−
area increase	+	−

Table 1. Velocity changes due to source terms in the flow: + denotes acceleration, - denotes deceleration

In the absence of source terms representing ionization effects, a continually decelerating flow satisfying (4) must have dA switching sign across the critical point M = 1. This is also true for a comet where the main contribution in the supersonic regime is a mass source (1) representing new ions: the increase in streamtube area in the subsonic continuation is induced by a curved bow shock of appropriate strength as in Fig.1.

An approximate analytic method confirmed by computations (Wallis 1971, 1973b; Brosowski and Wegmann 1972) is available for calculating solutions with a mass source alone. The shock is weak, the Mach number of the flow ahead of it decreasing according to (5) to $M \simeq 2$. The shock Mach number does not depend on the downstream boundary condition as in ordinary channel flows, because the inner dense part of the comet is relatively tiny in size. Instead, the shock is of such strength as to increase the streamtube cross-section A sufficiently to accommodate the continuing mass increase.

In the general comet case, plasma-gas collisions represented by energy and momentum loss terms also enter via equations (2) and (3), their forms depending to some extent on the degree of thermalization. In particular, collisions resulting in cooling of the plasma in the subsonic regime (cf. Table 1) cause weakening of the shock. But energy loss in the significant case of charge exchange interactions cannot completely eliminate the shock, for the basic reason that the critical point $M=1$ is mathematically always a 'saddle' (cf. Wallis and Ong 1975). This means too that the shock position varies considerably with slight changes in pressure, so in practice with fluctuating parameters one expects the bow shock to oscillate wildly. It may be that dissipation associated with the general level of plasma turbulence is sufficient to eliminate the shock (as a structure with steep gradients inducing sufficient intrinsic turbulence and dissipation). The 'shock' may thus be a mathematical artifact rather than a real structure to be detected by a space-probe.

On the simplest view, the shock being weak means that the electrons are simply heated adiabatically (magnetic moment conserved) up to and through the shock, the extra resistive heating amounting to only O(10eV) (Wallis 1973b): their subsequent cooling should be considered separately from that of the ions and will occur for higher energies through collisional ionization and excitation and for lower energies through vibrational and rotational excitations of H_2O and other molecules (Wallis and Ong 1976). The ions cool mainly through charge exchange processes in the subsonic region which substitute low energy for high energy ions. Despite the different nature of these cooling processes, they turn out conveniently to have very similar length scales, so the single fluid model is reasonable.

A rough hydrodynamic solution can thus be obtained along the lines used earlier (Wallis 1973b), taking A = const. in the supersonic flow, but p = const., $A \neq$ const. in the subsonic regime. However, no simple integrals exist and, for realistic forms of the momentum and energy loss functions (2) and (3), computation of the rough solutions has little merit. Also, it is not clear whether application of the Chapman-Jouguet hypothesis (cf. Wallis and Ong 1975) is not more appropriate than a somewhat arbitrary assumption of shock curvature.

IONOSPHERE

In the inner coma - within 10^3 km in the present example - the gas and plasma are strongly coupled by binary collisions. They expand together radially outwards from the nucleus. Photo-ionization processes with release of energetic electrons are counteracted by electron collisional cooling and dissociative recombinations. With the expansion velocity varying little with radius and the chemical equilibrium time shorter than the hydrodynamic time (local thermodynamic equilibrium), the plasma and neutral gas densities vary roughly with radius R as

$$n \sim N^{\frac{1}{2}} \sim R^{-1}. \tag{6}$$

Computations of the fluid and rate equations by Ip and Mendis (1976) confirm this dependence, though their representation of heating by a polytropic relation is artificial. Shimizu (1976) calculates that theoretically the gas should cool to a few ^{0}K by 300km radius. However, direct observations of collisional effects on cometary molecular spectra reveal several hundred ^{0}K (Arpigny 1976), which presumably confirms the net heating through photo-dissociative processes (Wallis 1974).

IONOSPHERE TRANSITION

The question arises, is there a tangential discontinuity where the pressure of the incoming plasma is balanced by that of purely cometary origin ('ionosphere discontinuity' of Fig.1)?

The cometary ions brought in by the solar wind and which carry much of its energy undergo charge-exchanging collisions with neutral molecules; with decreasing plasma flow velocities and with penetration into ever-increasing neutral densities, the transition to temperatures comparable to that of the neutral gas is exponentially sharp. The electrons cool through collisional excitations on a very similar length scale, while with plasma pressure near constant the less rapidly cooling component controls the density. Generalizing the earlier result (Wallis and Ong 1976), the scale radius of the sharp transition is found as

$$R_{trans} \simeq Q\, r \times 10^4 \text{ km} \qquad (7)$$

at heliocentric distance r AU, for a comet of size $Q \times 10^{29}$ molecules of H_2O ster^{-1} s^{-1}. The variation with r covers the change in ionization rates ~ r^{-2}. The ionospheric plasma has density

$$n \simeq 6\times10^2\, Q^{\frac{1}{2}} / r\, R[4] \quad \text{cm}^{-3} \qquad (8)$$

where R[4] denotes radial distance in units of 10^4 km. The ionospheric electron contribution to the pressure is small; the stagnation pressure of the ions is calculated assuming slow increase with comet size due to the photo-dissociative heating. Comparing it with the solar wind stagnation pressure from a flux of $3\times10^8\, r^{-2}$ ions cm^{-2} s^{-1}, one finds

$$p_{ionos} / p_{solar\ wind} \simeq 3.5\times10^{-2}\, Q / R[4]. \qquad (9)$$

At the transition position (7), the ionosphere pressure is thus

$$p_{ionos} \simeq 3.5\times10^{-2}\, r^{-1}\, p_{solar\ wind}. \qquad (10)$$

Due to the ionizing interaction, the plasma's pressure is at least a factor 2 less than that of the free solar wind after passing through the bow shock (Wallis 1973b); one concludes that the ionospheric pressure is a factor 10r too small to hold off the incoming plasma, that for an H_2O comet under normal steady solar conditions there can be no 'ionopause', except at very small heliocentric distance r.

This conclusion might be invalidated in a number of ways:
1) increased ionization rate ν - if this exceeds the 0.5×10^{-6} s^{-1}

assumed for (8), the numerical factor is increased ($n \simeq V^{1/2}$). Solar UV ionization rates cannot be higher by as much as a factor 100, but electrical discharge ionization suggested by Ip and Mendis (1976) to give a rate 10^{-4} s^{-1} would suffice;
2) reduced electron cooling and recombination rate - if the gas is largely atomic (O for example), recombination may become unimportant:

$$R^{-1}\, d(nVR^2)/dR = Q/VR^2 \;\rightarrow\; n \simeq 5\times10^3\; Q/r^2\; R[4] \quad cm^{-3}\,. \tag{11}$$

If this value is used in (10), the extra factor $10Q^{\frac{1}{2}}/r$ would increase the ionospheric pressure sufficiently to exceed the solar wind pressure for larger comets. Though OH is indeed observed to be destroyed rather more quickly than shown on Fig.1, on a scale 2×10^5 km (Keller and Lillie (1974), the coma of an H_2O comet is still unlikely to be sufficiently dissociated to allow (11) to hold at small enough R;
3) reduced solar wind pressure - as the pressure is often small by a factor 3 than the mean 2×10^{-8} dynes cm^{-2} used for (9), but not more than a factor 10 smaller, this reduction is too little;
4) reduced plasma pressure in the incoming subsonic flow - because of the ionizing interaction, the momentum flux of the outflowing ionized cometary gas may virtually balance the solar wind pressure. This depends on how effectively the incoming plasma is cooled, being far from true for incompressible flow (Wallis and Dryer (1976), but no representative solution has yet been given.

There are observational grounds for a layer of denser plasma on the sunward side of comets (at least, for the major comets) which encourages ideas of tangential discontinuity and pressure balance. Schematically, the flow pattern would appear as in Fig.2; in cases 1) or 2) the visible structure would be in region C1 to the border with C2, whereas in case 4) the visible plasma would be in region C2 just outside the discontinuity.

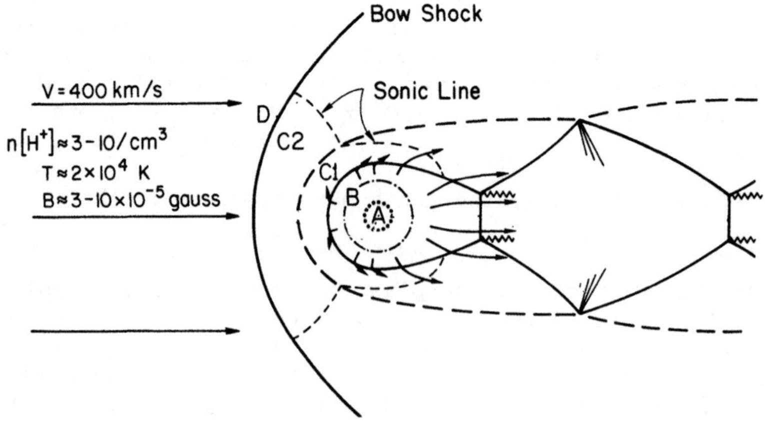

Fig.2. Schematic flow regions for the comet-solar wind interaction. Region A is subsonic flow near the nucleus, region B is supersonic with transition to collision-free regime.

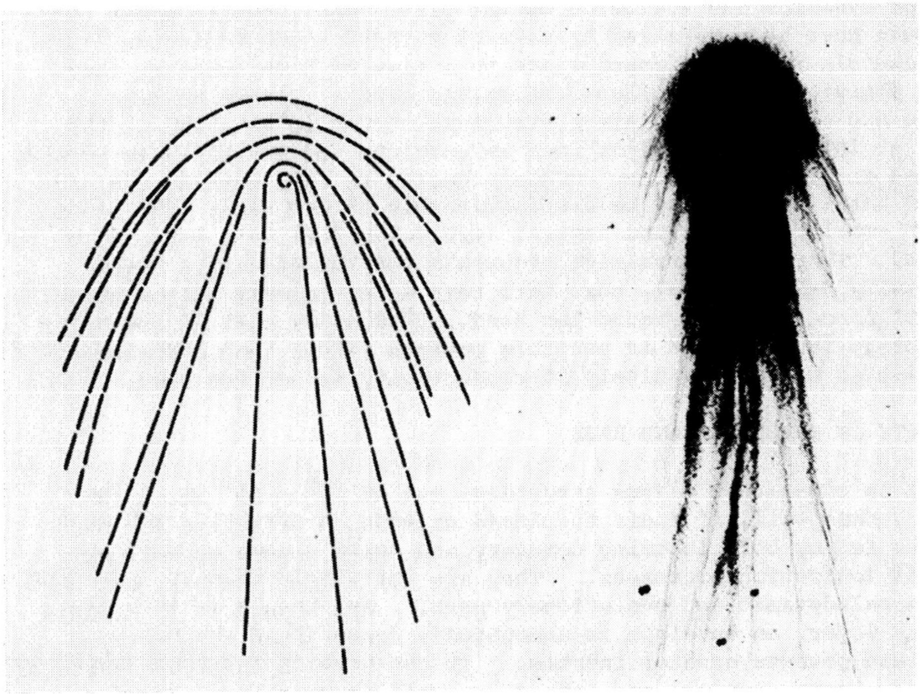

Fig.3. Plasma envelopes in comet Morehouse (1908III)

PLASMA STRUCTURES AND ENVELOPES

Structures sunward of the comet nucleus that one can identify with reasonable confidence as composed of plasma, have been observed in major comets such as Bennett and Humason. But one must go back to comet Morehouse of 1908, studied in detail by Eddington (1910), for pictures with the least contamination and adequate time sequences. One explanation advocated consistently and primarily by Wurm (e.g.1974) is that the 'jets' and 'envelopes' originate from internal sources close to the nucleus (within a few hundred km in particular cases). The alternative view is of a saucer-shaped plasma region sunward of the nucleus at several thousand or more km, which divides into rays on flowing into the tail.

A main difficulty with Wurm's description is that collisional processes and particularly dissociative recombinations occur rapidly at the high gas densities now known to be present in the inner coma. With a rapid enough ionization mechanism, as the current discharge suggested by Ip and Mendis (1976), fully-ionized structures capable of emerging from the collisional region are conceivable. But the production of narrow jets and the formation of successive envelopes by such a mechanism is less easy to envisage.

The dynamics and evolution of the plasma envelopes in comet Morehouse have been detailed by several authors since Eddington (1910). The principle features, appropriate to a time of high cometary (and solar) activity are as follows (cf.Wallis 1968): plasma becomes visible and consolidates into a structure around 10^5 km ahead of the comet in 10^3 sec; it extends into an envelope and contracts in towards the head in 10^4 secs; a new envelope begins to form outside the old one and three or more may be distinguishable at one time; the plasma of each envelope flows into several (typically 4-6) tail rays, at up to 50 km/s. There are persuasive arguments for the structure being essentially 3-dimensional, that tail rays do not simply follow magnetic lines of force wrapping round the head. The major part of the solar wind energy is of course in particle motions rather than magnetic field - the latter behaves passively if conductivity is not too high.

STABILITY OF ENVELOPES AND RAYS

It is clear that plasma structures out at 10^4 - 10^5 km in the cometary coma will, if their thickness exceeds an effective ion mean path, be fed by both incoming cometary and solar plasma as well as internal ionization processes. They are not simply tracers, but have an internal dynamic and evolutionary path. And though it is a fluid boundary layer, an envelope is undoubtedly denser than the nearby plasma and possess greater inertia. It can have an internal source of material, energy sources and losses, as well as momentum sources.

There are several possibilities for new hydrodynamic properties. In particular, it is clear that a stagnation point in the plasma flow with relatively moving gas is singular (Wallis and Dryer 1976). The plasma-gas interaction is enhanced near the stagnation point and the flow is unstable if the diffusion path is much smaller than the flow scale. Qualitatively, a stationary saucer-shaped envelope will tend to burst sunwards from its centre (Fig.4, first picture) due to the enhanced contribution of momentum from newly ionized neutral particles in this region. While the incompressible (uncooled) description of Wallis and Dryer (1976) over-estimates diffusion effects, even this indicates that diffusion does not stabilize envelopes in larger comets, outside 10^4 km.

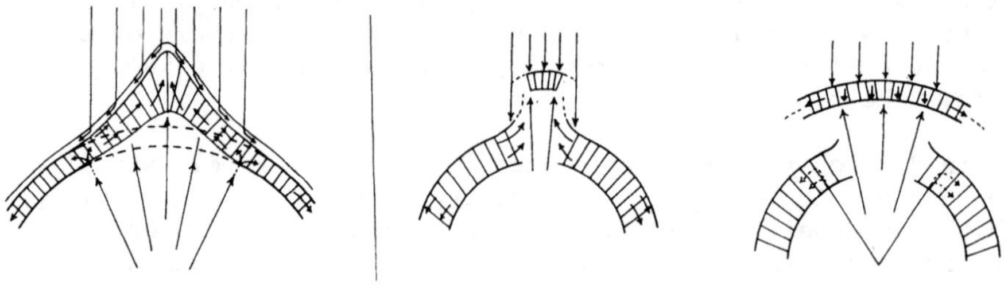

Fig. 4. Hydrodynamic outbursting instability of an ionizing plasma envelope, leading to periodic formation of new envelopes.

Also, chemical reactions and particularly collisional ionization processes can combine with fluid motions to give instability. If the ionization with net energy loss gives increased density and increased ionization rate in an element of plasma, then density fluctuations will grow. The electron collisional ionization process originally suggested (Wallis 1968) probably fails because the electrons are little heated, but charge exchange interactions can work similarly in far subsonic flow. This condensation process can explain the formation of 'knots' and similar irregularities in the comet head, but not such features that first appear in the tail (one outstanding example of a tail feature appeared as a sub-comet in the tail of comet Burnham (Beyer 1962) and may have originated from a solid fragment - if so, this may be common on a smaller scale).

Standard MHD instabilities of an interface may also be relevant. If a magnetic field lies along the envelope and is not purely passive, interchange or flute instabilities would develop at the envelope apex. They would enhance (weakly) the hydrodynamic outbursting instability. In the lateral parts of the envelope, accelerated back into the tail, the Taylor instability will operate, which perhaps underlies the division of the envlope into tail rays (Wallis 1968). Under tangential streaming, the interface is unstable to Kelvin-Helmholtz perturbations, very plausibly resulting (Ershkovich et al. 1971) in finite amplitude waves. Narrower tail rays are apparently not susceptible to these instabilities, presumably because of transport effects. An important deduction from the observed wavelength is that the magnetic field is weak and probably of external (interplanetary) origin (Lüst 1962; Ershkovich 1976).

COMET VARIABILITY

The intrinsic tendencies to instability outlined above serve to obscure the responses of comets to solar flux changes and distort statistical correlations (Sekanina 1959). It is clear that solar UV radiation is more important than the solar wind: for example, photo-dissociation of H_2O alone (at 10^{-5} s^{-1}, releasing 1.8eV per dissociation) in the $Q = 10^{29}$ molecules $ster^{-1}$ s^{-1} comet releases more energy within 5000km radius than the total solar wind energy flux through an area of the same radius. Adding to this the fact that only a fraction of the solar wind flux could conceivably penetrate the inner coma, it is clear that solar radiation variations are more important in changing evaporation and chemical interaction rates. Variations in total brightness and in dust production, which correlate to some extent with solar activity (Sekanina 1959; Miller 1976), are much more likely to be radiation induced.

Solar wind responses are of course apparent in causing tail fluctuations, but the rather poor success of specific correlations shows that quite small changes in solar wind may affect the tail visibility. On the other hand, the most serious effect - a complete disruption of the tail - does correlate with solar flares. There are several examples of such disruptions, in comet 1939III (Guigay 1963; Richter 1939), comet Humason (Biesbroeck 1963) and Ikeya-Seki (Churumov and Demenko 1970). From a solar flare, the solar wind momentum flux increases typically by a factor 10-100 within 100 s. This is far faster than the rate of evolution of plasma envelopes. Visually, the envelope is swept back into the tail region, the head is swept clean of plasma and new production is delayed for some hours - presumably the time for a new envelope to be generated at the new position of momentum balance. It is not clear how far the qualitative discussion of shock-comet interaction of Dryer et al. (1976) applies to the highly inhomogeneous cometary plasma situation. Quantitative discussion must await detailed models of the plasma envelope and treat the comet as a self-regulating if delicately-balanced system.

REFERENCES

Arpigny, C.: 1976, in *The Study of Comets*, NASA SP-393, 797.

Beyer, M.: 1962, *Astr. Nachr. 286,* 211.

Biesbroeck, G. van: 1963, *Astro. J. 68,* 733. (See also *Sky Teles. 24,* 124, 216.

Bogdanov, A. V., Vaisberg, O. L., Kalinin, A. P., and Smirnov, V. N.: 1975, *Dokl. ANSSSR 225,* 1284 (= *Sov. Phys. Doklady 20,* 791).

Brosowki, B., Wegmann, R.: 1972, *MPI/PAE-Astro* 46, *M.P.I. Astrophys.,* Munich.

Churumov, K. I., and Demenko, A.A.,: 1970, *Astromet. i Astrofiz.* no. 11, 55 (Kiev).

Dryer, M., Ershkovich, A. I., and Shen, W. -W.: 1976, *J. Geophys. Res. 81,* 6184.

Eddington, A. S.: 1910, *Mon. Not. Roy. Astron. Soc. 70,* 442.

Ershkovich, A. I.: 1976, *Mon. Not. Roy. Astron. Soc. 177,* 649.

Ershkovich, A. I., Nusinov, A. A., and Chernikov, A. A.: 1971, preprint D-86, Space Res. Inst., Moscow.

Guigay, G.: 1963, *J. d. Observateurs 46,* 181.

Ip, W. -H., and Mendis, D.: 1976, *Icarus 28,* 329, *29,* 147.

Keller, H. U., and Lillie, C. F.: 1974, *Astron. Astrophys. 34,* 187.

Lust, Rh.: 1962, *Z. Astrophys. 54,* 67.

Mendis, D., and Ip, W. -H.: 1977, *Space Sci. Rev.* (in press).

Miller, F. D.: 1976, *Space Sci. Rev. 19,* 739.

Richter, N.: 1939, *Astr. Nachr. 269,* 103.

Sekanina, Z.: 1959, *Bull. Astron. Inst. Czech. 10,* 103.

Shapiro, A. H.: 1953, *The Dynamics and Thermodynamics of Compressible Fluid Flow, 8,* New York.

Shimizu, M.: 1976, *Astrophys. Space Sci. 40,* 149.

Vaisberg, O. L., Romanov, S. A., Smirnov, V. N., Karpinsky, I. P., Khazanov, B. I., Polenov, B. V., Bogdanov, A. V., and Antonov, N. M.: 1976, *Kosmich. Issled. 14,* 827.

Wallis, M. K.: 1968, *Planet. Space Sci. 16,* 1221.

Wallis, M. K.: 1971, *TRITA-EPP-72-02, R. Inst. Tech.,* Stockholm.

Wallis, M. K.: 1973a, *Astron. Astrophys. 29,* 29.

Wallis, M. K.: 1973b, *Planet. Space Sci. 21,* 1647.

Wallis, M. K.: 1974, *Mon. Not. Roy. Astron. Soc. 166*, 181.
Wallis, M. K., and Dryer, M.: 1976, *Astrophys. J. 205*, 895.
Wallis, M. K., and Ong, R. S. B.: *Planet. Space Sci. 23*, 713.
Wallis, M. K., and Ong, R. S. B.: in *The Study of Comets*, NASA SP-393, 856.
Wu, C. S., and Davidson, R. C. : 1972, *J. Geophys. Res. 77*, 5399.
Wu, C. S., and Hartle, R. E.: 1974, *J. Geophys. Res. 79*, 283.
Wurm, K.: 1974, *Astrophys. Space Sci. 27*, 211.

DISCUSSION

Grzedzielski: What is the cooling mechanism that allows the solar wind plasma to cool fast enough in order to be able to pile up in the vicinity of the stagnation point? Is the cooling determined by the cometary matter picked up by the solar wind plasma on the way toward the stagnation point?

Wallis: The cooling mechanisms are several; the ions cool through charge exchange with neutral cometary particles - resonant charge exchange cross-sections are high, as are also cross-sections for protons with O or H_2O. Higher energy electrons cool through collisional ionization and excitation of neutrals, but below 10 eV may cool through vibrational and rotational excitations of molecules. Cooling rates depend on the particular comet constituents, but the scale distance is little changed. Cooling of incoming protons, cometary ions and electrons is found to occur "explosively" (i.e., exponentially with distance) close to 10^4 km radius ahead of the nucleus.

THE COMET-SOLAR WIND INTERACTION

D. A. MENDIS
Department of Applied Physics and Information Science
University of California, San Diego
La Jolla, California 92093 U. S. A.

ABSTRACT

The important roles of comets as natural probes of the solar wind, particularly at high heliographic latitudes and small heliocentric distances, unattained thus far by artificial space probes, is stressed. It is becoming clear the solar wind is not merely responsible for shaping and maintaining the cometary plasma tail, but is also indirectly responsible for the rapid ionization processes in the coma. Despite this, however, the detailed interaction of the solar wind with comets, which is underlined by the complex and time-varying morphology observed in the cometary head and the tail, constitutes a formidable problem, and is only very imperfectly understood at present. The current views are critically reviewed and an attempt is made to identify the dominant physical mechanisms that are involved.

INTRODUCTION

In a symposium devoted to the study of travelling interplanetary phenomena it is appropriate to begin by stressing the historic as well as the continuing role of comets as probes of the interplanetary medium. Indeed it was the interpretation by Biermann (1951) of the observed orientation of the plasma tails of comets in terms of dynamical aberration that provided the earliest unambiguous information about the continuous outflow of corpuscular radiation from the sun. In this important role comets have not been entirely superseded by the advent of artificial space probes, because while the latter are confined to regions close to the ecliptic plane, long period comets approach the sun at all

M. A. Shea et al. (eds.), Study of Travelling Interplanetary Phenomena/1977, 291–303. All Rights Reserved.

inclinations and a few of them get closer to the sun than any artificial space probe has or will in the foreseeable future.

The technique of using this dynamical aberration of the plasma tails of comets as a means of delineating the time averaged global velocity field of the solar wind over a period of several solar cycles has been used to good advantage over the years, particularly by Brandt and his coworkers (e. g., Brandt et al., 1975; Brandt and Rothe, 1976). Likewise sudden changes in the orientation or structure of the plasma tail may indicate its response to discontinuities in the solar wind flow (e. g., see Brandt and Mendis, 1977, for a recent review).

Thus while we can make use of the plasma tails of comets as tracers of the solar wind flow with some degree of confidence at the present time, much less satisfactory is our understanding of the tracer itself (which is characterized by a complicated and rapidly varying fine structure) from the point of view of its interaction with the ambient medium.

Not only is the nature and observed motions of these complicated structures obscure, but even such basic questions as the dominant process of ionization of cometary gases, the mechanism of coupling between the cometary plasma and the solar wind, and the possible existence of significant cometary magnetic fields remain, to a lesser or greater degree unresolved.

I would like here to address the attention of the solar wind community to some of these questions, stressing mainly the physical processes that are likely to be involved.

THE FLUID INTERACTION

If the comet is treated as a source of gas within an ionizing flow of magnetized plasma, the basic features of the plasma flow on the windward side of the comet are shown schematically in Fig. 1, where the original picture of Biermann et al. (1967) has been embellished by the addition of the interplanetary magnetic field and an internal shock. The more recent calculations by Brosowski and Wegmann (1972) and Wallis (1973) seem to indicate that the outer shock is much weaker than thought earlier, being typically about Mach 2.

The position of the tangential discontinuity is determined by the pressure balance between the inflowing solar wind and the outflowing cometary plasma. Wallis and Ong (1976) have argued that, due to the

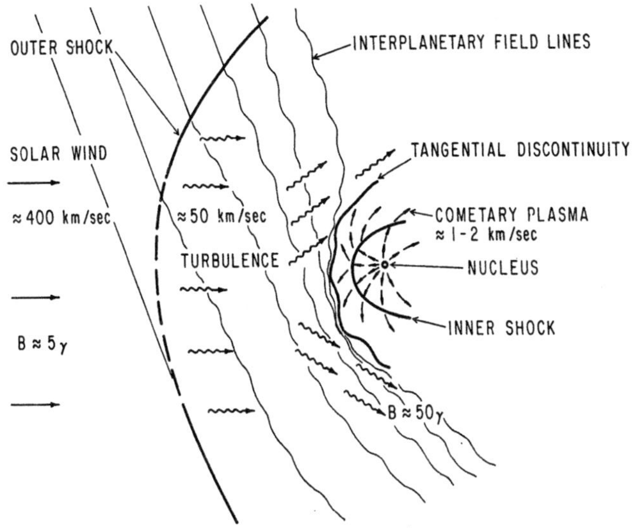

Fig. 1. Schematic representation of the fluid-like interaction of the solar wind with a comet.

strong cooling of the cometary plasma, its pressure is low and consequently that the solar wind can penetrate very close to the nucleus before pressure equilibrium occurs. They also point out that what we have in this situation is not the separation of distinct plasmas but rather the continuous transition from the contaminated solar wind plasma to the cometary plasma, and that therefore the concept of a tangential discontinuity is rather dubious.

While the temperature of the cometary ions is quite uncertain, one expects these ions to be strongly coupled to the neutral species within $r \leqslant 10^4$ km from the nucleus and to be expanding supersonically there (Ip and Mendis, 1976a, 1977). The ram pressure of the quiet solar wind at 1 AU $\approx 10^{-8}$ dyne cm^{-2}. For a medium bright CO^+-rich comet like Kohoutek (1973f) with a production rate $\approx 10^{29}$ mols s^{-1} at 1 AU and an ionospheric expansion velocity ≈ 1 km s^{-1}, this pressure is equalized by the ram pressure of the cometary ionospheric plasma at $r = r_c$. While $r_c \leqslant 10^3$ km if the only sources of ionization are photoionization and charge exchange ionization, both with time scales of about 10^6 s, $r_c \approx 10^4$ km if ionization takes place with a much shorter time scale of about 10^4 s, as is suggested both by observation, as well as by theoretical considerations (Ip and Mendis, 1977) which will be discussed in the next section. The brightness profile of the 8-10 band of H_2O^+ in the medium bright comet Bennett (1970II) at a heliocentric distance of ≈ 0.84 AU, recently published by Delsemme and Combi (1976) suggests that a tangential discontinuity, if present, must necessarily be at a nuclear distance $r \leqslant 2 \times 10^4$ km. On the other hand, the stability of such a surface against various hydrodynami and hydromagnetic instabilities remains an open question (e.g., Ioffe, 1970; Dobrowolny and D'Angelo, 1972; Ershkovich and Chernikov, 1973; Mendis and Ip, 1977).

The presence of a magnetosheath of compressed interplanetary magnetic field, as indicated in Fig. 1, is also an open question. Arguments based on the long time of diffusion of the interplanetary field lines through the cometary ionosphere suggest an enhancement of the interplanetary field there up to a value set essentially by the ram pressure of the solar wind (e.g., Dessler, 1968; Mendis and Ip, 1977). On the other hand, currents set up by the photo-production of cometary ions in the post-shock region as well as by the collisions of solar wind ions with cometary neutrals may reduce this field substantially, as has been suggested for the case of Venus and Mars (Cloutier, 1970; Cloutier and Daniell, 1973).

THE PRODUCTION OF COMETARY IONS

Spectroscopically the ionospheres and plasma tails of comets are observed mainly by the emission bands of molecular ions, while the emission lines of a few atomic ion species have been observed in a few cases. A complete list of the ion species observed so far is given in Table 1.

Table 1. Cometary Ions

H_2O^+	CO_2^+	CO^+	OH^+	CH^+	CN^+	N_2^+	Ca^+	C^+	Si^+

The species C^+, Si^+ and CN^+ have been added recently by the ultraviolet observations of comet West (1975n) (Feldman and Brune, 1976; Smith et al., 1976).

Even a casual inspection of a photograph of a cometary head, which is not overexposed, shows that the source of the tail ions is confined to a rather small region around the nucleus (typically $\leqslant 10^4$ km for a medium bright comet at a heliocentric distance ≈ 1 AU). This point was strongly emphasized by Wurm (e.g., 1943, 1963) who also gave a circumstantial argument for a short time scale of ionization of CO ($\sim 10^{3.5}$ s at 1 AU) based on the time scale of the observed fluctuation in the intensity of CO^+ features in several comets.

While Wurm's arguments for a short time scale for CO is by no means conclusive (e.g., see Mendis and Ip, 1977), the idea of a short time scale for the production of cometary ions has been strengthened by more recent observations of the distribution of H_2O^+ in comets Bennett (1970II) and Kohoutek (1973f) (Delsemme and Combi, 1976; Wyckoff and Wehinger, 1976). With typical expansion speeds of $H_2O^+ \approx 1$ km s^{-1} (Ip and Mendis, 1976a), these observations require an ionization time scale for H_2O at 1 AU of $\leqslant 10^4$ s.

If one accepts these short time scales for the ionization of the cometary species one is faced with a problem. The reason being that the immediately obvious sources of ionization, viz. (i) photoionization by the ultraviolet solar radiation and (ii) charge exchange ionization by the solar wind protons, for molecules such as H_2O, CO and N_2 etc., give time scales in excess of 10^6 s at 1 AU. Axford (1964) and Beard (1966) have stressed the role of collisional ionization by energetic electrons produced at the outer shock front. If energetic electrons in the keV range are produced as suggested by Axford (1964), multiple ionization by these could reduce the time scale for ionization to $\leqslant 10^5$ s. However, Wallis (1973) has pointed out that if the upstream shock is subcritically weak (i.e., Mach number $\leqslant 3$) as is likely to be the case, the shock induced heating of the electrons is only about 10 eV rather than 1 keV, and this is inadequate to produce the enhanced ionization.

The alternative is to look for the so-called 'internal' or 'intrinsic' ionization processes suggested by Wurm (1966, 1974). One possibility is associated with fast ion-molecule chain reactions believed to occur in the collision dominated inner coma (Oppenheimer, 1975; Shimizu, 1975; Huntress, 1977). In this scheme the ionization of the observed ion in question is not produced directly by either solar electromagnetic or solar corpuscular radiation, but is achieved via ion exchange and charge exchange reactions involving other ions, which are themselves produced by the direct ionization process. The catch in this proposal is that these latter ions must themselves be produced quickly ($\tau_i \leqslant 10^4$ s), be abundant, and also have ionization potentials comparable to those of the ions they eventually produce. A way around this difficulty is to produce the intermediate 'donar' ion by the fast photoionization of an atom or radical, which has been left in a high excited metastable state during its production by photodissociation. A plausible scheme along these lines has recently been proposed by Delsemme and Combi (1976) for the production of H_2O^+ with $\tau_i \approx 5 \times 10^4$ s in comet Bennett (1970II) using C^+ as the intermediate ion.

Ip and Mendis (1976b), on the other hand, have associated

Wurm's 'internal ionization process' with large currents of energetic electrons closing through the inner coma from the tail. These authors proposed that the folding of the interplanetary magnetic field lines into the tail of the comet as the solar wind sweeps past generates a cross-tail current $I_t \approx cB_t \ell_t / 2\pi \geqslant 10^8$ A, separating the two tail lobes of opposite polarity. Here ℓ_t = length of the tail current-sheet and B_t = the tail field just outside the neutral sheet. These authors have further argued that this current will be eventually disrupted due to the onset of the ion-sound current instability which will cause a substantial fraction of these current electrons (which are energized to the Alfvén energy $E_A \approx B_t^2 / 4\pi n_e \approx$ 1-10 keV) to flow along the field lines and close through the inner coma, causing ionization with a time scale $\leqslant 10^4$ s. It may also be pointed out that such large currents require plasma densities $\approx 10^2$ cm^{-3} in the tail plasma sheet for their maintenance, and such values may be attained during the folding of tail rays of enhanced plasma density into this sheet. This leads to a natural explanation of the observed time scales for outbursts of rapid ionization ($\approx 10^4$ s). Also the field aligned currents will probably flow along sheets or filaments and close through the sunward side of the coma. Furthermore, they will ionize most efficiently towards the ends of their paths when their energies are $\leqslant 100$ eV as in the case of the terrestrial ionosphere. The observed beginning of tail rays in the sunward hemisphere at nuclear distances of about 10^3 km, as well as the shell structure observed in front of the nucleus may conceivably be explained along these lines.

Detailed hydrodynamic models of expanding, H_2O-dominated as well as CO-rich, ionospheres have been computed (Ip and Mendis, 1976a, 1977). It was found that the observed H_2O^+ ion distribution in comet Kohoutek (1973f), which was restricted to a region $r \leqslant 5 \times 10^3$ km in the sunward direction (Wyckoff and Wehinger, 1976), can be reproduced satisfactorily only when such a rapid ionization source (with $\tau_i \leqslant 10^4$ s) was included. The computed distributions of column densities of the two ions H_2O^+ and CO^+ in the sunward direction of an H_2O dominated comet (with $\gamma = n_o(CO)/n_o(H_2O) \approx 0.1$ and nuclear radius r_n = 1 km) both with and without this "internal" ionization source is shown in Fig. 2.

THE FINE STRUCTURE OF THE PLASMA TAIL

It has already been stressed that the plasma in the cometary tail is far from homogeneous. It is characterized by a discrete and time varying structure. The most dominant feature observed in weakly exposed photographs with high surface resolution near the 'head' of the

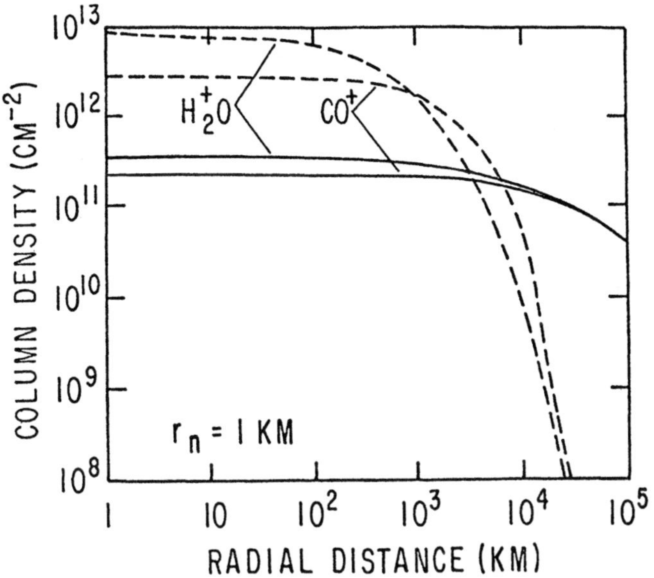

Fig. 2. Distribution of the column densities of H_2O^+ and CO^+ ions in the sunward direction of a H_2O dominated comet with $\gamma = n_o(CO)/n_o(H_2O) = 0.1$, (——) without 'internal' ionization source, (-----) with 'internal' ionization source (γ_n = 1 km) (from Ip and Mendis, 1976a).

tail is a fan like structure of straight and narrow 'rays' and bundles of rays called 'streamers' which start at a large angle to the tail axis and continuously turn on to the tail axis rather like a 'folding umbrella.' The distant plasma tail is further characterized by a complicated structure involving 'knots,' 'helicies' and large diffuse and irregular structures, all of which propagate down the tail often with large velocities and accelerations. A detailed description and analysis of these structures is given in a recent review by Brandt and Mendis (1977). Here I will merely confine myself to a few brief comments.

As regards the ray structure, it is difficult to avoid the conclusion, following Alfvén (1957), that they are magnetic flux tubes. Their umbrella folding motion is then simply their drift in crossed electric and magnetic fields. While the extreme straightness and stability of the rays could conceivably be merely a consequence of hypersonic flow (Wallis, 1967), a purely hydrodynamic model does not address itself to the evolution of the rays, particularly their umbrella folding geometry.

With regard to the other structures, observed down the tail, let us consider their motions first. The observed accelerations of these structures are typically 10-200 cm s^{-2} during quiet solar wind conditions, while occasionally attaining values as large as 10^4 cm s^{-2} during disturbed solar wind conditions.

It is now generally accepted that despite the large particle collision mean free paths, an essentially fluid-like interaction takes place between the solar wind and the cometary plasmas. According to one point of view this is brought about by the interplanetary magnetic field which enhances the coupling between the two plasmas essentially by reducing the proton-ion interaction distance (Harwit and Hoyle, 1962).

In a simple hydromagnetic description given by Ioffe (1968) the observed accelerations are a result of the plasma being squeezed out of flux tubes by magnetic pressure rather like 'toothpaste out of a tube due to lateral pressure.' This 'toothpaste tube' model predicts acceleration

$$a \approx \frac{B^2}{2\pi \rho_i \ell}$$

for a fluid element at a distance ℓ from the nucleus, down the tail. Taking $n_i \approx 10^2$ cm^{-3}, $m_i \approx 28$ amu, $B \approx 100\ \gamma$ and $\ell \approx 5 \times 10^5$ km, one obtains $a \approx 200$ cm s^{-2} which is in agreement with observation. However, this acceleration mechanism, while likely to be effective close to the head where the field lines have a fan like configuration, is unlikely to be so, far down the tail ($\ell \gtrsim 5 \times 10^6$ km) where the field lines are approximately parallel to the axis.

According to the second point of view the enhancement of the momentum transfer rate is brought about by plasma turbulence excited in the tail via some instability. Chernikov (1975) has considered this question in considerable detail and has concluded that of various instabilities that can occur, the ion-sound instability has the greatest growth rate both in the absence of magnetic fields as well as in the presence of fields that are likely to be present. If the relative velocity between the solar wind and cometary plasma is larger than a critical value of about 200 km s^{-1}, accelerations around 100 cm s^{-2} may be achieved with reasonable values for the cometary plasma parameters. Inherent in this analysis, however, is the supposition that the solar wind and cometary plasmas are freely interpenetrating, which of course may be prevented if a stable tangential discontinuity surface separating these two plasmas exists.

Theoretical discussion of the stability of this discontinuity surface (e.g. Dobrowolny and D'Angelo, 1971; Ershkovich et al., 1972;

Ershkovich, 1976; Mendis, 1977) has hinged on the strength of the cometary tail magnetic field, B_t, with values typically $\gtrsim$ 25-50 γ quenching the Kelvin-Helmholtz instability that would arise for smaller values of B_t. It also needs to be pointed out that these theoretical considerations to date are based on highly oversimplified models of the tail field configuration, which correctly consists of the vector sum of the ambient field and the tail field generated by the neutral sheet current. On the other hand, although one occasionally sees the distant cometary tail lose its regular structure and turn into a turbulent wake (e.g. comet Kohoutek (1973f)), it is not a regular occurrence.

Finally, the interesting possibility that large amplitude Alfvén waves propagating down the tail may themselves be responsible for the local heating and acceleration must be mentioned. That outward propagating Alfvén waves are most prominent in hot high velocity solar wind streams has already been noted by Belcher and Davis (1971), who suggest that these waves may play an important role in the dynamics of these high velocity streams.

If large amplitude Alfvén waves ($\delta B/B \gtrsim 1$) are generated by the classical kink instability when parallel current begins to flow (e.g., see Alfvén and Fälthammar, 1963), their dissipation may be large due to nonlinear mode coupling mechanisms which feed energy from the Alfvén mode to the magnetosonic modes, which are strongly damped (e.g., Belcher and Davis, 1971; Barnes, 1966). Then using a simple 1-D, isothermal, supersonic flow model (which seems appropriate for the comet tail) with this extended source of heat input, it is easily shown that the bulk velocity, V_i down the tail tends asymptotically to $V_{i\infty} \approx C_o \chi$, while the Alfvén velocity, V_A, tends to $V_{A\infty} \approx V_{Ao}\sqrt{\chi}$ where C_o and V_{Ao} represent the initial sound speed and Alfvén spee respectively, and

$$\chi \approx 3 + \left(\frac{\delta B}{B}\right)_o^2 \left(\frac{V_{Ao}}{C_o}\right)^2$$

Assuming $(\delta B/B)_o \gtrsim 1$, and taking $T_o \approx 10^4$ K, $B_o \approx 100\ \gamma$, $n_o \approx 5\times 10^2$ cm^{-3} and $m \approx 28$ amu, we find $V_{i\infty} \gtrsim 100$ km s^{-1} and $V_{A\infty} \gtrsim 100$ km s^{-1}, giving the observed wave velocity with respect to the nucleus far down the tail $= V_{i\infty} + V_{A\infty} \gtrsim 200$ km s^{-1}, which is certainly consistent with the observations of such wave-like features in comet Kohoutek (1973f)(Hyder et al., 1974).

As regards the nature of these distant tail structures our

understanding at the present time is very poor. The helical feature observed in comet Kohoutek (1973f), as well as in other comets, has been identified as a wave excited by the classical kink instability produced by a parallel current by Hyder et al. (1974). It has been identified as a wave also by Ershkovich (1976), who, however, suggests that it is excited by the Kelvin-Helmholtz instability down the tail, and thereafter nonlinearly stabilized.

While the bends and kinks in the tail can be associated with sudden changes of the solar wind flow (e.g., Jockers and Lust, 1973), the large diffuse, irregular features far down the tail seen in a few comets (e.g., Kohoutek (1973f), Morehouse (1908III), Humason (1962VIII)) gives the impression of a highly turbulent wake, produced perhaps by the unquenched growth of waves of one sort or another.

The small knots mentioned earlier are a common cometary tail phenomenon. However, once again we have very little understanding of their formation. Wallis and Ong (1976) have suggested that these knots may be caused by the recombination instability which has been investigated by D'Angelo (1967). However, this requires a number of conditions which include, besides a strong temperature dependence of the volume recombination coefficient, a negligible heat conduction and an electron temperature $T_e \gg$ the ion temperature T_i. Whether these conditions are met in the cometary plasma tail remains unclear. Furthermore, recombination far down the tail where such knots are observed, is negligible.

As an alternative it is tempting to suggest that these may result from charge sheet and current sheet instabilities which produce vortex like forms in the laboratory, and which have been invoked to explain similar forms in the aurora (Webster and Halliman, 1973).

CONCLUSION

As pointed out at the outset, there is little doubt that the solar wind with its associated magnetic field is responsible for the complicated plasma dynamics of the cometary head and tail. We have further suggested that it may also be indirectly responsible for the bursts of rapid ionization observed in the inner coma. However, the detailed understanding of the interaction of the solar wind with cometary atmospheres is in a very preliminary stage. Future cometary missions, like the one under consideration for comet Halley in 1986, carrying in situ measurements of the particles and field environment would undoubtedly produce a dramatic advance. Such questions as the existence of an outer and inner shock, a tangential discontinuity, an amplified magnetic field and energetic particles may then be settled. More cometary ions including H_3O^+ and HCO^+, which are predicted by theoretical model ionospheres (e.g., see Ip and Mendis, 1976a, 1977) to be the most dominant species in the innermost regions, will hopefully be detected.

In the meantime, however, much work remains to be done by ground based observations. Of primary importance are more detailed observations of the morphology and time variation of the complicated fine structure. Both accurate drawing by visual observers of features close to the head as well as kinematographic studies such as the one successfully carried out by Jockers et al. (1976) for comet Kohoutek (1973f) are in order.

There is also the pressing need for the direct measurement of Doppler shifts of ions far down the tail. Recently, Huppler et al. (1975) obtained Doppler shift measurements of H_2O^+ using high resolution Fabry-Perot spectrometry. This was, however, confined to regions close to the nucleus ($r \leqslant 2 \times 10^5$ km), where the emission is strong. Every effort should be made to extend these to the faint structures far down the tail. This is the ultimate way to settle the vexing question as to whether the motions observed down the tails are pure wave motions or are associated with bulk motions. This will also have a bearing on the important question regarding the morphology and strength of the magnetic field in the tail. With regard to the latter question, however, direct information can come only from observations of Zeeman splitting. The observed sharp 1665 MHz hyperfine component of the OH Λ doublet is an obvious candidate for this purpose.

Further laboratory studies and computer simulations pertinent to the plasma chemistry in the inner coma are also in order and these need to be supplemented by detailed mechanistic studies.

The need for the development of more reliable theoretical models of cometary ionospheres and plasma tails, while paying greater attention to the detailed plasma processes involved, hardly bears statement.

ACKNOWLEDGMENTS

This work is supported by the following grants: NASA-NSG 7102 of the NASA Planetary Geology Program, NASA-NGR-05-009-110 of the NASA Cometary Science Program and NSF-MPS74-23501 of the NSF Solar System Astronomy Program.

REFERENCES

Alfvén, H.: 1957, Tellus 9, 92.
Alfvén, H. and Fälthammar, C.-G.: 1963, Cosmical Electrodynamics, Oxford Clarendon Press, 120.
Axford, W. I.: 1964, Planetary Space Sci. 12, 719.
Barnes, A.: 1966, Phys. Fluids 9, 1483.
Beard, D. B.: 1966, The Solar Wind (R. J. Mackin, Jr. and M. Neugebauer, Eds.), Pergamon Press, N.Y., 373.
Belcher, J. W. and Davis, L.: 1971, J. Geophys. Res. 76, 3534.
Biermann, L.: 1951, Z. Astrophys. 29, 274.
Biermann, L., Brosowski, B., and Schmidt, H. U.: 1967, Solar Phys. I, 254.
Brandt, J. C., Harrington, R. S., and Roosen, R. G.: 1975, Astrophys. J. 196, 887.
Brandt, J. C. and Rothe, E. D.: 1976, The Study of Comets, NASA-SP-393 (B. Donn et al., Eds.), 878.
Brandt, J. C. and Mendis, D. A.: 1977, Solar System Plasma Physics: A Twentieth Anniversary Review (C. F. Kennel, L. J. Lanzerotti, E. N. Parker, Eds.), North-Holland Pub. Co. (in press).
Brosowski, B. and Wegmann, R.: 1972, Max-Planck-Institut, Astrophysik Publ., MPI/PAE-Astro 46.
Chernikov, A. A.: 1975, Soviet Astron. 18 (4), 505.
Cloutier, P. A.: 1970, Radio Science 5, 387.
Cloutier, P. A. and Daniell, R. E., Jr.: 1973, Planetary Space Sci. 21, 463.
D'Angelo, N.: 1967, Phys. Fluids 10, 719.
Delsemme, A. H. and Combi, M. R.: 1976, Astrophys. J. 209, L153.
Dessler, A. J.: 1968, The Atmospheres of Venus and Mars (J. C. Brandt and M. B. McElroy, Eds.), Gordon and Breach Science Pub., Inc., N.Y., 241.
Dobrowolny, M. and D'Angelo, N.: 1971, Cosmic Plasma Physics (K. Schindler, Ed.), Plenum Press, N.Y., 149.
Ershkovich, A. I.: 1976, Planetary Space Sci. 24, 287.
Ershkovich, A. I. and Chernikov, A. A.: 1973, Planetary Space Sci. 21, 663.
Ershkovich, A. I., Nusinov, A. A., and Chernikov, A. A.: 1972, Planetary Space Sci. 20, 1235.
Feldman, P. D. and Brune, W. H.: 1976, Astrophys. J. 209, L45.
Harwit, M. and Hoyle, F.: 1962, Astrophys. J. 135, 875.
Huntress, D.: 1977, Jet Propulsion Laboratory Preprint.
Huppler, D., Reynolds, R. J., Roesler, F. L., Scherb, F., and Tranger, J.: 1975, Astron. J. 202, 276.

Hyder, C. L., Brandt, J. C., and Roosen, R. G.: 1974, Icarus 23, 601.
Ioffe, Z. M.: 1968, Soviet Astron. 11, 1044.
Ioffe, Z. M.: 1970, Soviet Astron. 13, 1042.
Ip, W.-H. and Mendis, D. A.: 1976a, Icarus 28, 389.
Ip, W.-H. and Mendis, D. A.: 1976b, Icarus 29, 147.
Ip, W.-H. and Mendis, D. A.: 1977, Icarus 30, 377.
Jockers, K. and Lust, Rh.: 1973, Astron. Astrophys. 26, 113.
Jockers, K., Roosen, R. G., and Cruikshank, D. P.: 1976, The Study of Comets (B. Donn et al., Eds.), NASA-SP-393, 370.
Mendis, D. A.: 1977, Monthly Notices Roy. Astron. Soc. 179, 5p.
Mendis, D. A. and Ip, W.-H.: 1977, Space Sci. Rev. (in press).
Oppenheimer, M.: 1975, Astrophys. J. 196, 251.
Shimizu, M.: 1975, Astrophys. Space Sci. 36, 353.
Smith, A. M., Bohlin, R. C., Stecher, T. P.: 1976, paper presented at the Comm. 15 Meeting XVIth IAU Gen. Assembly (Grenoble, France, Sept. 1976).
Webster, H. F. and Hallinan, T. J.: 1973, Radio Sci. 8, 475.
Wallis, M. K.: 1967, Planetary Space Sci. 15, 137.
Wallis, M. K.: 1973, Planetary Space Sci. 21, 1647.
Wallis, M. K. and Ong, R. S. B.: 1976, The Study of Comets, NASA-SP-393 (B. Donn et al., Eds.), 856.
Wurm, K.: 1943, Mitt. Hamburger Sternw. 8, No. 51.
Wurm, K.: 1963, The Moon, Meteorites and Comets (B. M. Middlehurst and G. P. Kuiper, Eds.), Univ. of Chicago Press, 573.
Wurm, K.: 1966, Icarus 5, 219.
Wurm, K.: 1974, Astrophys. Space Sci. 27, 211.
Wyckoff, S. and Wehinger, P.: 1976, Astrophys. J. 204, 616.

SOLAR WIND INTERACTION WITH TYPE-1 COMET TAILS

A.I. ERSHKOVICH
Department of Geophysics and Planetary Sciences
Tel-Aviv University, Ramat Aviv, Israel

ABSTRACT

A comet tail is considered as a plasma cylinder separated by a tangential discontinuity surface from the solar wind. Under typical conditions a comet tail boundary is shown to undergo the Kelvin-Helmholtz instability. With finite amplitude the stabilizing effect of the magnetic field increases, and waves become stable. The proposed model supplies the detailed quantitative description of helical waves observed in type-1 comet tails.

This theory enables us to evaluate the comet tail magnetic field by means of the observations of helical waves. The magnetic field in the comet tail turns out to be of the order of the interplanetary field. This conclusion seems to be in accordance with Alfvén's idea that the magnetic field in type-1 comet tails is a captured interplanetary field.

1. INTRODUCTION

Alfvén (1957) was the first to suggest that the wave motions which have long been observed in type-1 comet tails are magnetohydrodynamic waves (see also Marochnik, 1964). According to Ioffe (1970) these waves arise because of the Kelvin-Helmholtz instability in the comet head, whereas the tail was assumed to be stable. Ershkovich et al. (1971, 1972) proposed a model of the type-1 comet tail as a plasma cylinder separated from the solar wind by a magnetohydrodynamic tangential discontinuity surface, and showed that this interface is unstable under typical conditions. A number of characteristics of the observed helical waves were interpreted therein to be a result of the Kelvin-Helmholtz instability generated by the solar wind interaction with the comet tail (rather than the coma). Somewhat later Dobrowolny and D'Angelo (1972) using different considerations (kinetic rather than magnetohydrodynamic approach, plane geometry, finite velocity and density gradients instead

M. A. Shea et al. (eds.), Study of Travelling Interplanetary Phenomena/1977, 305–319.

of the tangential discontinuity) also came to the conclusion that the comet tail boundary undergoes the Kelvin-Helmholtz instability.

This paper is devoted mainly to the analysis of the stability of the comet tail boundary, which explains the wave motions observed in type-1 comet tails and enables us to evaluate the comet tail magnetic field. Other aspects of the solar wind interaction with the comet plasma have been considered in the fundamental papers by Biermann et al. (1967, 1974), Brosowski and Wegmann (1972) and Wallis (1967a, 1967b, 1973a, 1973b).

2. WAVES OF INFINITESIMAL AMPLITUDE IN INCOMPRESSIBLE PLASMA

Let us consider a comet tail as a plasma cylinder with a free boundary with the solar wind flowing around it. The plasma is supposed to be uniform, ideally conducting and incompressible, and both the unperturbed magnetic field $\vec{B}$ and plasma velocity $\vec{V}$ are uniform and parallel to the cylinder axis z. Indeed, observations show that the central tail is cylindrical, the radius R of the tail does not vary systematically with the distance z from the nucleus, and the direction of the mass motion in the central parts is approximately parallel to the main axis (Biermann et al., 1974). The aberration phenomenon undoubtedly shows that the solar wind velocity is also parallel to the tail axis. The assumed configuration of the magnetic field is compatible with the well-known picture by Alfvén (1957), which describes the origin of the comet tail field (a neutral sheet which has to appear in such a picture will be shown not to be significant in the problem under consideration). The undisturbed values in this model, of course, satisfy the stationary equations of magnetohydrodynamics.

The proposed simplified model intends to describe the most general properties of the observed wave motions. It does not allow for a fine structure of the comet tail, such as streamers or rays and the details of their activity (Wurm, 1975). In addition, the condition of incompressibility does not hold: plasma could be considered as incompressible if the sound velocity were large in comparison with other characteristic velocities, whereas in our case the opposite holds. However the analysis of incompressible fluid is very instructive, and the allowance of the plasma compressibility will be shown not to alter its main results.

A free boundary of the tail is supposed to be a discontinuity. Since in our model the component B_n of the magnetic field normal to the comet tail boundary is zero, this interface is a tangential discontinuity. A model of the tail boundary as a contact discontinuity ($B_n \neq o$), which may correspond to reconnection between the interplanetary and the comet tail magnetic fields, necessarily results in the continuity of the plasma bulk velocity (Landau and Lifshitz, 1963). However observations show that the velocity of the plasma outflow in the tail is substantially less than the solar wind velocity (Jockers et al., 1972).

The linearized system of magnetohydrodynamics equations in our case is:

$$\frac{\partial \vec{b}}{\partial t} = B \frac{\partial \vec{v}}{\partial z} - V \frac{\partial \vec{b}}{\partial z} , \tag{1a}$$

$$\rho\left(\frac{\partial \vec{v}}{\partial t} + V \frac{\partial \vec{v}}{\partial z} \right) = - \nabla\Phi + \frac{B}{4\pi} \frac{\partial \vec{b}}{\partial z} , \tag{1b}$$

$$\text{div } \vec{b} = \text{div } \vec{v} = o, \tag{1c}$$

where ρ is the density, $\vec{v}$, $\vec{b}$ and Φ are the disturbances of the plasma velocity, magnetic field and total pressure (gaskinetic and magnetic), respectively. We take all the disturbances as cylindrical waves proportional to $\exp[i(kz + m\phi - \omega t)]$, where ϕ is an azimuthal angle in a cylindrical coordinate system (r,ϕ,z) aligned with the tail axis z, ω is the frequency, k is the wave number, m is an integer. Taking divergency of (1b) and using (1c) one obtains Laplace equation for Φ, the solutions of which limited both inside and outside the cylinder are:

$$\Phi_i = A_i I_m(kr)\exp[i(kz + m\phi - \omega t)], \tag{2a}$$

$$\Phi_e = A_e K_m(kr)\exp[i(kz + m\phi - \omega t)]. \tag{2b}$$

Here A is a constant, I_m and K_m are modified Bessel functions, indices i and e refer to the internal and external fluids, respectively. Both the total pressure and the component of the plasma velocity normal to the disturbed interface must be continuous across the tangential discontinuity. After linearization these boundary conditions are:

$$\Phi_i = \Phi_e, \tag{3a}$$

$$\frac{v_{ri}}{\omega - kV_i} = \frac{v_{re}}{\omega - kV_e} \tag{3b}$$

on the undisturbed cylinder surface r = R. Since $\vec{v}$ could be expressed through $\nabla\Phi$ by using (1a) and (1b), one obtains from (2) and (3) a homogeneous algebraic system for coefficients A_i and A_e. The condition of existence of non-trivial solutions of this system is (Ershkovich and Nusinov, 1970, 1972):

$$\frac{\omega}{k} = \frac{\rho_i V_i - L_m\rho_e V_e \pm \left[\frac{1}{4\pi} (\rho_i - L_m\rho_e)(B_i^2 - L_m B_e^2) + L_m\rho_i\rho_e(V_i - V_e)^2\right]^{\frac{1}{2}}}{\rho_i - L_m\rho_e} \tag{4}$$

where

$$L_m = \frac{I'_m(kR)K_m(kR)}{I_m(kR)K'_m(kR)} , \tag{5}$$

and prime means the derivative by argument kr. With $B_i = B_e = o$ the expression under the radical in (4) is always negative since $o > L_m > -1$ in the interval $o < kR < \infty$ (see Figure 1 of Ershkovich and Chernikov,

1973), and a well-known result follows from (4): in usual hydrodynamics a tangential discontinuity surface is always unstable (Landau and Lifshitz, 1959). The physical meaning of this instability, named the Kelvin-Helmholtz instability, is as follows. A sporadic disturbance, e.g. an inflation, appears on the interface. The velocity of the fluid, flowing around the inflation, increases. According to Bernoulli's law the pressure decreases there, and the inflation would grow further, being convected downstream. The energy of the instability is the kinetic energy of the relative motion of both fluids, with the velocity $V_i - V_e$. Magnetic field stress, acting like a surface tension, impedes the instability (cf. Chandrasekhar, 1961).

The function $L_m(kR)$ tends to -1 for short wavelength $kR >> 1$ (for such disturbances a cylindrical interface may be considered as plane), and also for long wavelength ($kR << 1, m \neq o$). Then the instability criterion, derived from (4), becomes:

$$\frac{B_i^2 + B_e^2}{4\pi} < \frac{\rho_i \rho_e}{\rho_i + \rho_e} (V_i - V_e)^2. \tag{6}$$

This criterion was first obtained by S.I. Syrovatskii in 1953 (see Landau and Lifshitz, 1963) for plane tangential discontinuity. It could be seen from (6) that the stability conditions in comet tails (where $\rho_i >> \rho_e$) and the Earth's magnetotail (where $\rho_i << \rho_e$) are just opposite. Hence one may expect that under typical conditions comet tails undergo the Kelvin-Helmholtz instability, whereas the geomagnetic tail is usually stable.

Ioffe (1970) drew attention to the fact that the Kelvin-Helmholtz instability may develop in the comet head since the expected configuration of the magnetic field could be favourable for the instability (in particular, the interface with $\vec{B}_i || \vec{B}_e \perp (\vec{V}_i - \vec{V}_e)$ is always unstable (Chandrasekhar, 1961; Landau and Lifshitz, 1963)). Dryer et al. (1976) by using criterion (6) have shown that even in the case when $\vec{B}_i || \vec{B}_e || (\vec{V}_i - \vec{V}_e)$ (such a configuration is the most stable) a substantial part of the sunlit cometary ionopause, including flanks, should undergo the Kelvin-Helmholtz instability under typical conditions. The same instability must also develop at the flanks of the ionopause of Venus.

With $\rho_i >> \rho_e$ the dispersion equation (4) becomes as follows:

$$\frac{\omega}{k} = V_i - L_m \frac{\rho_e}{\rho_i} V_e \pm \sqrt{\frac{B_i^2 - L_m B_e^2}{4\pi\rho_i} + L_m \frac{\rho_e}{\rho_i} (V_i - V_e)^2}. \tag{7}$$

The dispersion equations (4) or (7) do not depend on the direction of the magnetic field. Thus our model may also include a neutral sheet (or a number of neutral sheets - cf. Ness and Donn, 1966), separating magnetic fields of equal magnitude and opposite direction. These equations could also be used for the stability analysis of an individual ray in the comet tail. However we have to keep in mind that if the relative velocity $|V_i - V_e|$ is small enough to produce the Kelvin-Helmholtz instability, other instabilities, e.g. flute instability, could also be effective.

The function $L_m(kR)$ tends to zero with $kR \ll 1$, $m = o$, i.e. longwave symmetric (sausage-like) modes are always stable, and cannot reach large enough amplitudes to become visible. Perhaps because of this reason they are never observed. However it should be noted that symmetric disturbances require symmetric conditions of excitation, and therefore they should occur rarely.

3. WAVES OF INFINITESIMAL AMPLITUDE IN COMPRESSIBLE PLASMA

The solution of this problem for compressible plasma yields:

$$\frac{\rho_i[(\omega-kV_i)^2 - k^2C_{Ai}^2]}{\rho_e[(\omega-kV_e)^2 - k^2C_{Ae}^2]} = \frac{\alpha_i I_m'(\alpha_i R)K_m(\alpha_e R)}{\alpha_e I_m(\alpha_i R)K_m'(\alpha_e R)}, \tag{8}$$

where α satisfies the following equation (for each medium):

$$\alpha^2 = \frac{[(\omega-kV)^2 - k^2C_A^2][(\omega-kV)^2 - k^2C_s^2]}{k^2C_A^2C_s^2 - (\omega-kV)^2(C_A^2+C_s^2)}, \tag{9}$$

and C_A, C_s are the Alfvén velocity and the sound velocity, respectively. With $V_i = o$ this dispersion equation was obtained by McKenzie (1970) (see also Ershkovich and Nusinov, 1972) for the stability problem of the Earth magnetotail; generalization for the case $V_i \neq o$ is obvious. With $C_s^2 \gg C_A^2$; $C_s^2 \gg (\frac{\omega}{k} - V)^2$ one obtains from (8)-(9) the dispersion equation (4) of incompressible plasma. Ershkovich and Heller (1977) investigated the transcendental equation (8) numerically, with plasma parameters from Tables I and II, which seem to be typical for the solar wind interaction with the tails of Comets Kohoutek, Arend-Roland and Morehouse. The choice of the parameters was discussed in detail ibid. The magnetic field B_i in the comet tail was evaluated from the pressure balance across the tangential discontinuity.

Unlike the case of the Earth-solar wind coupling, the solar wind interaction with comets is controlled by ionization processes (Biermann et al., 1967, 1974; Brosowski and Wegmann, 1972; Wallis, 1967a, 1967b, 1973a, 1973b). Neutral molecules emitted by the nucleus are ionized by various processes (photo-ionization, charge-exchange, electron impact). Heavier cometary ions join the solar wind plasma far outside the tangential discontinuity surface, and effectively change the total mass, momentum and energy fluxes. As a result, the Mach number in front of the bow shock decreases substantially, the shock becomes weak, and the interaction occurs over an extended region, with gradual change of the

TABLE I

Typical solar wind parameters used for calculations

Heliocentric distance (AU)	Number density (cm^{-3})	Effective temperature (K)	Interplanetary magnetic field B_e (gammas)	Solar wind velocity V_e (kms^{-1})	Sound velocity C_{se} (kms^{-1})	Alfvén velocity C_{Ae} (kms^{-1})
0.58	20	2.6×10^5	23.0	540	63	115
0.80	10	2.0×10^5	8.0	500	52	55
1.40	1	1.4×10^5	3.5	300	44	76

TABLE II

Parameters for comets

Comet	Heliocentric distance (AU)	Number density (cm^{-3})	Temperature (K)	Mean mass (largely CO^+)	Radius of the tail R(km)	Wavelength λ(km)	kR	Bulk velocity V_i (kms^{-1})	Magnetic field B_i (gammas)	Sound velocity C_{si} (kms^{-1})	Alfvén velocity C_{Ai} (kms^{-1})
Kohoutek (1973f)	0.58	10	10^4	28	2.3×10^5	1.4×10^6	1.0	235	26.8	2.2	41.3
Arend-Roland (1957III)	0.80	30	10^4	28	1.0×10^5	5.0×10^5	1.25	70	11.7	2.2	8.8
Morehouse (1908c)	1.40	30	10^4	28	5.0×10^5	1.3×10^6	2.4	40	2.6	2.2	2.0

flow parameters. Therefore the plasma parameters just outside the tangential discontinuity surface differ from the undisturbed pre-shock values.

However wave motions are observed in the distant tail, at several millions km from the cometary nucleus (0.1 AU in the case of Comet Kohoutek), where these effects seem to be not so important. Biermann et al. (1974) calculated the distribution of molecular ions for the production rate of the cometary nucleus of $10^{30}sec^{-1}$ and an outflow velocity of neutral molecules of 1 kms^{-1}. Extrapolation of Figure 5 ibid. to a distance of 3-4 millions km (where helical waves have been observed in the cases of Comets Arend-Roland and Morehouse) gives the density just outside the tangential discontinuity surface to be of the order of 10 $m_p cm^{-3}$ (m_p is the proton mass). An alternative estimate could be derived from the observational fact that the visible boundary of the brightness is rather sharp, whence one may conclude that the number density of molecular ions just outside the boundary is substantially less than that inside the tail. Thus we come to the following conclusion: for Comet Arend-Roland and particularly for Comet Kohoutek the effects under consideration are negligible, and the density ρ_e equals approximately the pre-shock solar wind density represented in Table I. However in the case of Comet Morehouse the actual density ρ_e could be an order of magnitude more than that in the undisturbed solar wind. Increase of the density ρ_e results in the increase of the instability growth rate and barely affects the phase velocity (cf. (7)).

In the case of instability, the arguments of the Bessel functions in (8) are complex, and we have two equations (for both real and imaginary parts), which determine two unknown quantities. Considering ω as real and k as complex (which is typical for boundary value problems) we choose ω and Imag k as such unknown quantities, supposing all the remaining parameters in (8) to be known (see Tables I and II). The interface is obviously unstable if Imag k/Rek < o. Numerical calculations give Imag k/Rek to be -0.30, -0.54, -0.17 for Comets Kohoutek, Arend-Roland and Morehouse, respectively. Consequently, with typical plasma parameters a comet tail boundary is subject to the Kelvin-Helmholtz instability. One may determine a critical magnetic field $B_{i,c}$ which corresponds to marginal stability (when Imag k = o). For Comets Kohoutek, Arend-Roland and Morehouse we obtained $B_{i,c}$ = 49γ (39γ), 65γ (43γ) and 15γ (8γ), respectively (the values in parentheses correspond to the model of incompressible plasma). The comet tail field estimated from the pressure balance across the tangential discontinuity (see Table II) turns out to be much less than $B_{i,c}$.

It is seen that plasma compressibility always results in an increase of the magnetic field required for stabilization of the interface. This important conclusion has been verified with the plasma parameters being allowed to vary over a wide range. Thus a simple instability criterion for incompressible fluid (viz. the expression under the radical in (4) or (7) should be negative) is also a sufficient (but not necessary) condition for the instability of a cylindrical tangential discontinuity surface in a compressible fluid with the same parameters B, ρ and V and for disturbances with the same wavelength.

Numerical analysis has shown that the phase velocity of helical waves approximately equals the Alfvén speed in the tail, in the reference system moving with the bulk velocity V_i of the plasma outflow:

$$\frac{\omega}{\mathrm{Rek}} \approx V_i \pm C_{Ai} \tag{10}$$

with an error of less than 20% with 300 $\mathrm{kms}^{-1} \lesssim V_e \lesssim$ 800 kms^{-1}; 1 $m_p\mathrm{cm}^{-3} \lesssim \rho_e \lesssim$ 100 $m_p\mathrm{cm}^{-3}$. This result is a consequence of the condition $\rho_i \gg \rho_e$ and is not obvious a priori, since the waves under consideration are surface waves propagating in both media with the same phase velocity which depends on the parameters of both fluids. The computed phase velocity equals 240 kms^{-1}, 85 kms^{-1} and 45 kms^{-1} for Comets Kohoutek, Arend-Roland and Morehouse, respectively. These values are in a good agreement with measurements. According to observations by Hyder et al. (1974) the phase velocity of helical waves in the tail of Comet Kohoutek was 235 kms^{-1} (with an error of 10%), being approximately coincident with the velocity of an irregular Swan-like cloud. The phase velocity measured by Miller (1969) in the tail of Comet Arend-Roland was 95 ± 16 kms^{-1}, whereas the velocity of an irregular structure ("hockey stick") was about 70 kms^{-1}. We believe that such irregularities represent bulk motion of material with the velocity V_i (see also Miller, 1969; Jockers et al., 1972; Biermann et al., 1974), although there is an alternative point of view (Hyder et al., 1974). These data indicate that the observed phase velocity differs little from the velocity V_i of the plasma outflow. Then (10) leads to the conclusion (which is important for the evaluation of the comet tail magnetic field) that the Alfvén speed should be small: $C_{Ai} \ll \omega/\mathrm{Rek} \approx V_i$.

Experimental data for Comet Morehouse are also consistent with formula (10). Indeed, according to Wolf (1909) both wavelength λ and plasma velocity V_i increased almost linearly versus the distance z along the tail of Comet Morehouse (see also Figures 2 and 3 of Ershkovich et al., 1972): $\lambda(z)/V_i(z) \approx$ Const $\approx$ 8.8 hours. Thus the wavelength grew approximately proportionally to the plasma velocity, in accordance with (10).

Observations of wave amplitudes in the tail of Comet Morehouse enable us to compare the computed instability growth rate with measurements. According to observations by Wolf (1909) the amplitude increased from 19" to 100" between z = 40' and z = 78' (1' corresponded to the distance of 7×10^4 km). Hence one obtains Imag k = -0.62×10^{-6} km^{-1}, supposing the amplitude grows exponentially ~ exp(-Imag k · z). Numerical solution of the dispersion equation (8) with plasma parameters, just corresponding to the centre of the interval under consideration, z = 60' (see Table II), yields the instability growth rate Imag k = $-0.85 \times 10^{-6}\mathrm{km}^{-1}$ and $-0.33 \times 10^{-5}\mathrm{km}^{-1}$ for the external density ρ_e = 1 $m_p\mathrm{cm}^{-3}$ and 10 $m_p\mathrm{cm}^{-3}$, respectively. The agreement is much better with the external density ρ_e taken to be equal to the density of the undisturbed (pre-shock) solar wind (cf. Table I).

Numerical investigation of the dispersion equation (8) by Ershkovich and Heller (1977) indicates that a simple model of incompressible fluid is a rather good approximation.

4. WAVES OF FINITE AMPLITUDE IN INCOMPRESSIBLE PLASMA

The observed amplitudes δ in the tails of Comets Kohoutek and Arend-Roland were large: $\delta/R \gtrsim 0.3$ (Hyder et al., 1974; Ershkovich, 1976a) and $\delta/R = 0.3$ (Miller, 1969), respectively, and they did not grow with distance z. In the case of Comet Morehouse the wave amplitude was small (e.g. $\delta/R = 0.044$ with $z = 40'$ (Wolf, 1909)), and increased with z. The instability growth rate for Comets Kohoutek and Arend-Roland, computed from (8), turned out to be substantially greater than that for Comet Morehouse. One may suppose that with a large instability growth rate the amplitude rapidly reaches a finite value, and then a non-linear stabilization of the Kelvin-Helmholtz instability occurs, whereas in the case of Comet Morehouse such a stabilization takes place only at the very end of the visible wave structure in the tail.

Indeed, with finite amplitude the stabilizing effect of the magnetic field increases. The magnetic force $1/4\pi$ rot $\vec{B} \times \vec{B}$ has a component directed towards the centre of curvature of the field line, viz. $B^2\vec{n}/4\pi R_c$, where $\vec{n}$ is a unit vector of the principal normal to the field line, R_c is the radius of curvature. There is an obvious analogy with the effect of surface tension, which results in an additional pressure on the curved interface in the fluid: $p = 2\sigma/R_c$, where σ is the coefficient of surface tension. Consequently, with the disturbance (e.g. an inflation on the interface) growing, the radius R_c of the inflation decreases, and stabilizing effect of the magnetic field is enhanced.

The existence of a non-linear stabilization of the Kelvin-Helmholtz instability at finite amplitudes has been demonstrated by Ershkovich and Chernikov (1973)(see also Ershkovich, 1976a). The dispersion equation for non-linear helical waves obtained therein has the same form as (4) or (7) but the function $L_m(kR)$ is replaced by $L_m^* = L_m[1-(\delta/R)^2\Lambda_m]$, where the function $\Lambda_m(kR)$ is represented on Figure 2 of Ershkovich and Chernikov (1973). With

$$L_m^* = -\frac{B_i^2}{4\pi\rho_e(V_i-V_e)^2 - B_e^2} \tag{11}$$

the radical in the dispersion equation vanishes, whence one obtains the critical value δ_c of the wave amplitude which corresponds to marginal stability. With typical plasma parameters (see Tables I and II) the denominator in (11) is positive. Then the helical waves are unstable if $\delta < \delta_c$, and are stable if $\delta \geqslant \delta_c$ (Ershkovich, 1976a).

Waves of large amplitude in comet tails could be excited because of the interaction with interplanetary shock waves. A correlation between a comet tail event and interplanetary shock waves was discussed by Jockers et al. (1972), Jockers and Lüst (1973); the observed kinks in the comet tails were attributed to a sudden shift in solar wind direction An abrupt change of the value of the solar wind velocity (e.g. due to the arrival of a shock front) also gives rise to similar effects via the aberration phenomenon (Dryer et al., 1976). Other aspects of the interaction of interplanetary shock waves with comets have been considered by Eviatar et al. (1970), Dryer et al. (1975, 1976), Miller (1976).

However helical waves in the tails of Comets Kohoutek and Arend-Roland have been observed under rather quiet conditions, when there were no notable solar flares (Miller, 1969; Hyder et al., 1974). In such a case waves of large amplitude could develop only as a result of the Kelvin-Helmholtz instability, and the amplitude of stabilized helical waves should be equal to the value δ_c. This prediction is in very good agreement with observations. Indeed, the critical amplitude δ_c, computed from (11) with plasma parameters taken from Tables I and II and $m = 1$, turns out to be 0.30R and 0.34R for Comets Kohoutek and Arend-Roland, whereas the observed values are $\delta \geqslant 0.3R$ and 0.3R, respectively. It is interesting, that the maximum amplitude measured by Wolf (1909) at the very end of the wave structure in the tail of Comet Morehouse ($z = 105'$) was 150", i.e. $\delta \approx 0.35R$, whereas formula (11) yields $\delta_c = 0.33R$.

The tail diameter is the only intrinsic scale which could determine the wavelength if helical waves are excited as a result of the Kelvin-Helmholtz instability under quiet conditions. Moreover, the amplitude δ_c of stabilized kink-modes ($m = 1$), calculated from (11), is maximum with $kR \sim 1$ (this result follows from the fact that the function $\Lambda_1(kR)$ has a sharp minimum with $kR \sim 1$). Thus the wavelength of visible disturbances should be of the order of cylinder circumference. This conclusion is also in good agreement with observations (see Table II).

Substituting (11) into the dispersion equation one obtains the phase velocity of stabilized helical waves in the form:

$$\frac{\omega}{k} = V_i + \frac{C_{Ai}^2 V_e}{(V_i - V_e)^2 - C_{Ae}^2} \quad . \tag{12}$$

The second term on the right-hand side of expression (12) is always small in comparison with the first term. Thus non-linear evolution of the instability does not result in a substantial change in phase velocity, which remains approximately equal to the velocity of plasma outflow in the tail when the amplitude varies over a wide range from infinitesimal up to finite values (cf. (7)).

Sausage-like disturbances ($m=o$) require symmetric conditions of excitation, and therefore will seldom occur. If, however, they develop, their amplitudes should be substantially greater than those for kink-modes the amplitude δ_c, derived from (11), reaches $0.5R \div 0.6R$ with $m = o$ and $kR \sim 1$. Perhaps, the break-up of the tail or the cometary ray into individual clouds with the characteristic size of the order of cylinder diameter (which is sometimes observed-see e.g. Dobrovolskij (1966)) could be associated with the non-linear evolution of sausage-like modes.

5. ESTIMATION OF THE COMET TAIL MAGNETIC FIELD

It is difficult to imagine an astrophysical plasma without immersed magnetic fields. Undoubtedly, there is a magnetic field in the cometary tails too, although it has not been measured yet. The filamentary structure appears to be compelling evidence for magnetic fields in comet tails (Brandt, 1968). Alfven (1957) pointed to the mechanism of a "capture" of the interplanetary field by comets. Although details of such a capture are not yet clear, the Kelvin-Helmholtz instability should undoubtedly facilitate the mixing between the interplanetary magnetic field and the cometary plasma.

The Larmour radius of CO^+ ions, obviously, could not exceed the visible radius of a comet ray ($\lesssim$ 3000 km), whence one obtains $B_i \gtrsim 0.2 \div 0.8\gamma$ with a thermal energy of CO^+ ions of $1 \div 10$ ev. There is a suggestion that the comet tail field could be as large as 100γ or even 1000γ. Hyder et al. (1974) identified the phase velocity of helical waves in the tail of Comet Kohoutek with the Alfvén speed (cf. (10)), whence the magnetic field $B_i \gtrsim 150\gamma$ was obtained. However such an interpretation does not take into account the bulk motion of the plasma outflow (Ip and Mendis, 1975a; Ershkovich, 1976a). Ip and Mendis (1975a, 1975b) derived very high values $300\gamma \lesssim B_i \lesssim 500\gamma$ (with $z \lesssim 6 \times 10^5$ km) from scintillation observations by Ananthakrishnan et al. (1975), although the experimenters proposed another interpretation (through local eddies and reverse velocity fields in the tail). Ip and Mendis (1976) supposed that the magnetic field of 1000γ could exist near the tail axis, just outside the neutral sheet, decreasing outwards up to the interplanetary values near the tail boundary. Such an assumption seems to be unlikely. A uniform and infinite current sheet yields uniform magnetic fields with opposite directions on both sides of the sheet. It may be shown that the finite scale of the current sheet results in a negligible decrease of the magnetic field towards the tail boundary. Analogy with the Earth magnetotail also points to the fact that the magnetic field is approximately uniform throughout the tail cross-section (except for thin neutral sheet).

Our theory enables us to evaluate the comet tail magnetic field by means of the observations of helical waves in the comet tails. Indeed, the comparison between the theory and observational data on the phase velocity, instability growth rate, etc. has demonstrated that the model of tangential discontinuity is a rather good approximation. If so, the magnetic field B_i in a comet tail may be estimated from the pressure balance across the tangential discontinuity surface. Such an estimate results in B_i-values of the same order of magnitude as the interplanetary magnetic field B_e (cf. Tables I and II). While the number density and the temperature in the tail (which participate in the pressure balance) could be unknown, one may evaluate the magnetic field B_i from (11) by using values of the amplitude δ_c of stabilized non-linear helical waves and the corresponding wavelengths (Ershkovich, 1976a). The magnetic field B_i, computed from (11), turns out to be very sensitive to the variations of the amplitude δ_c. Therefore rather accurate measurements of the amplitude are required. Another method utilizes simultaneous observations of the phase velocity of helical waves and bulk velocity of the plasma outflow in the tail, according to (10). More accurate estimation of the magnetic field B_i requires a numerical solution of the dispersion equation (8) (Ershkovich, 1976b). Different considerations lead to the same conclusion: the magnetic field in the comet tail is of the order of the interplanetary field (Ershkovich, 1976b). This conclusion seems to be in accordance with the idea that the magnetic field in type-1 comet tails is a "captured" interplanetary field (Alfvén, 1957) If the comet tail field were of internal origin, it would decrease with distance z away from the nucleus, and the comet tail would increase in cross-section because of the flux conservation. However the radius of the tail does not vary systematically with distance z (Biermann et al., 1974; see also data by Miller and Dennis, cited in Ershkovich, 1976b), whence one may conclude that the magnetic field in the comet tails is of external origin.

6. CONCLUDING REMARKS

The proposed model of a comet tail boundary as a tangential discontinuity surface was shown to be in accordance with a number of characteristics of the observed helical waves, viz.: the phase velocity, the instability growth rate, the wavelength ($kR \sim 1$) and particularly the amplitude δ_c of stabilized non-linear waves. The theory also explains qualitatively the following observational facts: the generation of helical waves under quiet conditions; the absence of long wavelength sausage-like modes with finite amplitudes; the increase of the wavelength along the tail axis proportionally to the mass velocity V_i (see (10)); the enhancement of comet tail activity with the motion of the comet towards the sun (Vsekhsvjatsky, 1967). The last-mentioned effect could be associated with the increase of the instability growth rate because of the growth of the relative velocity $V_e - V_i$ (see (7)).

The most important prediction refers to the magnetic field: the Alfvén velocity should be small as compared with the phase velocity: $C_{Ai} \ll \omega/k \approx V_i$, and the magnetic field in the tail should be of the order of the interplanetary field. Different methods of the evaluation of the magnetic field by means of observations of helical waves were presented.

The comet tail boundary considered as a tangential discontinuity surface was shown to be unstable under typical conditions. In such a case a question arises: why are helical waves in type-1 comet tails seen only occasionally and not at all times? The only plausible explanation is that the model of tangential discontinuity is not always applicable. A transition layer between two fluids with different parameters may be considered as a discontinuity surface if the thickness d of the layer is small in comparison with the wavelength: $kd \ll 1$, or $d \ll R$ for the observed helical waves ($kR \sim 1$). The visible boundary of the luminosity is rather sharp, and the characteristic scale d for the density variation satisfies this condition. However the variation of the plasma velocity could occur more gradually (Biermann et al., 1967; Brosowski and Wegmann, 1972; Wallis, 1973a, 1973b). According to Jockers et al. (1972) the plasma velocity in the tail of Comet Tago-Sato-Kosaka varied from $40 \div 100$ kms^{-1} near the tail axis up to 300 kms^{-1} in the outer parts of the tail. The solar wind velocity was of $300 \div 400$ kms^{-1}. Consequently, the condition $d < R$ holds in any case. Our estimations show that with $d < R$ the Kelvin-Helmholtz instability could sometimes develop, although the smooth velocity profile $V(r)$ certainly exerts a stabilizing effect (cf. Chandrasekhar, 1961). With $d < R$ (instead of $d \ll R$) the stability conditions in the comet tails become marginal (cf. Dobrowolny and D'Angelo, 1972), and one may expect only qualitative (rather than quantitative) agreement between the model of tangential discontinuity and observations. However there is no doubt that this model supplies the detailed quantitative description of helical waves (provided they develop). Hence we come to the following conclusion: the generation of helical waves in type-1 comet tails under quiet conditions occurs as a result of the Kelvin-Helmholtz instability, which arises only when the gradient $\partial V/\partial r$ of the plasma velocity becomes large enough (in order to consider the comet tail boundary as a tangential discontinuity surface). The mechanism which controls the gradient $\partial V/\partial r$ is obviously associated with anomalous viscosity, and requires special investigation.

ACKNOWLEDGMENTS

I am grateful to the participants of the seminar under the guidance of Prof. L. Biermann and Dr. Rh. Lüst, especially to Drs. H.U. Schmidt and K. Jockers, for fruitful and stimulating discussion.

This research was supported by a grant from the United States-Israel Binational Science Foundation (BSF), Jerusalem, Israel.

REFERENCES

Alfvén, H.: 1957, *Tellus* 9, 92.
Ananthakrishnan, S., Bhandari, S.M., and Pramesh Rao, A.: 1975, *Astrophys. Space Sci.* 37, 275.
Biermann, L., Brosowski, B., and Schmidt, H.U.: 1967, *Solar Phys.* 1, 254.
Biermann, L., Lüst, R., and Wegmann, R.: 1974, *Astronaut. Acta* 18 (*Suppl.*), 113.
Brandt, J.C.: 1968, *Ann. Rev. Astron. Astrophys.* 6, 267.
Brosowski, B. and Wegmann, R.: 1972, '*Numerische Behandlung Eines Kometenmodells*', Max-Planck-Inst. für Physik und Astrophysik, München, preprint MPI/PAE - Astro. 46.

Chandrasekhar, S.: 1961, *Hydrodynamic and Hydromagnetic Stability*, Clarendon Press, Oxford, Ch. XI.
Dobrovolskij, O.V.: 1966, *Komety (Comets)*, Nauka, Moscow, Ch. 6.
Dobrowolny, M. and D'Angelo, N.: 1972, in K. Schindler (ed.), *Cosmic Plasma Physics*, Plenum Press, New York-London, p. 149.
Dryer, M., Ershkovich, A.I., and Shen, W.-W.: 1976, *J. Geophys. Res.* *81*, 6184.
Dryer, M., Eviatar, A., Frohlich, A., Jacobs, A., Joseph, J.H., and Weber, E.J.: 1975, *J. Geophys. Res.* *80*, 2001.
Ershkovich, A.I.: 1976a, *Planetary Space Sci.* *24*, 287.
Ershkovich, A.I.: 1976b, *Monthly Notices Roy. Astron. Soc.* *177*, 649.
Ershkovich, A.I. and Chernikov, A.A.: 1973, *Planetary Space Sci.* *21*, 663
Ershkovich, A.I. and Heller, A.B.: 1977, 'Helical Waves in Type-1 Comet Tails', *Astrophys. Space Sci.* (in press).
Ershkovich, A.I. and Nusinov, A.A.: 1970, 'Earth's Magnetic Tail Oscillations', Rept. at Int. Symp. on Solar-Terr. Phys. in Leningrad, May 1970 (Preprint No. 37, Space Res. Inst., U.S.S.R. Acad. of Sci., Moscow).
Ershkovich, A.I. and Nusinov, A.A.: 1972, *Cosmic Electrodyn.* *2*, 471.
Ershkovich, A.I., Nusinov, A.A., and Chernikov, A.A.: 1971, 'Oscillations of Type-1 Comet Tails', Rept. at XV General Assembly of IUGG in Moscow, August 1971 (Preprint No. D-86, Space Res. Inst., U.S.S.R. Acad. of Sci., Moscow).
Ershkovich, A.I., Nusinov, A.A., and Chernikov, A.A.: 1972, *Planetary Space Sci.* *20*, 1235.
Eviatar, A., Joseph, J.H., and Dryer, M.: 1970, *Cosmic Electrodyn.* *1*,239
Hyder, C.L., Brandt, J.C., and Roosen, R.G.: 1974, *Icarus* *23*, 601.
Ioffe, Z.M.: 1970, *Soviet Astron.* *13*, 1042.
Ip, W.-H. and Mendis, D.A.: 1975a, *Icarus* *26*, 457.
Ip, W.-H. and Mendis, D.A.: 1975b, *Astrophys. Space Sci.* *35*, L1.
Ip, W.-H. and Mendis, D.A.: 1976, *Icarus* *29*, 147.
Jockers, K. and Lüst, R.: 1973, *Astron. Astrophys.* *26*, 113.
Jockers, K., Lüst, R., and Nowak, T.: 1972, *Astron. Astrophys.* *21*, 199.
Landau, L.D. and Lifshitz, E.M.: 1959, *Fluid Mechanics*, Pergamon Press, London, Ch. III.
Landau, L.D. and Lifshitz, E.M.: 1963, *Electrodynamics of Continuous Media*, Pergamon Press, Oxford, Ch. VIII.
Marochnik, L.S.: 1964, *Soviet Phys. Usp.* *7*, 80.
McKenzie, J.F.: 1970, *J. Geophys. Res.* *75*, 5331.
Miller, F.D.: 1969, *Astron. J.* *74*, 248.
Miller, F.D.: 1976, *Space Sci. Rev.* *19*, 739.
Ness, N.F. and Donn, B.D.: 1966, in *Nature et Origine des Comètes*, 13th Liège Symposium, Cointe-Sclessin, Belgium, Inst. d'Astrophys., p. 343.
Vsekhsvjatsky, S.K.: 1967, *Komety (Comets) 1961-1965*, Nauka, Moscow, p.58.
Wallis, M.K.: 1967a, *Planetary Space Sci.* *15*, 137.
Wallis, M.K.: 1967b, *Planetary Space Sci.* *15*, 1407.
Wallis, M.K.: 1973a, *Astron. Astrophys.* *29*, 29.
Wallis, M.K.: 1973b, *Planetary Space Sci.* *21*, 1647.
Wolf, M.: 1909, *Astron. Nachr.* *180*, 1.
Wurm, K.: 1975, *Astrophys. Space Sci.* *32*, L19.

DISCUSSION

Wallis: Would you detail why you expect the waves to be more active when the comet is approaching the sun?

Ershkovich: Our dispersion equation (7) is written in the reference system connected with the cometary nucleus. In this system the external velocity, V_e, increases when the comet moves towards the sun (and decreases with the motion in the opposite direction), whereas the internal velocity, V_i, is fixed. Thus the instability growth rate (which is approximately proportional to the relative velocity V_e-V_i) increases in this case. Although the increase of the growth rate is of several tens percent only, the wave amplitude could grow significantly.

Wallis: At what position relative to the comet head do you predict that waves should appear?

Ershkovich: Under quiet conditions helical waves of infinitesimal amplitude could appear in the region where the tail becomes cylindrical, i.e., at the distances of several hundred km from the nucleus downstream. With the computed values of the instability growth rate they reach large enough amplitudes (to become visible) at the distances of several millions km from the nucleus (i.e., at several wavelengths).

Grzedzielski: I would rather support the point made by Max Wallis. One could expect that due to viscous interaction the gas in the tail would acquire a velocity of the order of the solar wind velocity, no matter what is the relative velocity of the nucleus and the solar wind. The situation could be much alike to the terrestrial case in the distant magnetotail.

Ershkovich: According to observations, the internal velocity, V_i, is substantially less than the solar wind velocity, V_e, although it increases with distance, z, from the nucleus due to the viscous interaction. Of course, the momentum transfer from the solar wind to the comet tail plasma results in the increase of the internal velocity (see the contributed paper by Perez-de-Tejada, et al. (1977). But at the distances where the helical waves are observed, the difference V_e-V_i is still large enough to produce the Kelvin-Helmholtz instability (see our Tables I and II for Comets Kohoutek, Arend-Roland, and Morehouse).
(Perez-de-Tejada, H., Orozco, A., Ershkovich, A., and Dryer, M.: 1977, in M. A. Shea, D. F. Smart, S. T. Wu (eds.), Contributed Papers to the Study of Travelling Interplanetary Phenomena/1977; Air Force Geophys. Lab., Bedford, Massachusetts, U.S.A.)

PART V

SOLAR COSMIC RAY PROPAGATION, MODULATION, AND INTERPLANETARY ACCELERATION

SIGNATURES OF SOLAR COSMIC RAY EVENTS AND THEIR RELATION TO PROPAGATION AND ACCELERATION PROCESSES

G. WIBBERENZ
Institut für Reine und Angewandte Kernphysik,
Olshausenstr. 40, 23 Kiel, FR-Germany

ABSTRACT

Significant improvements have been made during recent years in the study of solar cosmic rays. Simultaneous observations from spacecraft distributed in solar longitude and radial distance from the sun can be used to separate solar and interplanetary propagation processes. Combined with statistical studies they define the average characteristics of the interplanetary propagation process. Measurements with the Pioneer 10/11 spaceprobes out to beyond 5 AU are of particular importance to reveal the average propagation characteristics of the interplanetary medium. For individual events, anisotropy measurements allow to draw conclusions on the interaction with the interplanetary medium as well as on the solar injection process. Combination of the various studies allows to derive the dependence of the average mean free path on particle rigidity. It shows an anomalous behaviour in the sense that $\partial\lambda/\partial P < 0$ in the rigidity range 20 - 200 MV which is difficult to reconcile with the results of contemporary scattering theories. Observations within the fast propagation region of a particle flare may lead to finite rise and decay times of the solar injection process. On the average, the decay times decrease with increasing proton energy. For a fixed energy, the decay times may vary from one event to the other. The $\approx$ 1 MeV protons occasionally behave different from higher energy protons and relativistic electrons, showing a long-lasting rise phase. Measurements with the Helios 1/2 spaceprobes within about 0.5 AU are particularly suited to study the solar injection process because of the reduced influence of interplanetary scattering. Under undisturbed conditions, details of the angular distribution can be used to study the pitch angle scattering mechanism. Local disturbances in the interplanetary magnetic field and changes in the degree of the field turbulence lead to changes in the angular distribution and may also influence the intensity patterns. This will make it difficult to find clear cases for the proposed non-diffusive propagation schemes, as coherent propagation or collimated convection. One example for the coherent mode of propagation is presented.

M. A. Shea et al. (eds.), Study of Travelling Interplanetary Phenomena/1977, 323–342.

1. INTRODUCTION

Individual solar particle events vary greatly in their size, duration, structure, chemical composition and other signatures. Some of these differences are related to the variable location of an observer with respect to the parent flare, but it has also become more and more apparent, that near-sun conditions greatly influence the appearance of an event. The injection of energetic particles from the sun into the interplanetary medium is determined by the combination of acceleration, storage, propagation, and release. Variations in one or more of these processes cause a large part of the observed differences. In addition, there are variations in the state of the interplanetary medium. The scattering efficiency of the interplanetary medium varies with the position in space and with time, and for particles of different type and energy, so that for a given solar injection profile we may obtain qualitatively different propagation schemes, ranging from ordinary diffusion to almost scatter-free propagation.

It is the purpose of this paper to summarize some of the recent advances. This brief review is far from being complete, and the papers referenced should be taken as examples which type of information can be obtained e.g. by observations at different radial distances, by detailed angular distribution measurements, and from the study of proton-to-alpha ratios. For related topics, the reader is referred to other reviews presented during this symposium (R.P. Lin; E.C. Roelof and S.M. Krimigis).

2. AVERAGE INTERPLANETARY PROPAGATION PARAMETERS

The propagation of solar energetic particles is essentially determined by a random walk process along the interplanetary magnetic field (imf). Even in cases when a simple diffusion model is not adequate to describe the particle behaviour, it is still useful to assign a "mean free path" λ to the particle motion. A "large" mean free path then means that propagation concepts different from ordinary diffusion have to be used. Some of these concepts will be referred to in more detail below.

The mean free path λ parallel to the average imf depends on particle rigidity P only and may vary with distance r from the sun, $\lambda_\parallel = \lambda_\parallel(r,P)$. In this section we shall concentrate on the dependence $\lambda_\parallel(P)$ at $r = 1$ AU. When a diffusion coefficient $\kappa_\parallel$ is obtained from fits to a diffusion model, $\lambda_\parallel$ is related to it by $\lambda_\parallel = 3\kappa_\parallel/v$ (v = particle velocity). In many cases a "radial" mean free path λ_r is determined. In accordance with the average conditions near 1 AU we then have $\lambda_\parallel \approx 2\lambda_r$, since $\lambda_r = \cos^2\phi \times \lambda_\parallel$, $\phi \simeq 45^o$ is the average angle between the radius vector to the sun and the imf. Here it is assumed that diffusion perpendicular to the imf can be neglected; at least close to 1 AU this assumption is justified by a number of observations. Note that a certain radial dependence $\lambda_r(r)$ leads to a different dependence

$\lambda_{\parallel}(r)$, since ϕ varies with distance. For previous summaries of $\lambda_r(P)$, $\lambda_{\parallel}(P)$ the reader is referred to Wibberenz (1974, 1976a,b).

Figure 1 contains in condensed form some of the previous results, complemented by more recent observations. Three points are taken from Wibberenz (1976b). Dashed lines connect "scatter free" protons $\gtrsim$ 0.3 MeV (Roelof and Krimigis, 1973), (the mean free path has somewhat arbitrarily been put equal to 1 AU), the average values for 10-60 MeV solar protons (Reinhard and Wibberenz, 1974) and the value derived from the integral gradient of galactic cosmic rays (McKibben et al., 1975). Note that in the horizontal bar according to Reinhard and Wibberenz (1974) effects of coronal propagation - which may also be important for flares on the western hemisphere of the sun - have been subtracted. It had been assumed that the mean free path λ_r does not vary with distance from the sun, i.e. b = 0 in a power law dependence of the mean free path, $\lambda_r(r) \sim r^b$.

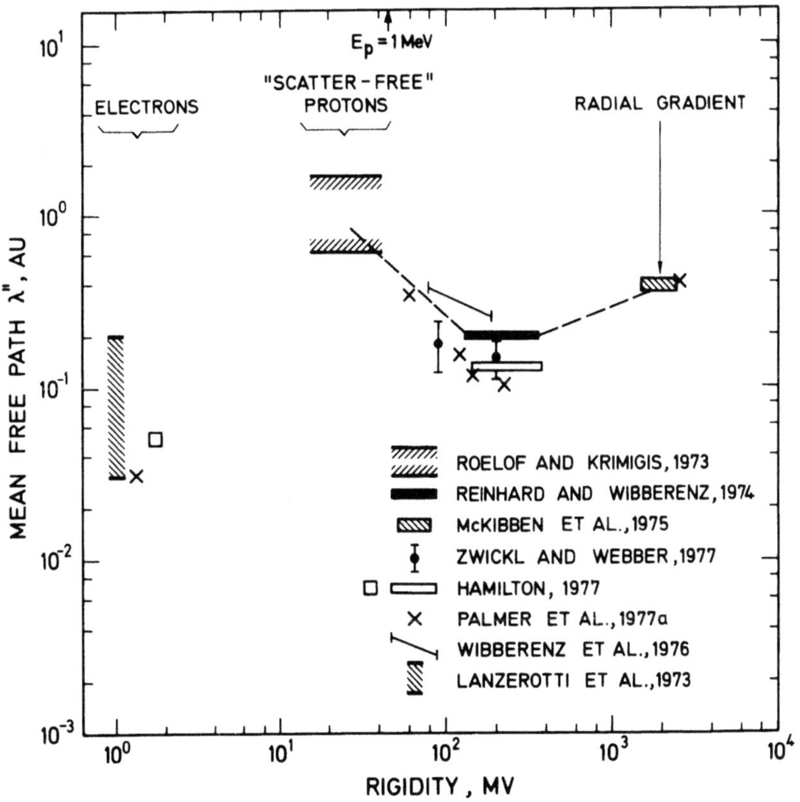

Figure 1: Parallel mean free path $\lambda_{\parallel}$ at 1 AU as a function of magnetic rigidity P. Symbols are listed (see insert) in the sequence as they are discussed in the text.

Let us now turn to additional information. The main improvements result from measurements taken over large radial distances with the Pioneer 10/11 spaceprobes and from simultaneous intensity and anisotropy analyses. Observations from the Pioneer 10/11 measurements are contained in papers by Zwickl and Webber (1977) and Hamilton (1977), leading to slightly different results. Zwickl and Webber (1977) have analysed a number of solar events observed on Pioneers 10 and 11 out to distances of 5 AU from the sun and compared the results with simultaneous observations on the near-earth satellites IMP 5 and 7. The authors derive propagation parameters for each event and subsequently perform a statistical analysis. They state "... that a numerical diffusion model which includes a nonimpulsive solar injection profile, a spectral index that increases with energy, along with convection and adiabatic deceleration can adequately explain solar particle events - at least out to 5 AU." The variation of the radial mean free path with distance r from the sun and magnetic activity P is described as $\lambda_r(r,P) = \lambda_0\, r^b\, P^\alpha$. Zwickl and Webber (1977) obtain $\alpha = -0.15 \pm 0.20$. Conversion to the parallel mean free path at 1 AU gives

$$\lambda_{\parallel}(1\ \text{AU}) = 0.18 \pm 0.06\ \text{AU for 3.4 to 5.2 MeV protons}$$
$$\lambda_{\parallel}(1\ \text{AU}) = 0.15 \mp 0.04\ \text{AU for 24-30 MeV protons}$$
$$b = 0.0 \mp 0.3$$

These data points have been added to Figure 1. The following results are noteworthy: (1) Solar particle observations at large distance from the sun are not very sensitive to details of the solar injection process, at least for the events considered in the study, which were restricted to those in the fast propagation region, where coronal transport times are relatively small. (2) Though the increase of the mean free path with decreasing rigidity from the two points is statistically not significant, Zwickl and Webber (1977) argue from a comparison with the large mean free path of ≃ 1 MeV protons (by reference to a private communication of Roelof and Gold) that one obtains the important result $\partial\lambda/\partial P < 0$ for rigidities below about 200 MV, which can at present not be explained by contemporary theories of cosmic ray scattering in interplanetary space. (3) The radial mean free path λ_r does not vary greatly with radial distance r, a result, which has been inferred more indirectly in earlier work and which has been used in many studies to derive the local value $\lambda_{\parallel}(1\ \text{AU})$ from the analysis of solar events. Recent theoretical studies (Morfill et al., 1976), on the basis of particle scattering by Alfvén waves, also result in little radial variation of λ_r.

A different method to describe solar particle events observed between 1 and 6 AU has been used by Hamilton (1977). He has fitted the intensity profiles observed at different radial distances to full solutions of the diffusion-convection model, using also a power-law dependence of the mean free path, $\lambda_r(r) \sim r^b$. Hamilton takes into account disturbances of the profiles related to solar wind dwells or shifts and to crossings of solar sector boundaries, which may lead to delayed onset phases. Converting the results of Hamilton (1977) to the parallel mean free path at 1 AU we obtain as an average over all events fitted:

$$\lambda_{\parallel}(1\ \text{AU}) = 0.13 \pm 0.02\ \text{AU for protons from 11 to 67 MeV}$$
$$b = 0.4 \pm 0.2.$$

This result has also been inserted into Figure 1. If one neglects convective effects, the time t_m of the intensity maximum varies as $t_m \sim r^{2-b}$. Figure 2 shows the experimental points $t_m(r)$ (Hamilton, 1977). The straight line corresponds to b = 0.4. (The slope of the curve changes only insignificantly, if convection and adiabatic deceleration are taken into account, Hamilton (1977)). Within the quoted errors the b-values obtained by Zwickl and Webber (1977) and Hamilton (1977) are consistent with each other. It could be, however, that the t_m-values by Hamilton near 1 AU have been overestimated. If one takes coronal propagation effects into account, one obtains the following average t_m-values at 1 AU (after Reinhard and Wibberenz, 1974): t_m = 4.2 hr for ≈ 13 MeV protons, t_m = 2.5 hr for ≈ 40 MeV protons. If one starts from these points at 1 AU in Figure 2 and draws the relation $t_m(r)$ corresponding to b = 0, the agreement with the remaining experimental points is rather good.

Hamilton (1977) also presents one example where 1 - 2 electrons are observed at different radial distances. The resulting mean free path of 0.05 AU has been added to Figure 1.

Let us now turn to the second method which allows to derive a mean free path, namely the interpretation of simultaneous intensity and anisotropy measurements. Figure 3 gives an example presented by Schulze et al. (1977). The upper part shows the intensity-time profile for 21 to 60 MeV protons on Pioneer 9. The overall profile can in principle be described by pulse-like solar injection and a relatively small diffusion coefficient. However, the experimentally observed anisotropies late in the event are much too large to be consistent with delta-function injection. The analysis reveals that an effective injection duration of the order of 10 hours combined with a larger diffusion coefficient is required. It should be noted that the drop in intensities and anisotropies around 2000 UT on Nov. 18 followed by a subsequent increase is not related to a new solar injection, but to a large scale disturbance in the interplanetary magnetic field (Wibberenz et al., 1976). This example is also shown to point out that a reduction of the observed anisotropy can be a local effect and does not necessarily mean that the further supply of particles from the sun has stopped. In a systematic and careful study, Zwickl and Webber (1976) have treated a number of additional effects related to the measurement of angular distributions. The recipes given for the correct interpretation of anisotropy measurements are highly recommended.

Complementary to the study of average propagation parameters which

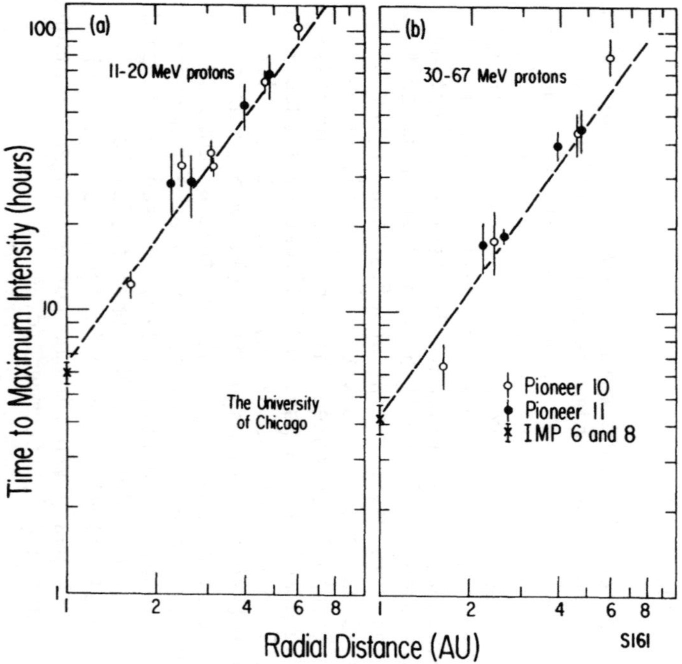

Figure 2. The radial dependence of the time of maximum intensity as a function of distance from the sun for a number of solar particle events (1972-1974), compared with predictions of the diffusion-convection model (b=0.4). (After Hamilton, 1977.)

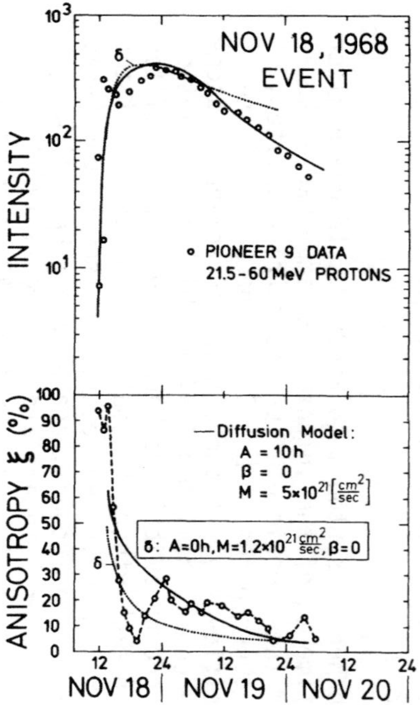

Figure 3. Hourly averages of 21.5-60 MeV proton intensities and anisotropies from the UTD experiment on board Pioneer 9 are compared with model calculations. An injection period A=10 hours is needed to fit both measurements simultaneously. (After Schulze et al., 1977.)

we have presented in the first part of this section are analyses of individual events over a large energy range. Palmer et al. (1977), for the solar event of 20 April 1973, discuss the time profiles of intensity and anisotropy for relativistic electrons and for protons over a wide range of energy from 1 MeV to a few Gev. As tools they have used diffusion theory and Monte Carlo technique, depending on the absolute value of the mean free path. The authors find large differences in both solar injection and interplanetary mean free path between the various particle groups. The deduced rigidity dependence of the mean free path has been inserted into Figure 1. The overall agreement with the "average" mean free path is surprisingly good. Note the minimum in $\lambda_{\parallel}$ near 30 MeV protons. Palmer et al. (1977) also point out that the dependence of $\lambda_{\parallel}(P)$ is anomalous according to current theories of particle scattering. They suggest that for decreasing proton rigidity propagation changes from diffusive to a "supercoherent" mode, where focussing by the diverging IMF is dominant (Earl, 1976a).

The 0.5 MeV electrons do not fit into a scheme where scattering efficiency decreases with decreasing rigidity (see Figure 1). It is interesting to note that if we return to the diffusion coefficient $\kappa_{\parallel} = v\lambda_{\parallel}/3$ instead of the mean free path, we see that κ varies only between 2.3 and 4.1×10^{21} $cm^2 s^{-1}$ over the rigidity range 1-200 MV. We leave it open whether or not this is significant. Note that Fisk (1976) has proposed an unusual scheme for particle diffusion in the interplanetary medium. In this case particles do not diffuse as a result of their own motion among small-scale irregularities - which leads to the dependence of $\lambda_{\parallel}$ on P only, as stated in the beginning of this section- but rather in a process in which they are trapped between time-varying constrictions in the large scale field.

The result $\partial\lambda/\partial P < 0$ has also been found by Wibberenz et al. (1976) for the 18 Nov 1968 event. They obtain $\lambda_{\parallel}(P) = 0.34\ (P/100)^{-0.44}$ (P in MV, $\lambda_{\parallel}$ in AU) for the rigidity interval 75-185 MV (see the straight line in Figure 1 for this interval). Since this result was obtained from the velocity dispersion in the early phase of the event, it is important that it can be confirmed by an independent observation. Wibberenz and Witte (1970) have pointed out that the study of the proton-to-alpha ratio for constant energy per nucleon as a function of time can be used to draw conclusions on $\lambda(P)$. In particular, one obtains $\partial(p/\alpha)/\partial t < 0$ as a consequence of $\partial\lambda/\partial P < 0$. A decrease of the p/α-ratio consistent with the above dependence $\lambda_{\parallel}(P)$ was indeed found for the 18 Nov 1968 event (Wibberenz et al., 1976).

Let us close this section by some additional remarks. The first relates to the fluctuations in $\lambda_{\parallel}$ from one event to another, which tends to obscure the overall behaviour summarized in Figure 1. For medium energy protons (≃10-60 MeV) deviations from the average conditions may amount up to a factor of 3 in both directions (Reinhard and Wibberenz, 1974; Ma Sung et al., 1975). In addition, there are a few cases when the "scatter-free" conditions are also observed for proton energies well above 1 MeV. Examples are the events of 24 March 1966 (Kahler et al., 1967), 20 April 1971 (Palmer et al., 1975), and 3 March 1975 (Kunow et al., 1977a). Cases where the interplanetary magnetic field is inspected siumultaneously suggest that this situation occurs when the imf is unusually quiet for a sufficiently long period of time.

The degree of scatter derived from solar electron events in the ≈ 1 MV region is also indicated in Figure 1, according to estimates from Lanzerotti et al. (1973). This still leaves us with the impression of the singular behaviour of the $\lesssim$ 1 MeV protons (as well as the $\gtrsim$ 40 keV electrons which are not discussed here, see Lin, 1974). This singular behaviour of the low energy protons has first been pointed out by Roelof and Krimigis (1973) and is meanwhile firmly established. The reliable determination of the absolute value of the mean free path is impeded by another peculiarity of the $\lesssim$ 1 MeV protons, namely their long-lasting solar injection (see section 3).

It is beyond the scope of this paper to discuss in detail the theoretical predictions of the mean free path based on knowledge of the interplanetary magnetic field fluctuations. It has been stated by many authors that theory yields values for the mean free path which are considerably smaller than those observed, in particular for protons below about 100 MeV. Some of the new approaches, in particular the results of non-linear methods, are discussed in a review by Völk (1975) and during a workshop on cosmic ray diffusion (Birmingham and Jones, 1975).

The variety of efforts to perform higher-order corrections to the quasi-linear approximation of the scattering problem has clearly brought one very important result. As pointed out by Wibberenz (1976) the discrepancy between theory and experiment is not removed by higher-order corrections (e.g. Goldstein, 1976) if the field fluctuations are described by the slab model, where all power is contained in wave vectors parallel to the mean field. This result is confirmed by the computer simulation of the velocity diffusion of charged particles in a static turbulent magnetic field (Kaiser et al., 1977). They follow particle orbits in a large ensemble of random magnetic field realizations representing the slab model. The pitch angle diffusion coefficient $D_\mu(\mu)$ remains finite for pitch angles close to 90° ($\mu = \cos\alpha \rightarrow 0$), in contrast to $D_\mu(\mu) \rightarrow 0$ in the quasi-linear approach. This means, however, that the resulting spatial diffusion coefficient turns out to be smaller than in the quasi-linear approximation, so that the discrepancy with the experimental results becomes even larger. Therefore, we need to reduce our ignorance about the actual form of the interplanetary magnetic field fluctuations.

Morfill et al. (1976) who take the wave vector distribution of Alfvén waves into account reproduce rather well the mean free path in the GeV range. However, the result $\partial\lambda/\partial P < 0$ in the lower rigidity range still requires a clarifying theoretical interpretation.

3. SOLAR INJECTION PROCESSES

Solar control of observed profiles has been demonstrated by a variety of different methods. Coronal propagation effects over large longitudinal distances ($\gtrsim 50^0$) are seen in statistical studies, and lead to a systematic increase of the time constants involved and a decrease of the event sizes with increasing longitudinal distance. The most remarkable feature of this "long distance" coronal propagation is its energy independence (see Wibberenz, 1976a,for a summary of experimental results).

The situation changes if we go to observations closer to the flare longitude. It is the main purpose of this section to indicate some of the observed variations of the solar injection time constants with energy and from one event to the other. This should allow to draw conclusions on the coronal propagation over small distances, on the release mechanisms, and hopefully, in some cases also on the acceleration process itself.

Figure 4 sketches some features of the injection process $I(t)$ which one tries to resolve from observations. We consider a fixed longitude Φ on the sun, so that I(t) represents the number of particles released into interplanetary space at this location as a function of time. For simplicity, only three types of profiles are considered in Figure 4. (a) represents the famous pulse-like injection, represented by a delta-function, (b) also has an abrupt onset, but is followed by a finite duration, (c) has a finite rise and decay.

The three cases are selected as examples of typically different solar conditions. Cases (b) and (c) are related to some kind of storage mechanism, by which the particles are not immediately released, in contrast to case (a). The difference in the rising portion of the event (cases (b) and (c)) may be partly related to the distance from the flare, but could also contain some hints to the acceleration process.

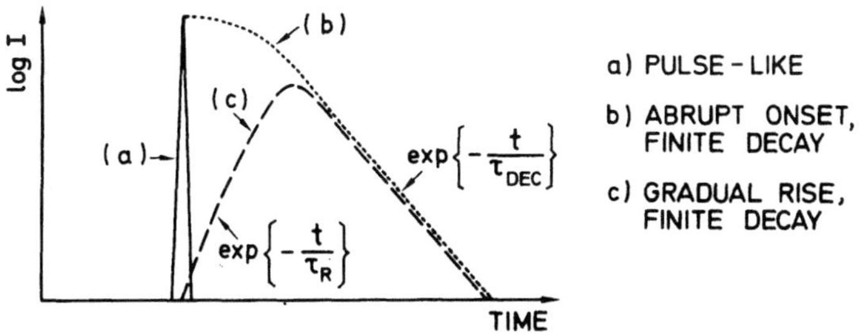

Figure 4. Types of solar injection profiles I(t) (schematic).

In deriving I(t), we are immediately faced with two difficulties.

(1) The influence of the interplanetary scattering does in general not allow to distinguish between various types of decay patterns. As shown by Schulze et al. (1977) one rather obtains an estimate of the effective "duration" of the injection. Differences in the onset phase will also be smeared out by interplanetary scattering, unless the mean free path is very large, as in the case of $\lesssim$ 1 MeV protons, or unless the observer is sufficiently close to the sun. Examples for the latter case will be presented below from the Helios 1/2 observations.

(2) Due to the rotation of the sun, an observer in space sees a sequence of particle populations injected at different solar longitudes. This problem is treated extensively by Roelof (1976). He concentrates on the variation of the coronal particle distribution with longitude. For many low energy particle events the fluxes mapped back to their high coronal connection longitude display at a fixed time a nearly exponential dependence on heliographic longitude, $S(\Phi)$ $\exp(-\Phi/\Phi_o)$. The e-folding angle Φ_o is typically of the order of 10°, but larger values up to $\simeq 30^\circ$ may also be found (Roelof, 1976).

A particularly simple situation occurs if the injection I(t) is independent of longitude, so that the coronal source distribution $S(\Phi,t)$ can be described by

$$S(\Phi,t) = \exp\{-\frac{\Phi}{\Phi_o}\} I(t) \tag{1}$$

The density profile U at distance r from the sun can be described by a convolution,

$$U(r,\Phi,t) = \int_0^t S(\Phi,t) f(r,t-\tau) d\tau \tag{2}$$

where f is the solution of the interplanetary transport for δ-function injection. From equ. (1) we obtain

$$U(r,\Phi,t) = \exp\{-\frac{\Phi}{\Phi_o}\} \int_0^t I(t) f(r,t-\tau) d\tau = \exp\{-\frac{\Phi}{\Phi_o}\} U_o(r,t) \tag{3}$$

with U_o independent of Φ. The field-aligned anisotropy is

$$\xi \sim \frac{1}{U}\frac{\partial U}{\partial r} = \frac{1}{U_o}\frac{\partial U_o}{\partial r} \tag{4}$$

which means that for a given (r,t) the observed anisotropy ξ is <u>independent</u> of Φ, and the time behaviour of the anisotropy can be used to draw conclusions on the injection process I(t), independent of the coronal distribution. This again demonstrates the importance of anisotropy measurements.

The relative importance of azimuthal and temporal effects in the density profiel $U(r,\Phi,t)$ can be estimated from eq. (3). For constant solar wind speed an observer moves by about 0.56^o/hr in solar longitude; this produces then a temporal change $\sim \exp(\pm t/\tau_{az})$, where $\tau_{az} \simeq (10^o)/(0.56^o/hr) = 18$ hr, and the sign depends on whether or not the observer moves towards or away from the center of the particle distribution. It depends on the time constants involved in I(t) (see Figure 4), whether we see essentially the injection profile for a fixed longitude on the sun, the longitudinal profile, or a combination of both.

Let us keep these difficulties in mind when we turn now to some examples. Selected results are summarized in Figure 5. Finite solar injections have been found for the Nov 18, 1968, event, which originated near W 90, by Wibberenz et al. (1976) and Schulze et al. (1977). The duration of the injection process increases systematically with decreasing proton energy: the duration is of the order 1 hr at neutron monitor energies, 10 hr for 21-60 MeV protons, 20 hr for 7-21 MeV protons; here the data are consistent with an abrupt onset of the injection. In contrast, a gradual increase of the solar injection for 0.3 - 1.0 MeV protons is observed, lasting at least for about 40 hours beyond flare maximum; after that time the event profile was distorted by superposition of an ESP event.

A strong dependence of the injection profiles has also been found by Palmer et al. (1977) for the event of April 29, 1973, caused by a 2B flare at (N14, W73). At neutron monitor energies the high degree of anisotropy for 2 - 3 hours after the flare suggest a finite length of injection (decay time ≃ 2 hours). A fast rise to intensity maximum accompanied by a fast decay in anisotropy is exemplified by the 0.41 - 1 MeV electrons and the 20 - 56 MeV protons. The 0.52 - 2.1 MeV protons show a much slower rise to maximum, accompanied by a high degree of anisotropy. The corresponding coronal rise time of ∼ 13 hr has been added to Figure 5.

Other points in Figure 5 have been taken from Reinhard (1975), Palmer et al. (1975), Gold and Roelof (1976), see also the discussion by Reinhard (1975). Estimates for finite solar injections from Zwickl and Webber (1977) are also inserted. We note the general decrease of the decay times with increasing energy, indicated by the dashed line. What is also peculiar is the general change in the nature of the profiles: in the cases discussed by Wibberenz et al. (1976) and Palmer et al. (1977), the ∼ 1 MeV protons exhibit a very slow increase of the solar injection profile (case (c), Figure 4), in contrast to the very abrupt rising phase for $\gtrsim$ 10 MeV protons (case (b)). Continuous acceleration of the $\lesssim$ 1 MeV protons is one of the explanations put forward to explain their characteristically different behaviour.

Let us now turn to differences between various events at the same energy. Here we take some examples from the Helios mission; when Helios is within say 0.5 AU, interplanetary scattering effects can be neglected to a large degree. Of particular importance are the two upper points at ∿ 5 MeV (Kunow et al., 1977b). The time constants differ by a factor of about 4, though both solar events originated in the same active region within 5 days. An even smaller decay time is found for the 3 March 1975 event (see the discussion in section 4 below).

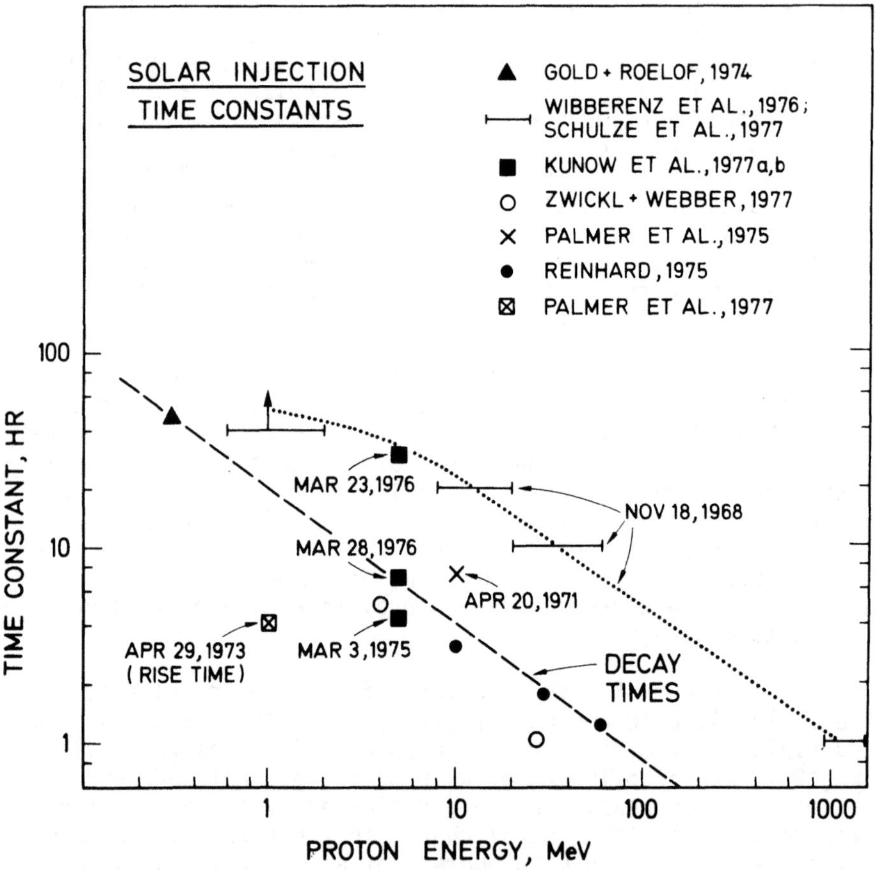

Figure 5: Solar injection time constants as a function of proton energy. The dotted curve connects points for the 18 Nov 1968 event which represent the effective duration of the injection. Note that the points at 1 MeV correspond to rise times of the events (lower limit in case of the 18 Nov 1968 event).

There are situations when the study of decay phases is rendered difficult due to repeated fresh injections from the sun. At large distances from the sun, interplanetary propagation tends to obscure this effect if the new injections occur sufficiently close to each other. Let us again take an example from Helios, where repeated injections can be resolved. Figure 6 shows in a schematic form the intensity-time profiles for three selected channels obtained with the University of Kiel cosmic-ray experiment on Helios 1. For the series of events starting on March 19, 1975, Helios 1 was at a distance of 0.32 AU from the sun, close to perihelion. The events were briefly discussed by Kunow et al. (1977a); Figure 6 is based on 5 minute averages, the curves have been drawn in such a way as to average out statistical fluctuations. The fine structure close to onset in the nucleon channels is partly related to local imf discontinuities; the remarkable feature is the different number of peaks for the three particle types: relativistic electrons show up only in the first peak; the second peak is seen only in the Helium nuclei. The small contribution of interplanetary propagation at these distances suggests that we see directly the

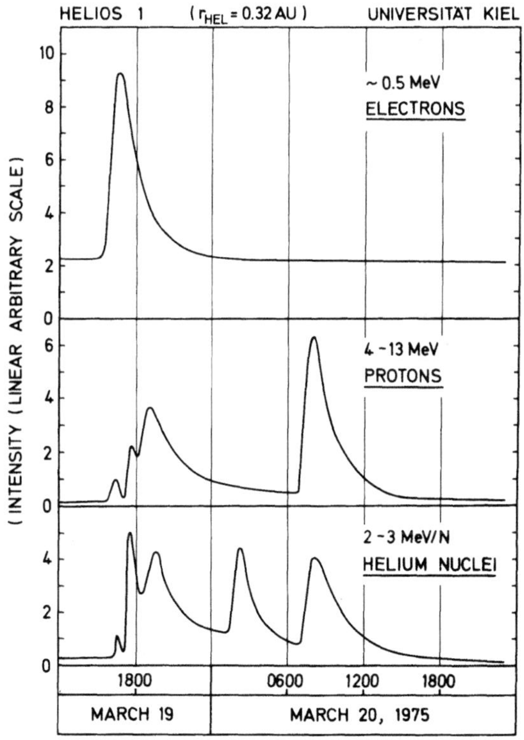

Figure 6. Time intensity profiles on a linear scale for three channels during the March 19/20, 1975 solar events, when Helios 1 was close to perihelion (see Kunow et al., 1977a).

solar injection profiles. The different composition for the three peaks is probably caused by different acceleration mechanisms. It is remarkable that these events are rich in He^3 (Kunow et al., 1977a); the ratio of He^3 to He^4 is of the order of 1 for all three peaks.

This example is shown to demonstrate the important role of the Helium nuclei in studying the solar acceleration process. The large variability in the proton/Helium ratio for solar particle events is well known; an important observation is the increase of the proton/Helium ratio with the size of the event (Van Hollebeke, 1975). There are more cases when the fresh injection of particles from the sun is mainly seen in the Helium channels (Witte et al., 1977).

Let us return to the discussion of the decay times. The time constants for the different particle species in Figure 6 are remarkably similar, corresponding to $\tau_{DEC} \simeq$ 2-3 hours (apart from the first proton peak where the decay phase might be disturbed by a statistically unresolved additional injection). If we interpret the decay phase as being solely due to solar processes, the values of τ_{DEC} are of the same order, but slightly smaller than those summarized in Figure 5 for $\simeq$ 5 MeV protons.

It should be understood that the interpretation of the observed decay phases as of solar origin requires a full understanding of the interplanetary propagation. The next section contains some remarks on the open questions in terms of proposed models.

4. DETAILS OF THE PITCH ANGLE SCATTERING PROCESS

In section 2 we had described the average propagation characteristics by a scattering mean free path $\lambda_{\parallel}$ parallel to the average imf. It was noted that the "anomalous" behaviour of $\lambda_{\parallel}(P)$ in the rigidity range $\simeq$ 1 - 200 MV is roughly identical with the constancy of the product $v \cdot \lambda_{\parallel}$ ($\sim$ diffusion coefficient). Assuming this as strictly valid and starting from λ = 0.15 AU for 20 MeV protons, we arrive at $\lambda_{\parallel}$ = 1.2 AU for 0.3 MeV protons.

This brings us to the interesting range of an almost "scatter-free" propagation and a breakdown of the diffusive approximation. For an observer within about 1 AU from the sun, this breakdown is not only related to the fact that the distance from the source is of the same order as the mean free path, but also to strong effects of adiabatic focussing (Roelof, 1969; Earl, 1976a). Earl (1976a) has introduced the term "supercoherent mode" for the case where the scale length $L_c = (1/B \times dB/dr)^{-1}$ of the diverging imf becomes small compared to the mean free path. Assume a situation where $L_c/\lambda_{\parallel}$ increases with increasing distance from the sun. This means that scattering becomes more and more important if the particles move further out along the field, until they are finally "scattered back", i.e. are moved across

90^0 pitch angle. This transition from the "supercoherent" to the "diffusive" mode of propagation with increasing distance from the sun is depicted schematically by Earl (1976b, see his Figure 2). This situation is not drastically different from the concept of "collimated convection" (Roelof, 1973; Nolte and Roelof, 1975); the main difference in the collimated convection model is the introduction of a fixed outer boundary where particle reflection occurs. Particles injected from this outer boundary are treated adiabatically in the inner (scatter-free) solar system.

Details of the pitch-angle distribution throughout an event are needed to decide if one of the two models describes the real situation. The absolute value of the pitch-angle diffusion coefficient and its functional dependence on μ determine the relative importance of adiabatic focussing and pitch angle scattering. Angular distribution measurements of energetic particles should help to resolve this question.

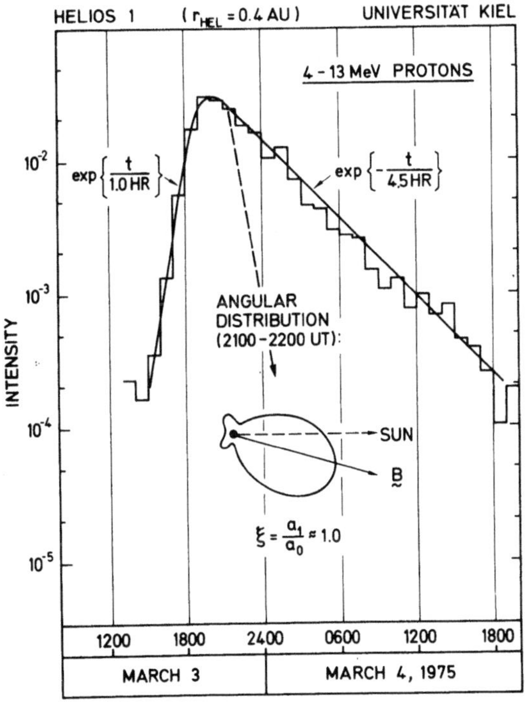

Figure 7. Hourly averages of 4 - 13 MeV proton intensities for the event of 3 March 1975 (upper part). Insert: Angular distribution derived from 8-sector measurements for the period 2100 - 2200 UT on 3 March (r = 0.4 AU for Helios 1).

Let us discuss one example which illustrates some of the potential results and the principle difficulties. Figure 7 shows in the upper part the 4 - 13 MeV proton intensities as observed on Helios 1 for the 3 March 1975 event, when Helios was at 0.4 AU (Kunow et al., 1977a). One of the essential features of this event is the highly peaked angular distribution for several hours around the intensity maximum. The insert shows for the time period 2100 - 2200 results of the 8-sector measurements represented by a Fourier expansion (Green, 1976), with a first order anisotropy of the order of 1. For pulse-like injection, the anisotropy should have dropped to about 0.14 at this time. Similar distributions are found for the whole period from 1900 - 2300 UT, i.e. around the intensity maximum.

By a lucky coincidence, Helios 1 is very close to the magnetic field lines connecting the sun and the earth at the time of the event (Kunow et al., 1977a). It is possible, therefore, to compare the intensity profile of Figure 7 with data from the Caltech Electron/Isotope Spectrometer on board IMP 8. The results are discussed in detail by Wibberenz et al. (1977). The essential point is that the shape of the intensity profile for the same proton energies near 1 AU is practically identical with that near 0.4 AU, but it is reduced in intensity by a factor determined by an r^{-2} - law and shifted in time roughly by $\Delta t = 2(\Delta s)/v$ (Δs = distance between Helios and the earth along the spiral, v = particle velocity). This corresponds to a propagation speed of the proton population equal to $v/2$, consistent with the coherent mode of propagation (Earl, 1974; Kunstmann and Alpers, 1976).

The observed profile (Figure 7) is therefore interpreted as being totally due to the solar injection process. The decay time constant τ_{DEC} = 4.5 hours for ≈ 5 MeV protons has been inserted in Figure 5. It should be mentioned that this small amount of interplanetary scattering in this energy range is not common. Inspection of interplanetary magnetic field data on Helios (courtesy F. M. Neubauer, G. Musmann) shows that the field is exceptionally quiet during the maximum phase of the event. As soon as the field becomes more disturbed, the anisotropy drops considerably. We feel that in many cases this is a local effect only which tends to obscure the continued delivery of particles from the sun.

This also means that it may be very difficult to extract very clear cases from observational results to test specific models with inherent simplifying assumptions. We can draw one preliminary conclusion from the results discussed previously: the intensity profiles are relatively smooth, whereas the anisotropy reveals sudden changes; this suggests that the "patchiness" of the field occurs rather along the average field than between adjacent field line bundles. The degree of disturbance of the IMF seems to change spatially on a scale which is large compared to the distance needed for the particles to establish a certain equilibrium pitch-angle distribution.

Let us close with a word of caution on the cosmic-ray scattering problem. So far it was tacitly assumed, by referring to the pitch-angle diffusion coefficient D_μ, that the scattering results from a superposition of many small-angle changes. The existence of an ensemble of magnetic field "blobs" (Barouch and Burlaga, 1975), of local constrictions in the field tubes (Domingo et al., 1976), and for isolated rotational discontinuities (Webb and Quenby, 1974) may lead to a sequence of large-angle scatterings, which cannot be treated by the usual Fokker-Planck formalisen for pitch-angle scattering.

ACKNOWLEDGEMENTS

It is a pleasure to acknowledge stimulating discussions with my colleagues from the Helios team at the University of Kiel, in particular G. Green, H. Kunow, and M. Witte, and their assistance in the preparation of Figures 6 and 7. I thank D.C. Hamilton, I.D. Palmer, W.R. Webber, and R.D. Zwickl that I could make use of their results prior to publication. This work was partly supplied by the Deutsche Forschungsgemeinschaft.

REFERENCES

Barouch, E. and Burlaga, L.F.: 1975, J. Geophys. Res. 80, 449.
Birmingham, T.J. and Jones, F.C.: 1975, Cosmic Ray Diffusion - Report of the Workshop in Cosmic Ray Diffusion Theory, NASA TN D-7873.
Domingo, V., Page, D.E., and Wenzel, K.-P.: 1976, J. Geophys. Res. 81, 43.
Earl, J.A.: 1974, Astrophys. J. 188, 379.
Earl, J.A.: 1976a, Astrophys. J. 205, 900.
Earl, J.A.: 1976b, Astrophys. J. 206, 301.
Fisk, L.A.: 1976, Astrophys. J. 206, 333.
Gold, R. E., and Roelof, E. C.: 1976, in M. J. Rycroft (ed.), Space Research XVI, Akademie-Verlag, Berlin, 791.
Goldstein, M.L.: 1976, Consequences of using nonlinear particle trajectories to compute spatial diffusion coefficientes, Preprint, (submitted to J. Geophys. Res.).
Green, G.: 1976, EOS (Trans. Am. Geophys. Union) 57, 979.
Hamilton, D.C.: 1977, The radial transport of energetic solar flare particles from 1 to 6 AU, Preprint, (submitted to J. Geophys. Res.)
Kahler, S.W., Primbsch, J.H., and Anderson, K.A.: 1967, Solar Phys. 2, 179.
Kaiser, T.B., Birmingham, T.J., and Jones, F.C.: 1977, Computer simulation of the velocity diffusion of cosmic rays, Preprint GSFC X-602-77/67.
Kunow, H., Witte, M., Wibberenz, G., Hempe, H., Müller-Mellin, R., Green, G., Iwers, B., and Fuckner, J.: 1977a, Cosmic ray measurements on board Helios 1 from December 1974 to September 1975: Quiet time spectra, radial gradients, and solar events, Preprint, (submitted to J. Geophys.).

Kunow, H., Wibberenz, G., Green, G., Müeller-Merlin, R., Witte, M., Hempe, H., and Fuckner, J.: 1977b, in M. A. Shea, D. F. Smart, S. T. Wu (eds.), Contributed Papers to the Study of Travelling Interplanetary Phenomena/1977; Air Force Geophys. Laboratory, Bedford, Massachusetts, USA.

Kunstmann, J.E. and Alpers, W.: 1976, Astrophys. J. 211, 587.

Lanzerotti, L.J., Venkatesan, D., and Wibberenz, G.: 1973, J. Geophys. Res. 78, 7986.

Lin, R.P.: 1974, Space Sci. Rev. 16, 189.

Ma Sung, L.S., Van Hollebeke, M.A.I., and McDonald, F.B.: 1975, Proc. 14th Internat. Cosmic Ray Conf. 5, München, 1767.

McKibben, R.B., Pyle, K.R., Simpson, J.A., Tuzzolino, A.J., O'Gallagher, J.J.: 1975, Proc. 14th Internat. Cosmic Ray Conf. 4, München, 1512.

Morfill, G.E., Völk, H.J., and Lee, M.A.: 1976, J. Geophys. Res. 81 5841.

Nolte, J.T. and Roelof, E.C.: 1975, Proc. 14th Internat. Cosmic Ray Conf. 5, München,1722.

Palmer, I.D., Palmeira, R.A.R. and Allum, F.R.: 1975, Solar Phys. 40, 449.

Palmer, I.D., Zwickl, R.D., Webber, W.R., and McDonald, F.B.: 1977, Spectrum of mean free paths near earth in the solar cosmic ray event of 29 April 1973, Preprint,(submitted to J. Geophys. Res.).

Reinhard, R.: 1975, Proc. 14th Internat. Cosmic Ray Conf. 5, München, 1687.

Reinhard, R. and Wibberenz, G.: 1974, Solar Phys. 36, 473.

Roelof, E. C.: 1969, in H. Ogelman and S. Wayland (eds.), Lectures in High Energy Astrophysics, NASA SP-199, chapter VII.

Roelof, E.C.: 1973, New Aspects of Interplanetary Propagation Revealed by > 3 MeV Solar Proton Events in 1967, Proc. Solar-Terrestrial Relations Conference, Calgary, Canada, August 1972.

Roelof, E. C.: 1976, in D. J. Williams (ed.), Physics of Solar Planetary Environments, Proc. of the International Symposium on Solar-Terrestrial Physics (Boulder), American Geophysical Union, 1, 214.

Roelof, E.C. and Krimigis, S.M.: 1973, J. Geophys. Res. 78, 5375.

Schulze, B.M., Richter, A.K., and Wibberenz, G.: 1977, Influence of finite injections and of interplanetary propagation on time-intensity and time-anisotropy profiles of solar cosmic rays, Preprint, (submitted to Solar Phys.).

Van Hollebeke, M.A.I.: 1975, Proc. 14th Internat. Cosmic Ray Conf. 5, München, 1563.

Völk, H.: 1976, Rev. Geophys. Space Phys. 13, 547.

Webb, S. and Quenby, J.J.: 1974, Solar Phys. 37, 235.

Wibberenz, G.: 1974, J. Geophys. 40, 667.

Wibberenz, G.: 1976a, in L. A. Fisk and W. I. Axford (eds.), Proceedings of the Symposium on the Study of the Sun and Interplanetary Medium in Three Dimensions, NASA X-660-76-53, 261.

Wibberenz, G., 1976b, in D. J. Williams (ed.), Physics of Solar Planetary Environments, Proc. of the International Symposium on Solar-Terrestrial Physics (Boulder), American Geophysical Union, 1, 346.

Wibberenz, G. and Witte, M.: 1970, in V. Manno and D.E. Page (eds), Intercorrelated Satellite Observations related to Solar Events, D. Reidel, Dordrecht, 499.
Wibberenz, G. Lanzerotti, L.J., and Venkatesan, D.: 1976, J. Geophys. Res. 81, 5807.
Wibberenz, G., Kunow, H., Green, G., Müeller-Mellin, R., Witte, M., Hempe, H., Mewaldt, R.A., Stone, E.C., Vogt, R.E., and Reinhard, R.: 1977, Radial development of a solar cosmic ray event between 0.4 and 1 AU on March 3, 1975, as observed from Helios 1 and IMP, Preprint.
Witte, M., Hempe, H., Green, G., Kunow, H., Müeller-Mellin, R., and Wibberenz, G.: 1977, in M. A. Shea, D. F. Smart, S. T. Wu (eds.), Contributed Papers to the Study of Travelling Interplanetary Phenomena/1977, Air Force Geophys. Lab., Bedford, Mass., USA.
Zwickl, R.D. and Webber, W.R.: 1976, Nucl. Instr. Meth. 138, 191.
Zwickl, R.D. and Webber, W.R.: 1977, Solar particle propagation from 1 to 5 AU, Preprint, (submitted to Solar Phys.).

DISCUSSION

Sarris: In many cases, as in corotating interaction regions, the magnetic field converges rather than diverges in the deep space. How would Earl's scheme describe solar events in such situations?

Wibberenz: The decay phase of solar events will certainly be influenced by this situation. In principle, in Earl's scheme this can be taken into account by inserting the correct scale length $\left(\frac{r}{B}\frac{\partial B}{\partial r}\right)^{-n}$ of the field. It will be interesting to see if characteristic differences in the decay phase will occur between compression and rarefaction regions.

Shea: Were the three helium increases shown for the event of 19-20 March 1975 each associated with a separate solar phenomena?

Wibberenz: The events were behind the west limb as seen from the earth. The only flare association could be obtained from the GSFC X-ray detector of Helios. The first event could not be related with an X-ray burst; unfortunately later on there was a data gap. It is possible, that these "unusual" events cannot be related to flaring activity.

Armstrong: Briggs reported at the 1977 Spring AGU meeting a study of ~ 35 flare events in 1972-75. He found that first flares from an active region were richer in alphas than subsequent flares. Your event having 3 alpha peaks may have involved more than one active region.

Wibberenz: At that time there was mainly one active region on the sun close to the Helios field line connection longitude. The He^3/He^4 - ratio might have slightly decreased in the third peak as compared to the first two peaks.

Morfill: The "anomalously high" mean free paths for MeV solar protons are difficult to calculate because the diffusion-convection picture has to be augmented by effects such as adiabatic deceleration, Fermi acceleration etc. This makes the transport theory much more difficult, and I wonder how reliable the estimates of the diffusion coefficient in this energy range are.

Wibberenz: The cosmic-ray diffusion coefficient is derived from solar events mainly during the initial phase when adiabatic deceleration plays a minor role, in particular if the resulting mean free path turns out to be large. In the ≥ 1 MeV range Fermi acceleration can very probably be totally neglected.

SOLAR ENERGETIC PARTICLES BELOW 10 MeV

E. C. Roelof and S. M. Krimigis
Applied Physics Laboratory
The Johns Hopkins University
Laurel, Maryland 20810 USA

ABSTRACT

The information on solar acceleration and coronal propagation contained in low energy solar particle observations must be extracted from the effects of propagation in a dynamic interplanetary medium and the proximity of the Earth's magnetosphere. The resulting separation reveals long-lived coronal injection and strong spatial ordering of coronal propagation.

INTRODUCTION

The basic theme of this paper is that solar energetic particles undergo many and diverse processes between their acceleration and their detection; consequently, we must carefully identify the signatures of these many processes and then compensate our data for their effects before we can discover the essential physics of solar particle acceleration and propagation. We stress the *compensation*, since we now know that, although coronal, interplanetary and magnetospheric processes *distort* the information contained in the solar particle fluxes, these processes often do not destroy the information beyond recovery. We shall give several examples which demonstrate how essential this approach is, drawn from studies of: (1) relativistic solar electron propagation near 1 AU; (2) solar **origin** of enhanced abundances of low energy heavy nuclei and $He^3 \geqslant 2$ MeV/nucleon; (3) interplanetary propagation of > 10 MeV solar protons near 5 AU; (4) coronal propagation of > 7.5 MeV protons in a flare event; and (5) the decay phase anisotropies of > 300 keV protons.

1. Relativistic Solar Electron Anisotropies

In Figure 1 we show the fluxes of > 220 keV electrons from the large flare event beginning late on day 253 (September 10), 1974, measured on the JHU/APL experiment on IMP-7. If we were to find large,

M. A. Shea et al. (eds.), Study of Travelling Interplanetary Phenomena/1977, 343–365.

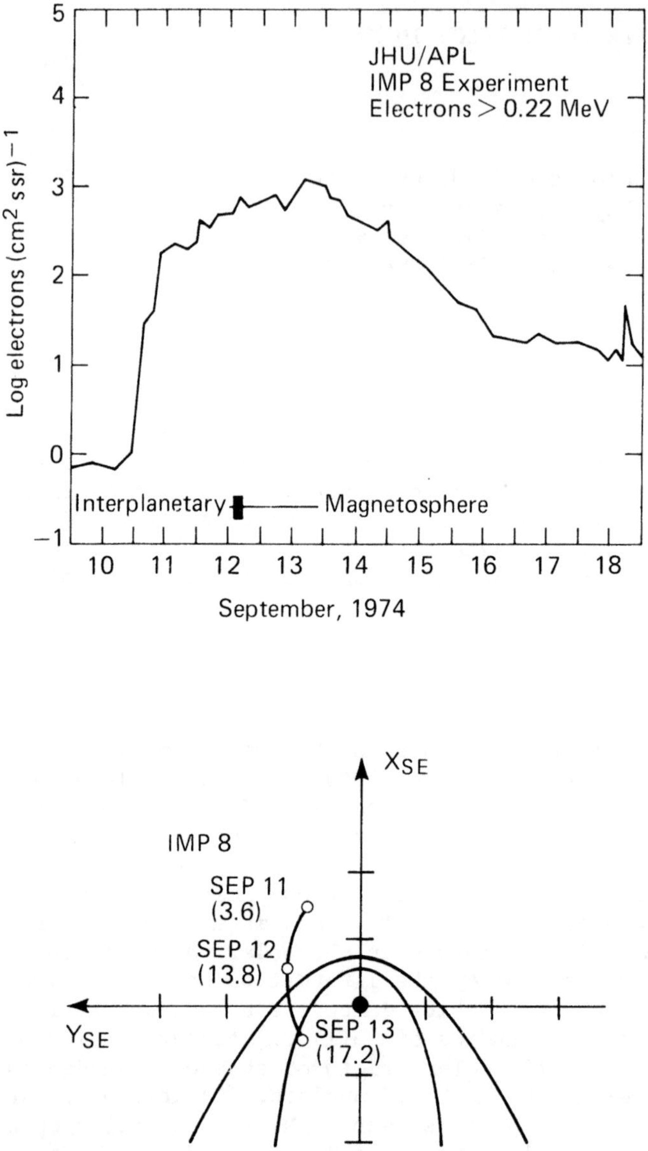

Figure 1. Relativistic electron intensity measured on IMP-8 from an eastern hemisphere flare 2121 UT, day 253 (September 10) 1974. Strong field-aligned outward anisotropies (distorted by reflection from the bow shock) persisted from onset until bow shock crossing. Distance of IMP-8 above ecliptic (Z_{SE}) is indicated on orbit plot.

persistent flux anisotropies in the early part of the event, that would be strong evidence for not only extended interplanetary injection, but also of a mean-free-path not much less than 1 AU. Past studies have given inconsistent results. *Lin (1970a,b)* identified > 45 keV electron events whose very prompt onsets and sharp rises suggest nearly scatter-free propagation. However, *Allum et al. (1971)* observed > 70 keV events which have no significant anisotropies after the first 10 minutes. We believe this apparent inconsistency is resolved by a recent analysis of > 220 keV electron events from 1972 to 1976 using IMP-7 and IMP-8 *(Ng and Roelof, 1977)*.

Figure 2 presents 5.5 minute averages of the first harmonic (C_1) of the electron anisotropy measured during the third through fifth hours of the September 10 event which we showed in Figure 1. The top panel shows the polar angle θ of the interplanetary magnetic field (IMF), courtesy of N. Ness and R. Lepping of NASA/GSFC. This is an important parameter, since the JHU/APL detector has a viewing cone of $22\frac{1}{2}°$ half-angle spinning in the ecliptic. Consequently for $|\theta| \geqslant 22\frac{1}{2}°$, the measured anisotropy will be an underestimate of the actual anisotropy. The lower panel of Figure 2 shows the distance L from the spacecraft to the bow shock along the instantaneous direction of the IMF measured on the spacecraft. When L becomes infinite, this means the extrapolated field line does not intersect the bow shock. The calculation is done in three-dimensions using the nominal bow shock configuration computed by *Fairfield (1971)*.

During the rise of the event, the IMF is quite variable, often swinging 90° in azimuth. The result is quite clear: when the IMF intersects the bow shock, the solar electrons are reflected and the anisotropy is greatly reduced. When the field line misses the bow shock, the true anisotropy is even larger (> 0.3) than that measured because the IMF is often well out of the ecliptic during unconnected periods. Consequently we conclude that the mean-free path is comparable to 1 AU in this event. A conservative estimate based on the anisotropy parameter $2\lambda/r$ introduced by *Roelof (1969)* and utilized by *Earl (1974)* gives $\lambda > 0.2$ AU. This value is smaller than those obtained in other events studied during 1972-1974.

One can readily see that if the IMF were to intersect the bow shock throughout a shorter electron event (as may have been the case in *Allum's* event), the measured anisotropy would be insignificantly small, even though the actual anisotropy on non-intersecting field lines could be so large as to approach the "scatter-free" propagation deduced by *Lin*. Thus, by use of IMF data on near-earth spacecraft we can identify the conditions under which strong electron anisotropies are reduced by reflection from the bow shock. Naturally, we also have learned something about the structure of interaction of the IMF and the magnetosheath.

The most striking characteristic of the September 10, 1974 electron event, revealed when we *compensate* for the bow shock effect, is

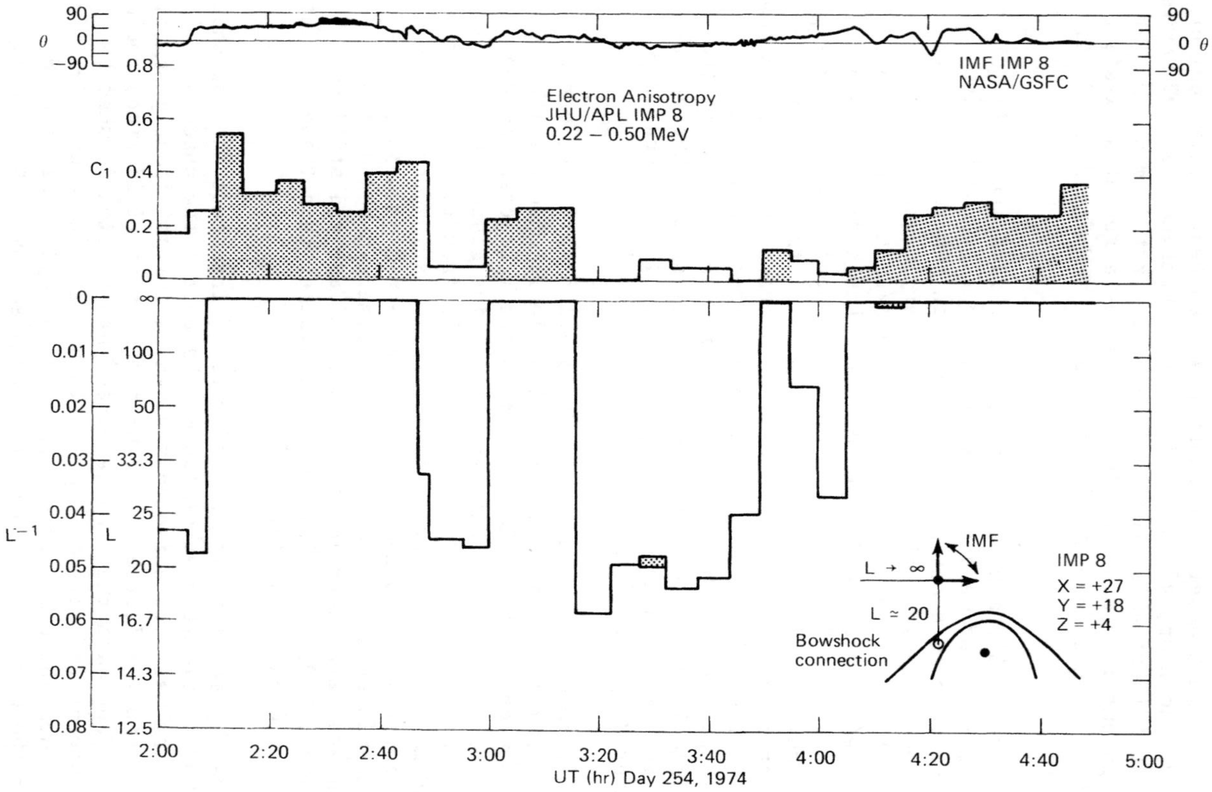

Figure 2. *Effect on relativistic electron anisotropy (C_1) of intersection of interplanetary magnetic field line with the bow shock. L is distance in earth radii from IMP-8 to bow shock along extrapolated field line. Shaded periods indicate no intersection ($L \to \infty$). From Ng and Roelof (1977).*

the duration of the relativistic electron anisotropy. The > 0.22 MeV electrons exhibit field-aligned outward anisotropies > 0.2 from the event onset (~ 2300, September 10) until IMP-8 crosses the dusk bow shock (~ 1800, September 12). Relativistic electrons are therefore being injected into interplanetary space for at least 36 hours after the original flare, and the strong anisotropies throughout this period imply that the intensity history in Figure 1 is essentially the coronal injection history for the first $1\frac{1}{2}$ days of the event. The slow rise then implies solar acceleration and/or storage for more than a day, since the flare (importance 2B) occurred at 2121, September 10 and was located in McMath Plage Region 13225 at N10, E61.

As we have pointed out in another study *(Sarris et al., 1976)*, the influence of the magnetosphere on the measurement of solar particles also appears directly through the injection of magnetospheric particles onto interplanetary field lines, both relativistic electrons and energetic protons (~ 1 MeV). With care, the magnetospheric particles can be identified, but if they are counted along with solar particles, anisotropies in particular will be severely distorted. *Marshall and Stone (1977)* have considered (but rejected) this possibility in their study of interplanetary 1.3-2.3 MeV proton anisotropies at moderate intensity levels.

2. Z ≥ 3 and He^3 Rich Solar Events

A most interesting class of small particle events has been identified during the past five years. These events are rich in heavy nuclei ⩾ 1 MeV/nuc and have He^3/He^4 ratios comparable to unity *(Price et al., 1971; Garrard et al., 1973; Hovestadt et al., 1975; Hurford et al., 1975a,b)*. Such an event measured by the JHU/APL experiment on IMP-8, is shown in Figure 3. The flux of Z ≥ 3 nuclei (⩾ 2 MeV/nuc) exceeds the 3 MeV proton flux and is comparable to the 3.5 MeV helium flux. There is a small electron event at about 14 hours on day 51 of 1974, but there is no flare of importance greater than a sub-flare on the visible sun. Magnetospheric electron and 0.4 MeV proton activity is appreciable only after day 54. This event is also rich in He^3 and Fe *(Hovestadt et al., 1975; Hurford, 1975a,b)*. The solar wind velocity (provided by A. Lazarus of MIT) is low until well into the decay of the particle event.

The first question to be asked is: Are these interesting events solar in origin? This can be answered in the affirmative by analyzing the flux anisotropies. In Figure 4 (from *Zwickl et al., 1977*) we show the 8-sector anisotropy measurements for the rise of four other events. Sector 5 views the sun while the ideal spiral direction falls in sector 7. The fluxes are clearly coming from the sun and cannot be magnetospheric or accelerated by some unspecified mechanism beyond 1 AU. In fact, since the solar wind speed (Figure 3 top) is low (< 350 km s^{-1}) and nearly steady during the event, it is unlikely that there is any stream-associated interplanetary acceleration inside 1 AU.

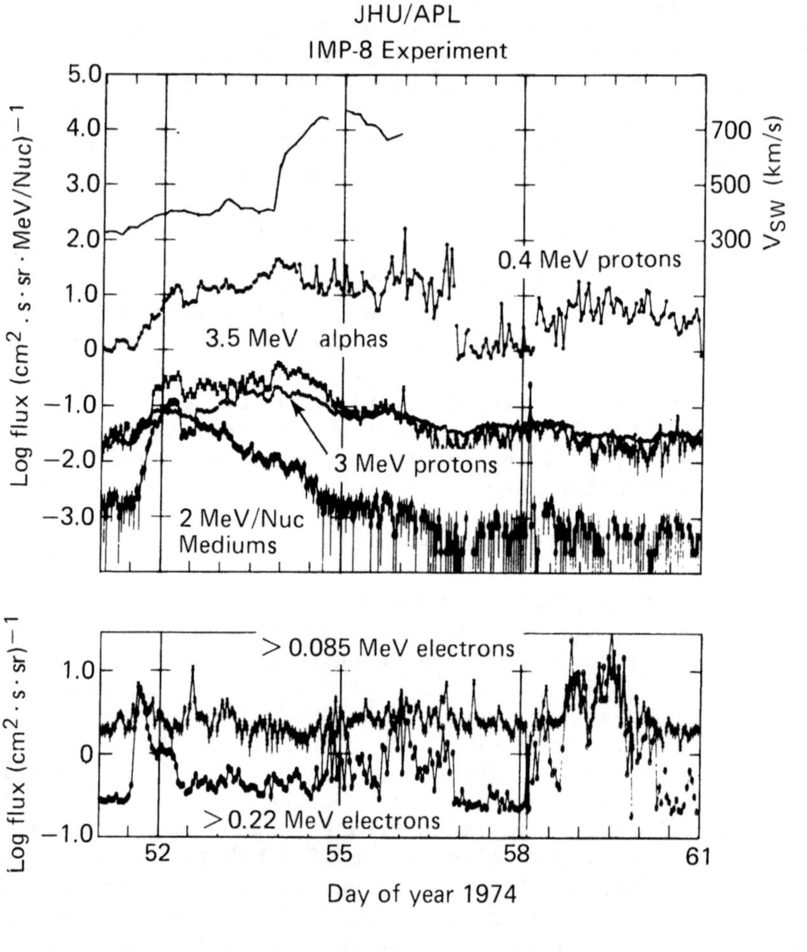

Figure 3. Intensities of ~ 1 MeV solar particles (JHU/APL) and solar wind velocity (MIT) measured on IMP-8 during a $Z \geq 3$-rich event (Zwickl et al., 1977).

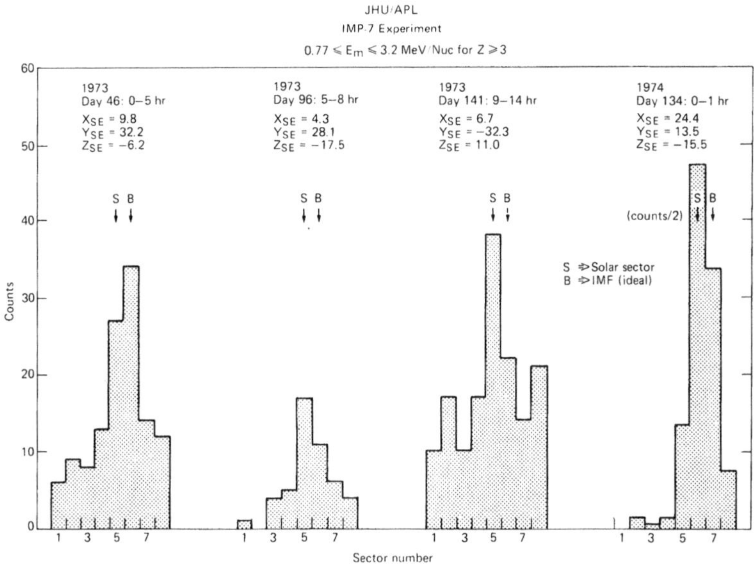

Figure 4. Eight-sector anisotropy histograms for four $Z \geq 3$-rich events showing strong outflow from the sun (Zwickl et al., 1977).

One then asks if these events originate from any characteristic type of region on the sun. A study of 10 events by *Zwickl et al. (1977)* is summarized in Table 1. The events overwhelmingly lack any association with significant flares, but 7 out of the 10 occur while solar wind velocities are quite low (compared to the > 700 km s^{-1} recurrent streams typical of these years). *Bame et al. (1975, 1977)*, have noted that the solar wind He abundance is most variable and that Si and Fe show correlated variations in low-speed solar wind. Since the solar wind O, Si and Fe abundances are, within error, the same as those determined for the lower corona and photosphere, the same $Z \geq 3$ rich regions in the lower solar atmosphere may serve as a common source for the energetic particles and low speed solar wind, even though the altitudes of the particle acceleration and plasma heating may still differ.

Thus by identifying the low-speed signature of the solar wind during these events, we may have extracted a significant clue as to the nature of their acceleration.

3. Interplanetary Propagation of > 10 MeV Protons: Coronal Connection Effects Near 5 AU

Before discussing the observations in this and the next section, we briefly illustrate in Figure 5 how solar wind stream structure affects the longitude at which the IMF line from earth (or any spacecraft) enters the corona. During the rise of the velocity at the stream leading edge,

TABLE 1. $Z \geq 3$ RICH EVENTS FROM SEPTEMBER 1972 - JANUARY 1976

Year	Event Duration	Flare Assoc.	Particle Data[1] $\gamma(P)$	$\gamma(A)$	$\gamma(Z)$	P/α	α/M	Solar Wind[2] (km/s)		Connection Longitude (°)	Enrichment[3]
72	302:00 - 304:00	?	2.4	2.0	1.4	25	3.0	360 - 410	S(i)	~ 40 → 15	Fe
73	46:00 - 47:18	?	2.7	3.2	2.0	5	7.1	325 - 380	D,S	30 → 10	Fe, ^3He
73	96:12 - 99:00	?	2.8	1.5	2.2	15	0.6	(?) - 425	D(i)	90 → 70	CNO
73	141:06 - 142.12	?	2.2	2.3	2.2	11	0.8	500 - 750	T	190 → 180	CNO
73	179:12 - 181:15	?	3.0	4.5	1.5	4	1.9	475 - 750	S,T	45 → 30	Fe, ^3He
73	248:15 - 250:12	?	2.5	3.6	2.9	4	6.2	375 - (?)	?(i)	250 → 230	Fe, ^3He
74	51:12 - 52:09	?	2.9	3.5	1.6	6	2.2	350 - 415	S	200 → 100	Fe, ^3He
74	127:08 - 130:15	1B,W38 ?	2.4	3.6	1.5	8	1.7	450 - 750	T	230 → 210	Fe, ^3He
74	133:15 - 136:12	2N,W66	3.6	3.6	1.2	25	0.8	350 - 410	S	180 → 140	Fe
74	335:12 - 336:12	?	3.1	—	1.6	34	0.3	320 - 350	S	60 → 40	Fe

1 Particle data are derived from ~ 6 hour average at t_{max}. $\gamma(P)$, $\gamma(A)$, $\gamma(Z)$ refer to spectral index for 1.85–7.8 MeV protons, 1.74–11.5 MeV/nuc alpha, and 0.77–3.2 MeV/nuc $Z \geq 3$, respectively. P/α is the proton to alpha ratio at 1.74 to 4.3 MeV, and α/M is the alpha to $Z \geq 3$ ratio at 1.6 to 4.3 MeV.

2
S = Solar wind stream onset and early portion of rise.
D = Solar wind stream decay with nearly constant high coronal connection longitude
T = Temporal solar wind stream superimposed on a large stream
i = Incomplete solar wind data

3 Fe and CNO rich designations are derived from scatter plots in Figure 2. ^3He designations are taken from *Hurford et al.* (1975a,b) and *Hovestadt et al.* (1975).

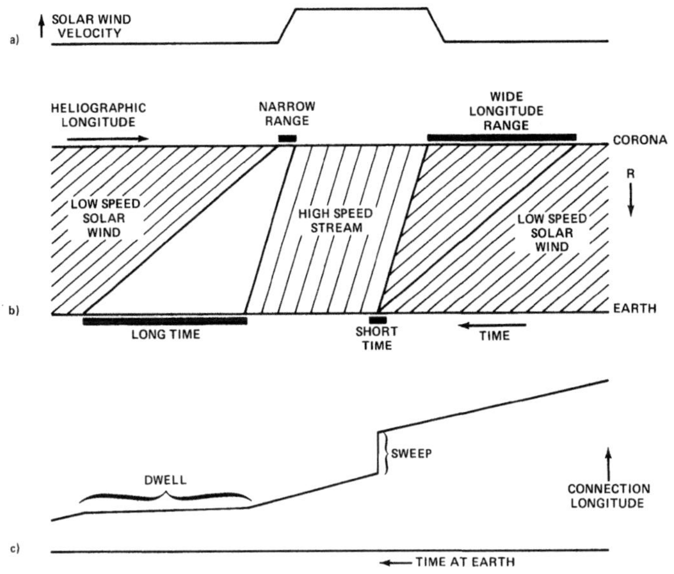

Figure 5. Effect of stationary solar wind stream structure on the coronal connection longitude of the interplanetary magnetic field lines at earth.

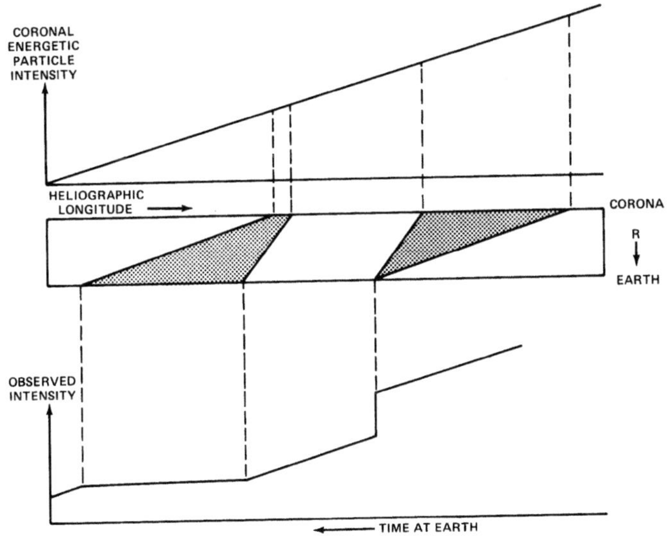

Figure 6. Reconstruction of a simple coronal longitude distribution from a time history of energetic particles at earth "deformed" by solar wind stream structure (Gold et al., 1977).

the source of the observed plasma (and hence the "frozen-in" IMF) "sweeps" rapidly over a large longitude interval to the east in the corona, whereas during the decay of the stream, the coronal connection "dwells" for days at almost the same heliolongitude. The "sweeps" in longitude cover an interval which is proportional to the distance of the observer from the sun. From earth, the sweeps jump ~ 30°, while at Jupiter, a sweep easily can jump half-way around the sun.

The drastic distortion of a very smooth coronal energetic particle injection longitude distribution by sweeps and dwells of the IMF connection longitude is illustrated in Figure 6 for the same solar wind stream structure depicted in Figure 5. One assumes, for quasi-stationary solar particle fluxes (as a first approximation), that the particle intensity gradients along the IMF are not large, as is consistent with multi-spacecraft observations *(Roelof, 1976)*.

The distortion effect is revealed dramatically in the recent analysis of Pioneer 10 and 11 observations by *Hamilton (1977)*. We are extending our domain of discussion above 10 MeV in order to include these most interesting observations. In Figure 7 (the first of three taken from his work), the coronal connection longitude of Pioneer 10 (pluses, bottom panel) during the September 7, 1973 event observed at 4.69 AU is nearly constant over 12 days, i.e., Pioneer 10 is in a lengthy solar wind "dwell". The 11-20 MeV (upper panel) and 30-67 MeV (middle panel) proton intensity histories are smooth and "classical". However, during the November 3, 1973 event shown in Figure 8, there is a pair of velocity increases at Pioneer 10 (4.92 AU), probably the forward and reverse shocks bounding an interplanetary magnetic "interaction region" *(Smith and Wolfe, 1976)*. The Pioneer 10 connection longitude "sweeps" eastward twice, first ~ 60° and then ~ 40°, each time bringing the connection longitude closer to the flare site. The 11-20 MeV proton fluxes also jump each time, just as expected for a smooth coronal gradient away from the flare site (compare with Figure 6). The same effect is apparent at 6.08 AU in the Pioneer 10 fluxes of both 11-20 MeV and 30-67 MeV protons in the November 5, 1974 event shown in Figure 9.

In both the November 3, 1973 and November 5, 1974 flare events, if one did not recognize the signature of the solar wind "sweeps", one would have assigned an erroneously late time of flux "maximum" to the Pioneer 10 protons. If further, one had forced a diffusion-model "fit" to the uncompensated data, the result would have been an erroneously small parallel diffusion coefficient.

The method for reconstructing the "coronal profile" of particle fluxes was first presented by *Roelof and Krimigis (1973)*. It consists simply of using the observed solar wind velocity to calculate the IMF coronal connection longitude and then assigning the particle flux observed at that time to that longitude. Thus, referring to Figure 10, if the solar wind stream structure in Figure 10(b) or 10 (c) distorted the simple coronal profile shown in 10(a), "mapping" the distorted flux histories using the observed solar wind velocities would re-create the

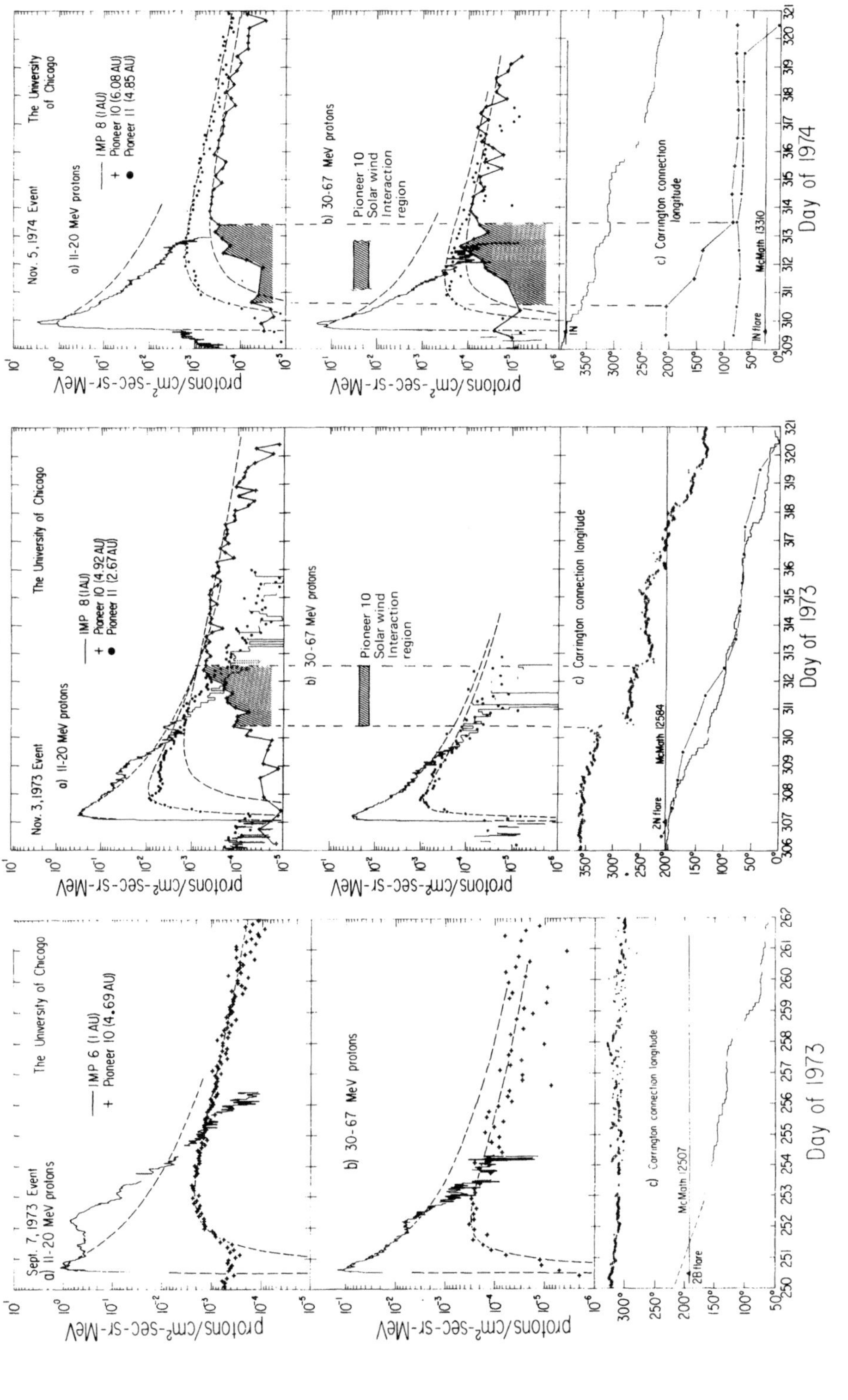

Figures 7, 8 and 9. Intensity histories of three solar proton events observed near earth and in deep space (adapted from Hamilton 1977). Connection longitudes are estimated from observed solar wind velocities; shading marks times of large "sweeps" in longitude which affect particle time histories.

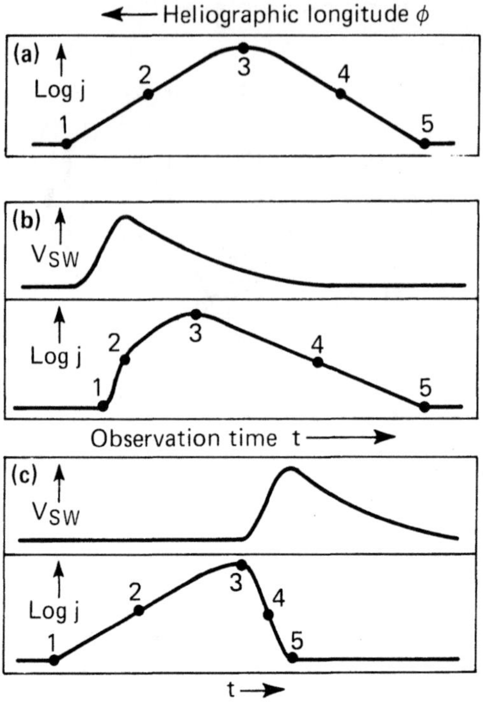

Figure 10. Examples of solar-wind/connection-longitude effects on a symmetric coronal particle injection profile: (a) velocity constant; velocity rise (b) early or (c) late in particle event. (Gold et al., 1977).

simple stationary profile. In other words, the computed connection longitudes provide the "key" to the coronal profile "encoded" by the varying solar wind velocity.

Gold et al. (1977) have carried out this "decoding" of particle events using the JHU/APL particle and the MIT solar wind measurements on IMP-7 and 8 from the launch of IMP-7 in 1972 into 1976. The result, shown in Table 2, of analyzing 1305 days of particle fluxes is that during an average of almost 70% of the time, particle flux histories at 1 AU are dominated by spatial distributions in the corona such as: rounded (quasi-stationary) longitude profiles; coronal gradients away from flare sites (established during the decay of flare event fluxes); and broad minima at longitudes between quasi-stationary distributions.

4. Coronal Propagation of > 7.5 MeV Flare Protons

Reconstruction of coronal longitude particle profiles using solar wind velocities from the same spacecraft requires special circumstances to separate spatial from temporal variations (e.g., solar wind sweeps

TABLE 2. Summary of Solar Particle Event Characteristics

Year	Data Days	Number of Events: Rounded Profiles	Decays Gradients	Broad Minima	Stream Association: Round	Minima	Spatial *: Days	(%)
1972	80	4	2	0	2	0	40	50
1973	365	17	6	0	16	0	255	62
1974	365	13	5	2	10	0	215	59
1975	360	16	3	6	13	5	267	74
1976	135	7	1	1	5	1	108	80
72-76	1305	57	17	9	46	6	892	68

* Total number of days (and fraction of observing period) on which rounded profiles, decay gradients and broad minima were identified.

identify longitudinal gradients, while solar wind dwells identify temporal flux changes at a constant coronal longitude). However, when observations of both energetic particles and solar wind velocity are available from well-separated spacecraft, much of the spatial/temporal ambiguities are removed.

Extraordinary circumstances existed in 1969, with Pioneers 6, 7, 8 and 9 distributed over nearly 180°. *McCracken et al. (1971)* and *Keath et al. (1971)* studied flare proton (> 7.5 MeV) events during this period, analyzing the events in terms of the spacecraft longitudes. Recently *Reinhard et al. (1977)* have added the JHU/APL > 10 MeV proton measurements from Explorer 34 and ordered the data in heliolongitude for the flare event of April 10, 1969 by calculating coronal emission longitudes from the solar wind measured on each spacecraft.

The time histories are shown in Figure 11 and one can note the considerable disparities among them. When they are "mapped back" as shown in Figure 12(a) for the initial three days of the event, the disparities order themselves into a remarkable transition (pointed out by *McCracken et al., 1971*), from intensities peaked over the flare site 7 hours after the flare, to a peak nearly two orders of magnitude ~ 100° west of the flare site. Beginning on April 13, the fluxes relax into a nearly perfect exponential longitude distribution with an e-folding scale ~ 35° late in the event. This is almost exactly the idealized longitude distribution shown in Figure 6.

The event becomes even more ordered when the time histories at constant coronal longitude are constructed in Figure 13 (a). These histories are as close as we can come with this technique to flux histories on the "same" field line discussed in idealized theoretical analyses such

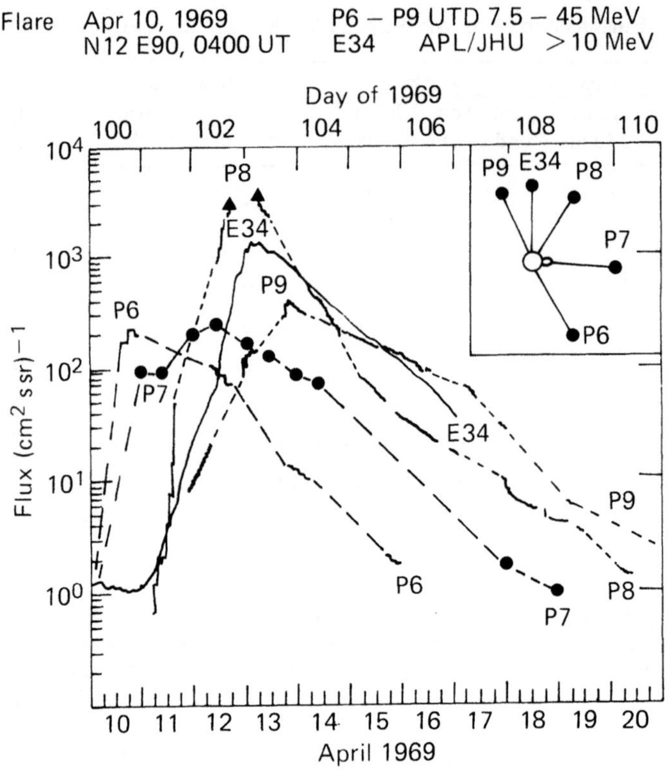

Figure 11. Intensity histories of a ~ 10 MeV solar proton event measured at five spacecraft over 180° longitude in April, 1969 (after McCracken et al., 1971).

as that of *Ng and Gleeson (1976)*. They clearly reveal a second flux increase at all longitudes from the flare site at 60° to 180°. The decay histories are similar, and moreover they are strikingly well fit at each longitude by the function $(t-t_o)^{-k}$ where t_o is 00 UT, April 11 and $k \simeq 4$ for all longitudes.

Comparison of the spacecraft time histories of Figure 11 with the constant helio-longitude histories of Figure 13 demonstrates the necessity of "decoding" the former to obtain the latter. Many portions of the spacecraft histories nearly could be fit piece-wise by exponential distribution in coronal longitude, but all the while the decay at any fixed heliolongitude was much closer to a power law in time.

McKibben (1972), in his compilation of multi-spacecraft observations, fit piece-wise exponential segments to the spacecraft histories. An example, the flare event of December 16, 1967 is shown in Figure 14. However, we note that the breaks in the flux history on December 17 and

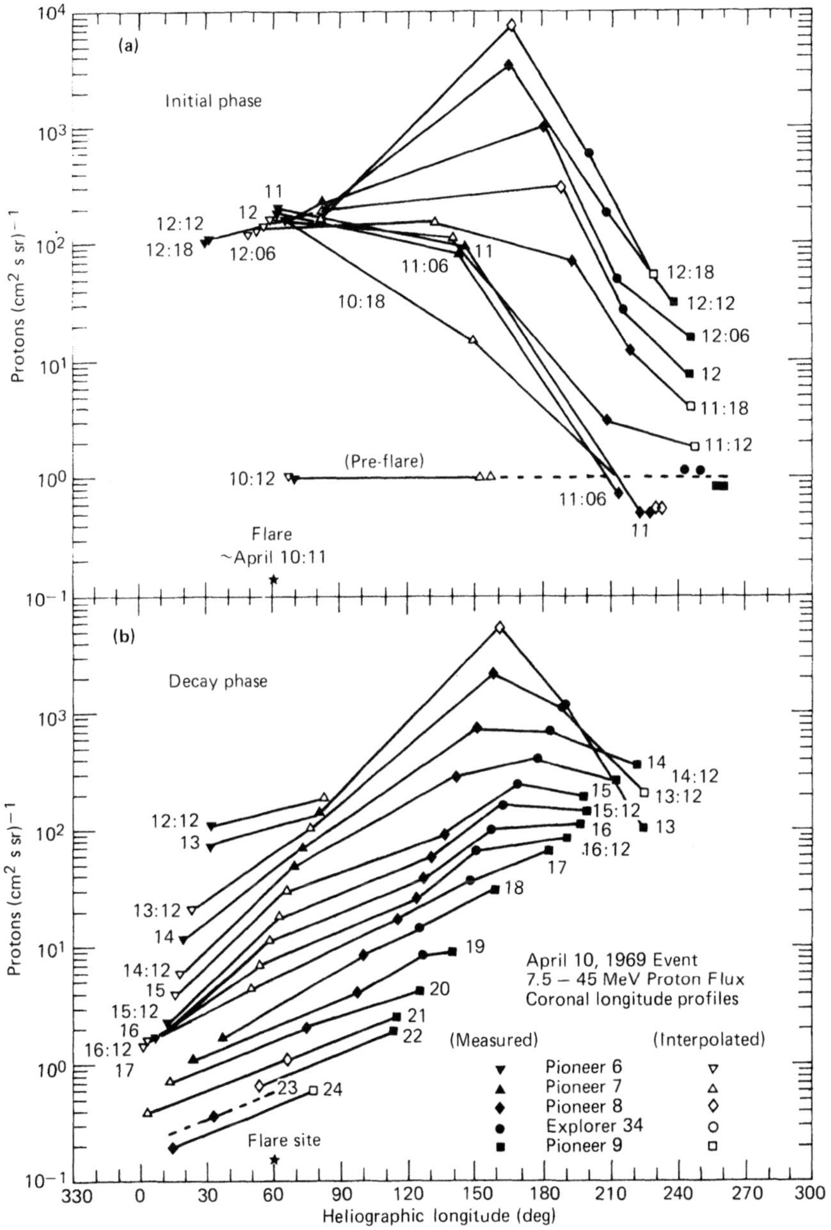

Figure 12. Coronal intensity profiles reconstructed from connection longitudes estimated from solar wind measurements on five spacecraft: (a) initial phase April 10-12, 1969; (b) decay phase April 13-24, 1969 (Reinhard et al., 1977).

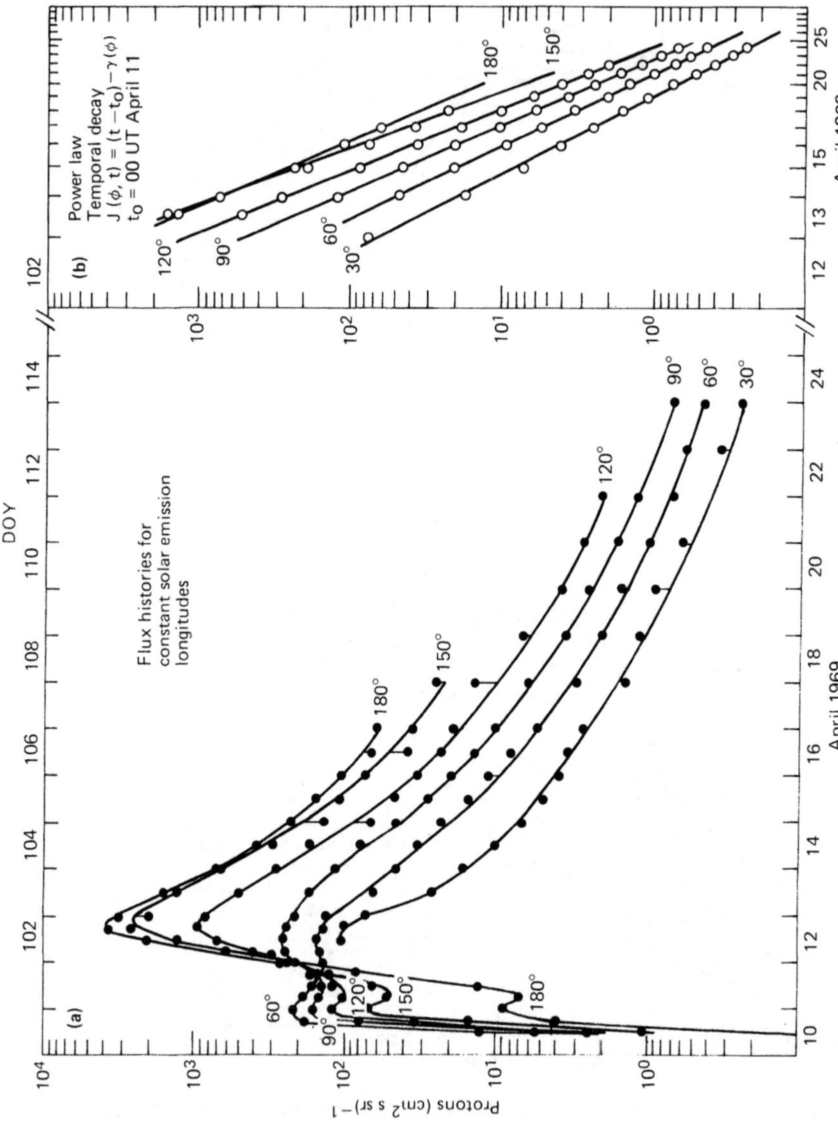

Figure 13. Coronal ~ 10 MeV proton intensity time histories (a) at fixed heliographic longitudes constructed from the coronal intensity profiles of Figure 12; (b) gives the same histories on a logarithmic time scale (Reinhard et al., 1977).

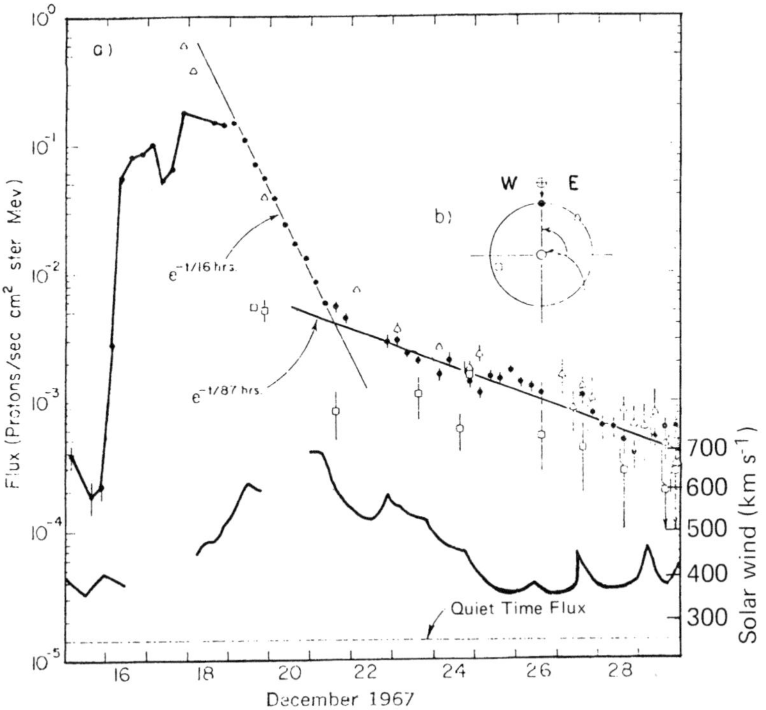

Figure 14. (a) Time-intensity profiles of 15 to 18.7 MeV proton fluxes observed by University of Chicago telescopes on Pioneer 6 (squares), Pioneer 7 (triangles), and IMP-4 (circles) during the December 16, 1967, solar-flare event. (b) Positions of the spacecraft relative to the flare and to the ideal interplanetary spiral field line (assuming a 400 km/sec solar wind) rooted at the flare site. Solar wind velocity (from NSSDC) rises during rapid drop in flux. After McKibben (1972).

December 21 correspond to the beginning and end of a solar wind sweep. We shall see in the next section that the solar wind stream structure has a more extensive effect on particle fluxes than simply changing connection longitudes.

5. Decay Phase Anisotropies of Solar Flare Events

We base this section on the extensive analysis of solar flare proton events ≥ 0.3 MeV from Explorer 35, 1967-1970, by *Innanen and Van Alle* *(1973)*. The statements concerning the two events we show here apply generally to all seven which they studied.

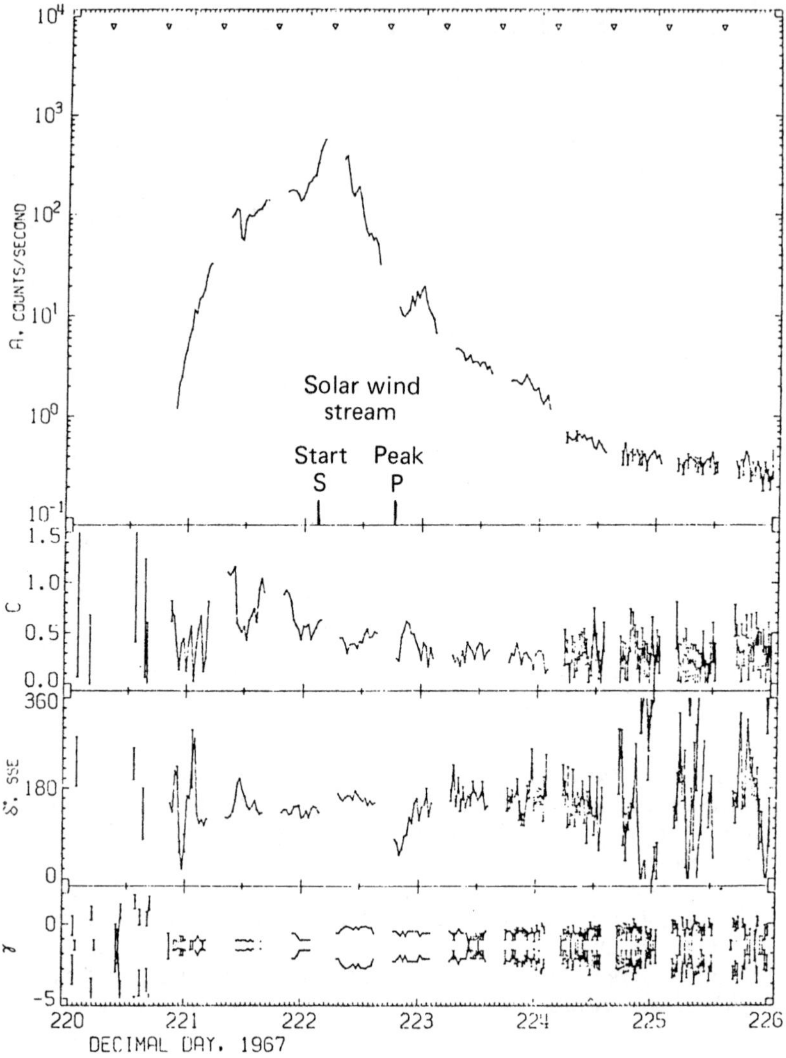

Figure 15. Intensity (A), anisotropy amplitude (C) and direction (δ), and differential spectral exponent (γ) for > 0.3 MeV protons from the University of Iowa detector on Explorer 35 for the flare event beginning August 9, 1967. Start (S) and peak (P) of the solar wind disturbance mark the discontinuity in the intensity history. After Innanen and Van Allen (1973).

In Figure 15, the flux, first harmonic anisotropy magnitude and direction and spectral index are shown for the event of August 9, 1967. We have indicated the start (S) and the peak (P) of the solar wind increase which reached earth August 11-12. Note that the sweep corresponds to the abrubt decrease in flux. Of greater interest is the effect upon the anisotropy. In Figure 16, Innanen and Van Allen's anisotropy analysis is presented for the same event of August 9, 1967 and also for the much-studied event of November 18, 1968. Not only does the "rise" phase of both events terminate at the start (S) of the solar wind increase but so does the "field-aligned" phase. The angle α between the 6-hour averaged IMF and flux first harmonic phase begins to (and continues to) deviate widely from its previous value near zero. Moreover, as can also be seen, the amplitude of the anisotropy is significantly reduced.

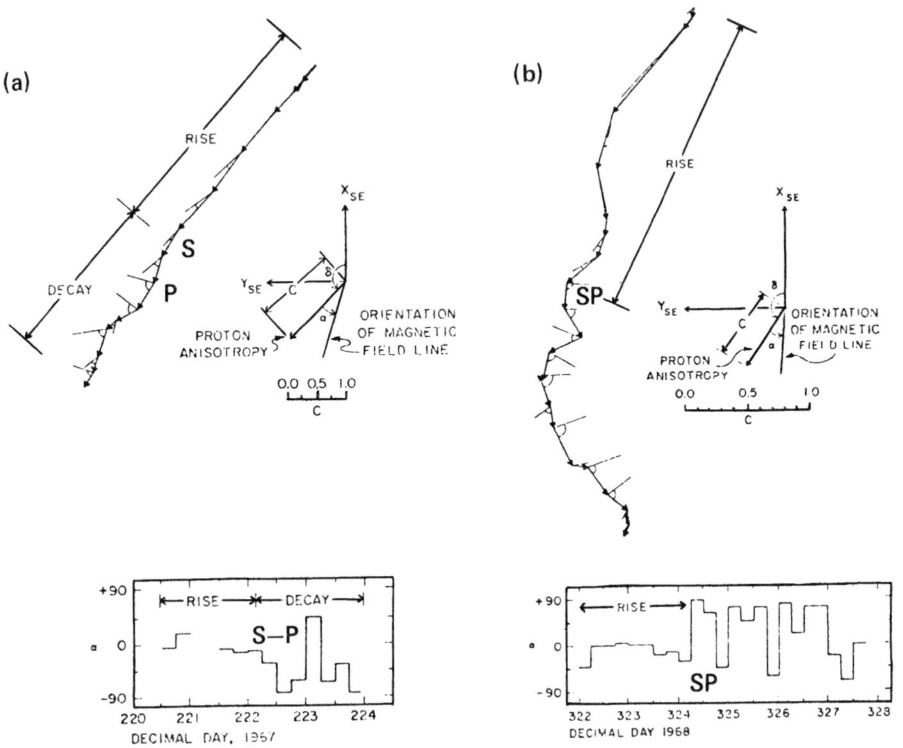

Figure 16. Anisotropy vector plot (6-hour averages in solar-ecliptic coordinates) of >0.3 MeV protons (see Figure 15) with start (S) and peak (P) of solar wind disturbance corresponding to end of field-aligned phase (α=0). Events (a) August 9, 1967 and (b) November 18, 1968. After Innanen and Van Allen (1973).

A similar association is found in all seven events between the arrival of the flare-heated plasma and the transition to the "decay" phase in which the particle anisotropy is reduced and ceases to be field-aligned.

The decay phase of solar events has been a great puzzle, and no satisfactory explanation has been offered for the tendency of the particle anisotropy to become radial (or sometimes more easterly as seen in > 7.5 MeV protons), and essentially insensitive to the local direction of the IMF *(Allum et al., 1974)*. What has emerged here is that the arrival of the flare-heated solar wind disturbance, which is a *concommitant* of most flare events, actually controls the end of the "field-aligned rise" phase of the event and the transition to the "radial" and later "field-independent" phase. It is therefore not surprising that the diffusion model "fits" to the events using an undisturbed medium could not simultaneously match the intensity and anisotropy histories *(Ng and Gleeson, 1971)*. What is actually happening is that the flare plasma stream-stream interaction with the slower ambient plasma ahead of it inhibits the propagation of particles through it, reducing the anisotropy behind it. The subsequent history of particles behind the interaction is often distorted by the change in flux when the fast solar wind "sweeps" the connection longitude to a region with different intensities than on those field lines in the preceding slow solar wind (see Figure 6).

CONCLUSIONS

There is little more to say; the data really tell the story. Solar energetic particles are so responsive to the dynamic state of the corona, the interplanetary medium and even the magnetosphere, that we make surer progress toward an eventual understanding of the origin of these particles by studying these concommitant interactions as carefully as the acceleration and "undisturbed" interplanetary propagation of the particles themselves.

ACKNOWLEDGEMENTS

This research was supported by the Atmospheric Research Section of the National Science Foundation (Grant ATM-76-23816) and by NASA and the Air Force Geophysics Laboratory under Contract N00017-72-C-4401 between The Johns Hopkins University and the Department of the Navy.

REFERENCES

Allum, F. R., Palmeira, R. A. R., Rao, U. R., McCracken, K. G., Harries, J. R. and Palmer, I.: 1971, *Solar Phys., 17,* 241.

Allum, F. R., Palmeira, R. A. R., McCracken, K. G., Rao, U. R., Fairfield, D. H. and Gleeson, L. J.: 1974, *Solar Phys., 38,* 227.

Bame, S. J., Asbridge, J. R., Feldman, W. C., Montgomery, M. D. and Kearney, P. D.,: 1975, *Solar Phys.*, *43*, 1975.
Bame, S. J., Asbridge, J. R., Feldman, W. C. and Gosling, J. T.,: 1977, *J. Geophys. Res.*, *82*, 1487.
Earl, J. A.: 1974, *Astrophys. J.*, *188*, 379.
Fairfield, D. H.: 1971, *J. Geophys. Res.*, *76*, 6700.
Garrard, T. L., Stone, E. C. and Vogt, R. E.: 1973, in R. Ramaty and R. G. Stone (eds.), *High Energy Phenomena on the Sun, NASA SP-342*, 341..
Gold, R. E., Krimigis, S. M. and Roelof, E. C.: 1977, *Proc. 15th International Cosmic Ray Conference (Plovdiv, Bulgaria)* in press.
Hamilton, D. C.: 1977, *University of Chicago Preprint*.
Hovestadt, D., Klecker, B., Vollmer, O., Gloeckler, G. and Fan, C. Y.: 1975, *Proc. 14th International Cosmic Ray Conference (Munich)*, *5*, 1613.
Hurford, G. J., Mewaldt, R. A., Stone, E. C. and Vogt, R. E.: 1975a, *Astrophys. J. Lett.*, *201*, L95.
Hurford, G. J., Stone, E. C. and Vogt, R. E.: 1975b, *Proc. 14th International Cosmic Ray Conference (Munich)*, *5*, 1626.
Innanen, W. G. and Van Allen, J. A.: 1973, *J. Geophys. Res.*, *78*, 1019.
Keath, E. P., Bukata, R. P., McCracken, K. G. and Rao, U. R.: 1971, *Solar Phys.*, *18*, 503.
Lin, R. P.: 1970a, *J. Geophys. Res.*, *75*, 2583.
Lin, R. P.: 1970b, *Solar Phys.*, *15*, 453.
Marshall, F. E. and Stone, E. C.: 1977, *Geophys. Res. Lett.*, *4*, 1977.
McCracken, K. G., Rao, U. R., Bukata, R. P. and Keath, E. P.: 1971, *Solar Phys.*, *18*, 100.
McKibben, R. B.: 1972, *J. Geophys. Res.*, *77*, 3957.
Ng, C. K. and Gleeson, L. J.: 1971, *Solar Phys.*, *20*, 166.
Ng, C. K. and Gleeson, L. J.: 1976, *Solar Phys.*, *46*, 347.
Ng, C. K. and Roelof, E. C.: 1977, *J. Geophys. Res.* (to be submitted).
Price, P. B., Hutcheon, I., Cowsik, R. and Barber, D. J.: 1971, *Phys. Rev. Lett.*, *26*, 916.
Roelof, E. C.: 1969, in H. Ogelman and J. R. Wayland (eds.), *Lectures in High Energy Astrophysics, NASA SP-199*, 111.
Roelof, E. C.: 1976, in D. J. Williams (ed.), *Physics of Solar Planetary Environments*, Proc. of the International Symposium on Solar-Terrestrial Physics (Boulder), *American Geophysical Union*, *1*, 214.
Roelof, E. C. and Krimigis, S. M.: 1973, *J. Geophys. Res.*, *78*, 5375.
Reinhard, R., Roelof, E. C. and Gold, R. E.: 1977, *J. Geophys. Res.*, (to be submitted).
Sarris, E. T., Krimigis, S. M. and Armstrong, T. P.: 1976, *J. Geophys. Res.*, *81*, 2341.
Smith, E. J. and Wolfe, J. H.: 1976, *Geophys. Res. Lett.*, *3*, 137.
Zwickl, R. D., Krimigis, S. M., Gold, R. E., Roelof, E. C. and Armstrong, T. P.: 1977, *Proc. 15th International Cosmic Ray Conference (Plovdiv, Bulgaria)*, in press.

DISCUSSION

Morfill: When you extract the longitudinal intensity variation from your data (during solar wind streams) is the result consistent with a longitude-independent diffusion coefficient (in interplanetary space)?

Krimigis: Referring to the April 1969 event (Figure 13(b)), the decay phases at a given heliographic longitude all show a power-law time dependence with approximately the same exponent over longitudes from 30° to 150°. Since the influence of coronal propagation is minimized in this late phase (> 3^d after the flare), the inference would be that there is no dependence on solar longitude for interplanetary propagation over these 120°.

Smart: In your Figure 3 you show a heavy ion event that occurs 1 or 2 days before the observation of a high speed solar wind stream. Is it possible the heavy ion event is the result of a high speed solar wind stream that "shorts out" the coronal propagation path (by producing a stream-stream interaction with the slower preceding solar wind), so that what you observed is the result of a solar wind switch?

Krimigis: We have given some thought to the possibility of such a "shorting out" of the connection longitude during an evolving flare-associated solar wind disturbance. However it is much less likely to occur in a co-rotating stream, and the velocity rise on day 53, 1974 (the event in Figure 3), is not flare-associated (Table 1). It is corotating, being part of the highly recurrent series which characterized 1974.

Wibberenz: The observer may be located at very different positions with respect to an interaction region, when a particle flare occurs. Why should this region hit the observer always late in the decay phase — in order to explain the easterly anisotropy?

Krimigis: By the "interaction region", I take it you mean the leading edge of the flare-associated solar wind disturbance. Since the flare also accelerated the energetic particles, the disturbance (which has a broad front in longitude), will usually reach earth 2 to 3 days after the particle increase, i.e., in the decay phase. The data reveal that it is at this time that the field-aligned anisotropy terminates; the 300 keV protons do not necessarily exhibit an "easterly anisotropy" after the solar wind disturbance.

Wibberenz: I would like to comment on the statement that the diffusion picture does not work if the anisotropies are large. For > 10 MeV protons the diffusion concept is generally valid, and large anisotropies merely indicate long-lasting solar injection. During an individual event the diffusion coefficient may vary by about a factor of 2.

Krimigis: Large anisotropies (> 0.50) cannot be consistent with diffusion-dominated transport. They imply that the time history is essentially the source injection history as long as the large anisotropy persists. We seriously question whether the diffusion concept has truly been proven valid for > 10 MeV protons since the severe distortions introduced by coronal injection gradients and solar wind disturbances (Figures 12 and 14) have not been properly considered in fitting diffusion parameters to event time histories.

ACCELERATION AND MODULATION OF ELECTRONS AND IONS BY PROPAGATING INTERPLANETARY SHOCKS

T. P. Armstrong and G. Chen
Dept. of Physics and Astronomy, U. of Kansas
Lawrence, Kansas 66045, U.S.A.

E. T. Sarris
Max Planck Institut für Aeronomie
D-3411 Katlenburg-Lindau 3, F.R.G.

S. M. Krimigis
The Johns Hopkins University Applied Physics Laboratory
Laurel, Maryland 20810, U.S.A

ABSTRACT

The theory and observations of spikes and discontinuities in charged particle fluxes associated with propagating interplanetary shocks are reviewed. The physical mechanisms that produce energy changes and modification of angular distributions are examined. The current status of in situ observations is reviewed.

I. INTRODUCTION

Shock spike events are one of the most thoroughly observed natural phenomena involving charged particle energization. Also, since this tends to be an isolated process occurring in comparatively simple electromagnetic fields, the physical models explain many of the significant characteristics of the energy spectra and angular distributions. Sensitive, high time and angular resolution particle and field measurements can be made in the midst of the acceleration region and thus shock spikes represent an excellent test of theories that may have extensions to other important astrophysical processes such as solar flares and super-nova explosions.

The shock spike process often occurs in a larger, longer duration, charged particle enhancement, the energetic storm particle (ESP) event. While it is clear that the ESP event is closely connected to the shock

M. A. Shea et al. (eds.), Study of Travelling Interplanetary Phenomena/1977, 367–389.

and may in fact be just the cumulative effect of the shock spike interaction acting for a prolonged period, we will not attempt to deal with ESP events here because to do so properly would require a review of interplanetary propagation. Therefore, in the limited space available here, we will attempt to review in a comprehensive manner the subject of shock spikes in theory and observation.

II. PHYSICAL CHARACTERISTICS OF CHARGED PARTICLE ORBITS THROUGH MHD DISCONTINUITIES

There are two qualitatively different regimes depending principally on the angle (β) between the magnetic field and the shock surface. If β is 0° or small $\leq 5^\circ$, all of the particles incident on the shock penetrate it and are transmitted into the post-shock regime. We will refer to this as the "perpendicular" (shock normal $\hat{n}$ perpendicular to magnetic field $\vec{B}$) regime. The regime $5^\circ \leq \beta \leq 10^\circ$ is characterized by the appearance of high energy reflected particles and will be referred to as "small angle oblique". The $10^\circ \leq \beta \leq 90^\circ$ is both fairly uncommon for interplanetary shocks and uninteresting in that it produces very little energy change of either transmitted or reflected particles. We will refer to this as the "oblique" regime.

The orbits of charged particles through perpendicular shocks, $\vec{B}\cdot\hat{n} = 0$, where $\hat{n}$ is the shock normal, have been studied previously (Parker, 1958; Shabanskii, 1962; Schatzman, 1963a,b; Hudson, 1965; Chen and Armstrong, 1972; Sarris and Van Allen, 1974).

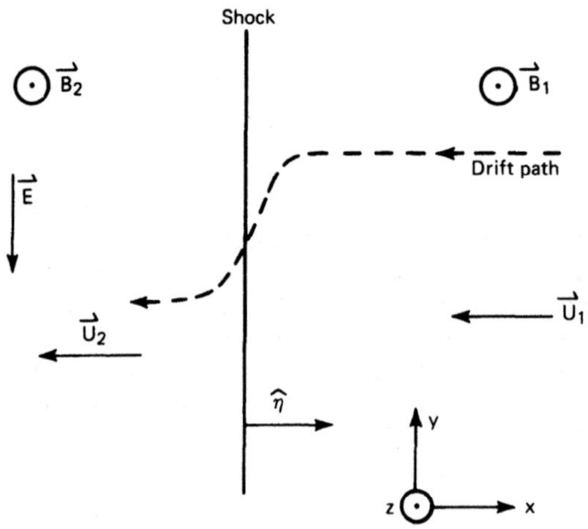

Figure 1. Schematic diagram of perpendicular shock geometry.

The rest frame of a perpendicular shock is shown in Figure 1 where $\vec{B}$, $\vec{U}$ are the magnetic field and plasma velocity. Subscript 1 refers to preshock and 2 to post-shock, $\hat{n}$ is the shock normal and points in the direction of shock propagation. The magnetic field is taken to be at rest with respect to the plasma and therefore the field lines move with respect to the shock surface. The effect of field line motion on particle trajectories can be represented in the shock rest frame by introducing an induced, spatially uniform, electric field $\vec{E} = -\vec{U} \times \vec{B}/c$. This field produces a drift of the particles equal to the plasma velocity in the shock rest frame that causes all particles to cross the shock. While the particle is within a gyroradius of the shock surface, it crosses and recrosses the shock thereby experiencing an effective "grad B" drift in the y-direction. On the average the particle gains an amount of energy in the transverse component of its motion given by $W_1/W_2 = B_2/B_1$ (Parker, 1958; Hudson, 1965; Chen and Armstrong, 1972; Sarris and Van Allen, 1974). This suggests that, averaged over phase, particles moving through a perpendicular shock tend to conserve the first adiabatic invariant.

We note that all of the energy appears in the transverse velocity component and that the perpendicular shock interaction should produce an anisotropy in the shock rest frame with its magnitude determined by the slope of the pre-shock energy spectrum. When the particle flux is transformed into the Earth-Sun rest frame the Compton-Getting effect (Ipavich, 1974) will amplify the flux coming from the solar wind direction and diminish it in the anti-solar wind direction. One of the indicators of the interaction with a perpendicular shock is the strong anisotropy perpendicular to the magnetic field in the post-shock regime.

Several other characteristics of the perpendicular shock interaction may be inferred from elementary considerations. At low energies where the drift speed is comparable to the particle speed, the particle may not cross the shock more than once and the "average" invariance of $\mu = W/B$ conservation will not be present.

The amount of energization depends on the number of times the particle crosses the shock. Now, in the case of perpendicular shocks the number of crossings is determined entirely by how rapidly the particle drifts through the region of interaction with the shock. In order to achieve greater energization in a perpendicular shock, some authors have considered stochastic scattering as a mechanism to prolong the particle residence time (Schatzman 1963a,b). In the case of small angle oblique shocks, however, there is a dynamical condition that allows for prolonged interaction without scattering.

The orbits of particles in oblique shocks have been investigated by many authors (Hudson, 1965; Chen, 1975; Chen and Armstrong, 1975; Terasawa, 1976) and we will outline here some of the main qualitative and quantitative results. The most significant difference between exactly perpendicular shocks, $\vec{B}\cdot\hat{n} = 0$, and small-angle oblique where $\vec{B}$

is inclined to the shock surface by 5 to 10 degrees, is the appearance of particles reflected into the pre-shock region. This has a simple interpretation.

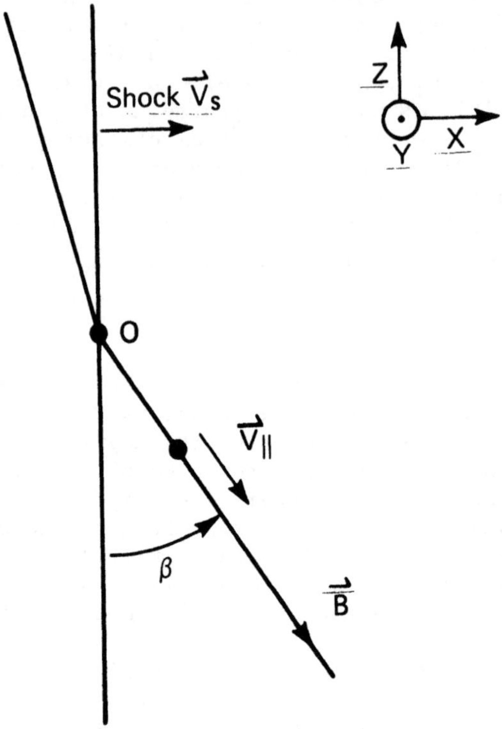

Figure 2. Schematic diagram of small angle oblique shock geometry.

Suppose that the geometry of an oblique shock is as shown in Figure 2. Now the particle moves along the field lines with a velocity $V_{||}$ and the shock moves to right (+ x direction) at some speed V_s with respect to the field lines. The point O at which a given field line enters the shock moves in the + x direction with a speed V_s or along the field line with a speed V_s csc β. If $\vec{V}_{||}$ for the particle points into the pre-shock medium and exceeds V_s csc β in magnitude, the particle can "out-run" the shock towards the upstream medium. For reflection, then, it is necessary for the particle to develop a large value of $\vec{V}_{||}$. Hence, at a given particle energy, the reflected particles in the pre-shock regime would tend to appear at small angles with respect to the magnetic field (Sarris and Van Allen, 1974; Chen and Armstrong, 1975; Chen, 1975) and one would expect a field aligned (in the shock frame) pre-shock particle anisotropy to result. Now a geometrical construction can be given that shows how $\vec{V}_{||}$ changes because of multiple shock crossings.

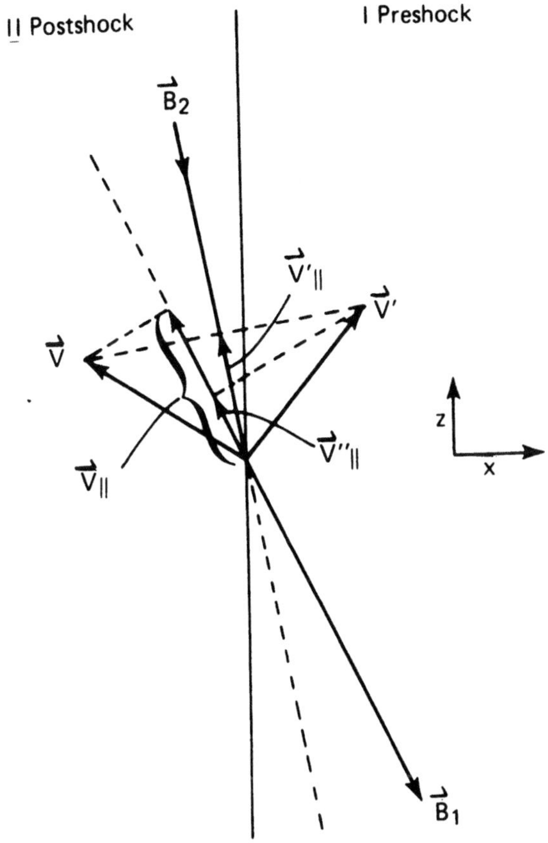

Figure 3. Velocity vector at successive crossings showing $V_{||}$

Figure 3 shows the X-Z plane containing the field lines for an oblique shock with the field line directions extended as dashed lines across the shock for reference. It is assumed that there is no Y or Z dependence of any variable and, although the particle moves in the Y and Z directions from crossing to crossing, the incidence and exit velocities can be referred to the same location. We consider first, for simplicity of visualization, the crossing geometry supposing the field lines to be stationary in the shock frame. Now $\vec{V}$ represents the X-Z plane projection of the particle velocity at some crossing (the Y component experiences no net change in a pair of crossings until field line motion is introduced and is not considered here) from region I, preshock, into region II, post-shock. $\vec{V}_{||}$ is the parallel component of the particle velocity referred to the pre-shock, $\vec{B}_1$, field before the first crossing. $\vec{V}_{\|}'$ is the parallel component referred to the post-shock field. Note the $\vec{V}_{\|}'$ is smaller in magnitude than $\vec{V}_{\|}$. When the particle re-enters region I, preshock, its X-Z velocity component is $\vec{V}''$ and the parallel component of velocity, referred to the preshock field is $\vec{V}_{\|}''$. Notice that the net change, $\Delta \vec{V}_{\|} = \vec{V}_{\|}'' - \vec{V}_{\|}$, of the parallel component of the velocity acts to increase the $\vec{V}_{\|}$ toward the pre-shock regime. This is the essential process that produces reflected particles. The motion of the field lines in the -X direction, however, limits the number of possible crossings. Hence, the ultimate fate of the particle, reflection or transmission, is determined by the competition between the drift due to field line motion and the evolution of $\vec{V}_{\|}$ due to multiple crossings.

A full examination of the geometry reveals (Chen, 1975; Terasawa, 1976) that particles entering the shock with very small pitch angles will easily penetrate and be transmitted while those with large pitch angles can be reflected. The illustration described above is only schematic; the actual orbit problem is more complex because field line motion and energy changes must be included. Under some circumstances it is possible to find a coordinate system wherein there is no magnetic field line motion (Hudson, 1965; Alekseyev and Kropotkin, 1970) and some aspects of the analytical formulation of the orbit problem are simpler. The essential difficulty remains, however, that repeated solutions of a transcendental equation are required to determine the particle orbit. Since this must be done numerically on a computer and the transformed coordinate system affords no improved economy, Chen (1975) used the shock frame of reference with the superposed field line motion in order to avoid the complication of applying the coordinate transformations.

We will now examine the salient features of particle orbits in oblique shocks using the results of Chen (1975). Dimensionless units were used in this simulation: time measured in $\omega_c^{-1} = mc/eB_1$ and velocity in units of shock speed, V_s, with respect to preshock plasma. Length units are then necesarily V_s/ω_c. For reasons that will be obvious later, the interesting regime of particle interaction with shocks includes the particle initial velocity range $10 \leq V_p < 100$, shocks of moderate strength $\vec{B}_2/\vec{B}_1 \sim 2$ with small normal component of magnetic field $B_x/B \sim 0.1$. Figures 4 and 5 show the X-Y particle coordinates versus time and kinetic energy versus time for a typical reflected and transmitted particle. First, note that there are a large number of crossings: 18 for the reflected particle and 9 for the transmitted one. Also, observe the displacement of the center of gyration of the particle in the -Y direction. The electrical field also points in the -Y direction and the source and mechanism of particle energization is clearly displayed. From the energy versus time plots it can be seen that the particle interacts with the shock for a long time, typically 5 gyroperiods for the transmitted particle and 10 for the reflected one. The duration of the shock interaction has been emphasized by Sarris and Van Allen (1974)

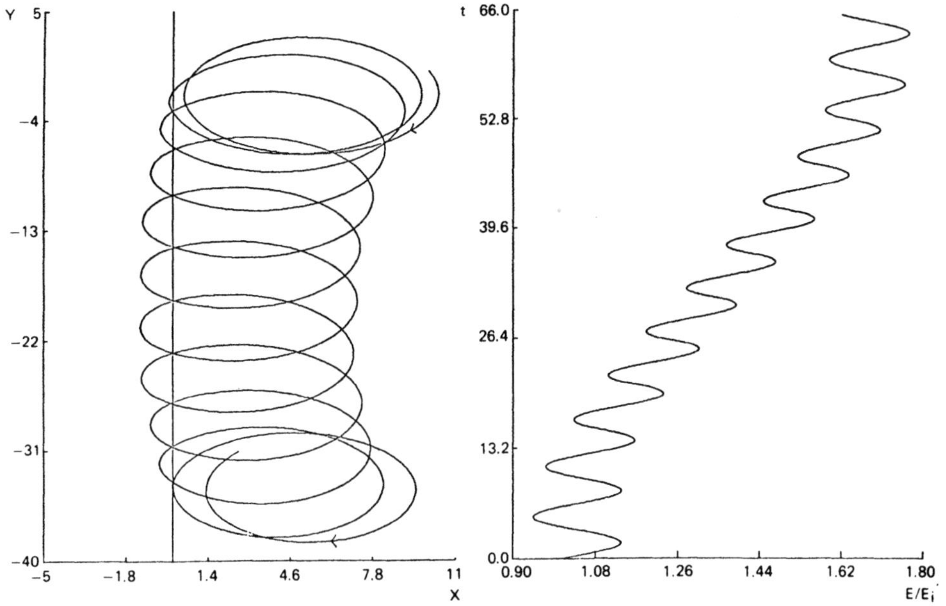

Figure 4. Orbit (a) and Energy (b) versus time for a typical reflected particle having $V_p/V_s = 10$, $B_2/B_1 = 2.$, and $B_x/B_1 = 0.1$. Left panel shows the motion in the x-y plane and the right the kinetic energy versus time.

and Chen and Armstrong (1975). For interplanetary shocks interacting with ∿1 MeV/nucleon ions this time scale is typically 3 to 5 minutes. This would be changed somewhat if evolution of the shock or scattering of particles from irregularities occured.

In order to characterize the change in a distribution of particles interacting with a shock, Chen used ensembles of particles with specified initial speed, each uniformly distributed on a given shell in velocity space. These particles were followed until the shock interaction was complete whereupon the post-interaction energy and pitch angle was determined. The initial phase space sphere was divided up into 4° strips of constant pitch and a number of particles was assigned to each strip according to the area of the strip. The particles were then spaced uniformly around in phase. This resulted in a close approximation to an initially isotropic distribution in the shock frame of reference. Ensembles of various sizes were tested and one with 1972 particles was selected for multiple runs to explore the dependences of final energies and pitch angle distributions on various parameters. Using more particles did not change the average energy gains and pitch angle distributions. A finer resolution study of the phase would yield slightly higher maximum energies.

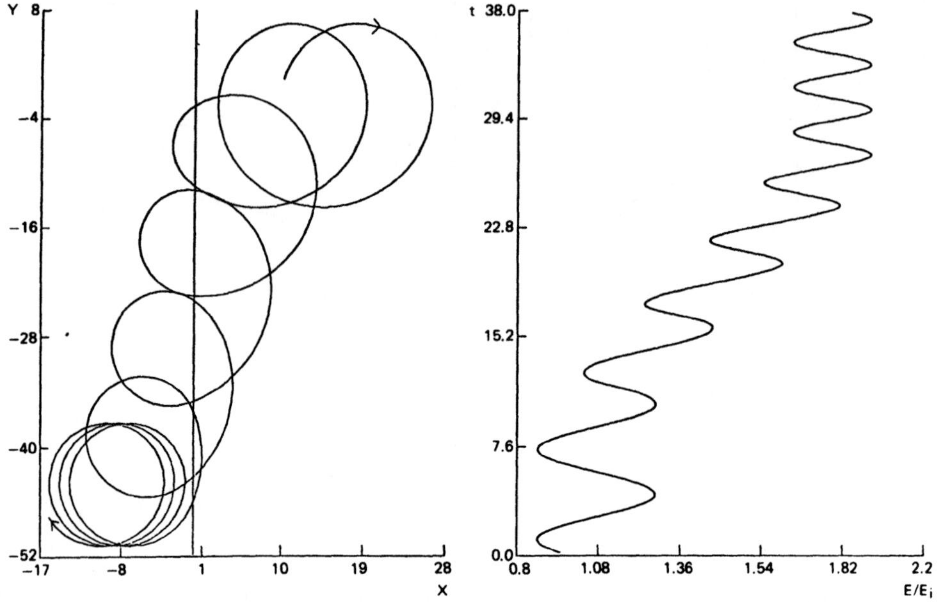

Figure 5. Orbit (a) and energy (b) versus time for a typical transmitted particle from the same ensemble as that shown in Figure 4. Only the initial pitch angle and phase angle are different.

Figure 6 summarizes the results for a run of 1972 particles of speed $V/V_s = 10$ in a shock of strength $\eta = B_2/B_1 = 2$ and normal component $B_x/B=0.1$ (preshock). The diagram shows a symbol for each particle at its initial pitch, θ, and phase angle ϕ. The symbols 0, 1, 2, ... stand for particles that were reflected from the shock with energies in the intervals $0<E/E_i<1$, $1<E/E_i<2$, ... respectively. The symbols +, A, B, C, ... are used for transmitted particles of energies $0<E/E_i<1$, $1<E/E_i<2$, etc. The asterisks denote particles that did not interact with the shock at all. The pitch angle θ is the angle between the velocity $\vec{V}$ and the forward directed $\vec{B}$ which point towards the preshock regime. The diagram of Figure 6 can be thought of as a map of the phase space with the energy gain or loss labelling each point. With this ensemble of 1972 particles, the resolution of this map is about 4°. By inspection of Figure 6 a number of important features of the shock interaction are evident. First, the highest energy particles, both transmitted and reflected, appear near the boundary in pitch and phase between the two regimes. Second, the reflected particles experience larger

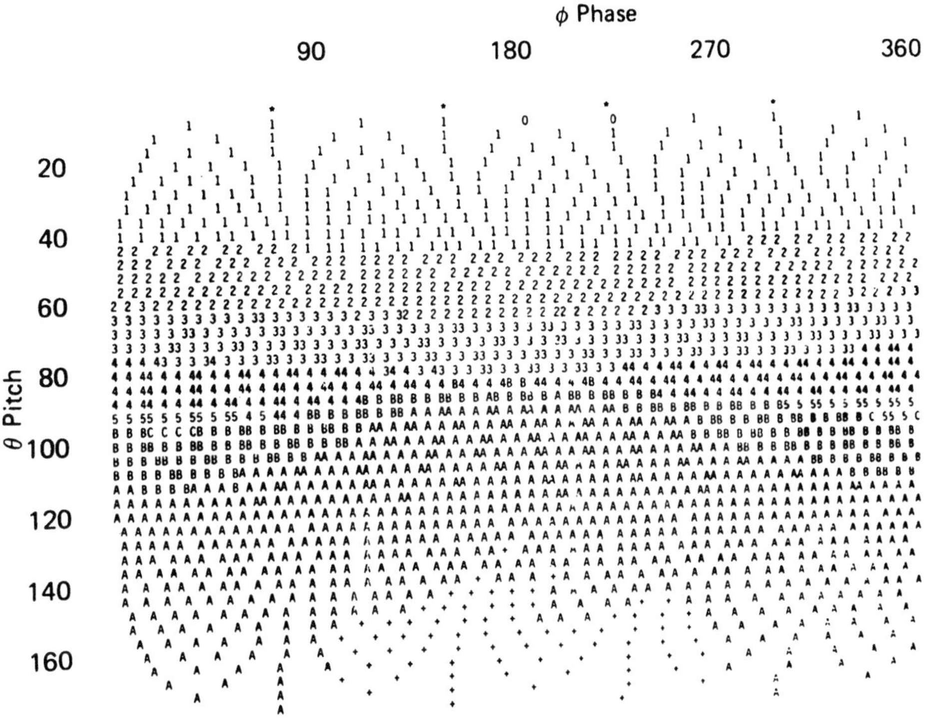

Figure 6. Distribution in initial pitch and phase for 1972 particles injected into a shock of strength 2, $B_x/B = .0.1$, and $V/V_s = 10$. See text for explanation of symbols. $\theta = 0^o$ points into the preshock regime.

energy gains. And third, the boundary is phase dependent so that both transmitted and reflected particles of high energy can appear at the same pitch angle. Hence, phase averaging that is used in many models as a simplification should be done with great care because much phase dependent structure appears for the highest energy particles. Before considering the resulting pitch angle distributions for this case, it is useful to examine the effect of scattering and magnetic irregularities on the shock interaction.

Chen (1975) has investigated irregularities by introducing random, small angle elastic scattering into the orbit solutions. The parameters of a typical scattering run were $\langle\omega_{SC}\rangle = 0.5\ \omega_C$ and $\langle\theta_{SC}\rangle = 10^o$, corresponding to frequent, small angle deflections of the particle velocity vector at randomly selected times. The phase space map of the result is shown in Figure 7 to which the description given previously for

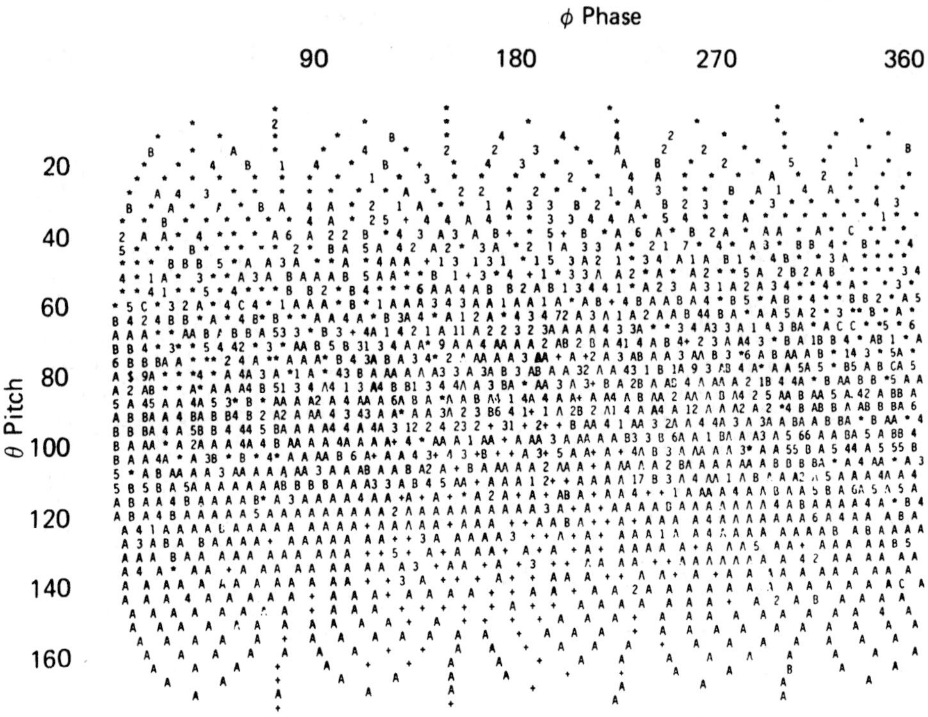

Figure 7. Same as Figure 5 but with random scattering $\langle\omega_{sc}\rangle = 0.5\omega_c$, $\langle\theta_{sc}\rangle = 10^{o}$ introduced.

Figure 6 also applies. Notice several features: one, disorder is introduced and the boundary between transmitted and reflected particles is no longer sharp; two, much higher energy reflected particles appear ($ refers to $10 < E/E_i < 11$); three, more non-interacting particles appear. The maximum energy found for a reflected particle with scattering is 10.3 times the initial energy compared to a maximum of 5.4 without. The physical reason for the greater energization of reflected particles in the presence of scattering is that the interaction time is prolonged. The scattering event itself, in this model, is elastic in the shock frame of reference. The deflection of the velocity vector leads to a displacement of the guiding center in the $\pm$ Y direction causing a loss or gain in particle potential energy. This causes a dispersion of the energies but no net gain. The most important effect of the scattering is to change the $V_{||}$ and when the change due to the scattering is opposite the change due to dynamical effects, it takes longer for the particle to acquire the $V_{||}$ necessary to escape the shock.

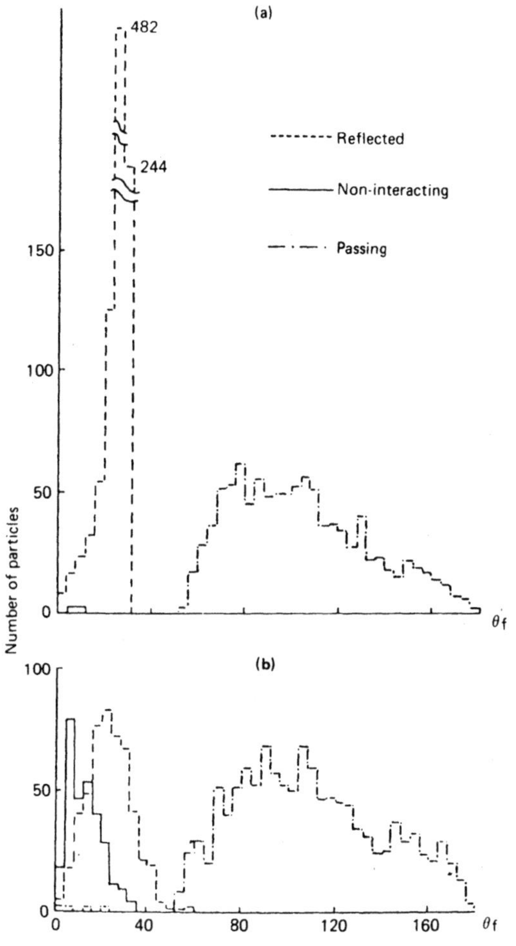

Figure 8. Post-interaction pitch angle distribution for a) without scattering, b) with scattering.

Now the next most obvious effect of the shock interaction is to modify the particle pitch angle distribution. Figure 8 shows the post-interaction pitch angle distribution for the "without scattering" and "with scattering" cases discussed above and shown in Figure 6 and 7. The diagrams show the number of particles versus pitch angle referred to the preshock $\vec{B}$ for reflected and noninteracting particles and to the post-shock field for the passing particles. The original, preshock distribution was isotropic in phase space and would go as sin θ on this diagram. Notice the large peak of reflected particles at $\sim 30^o$ to the preshock field in the unscattered (a) case. The effect of the scattering was to broaden the peak and reduce its height and to cause more noninteracting particles. The scattering does not change the post shock pitch angle distribution very much; notice that it peaks at 90° to the field in each case.

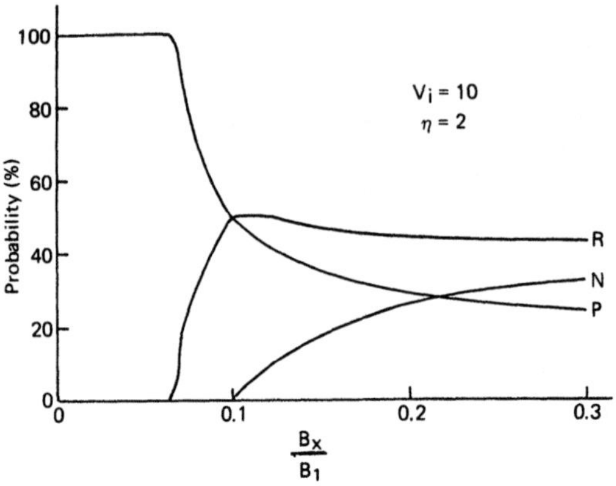

Figure 9. Probabilities for transmission (P) reflection (R) and non-interaction vs. B_x/B_1

In the studies of Chen (1975) and Sarris and Van Allen (1974) the important parameters determining the type of expected charged particle effects have been discussed. One of these is the angle, β_1, between the shock surface and magnetic field in the preshock regime. Figure 9 shows that phase-averaged probabilities for reflection, transmission and non-interaction versus $B_x/B_1 = \tan\beta_1$ for a shock of strength 2 and particles of speed $V/V_s = 10$. Notice that for $B_x/B_1 < 0.06$ ($\beta_1 < 3.4^o$) there are no reflected particles. This is a regime in which the main charged particle spike would be observed following the shock and would have an anisotropy perpendicular to the field. As B_x/B_1 increases from 0.06 to 0.1 ($\beta_1 = 5.7^o$) the reflected particles appear and rise to a maximum of 50% and the probability for transmission falls rapidly. One expects higher energy particles just ahead of the shock, moving along the field and lower energy particles behind the shock peaked perpendicular to the field. For large entry angles of the field $B_x/B_1 \geq 0.2$ ($\beta_1 \geq 11^o$) the fraction of noninteracting particles becomes large and the number of crossings for both transmitted and reflected particles diminishes rapidly thereby reducing the energy gains. In this regime, we do not expect to observe large shock spikes or modulation of particle intensities. Large angle oblique shocks should have minimal charged particle effects.

We can also describe, using the results of Chen (1975), the interval of incident particle speeds for which the shock spike interaction is large. The low speed cutoff is determined by several effects: one, below a minimum speed ($V/V_s = 6$ in the case of a shock with $\eta = 2$, $B_x/B_1 = 0.1$) there is no reflection - it is not dynamically possible; two, the particle drifts through the shock too fast and has a small number of crossings. The high speed cutoff is more gradual and is due to the fact that the fractional energy gain experienced by both transmitted and reflected particles falls off. Some effects may occasionally be observed up to speeds of several hundred times the shock speed, typically of the order of 10-20 MeV/nucleon for ions.

III. OBSERVATIONS OF SHOCK SPIKES

There have been many reported observations of shock spikes and abrupt intensity changes of low energy particles in intimate association with interplanetary shocks (Van Allen and Ness, 1967; Armstrong, et al., 1970; Singer, 1970; Lanzerotti, 1970; Lanzerotti and Soltis, 1970; Singer and Montgomery, 1971; Ogilvie and Arens, 1971; Palmeira, et al., 1971; Armstrong and Krimigis, 1973; Sarris and Van Allen, 1974; Gloeckler, et al., 1974; Sarris, et al., 1974; Sarris, et al., 1976; Richter, et al., 1976, Armstrong, et al., 1976). In addition, there has been one reported observation of a shock associated relativistic electron enhancement (Armstrong and Krimigis, 1976). Further, intensive searches of several years data has not yielded any comparable electron enhancements, although noticible SSC-associated discontinuities in the electron intensities have been identified (Decker and Armstrong, 1977). Events with significant energy changes of relativistic electrons seem to be much less common than those for low energy ions. We will concentrate here on reviewing the salient features of shock spikes for ions by considering the extent in energy, the species observed, the anisotropies, and the presence of reverse-shock-associated events.

The highest extent in energy for a shock spike event is well illustrated in Figure 10 (after Sarris, et al., 1976) where enhancements were observed for protons extending to above 25 MeV for the shock of 15 September, 1974. Notice that the spike intensities tend to be higher following the shock for the lowest energy particles and ahead of the shock for the highest energy particles. This seems to be a common feature of many shocks (cf. Sarris, et al., 1975; Gloeckler, et al., 1974; Armstrong and Krimigis, 1973) and is consistent with a picture of particle acceleration in nearly perpendicular oblique shocks (Chen and Armstrong, 1975; Sarris, et al., 1976).
at 50-100 MeV in the September 15, 1974, spike event showed no enhancement. This suggests that the relativistic proton events sometimes associated with terrestrial SSC's following solar flares (cf. Pomerantz and Duggal, 1974) may have some other mechanism acting, such as reflection of particles between a converging pair of shocks (Levy, et al., 1976).

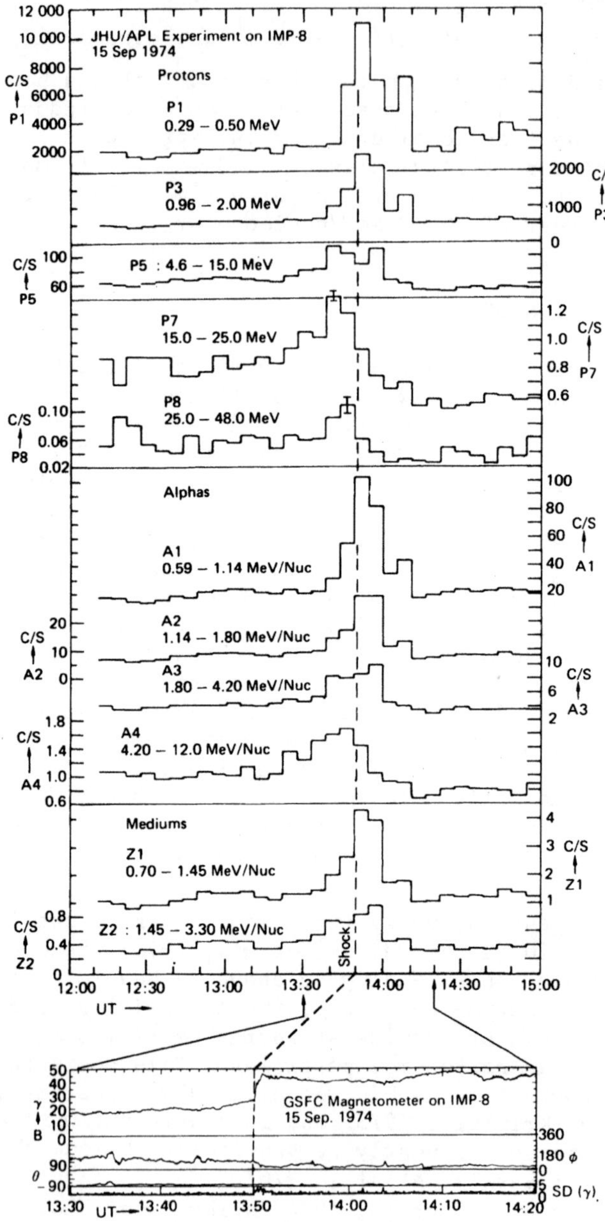

Figure 10. Intensity vs. time for a spike extending to high energies

The ionic species that have been observed to be affected by interplanetary shocks include iron group nuclei (Armstrong, et al., 1976), and the relative abundances of ionic species, p/α, α/Medium (C,N,O), and Fe/O have been observed to vary in intimate temporal association with interplanetary shocks (Lanzerotti and Robbins, 1969; Armstrong and Krimigis, 1973; Sarris, et al., 1976; Armstrong, et al., 1976).

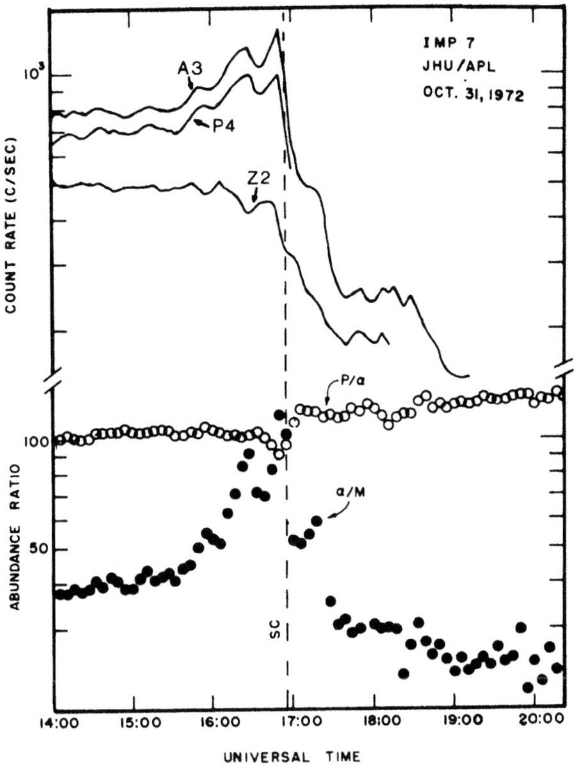

Figure 11. Plots of the 5.5 minute average count rate and abundance ratios for protons, alphas and medium nuclei from 14:00 to 20:00 U.T. on October 31, 1972. P4 is 1.85 to 4.5 MeV protons, A3 is 1.74. to 4.3 MeV/nucleon alphas, and Z2 is 1.6 to 3.2 MeV/nucleon mediums. P4, A3, and Z2 have been multiplied by .1, 10, and 100 respectively in order to plot them on the same decade. P/alpha refers to the interval 1.74 to 4.3 MeV/nucleon and alpha/M to 1.6 to 3.2 MeV/nucleon. The terristrial SSC at 16:54 U.T. is shown at the vertical dashed line.

An example of the variation of composition ratios around the time of a shock passage is shown in Figure 11 (after Armstrong and Krimigis, 1973). In this case all species, protons, alphas, and medium (CNO) nuclei have the same energy/nucleon, and we observe that the α/M shows a large enhancement at the SSC while p/α shows a small dip. This indicates that the shock acceleration process was more effective at this time for protons and alphas than for medium nuclei, despite the fact that the alphas and mediums have more nearly the same rigidity. The preshock energy/nucleon spectra had different slopes of a power law fit in this case with $\gamma \sim -4.3$ for alphas, $\gamma \sim -3.5$ for protons and $\gamma \sim -2.8$ for medium nuclei. As pointed out in Sarris and Van Allen (1974) the magnitude of the shock spike enhancement depends on the slope of the spectrum and is largest for steeply falling preshock spectra. This may explain some of the observed α/M abundance ratio enhancement in this case because one would expect on the basis of spectrum that the alphas would show the largest relative shock spike.

Another important characteristic of the shock interaction is its effect on the angular distributions of particles. Generally, the strongest anisotropies occur in the preshock regime where the fluxes tend to be field aligned coming away from the shock (Sarris and Van Allen, 1974; Sarris, et al., 1975; Armstrong, 1972). Behind the shock the anisotropy tends to switch to a peak perpendicular to the shock. The anisotropies can be very large; and example of a 1000:1 anisotropy for the shock of February 20, 1968, is given in Figure 12 from Singer and Montgomery (1971) wherein the anisotropy magnitude is indicated by the roll modulation of the fluxes. Several factors have made the study of angular distributions difficult such as the nearby presence of the magnetosphere with its uncertain effect and the large fluctuations of the magnetic field often present around the shock.

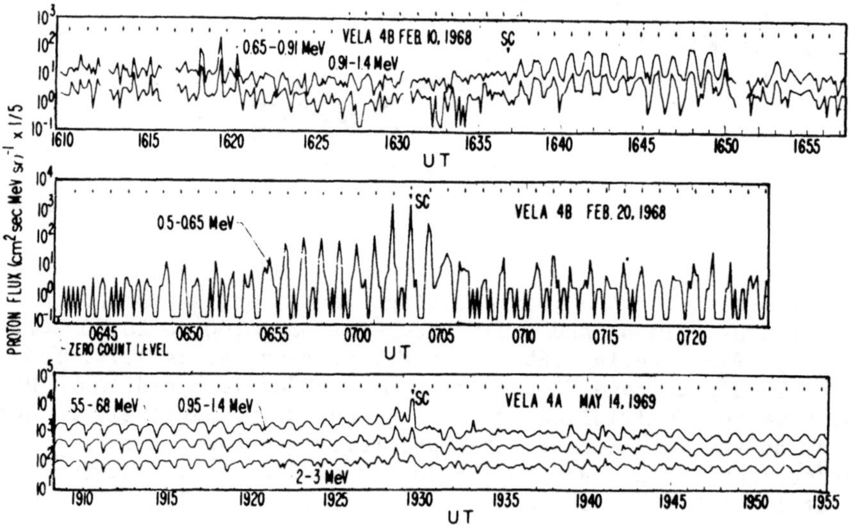

Figure 12. Spin modulated low energy proton intensities versus time from several Vela shock spike observations (Singer and Montgomery, 1971).

Further, the nearby presence of the magnetosphere as a source of energetic ion and electron bursts sometimes complicates the identification of the spikes themselves. Often, the SSC can trigger such bursts from the magnetosphere (Decker and Armstrong, 1977) and confusion can sometimes result.

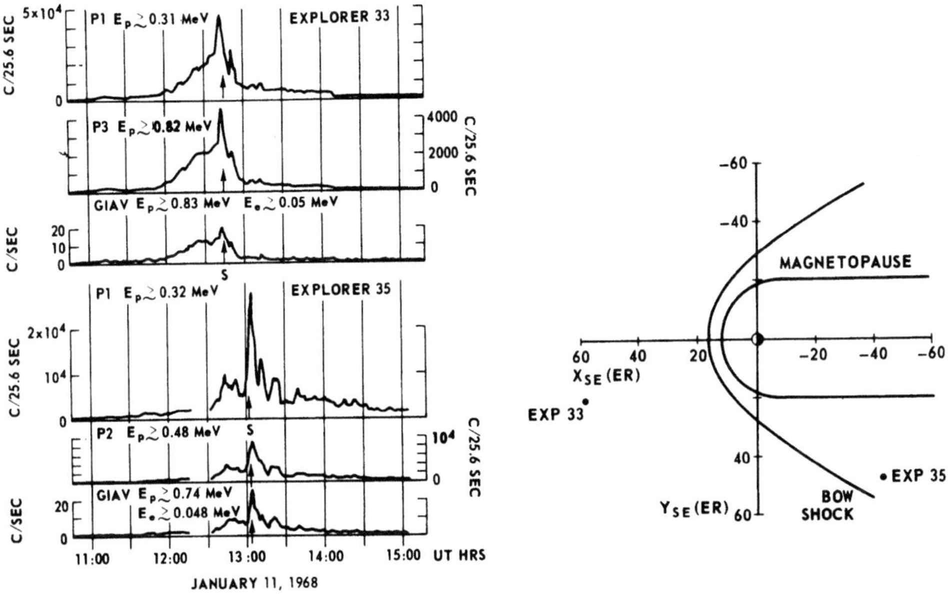

Figure 13. Plots of the counting rates from P1, P3 and G1AV (G.M. Tube rate due to particles, solar x-rays omitted) from Explorer 35 for the January 11, 1968 event. S denotes the shock time from magnetic field measurements. Explorer 35 is located at the largest radial distance from the sun.

There have been several multispacecraft observations of the shock spikes that show the propagation and evolution of particle intensities along with the shock (Armstrong et al., 1970; Sarris and Van Allen, 1974; Sarris, et al., 1975). One such case is illustrated in Figure 13 (after Armstrong et al., 1970) that shows a clear propagation time delay. Notice that while the peak structure is generally preserved, there is evidence of significant evolution of the details. It was pointed out in Sarris and Van Allen (1974) that the particle intensity could change over a few minutes due to changes in the shock normal-magnetic field angle. While this observation is complicated by the fact that Explorer 35 was located in the magnetosheath at this time it is useful because the two spacecraft are widely separated in the X_{SE} direction and it shows the shock spike behind the bow shock. The observations showed that the proposed mechanism of reflection of particles between the bow shock and the propagating interplanetary shock (Axford and Reid, 1963; Ogilvie and Arens, 1971) did not apply in this case.

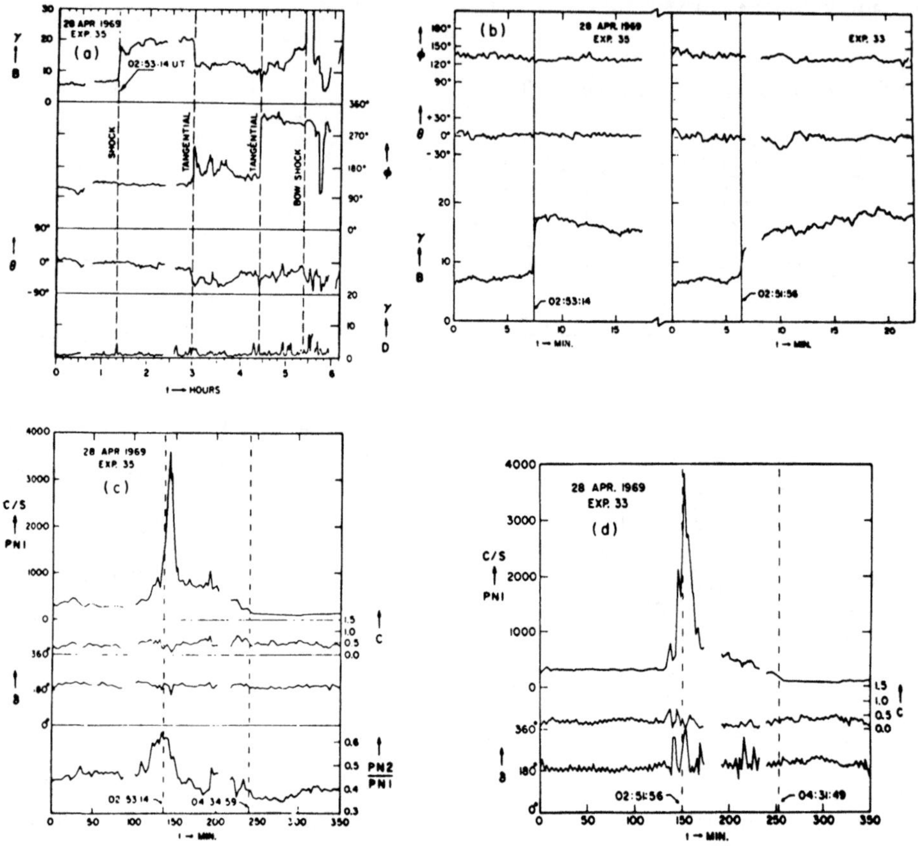

Figure 14. Plots versus time of > .3 MeV proton intensities and magnetic field data from Explorers 33 and 35 for the shock of 28 April 1969. (From Sarris and Van Allen, 1974.)

Further demonstration of the propagation of shock spikes is shown in Figure 14 from Sarris and Van Allen (1974) that give simultaneous observations from Explorers 33 and 35 for the April 28, 1968 event. Notice that the spike remains intimately associated with the magnetic discontinuity and fairly stable in shape over this short propagation interval.

Very little is known of the longer time scale evolution of shock spike events, although they are observed within 1 AU from Helios (Richter, et al., 1976) and some evidence for shock acceleration seems to be present from 1 to 5 AU in interplanetary active regions (IAR) (Pesses, et al., 1977) although the details are not yet well-established. There seem to be several possibilities, that shocks observed at 1 AU persist and systematically evolve in strength and geometry thus accounting for some IAR phenomena, or that new shocks are formed beyond IAU or some other process acts.

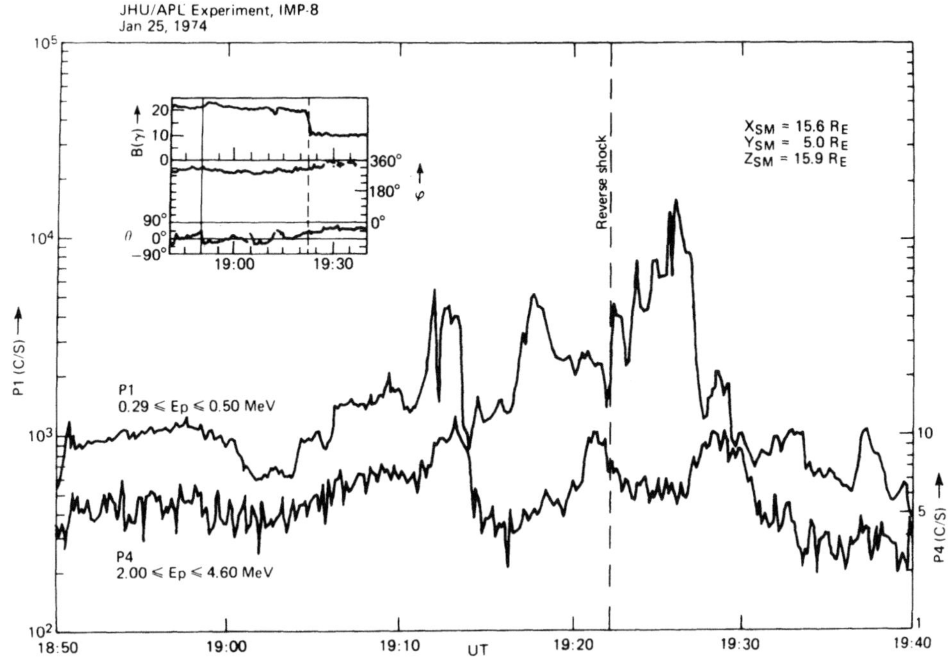

Figure 15. Observation from IMP 8 of spikes associated with a reverse shock. (Magnetic field data courtesy of Drs. Lepping and Ness, (GSFC).)

A characteristic feature of IAR's is the presence of a reverse shock on the trailing edge (Smith and Wolfe, 1976). A recent observation with IMP 8 at 1 AU, as illustrated in Figure 15, shows a reverse shock with multiple associated spikes, rapid temporal variations and characteristically hard energy spectra. We believe this to be a convincing case of acceleration at a reverse shock.

Finally, we show in Figure 16 a result from Sarris and Van Allen (1974) illustrating the absence of the shock spike effect for a large angle oblique shock. When the field direction changes by more than 10° or so across the shock this is the signature of a large angle oblique shock. There is no apparent effect of the shock on the particles in this case - a fact consistent with the theoretical framework outlined above.

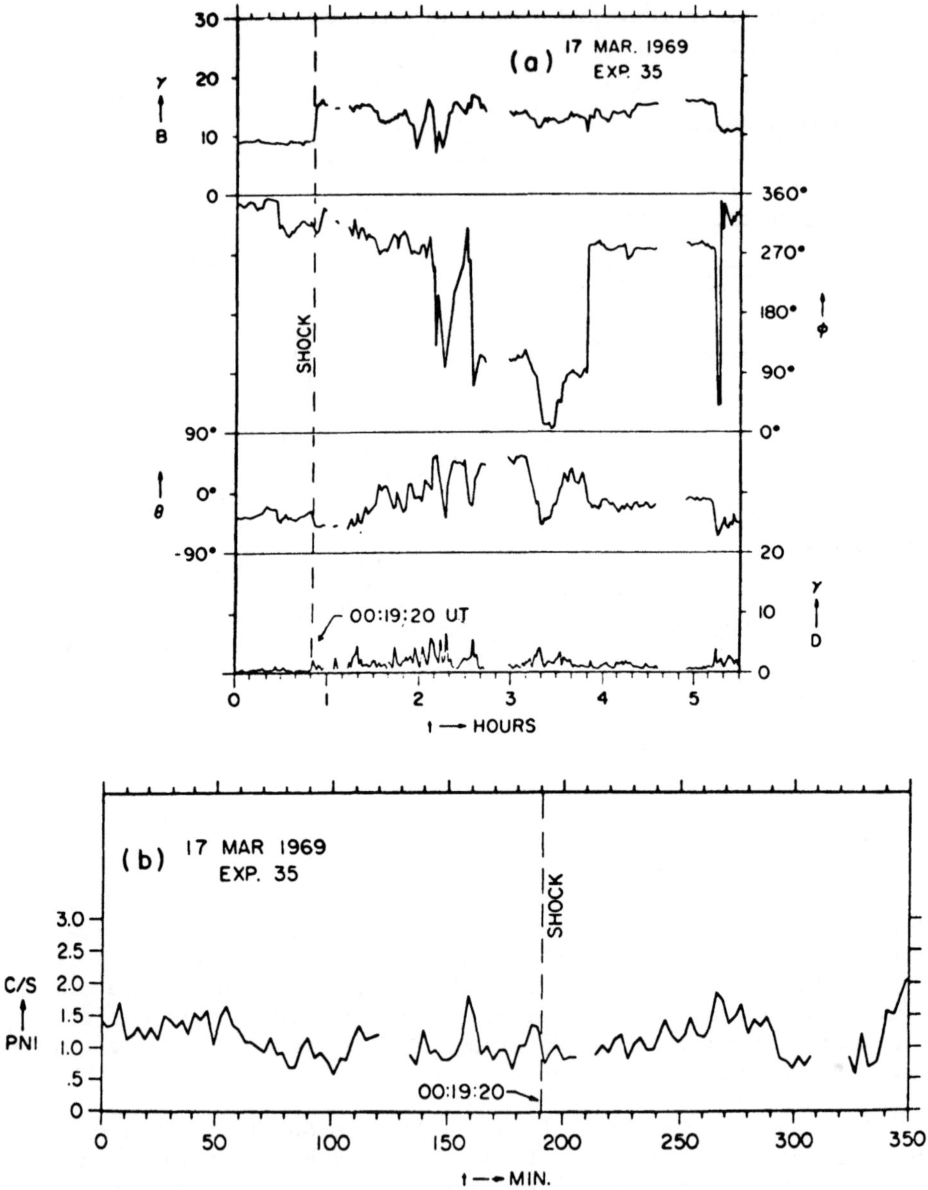

Figure 16. Plots of magnetic field and proton responses for a large angle oblique shock (Sarris and Van Allen, 1974).

SUMMARY

Summarizing the state of observations of shock spikes we have:

1) Sharp spikes in intensities, down to ion gyroradii dimensions, sometimes ahead, sometimes behind the shock with the highest energies ahead.
2) Generally larger, field-aligned ion anisotropies ahead of shocks rather than behind where the anisotropy tends towards two peaks perpendicular to $\vec{B}$.
3) Intensity modulation of protons, alphas, medium nuclei and iron have been observed. Composition ratios often change greatly during a shock spike.
4) Shock spike events show evidence of rapid evolution depending on the geometry of the shock.

There remain significant observational and theoretical areas to be investigated:

1) What is the large scale, three dimensional structure of the shock surface, particularly the magnetic field geometry and how does it influence the spike intensity and duration?
2) The models of the shock interaction need to include both a realistic pitch and phase dependent transmission probability and energy change and to correctly describe the particle transport into and away from the shock.
3) The presence of magnetic field gradients, curved lines, and possibly non-planar shock surface must be considered.
4) More sensitive and more detailed angular resolution particle flux measurements will allow the theoretical models to be more severely tested.

ACKNOWLEDGEMENTS

Portions of the work reported here were supported by NASA under Task I of Contract N00017-72-C-4401 between the Johns Hopkins University and the Department of the Navy. Some of the computing was performed at the University of Kansas computer center and one of (G.C.) did some of this work while supported by a General Research Fund grant from the University of Kansas. One of us (ETS) was supported by a JHU/APL post-doctoral fellowship at the Applied Physics Laboratory during some of this work. In the reduction and analysis of IMP-7 and 8 data used here, J. S. O'Donnell and J. L. Gunther of the Applied Physics Laboratory were of great assistance.

REFERENCES

Alekseyev, I., and Kropotkin, A.: 1970, Geomagn. and Aeronomy 10, 755.

Armstrong, T., Krimigis, S. and Behannon, K.: 1970, J. Geophys. Res. 75, 5980.

Armstrong, T: 1972, Proc. Conf. on Solar Terrestrial Relations, Calgary, 441.
Armstrong, T. and Krimigis, S.: 1973, Proc. 13th Int'l. Cosmic Ray Conf. (Denver) 2, 1504.
Armstrong, T. and Krimigis, S.: 1976, J. Geophys. Res. 81, 677.
Armstrong, T., Krimigis, S., Hovestadt, D., and Gloeckler, G.: 1976, Solar Phys. 49, 395.
Axford, I. and Reid, G.: 1963, J. Geophys. Res. 68, 1793.
Chen, G. and Armstrong, T.: 1972, Proc. of Conf. on Solar Terrestrial Relations, Calgary, 453.
Chen, G.: 1975, PhD thesis University of Kansas, Department of Physics and Astronomy.
Chen, G., and Armstrong, T.: 1975, Proc. 14th Int'l. Cosmic Ray Conf. (Munich) 5, 1814.
Decker, R. and Armstrong, T.: 1977, Trans. Am. Geophys. Union, 56, 467.
Gloeckler, G., Ipavich, F., Fan, C. Y., and Hovestadt, D.,: 1974, Geophys. Res. Letters 1, 65.
Hudson, P.D.: 1965, Monthly Notices Roy. Astron. Soc. 131, 23.
Ipavich, F.: 1974, Geophys. Res. Lett. 4, 149.
Lanzerotti, L. and Robbins, M.: 1969, Solar Phys. 10, 212.
Lanzerotti, L.: 1970, Upper Atmosphere Geophys. Report 9, World Data Center A, UAG-9, 34.
Lanzerotti, L. and Soltis, C.: 1970, Upper Atmosphere Geophys. Report 8, World Data Center A, UAG-8, 198.
Levy, E., Duggal, S., and Pomerantz, M.: 1976, J. Geophys. Res. 81, 51.
Ogilvie, K. and Arens, J.: 1971, J. Geophys. Res. 76, 13.
Palmeira, R., Allum, F. and Rao, U.: 1971, Solar Phys. 21, 204.
Parker, E.,: 1958, Phys. Rev. 109, 1328.
Pesses, M., Van Allen, J. and Goertz, C.: 1977, U. of Iowa Dept. of Physics Preprint 77-11.
Pomerantz, M. and Duggal, S.: 1974, J. Geophys. Res. 79, 913.
Richter, A., Keppler, E., Richter, K. D., Wilken, B., Umlauft, G.: 1976, Max Planck Institute für Aeronomie Preprint.
Sarris, E., and Van Allen, J.: 1974, J. Geophys. Res. 79, 4157.
Sarris, E., Krimigis, S., and Armstrong, T.: 1975, Proc. 14th Int'l. Cosmic Ray Conference, (Munich) 5, 1853.
Sarris, E., Krimigis, S., and Armstrong, T.: 1976, Geophys. Res. Letters 3, 133.
Schatzman, E.: 1963a, Ann. Astrophys. 26, 235.
Schatzman, E.: 1963b, Space Res. III, 709.
Shabanskii, V. P.: 1962, Soviet Phys. JETP 14, 791.
Singer, S.: 1970, in V. Manno and D. E. Page, Intercorreleated Satellite Observations Related to Solar Events, D. Reidel, Dordrect, Netherlands, 571.
Singer, S. and Montgomery, M.: 1971, J. Geophys. Res. 76, 6628.
Smith, E. and Wolfe, J.: 1976, Geophys. Res. Lett. 3, 137.
Terasawa, T.: 1976, Institute of Space and Astronautical Science, U. of Tokyo Research Note 22 (also Note 15).
Van Allen, J. A., and Ness, N. F.: 1967, J. Geophys. Res. 72, 935.

DISCUSSION

Gall: How would your results be affected if you take into account the width of the shock waves considered.

Armstrong: This question has been discussed by Terasawa (1976). There are some differences that appear if the shock width is not infinitesimal. Energy gains are about the same in an average sense.

Lin: Why is the range of energies limited to 100 keV to ~ 10 MeV for shock interaction?

Armstrong: The low energy limit is due to the particle drifting too rapidly through the shock and making too few crossings. At high energies the induced $\vec{E}$ seen by the particle is too small to change the particle energy significantly.

Morfill: You said that particles average over a large region of the shock surface, i.e., - since the interplanetary magnetic field fluctuations on all wavelength scales - they must average over a range of angles between B and the shock normal. Is it sufficient to take, for example, the Archimdean spiral field direction for this average in front of the shock, or does a correct sampling average over the known distribution of magnetic field directions give a significantly different answer?

Armstrong: Accounting for the evolution of the shock angle as seen by the particle is a necessary refinement of the present calculations. Changes of $\pm 5^{o}$ could have a substantial effect. The particle is with the shock ≈ 3 minutes and moves ~ 10 to 100 R_E so changes on those time or distance scales could be important.

Scholer: Is there a lower energy limit on the probability for reflection or transmission you get from the trajectory calculations that is approximately equal to the adiabatic reflection/transmission coefficient?

Armstrong: There is a regime of large angle oblique shocks for which adiabatic reflection applies. Terasawa (1976) has examined this question by simulation. Adiabatic estimate would not apply for small angle ($5\text{-}10^{o}$) shocks.

PART VI

STUDY OF TRAVELLING INTERPLANETARY PHENOMENA - STIP INTERVAL I AND STIP INTERVAL II

A SUMMARY OF SIGNIFICANT SOLAR-INITIATED EVENTS DURING STIP INTERVALS I AND II

M. A. SHEA and D. F. SMART
Air Force Geophysics Laboratory
Bedford, Massachusetts 01731, U.S.A.

H. E. COFFEY
World Data Center A for Solar-Terrestrial Physics
Boulder, Colorado 80302, U.S.A.

ABSTRACT

Using data available at the present time, a summary of the significant solar-terrestrial events that occurred during STIP Interval I (September-October 1975) and STIP Interval II (15 March-15 May 1976) has been prepared. The first STIP Interval was characteristic of typical conditions expected near solar minimum. In contrast, STIP Interval II was an active period. The principal events during STIP Interval II were associated with solar activity in a region at Carrington Longitude $\sim 45^{\circ}$ (designated McMath Plage 14143 during solar rotation 1639 and designated McMath Plage 14179 during solar rotation 1640). Among the events summarized are the following:

(1) Solar activity during the period 20-23 March from a region behind the eastern limb of the sun. This was followed by the observation, on 6 April, of an increase in solar wind velocity at $\sim$ 9.7 AU possibly associated with a solar event behind the eastern limb of the sun on 20 March;

(2) A large geomagnetic storm with intense mid-latitude auroral activity on 26 March;

(3) Moderate solar particle intensities at the earth from 26 March through 31 March; and

(4) A ground-level solar cosmic ray event on 30 April (12% increase at Inuvik, Canada) having a very hard spectrum and only moderate polar cap absorption (2.7 dB at Thule, Greenland) followed by a large geomagnetic disturbance on 2-3 May 1976.

M. A. Shea et al. (eds.), Study of Travelling Interplanetary Phenomena/1977, 393–420.

INTRODUCTION

During the 18th Plenary Meeting of COSPAR held in Varna, Bulgaria in 1975, members of the Study of Travelling Interplanetary Phenomena (STIP) Project recommended two intervals for special studies. STIP Interval I was scheduled for September and October 1975. STIP Interval II was planned for early 1976, the final dates of 15 March-15 May being selected after the successful launch of Helios 2 on 15 January 1976.

Both of these intervals were during solar minimum as illustrated by the graph of observed and "smoothed" sunspot numbers shown in Figure 1. In fact, statistical "sunspot minimum", defined by the minimum in a weighted 13-month average of the Zurich sunspot number, apparently occurred in March 1976, although the observed sunspot number for this particular month was the highest monthly value in a 17-month period (September 1975-January 1977).

In this paper we will present a summary of the significant solar-terrestrial phenomena that occurred during these two intervals.

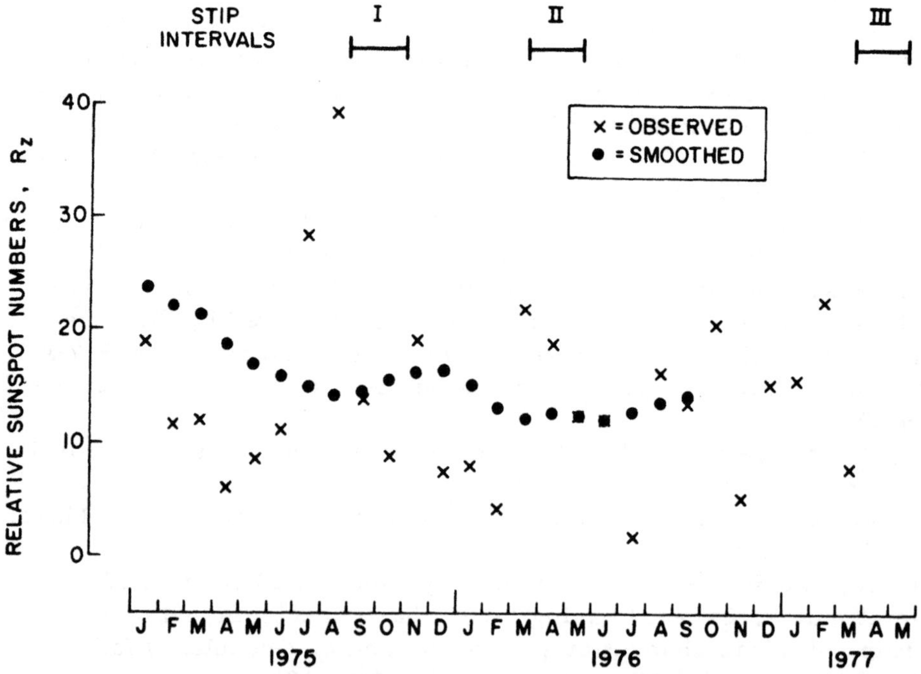

Figure 1. Observed and smoothed sunspot numbers for the period January 1975 through March 1977. The time periods of the STIP Intervals are indicated by the bars at the top of the figure.

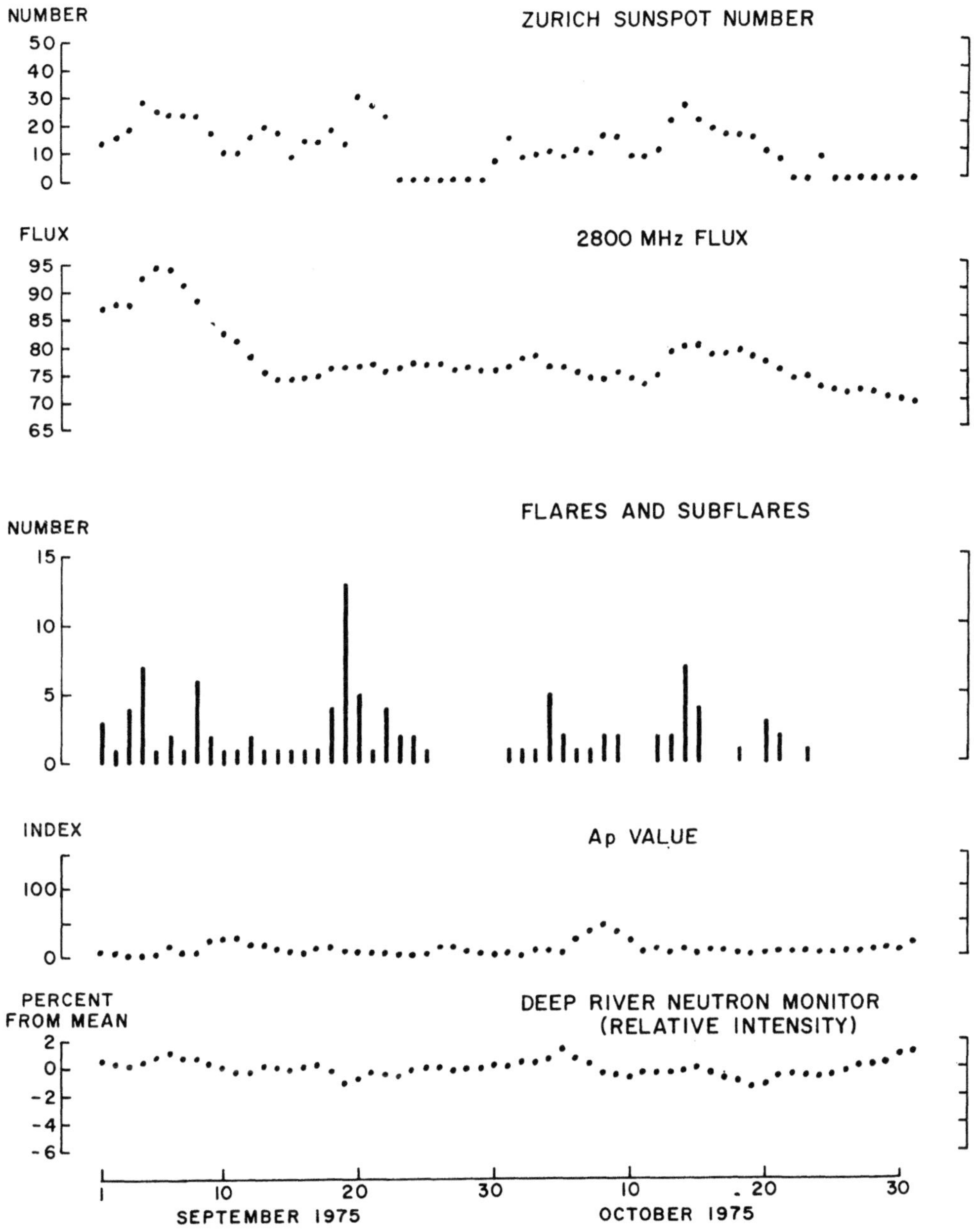

Figure 2. Graphical display of the daily values of the solar activity (as indicated by the Zurich sunspot numbers, the 2800 MHz flux, and the number of flares and subflares), and the geomagnetic activity (as indicated by the A_p index) during STIP Interval I. The relative cosmic ray neutron monitor intensity, as measured by the Deep River, Canada neutron monitor, is given at the bottom of the figure.

STIP INTERVAL I

STIP Interval I was characteristic of typical conditions expected near sunspot minimum with observed Zurich sunspot numbers of 14.1 and 9.0 for September and October 1975, respectively. Figure 2 is a graphical display of the daily values of solar activity, geomagnetic activity, and neutron monitor intensity for the entire period.

Although STIP Interval I was scheduled for the entire two months of September and October 1975, special emphasis was placed on the period 8-27 September when the Helios 1 space probe was radially aligned with the Pioneer 6, Pioneer 8, and Pioneer 9 space probes, respectively, as shown in Figure 3. In addition there were data from two Venera space probes during STIP Interval I. Venera 9, launched on 8 June 1975 and Venera 10 launched on 14 June 1975 arrived at the

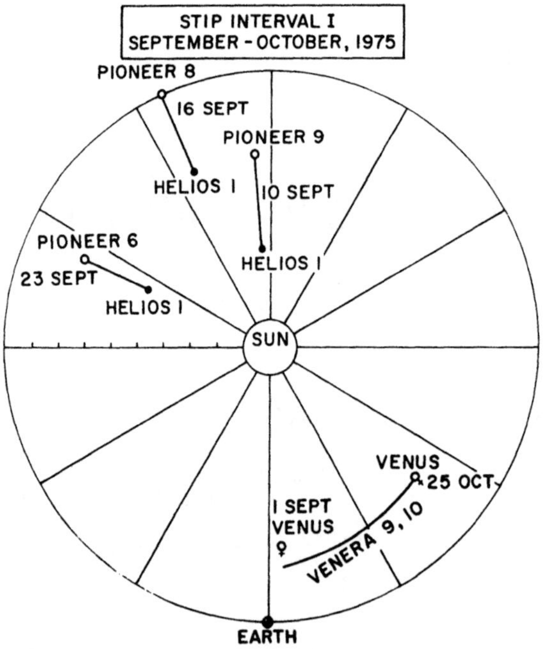

Figure 3. Relative positions of important spacecraft during STIP Interval I in a coordinate system based on a fixed sun-earth line. The dates and positions of radial alignments between the Helios 1 and the Pioneer 9, Pioneer 8, and Pioneer 6 space probes are indicated. The relative position of the planet Venus on 1 September and 25 October and the approximate trajectory of the Venera 9 and Venera 10 space probes for this same time period are indicated.

Table 1

Solar Flares With Importance ⩾ 1 or Time Associated With Type II Radio Bursts that Occurred During STIP Interval I

Date	Onset (UT)	Solar Coordinates	McMath Region	Imp.	Type II Onset (UT)
1 Sept 75	1507	N03, E52	13826	1N	
21 Sept 75	1515	N26, W71	—	1B	
8 Oct 75	2341	N32, W21	13875	sf	2342
13 Oct 75	0825	S11, E60	13892	1N	

planet Venus on 22 October and 25 October respectively. The positions of the planet Venus during STIP Interval I and the approximate trajectory of the Venera 9 and Venera 10 space probes with respect to a fixed sun-earth line are also shown in Figure 3.

No outstanding solar-terrestrial events occurred during this interval. Only three solar flares with importance equal to or greater than 1 were reported during this period, two of which (21 September and 13 October) were reported by only one station. Table 1 lists these three flares plus the one Type II radio burst (and associated flare) that occurred during this interval.

The solar particle observations by earth-orbiting spacecraft do not indicate any significant solar particle events during STIP Interval I. Representative data for electron and proton fluxes as measured by IMP 8 are illustrated in Figure 4. These fluxes are typical of observations during relatively quiescent solar conditions when high sensitivity low background instrumentation may observe co-rotating particle streams.

The solar wind bulk speed measurements made by earth orbiting spacecraft (IMP 7 and IMP 8) during STIP Interval I are illustrated in Figure 5. Geomagnetic activity was relatively low with the apparent initiation of a sequence of recurrent magnetic storms 9-12 September that returned 6-10 October and continued for the next five solar rotations as shown in Figure 6. The last remnants of this sequence may be discernible during the start of STIP Interval II (15-17 March 1976). A sudden commencement geomagnetic storm was reported at several stations at 1116 UT on 3 October with K_p increasing from a value of 1- prior to the sudden commencement to a maximum value of 4+ early on 4 October. No significant solar flares were reported for the four-day period preceding this sudden commencement. An examination of the available solar X-ray and solar radio data during this period does

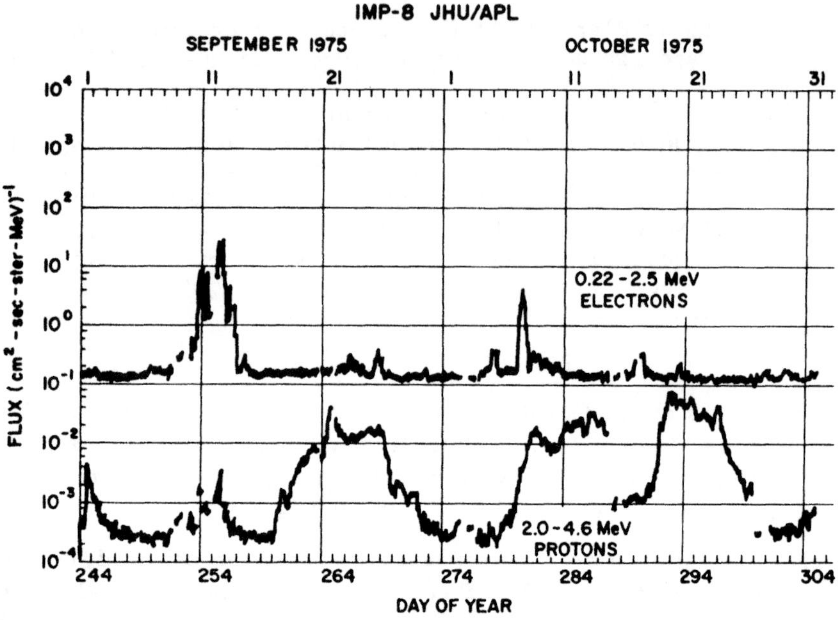

Figure 4. Three-hourly electron (0.22-2.5 MeV) and proton (2.0-4.6 MeV) fluxes from IMP 8 for the period 1 September-31 October 1975. Each tick mark represents one day with every tenth day designated by a vertical line.

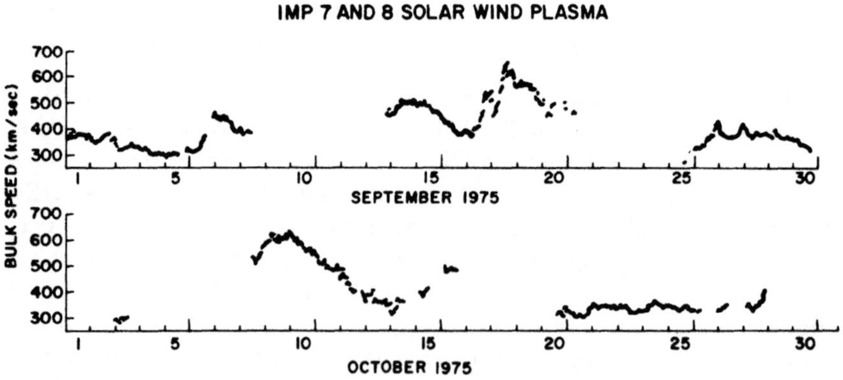

Figure 5. Solar wind bulk speed measurements observed by the earth-orbiting space craft IMP 7 and IMP 8 during STIP Interval I. These data, from the MIT plasma probe, were published in Solar-Geophysical Data.

not indicate the occurrence of significant activity on the visible hemisphere of the sun.

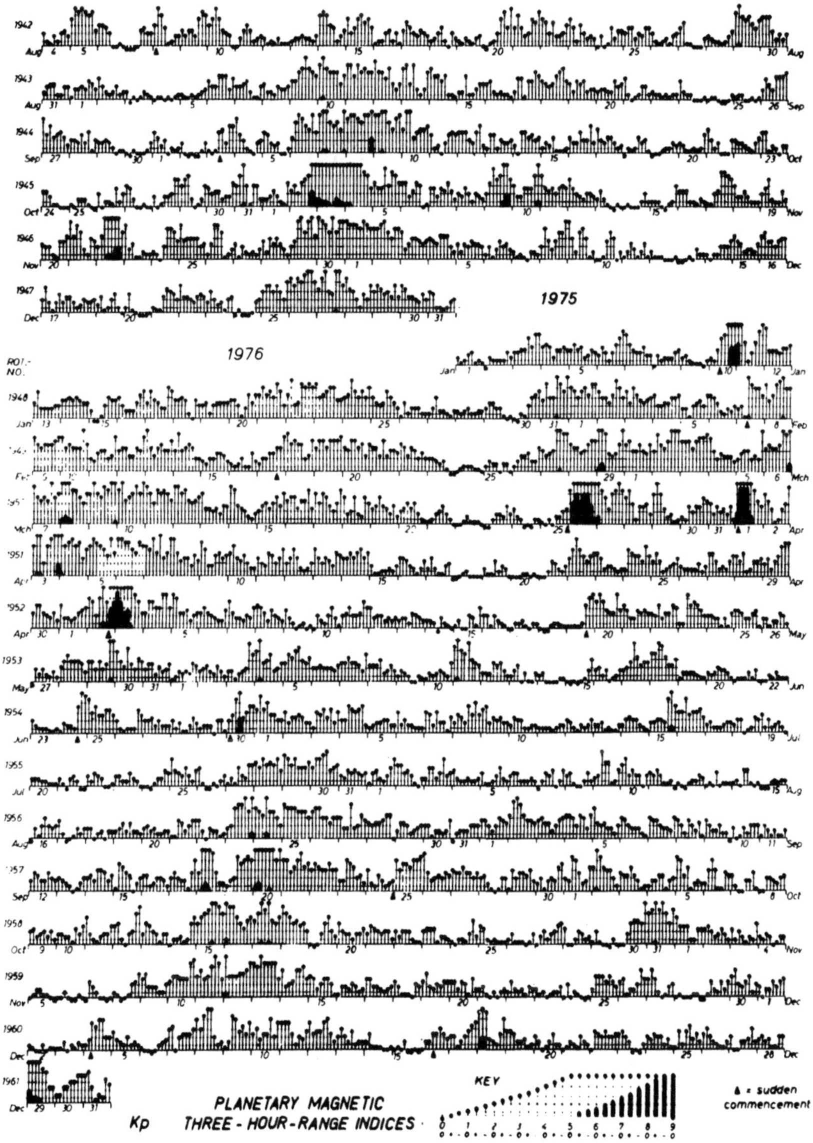

Figure 6. Planetary magnetic three-hour-range indices, K_p, for the period 4 August through 31 December 1975 and for the entire year of 1976. The rotation numbers given on the left hand side of this figure are Bartels rotation numbers and not the solar rotation numbers referred to in the text of this paper.

STIP INTERVAL II

The dates for STIP Interval II were selected primarily because of the possibility of obtaining extensive and hitherto unavailable particle, plasma, and magnetic field observations from two space probes close to the sun, Helios 1 and 2, both of which underwent perihelion during this interval. The orbits of these two probes are shown in Figure 7.

In contrast with the typical solar minimum conditions observed during STIP Interval I, STIP Interval II was a much more active period. Figure 8 is a graphical display of the daily values of solar activity, geomagnetic activity, and the neutron monitor intensity for this interval, plotted on the same scale as similar values in Figure 2.

Nine solar flares with importance equal to or greater than 1 were reported during STIP Interval II. The 1F solar flare on 9 April was a single station report. Table 2 lists these nine flares plus the Type II radio bursts (and associated flares, if identified) that occurred during this interval. Other flares, indicative of the solar activity during this period are also given.

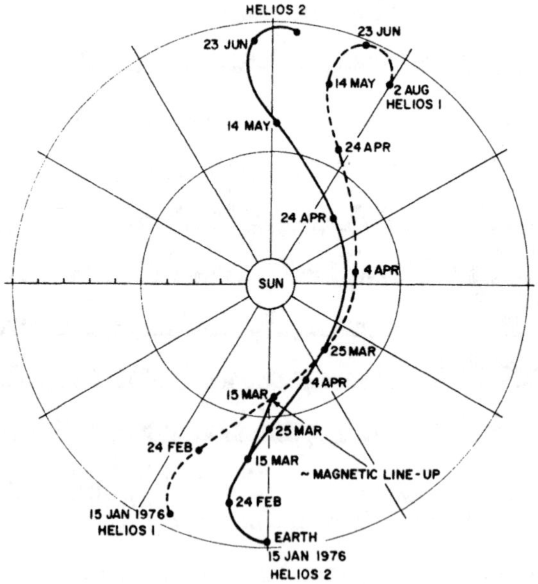

Figure 7. The orbits of the Helios 1 and 2 space probes with respect to a fixed sun-earth line. The time period is from 15 January to August 1976.

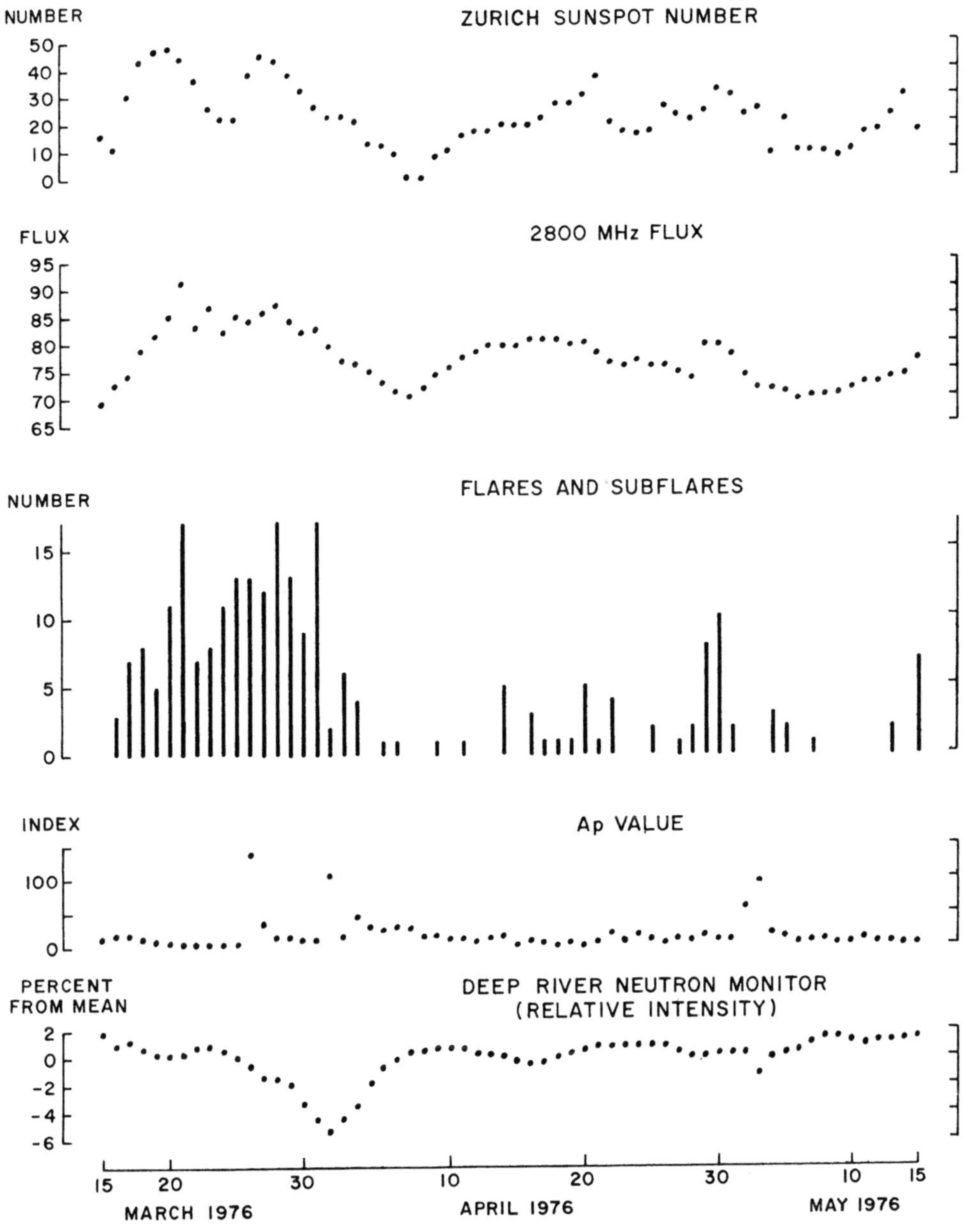

Figure 8. Graphical display of the daily values of the solar activity (as indicated by the Zurich sunspot numbers, the 2800 MHz flux, and the number of flares and subflares), and the geomagnetic activity (as indicated by the A_p index) during STIP Interval II. The relative cosmic ray neutron monitor intensity, as measured by the Deep River, Canada neutron monitor, is given at the bottom of the figure.

Table 2

Major Solar Flare Activity During STIP Interval II

Date	Onset (UT)	Solar Coordinates	McMath Plage	Imp.	Type II Onset (UT)	CFI
20 March 76					0203	2
20 March 76					2253	3
21 March 76	0750	N04, W29	14127	1B		5
23 March 76	0837	S05, E90	14143	sb	0841	12
	0907	S05, E90	14143	sn		
24 March 76	0011	N02, W66	14127	1N		5
25 March 76	1154	S06, E75	14143	sn		8
25 March 76	1305	S05, E69	14143	1N		9
26 March 76	1720	Type IV, No flare reported				3
27 March 76	1156	S10, E49	14143	1N		8
28 March 76	0546	S07, E37	14143	sb	0557	6
28 March 76	0922	S08, E31	14143	sn		5
28 March 76	1342	S06, E37	14143	sf		3
	1534	S06, E27	14143	sf		
	1625	S06, E25	14143	sf		
28 March 76	<1905	S07, E28	14143	1B	1921	10
31 March 76	1138	S07, W09	14143	1N		8
5 April 76		No flare patrol			2153	4
9 April 76	0019	N06, E80	14161	1F		1
20 April 76	1740	N03, W77	14161	sf	1803	6
29 April 76	1904	S08, W31	14179	sn		6
30 April 76	1242	S06, W41	14179	sn		7
30 April 76	2047	S08, W46	14179	1B	2107	11
1 May 76	2140	S08, W60	14179	sf	2206	7
15 May 76	1002	S14, E59	14211	1F		1

* CFI = Comprehensive Flare Index (Dodson and Hedeman, 1971)

This table includes:

(1) All flares with importance ≥ 1;
(2) All Type II radio bursts; and
(3) All "major" events identified by Dodson and Hedeman (1977).

Geomagnetic activity was very disturbed during portions of this interval, and very quiet at other times as shown in Figure 6. Three sudden commencement geomagnetic storms were reported: 0233 UT on 26 March, 0255 UT on 1 April, and 1829 UT on 2 May. Each of these sudden commencements was followed by a period of disturbed geomagnetic activity during which the K_p values reached 8 or greater. In general the geomagnetic activity was much higher during STIP Interval II than during STIP Interval I as illustrated in Figure 9.

Particulate radiation was also enhanced during this period as indicated by the 0.22-2.5 MeV electron flux and the 2.0-4.6 MeV proton flux observed by the IMP 8 satellite shown in Figure 10. In fact, preliminary evaluation of the solar-terrestrial activity that occurred during this period, together with the data acquired by various ground-based and satellite detectors, resulted in a request from the STIP

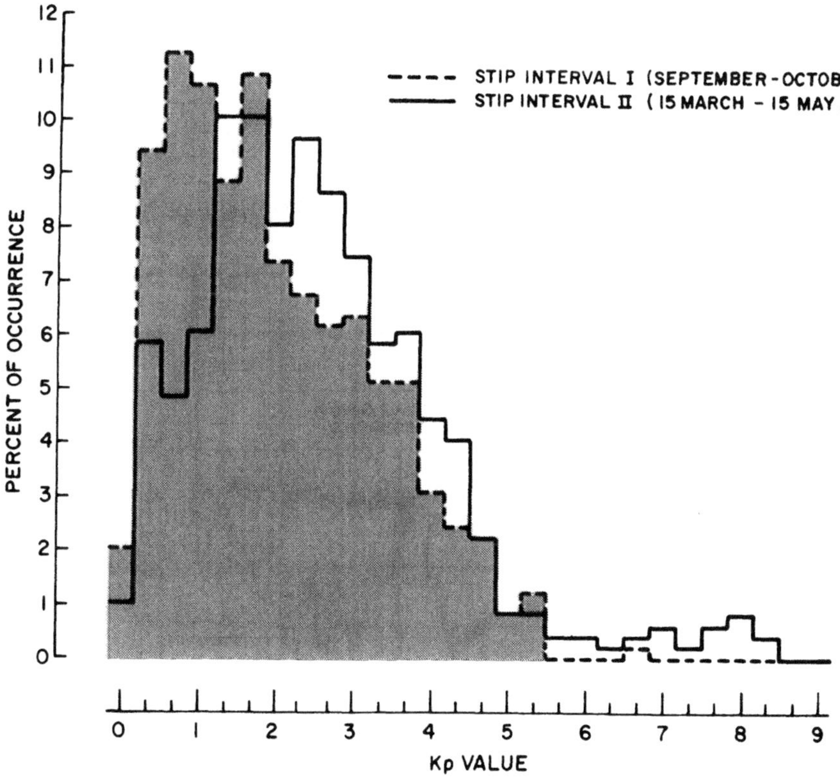

Figure 9. Comparison of the distribution of K_p values during STIP Interval I and STIP Interval II. The shaded area represents the K_p value distribution during STIP Interval I.

project that the period 20 March-5 May 1976 be designated as a "Retrospective World Interval". In conjunction with the formal declaration of this interval by the SCOSTEP Monitoring Sun-Earth Environment Committee, World Data Center A for Solar-Terrestrial Physics in Boulder, Colorado, U.S.A., undertook the publication of UAG Report No. 61 (Coffey and McKinnon, 1977) presenting significant solar-terrestrial phenomena during this period. Since the solar-terrestrial phenomena during STIP Interval II was dominated primarily by activity in a region near Carrington Longitude 45°, data contributions were solicited from the scientific community with the proviso that data to be published be confined to events believed associated with McMath Plage 14143 (solar rotation 1639) and its return as McMath Plage 14179 (solar rotation 1640). Table 3 presents a listing of some of the major solar-terrestrial phenomena that occurred between 20 March and 5 May 1976, with emphasis on those believed associated with McMath Plages 14143 and 14179.

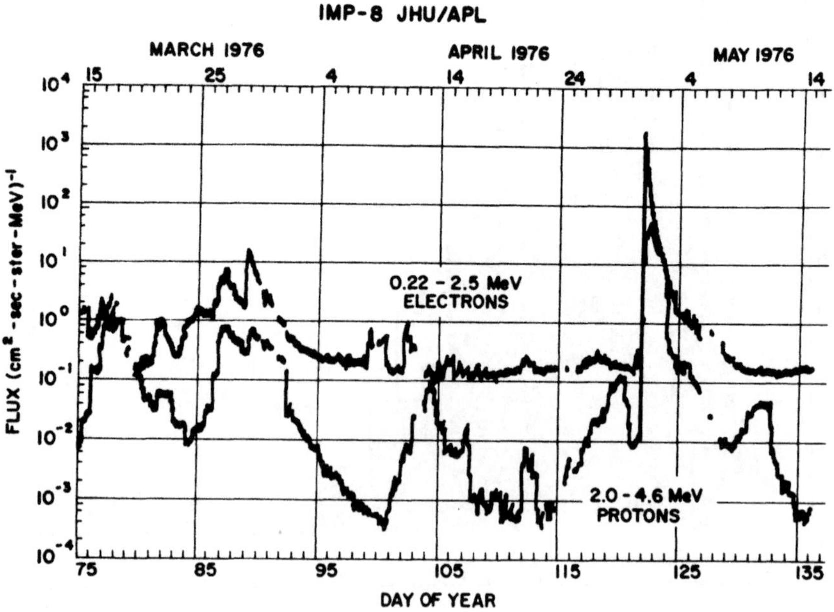

Figure 10. Three-hourly electron (0.22-2.5 MeV) and proton (2.0-4.6 MeV) fluxes from IMP 8 for the period 15 March-15 May 1976. Each tick mark represents one day with every tenth day designated by a vertical line.

TABLE 3

List of Significant Solar-Terrestrial Phenomena (20 March-2 May 1976)

20 March 1976		
0203	UT	Type II radio burst at the east limb (Culgoora)
2257	UT	Type II radio brust at the east limb (Culgoora)
Note:		Solar wind velocity enhancement on 9 April observed by Pioneer 10 (9.7 AU, 147° east of the sun-earth line).
23 March 1976		
0450	UT	X-ray enhancement, east limb (OSO-8)
0837-0841	UT	Solar flare (S05, E90; McMath Plage 14143; sb)
0840-1900	UT	X-ray enhancement (GOES-1)
0841	UT	Type II and Type IV radio bursts (IZMIRAN, Moscow)
0842	UT	Type IV radio burst (Durnten)
0907-0945	UT	Solar flare (S07, E90; McMath Plage 14143; sn)
1100	UT	Many loops observed in H-alpha, SE limb (Ramey)
26 March 1976		
0233	UT	Sudden commencement geomagnetic storm (A_p=138)
28 March 1976		
1840	UT	Onset of X-ray enhancement, 1-8Å (GOES-1)
1905-2021	UT	Solar flare (S07, E28; McMath Plage 14143; 1B)
1915	UT	Onset of 8800 MHz solar noise burst. Maximum at 1934 UT with flux density of 3719 flux units (Sagamore Hill)
1921-1950	UT	Type II radio burst (Harvard)
1925-1939	UT	Type IV radio burst (Sagamore Hill)
1931-2400	UT	Type IV radio burst (Boulder)
31 March 1976		
1108	UT	Onset of X-ray enhancement, 1-8Å (SMA/GOES)
1138-1350	UT	Solar flare (S07, W09; McMath Plage 14143; 1N)
1153-1512	UT	Type IV radio burst (Durnten)
1356-1407	UT	Solar flare (S08, W09; McMath Plage 14143; sf)
1437-1442	UT	Solar flare (S10, W03; McMath Plage 14143; sn)
1445-1508	UT	Solar flare (S11, W08; McMath Plage 14143; sf)
1 April 1976		
0255	UT	Sudden commencement geomagnetic storm (A_p=107)
		* * * * * * * * *
30 April 1976		
2043	UT	Onset of X-ray enhancement, 1-8Å (GOES-1)
2047-2218	UT	Solar flare (S08, W46; McMath Plage 14179; 1B)
2047	UT	Onset of 2800 MHz solar noise burst at Ottawa. Maximum at 2109 UT with flux density of 1670 flux units.
2103-2130	UT	Type IV radio burst (Sagamore Hill)
2107-2129	UT	Type II radio burst (Harvard)
2120-2125	UT	Onset of GLE (Inuvik, Canada neutron monitor)
2125-2130	UT	Onset of PCA (Thule, Greenland; Cape Zhelaniya and Dixon Island, USSR)
2130	UT	Onset of 6-10 MeV protons observed on SMS/GOES
1 May 1976		
0130	UT	Maximum PCA at Thule, Greenland (2.7 dB)
2 May 1976		
1829	UT	Sudden commencement geomagnetic storm (A_p=58 on 2 May; A_p=94 on 3 May)

SIGNIFICANT EVENTS DURING STIP INTERVAL II

Solar initiated event, 20 March 1976

Type II radio bursts were recorded by the Culgoora, Australia radioheliograph and radio spectrograph at 0203 UT and 2255 UT on 20 March 1976 which Nelson and McLean (1977) attribute to flares 55° and 44° behind the eastern limb of the sun, respectively. Events of this type have been observed before (e.g. Smerd, 1970) but not from such great distances behind the limb (Nelson and McLean, 1977).

During the previous solar rotation, McMath Plage 14118, located at Carrington Longitude 44°, was observed in X-rays from 6 March until this new region was ~1.5 days behind the western limb of the sun on

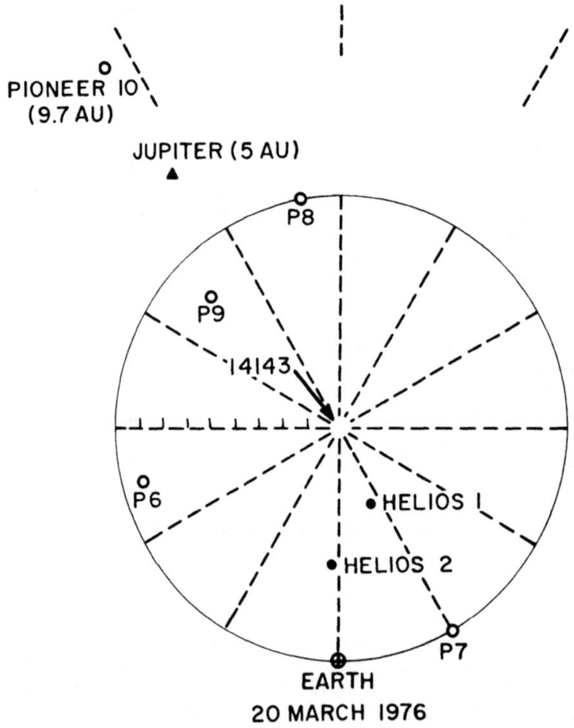

Figure 11. Positions of important space probes and the planet Jupiter relative to the sun-earth line on 20 March 1976. The area within one AU is drawn to scale. The radial positions of Jupiter and Pioneer 10 are not drawn to scale; however, the angular location is accurate. The Pioneer space probes within one AU are designated by the letter P and the appropriate number. The location of McMath Plage 14143 is indicated by the arrow.

11 March (Wolfson et al., 1977). On 23 March a very active region at the same solar location as McMath Plage 14118 was optically observed rotating over the eastern limb of the sun - presumably a continuation of the same active region that disappeared over the western limb ~13 days earlier. Since all plages are identified by an unique number for each solar rotation, this solar active region was renumbered as McMath Plage 14143 for this solar rotation. From these observations McMath Plage 14143 was considered to be the probable source of the Type II radio bursts on 20 March.

Figure 11 illustrates the position of various space probes, the planet Jupiter, and the location of McMath Plage 14143 on 20 March 1976. Unfortunately, there were no observational data from the Pioneer 9 space probe during this time period which could have confirmed the existence of an interplanetary shock associated with the Type II radio bursts on 20 March. However, there are two observations which can be interpreted as indicating the passage of a shock through the interplanetary medium. Non-Io associated radio emission at 17 to 24 MHz was observed from the planet Jupiter from 0234 to 0250 UT on 30 March (Warwick, private communication). This rare type of radio emission occurs only 2-3% of the time and falls within the time window expected if an interplanetary shock originated on 20 March from region 14143 and was propagating outward through the interplanetary medium with an average velocity of $\sim$865 km s^{-1} or $\sim$940 km s^{-1}

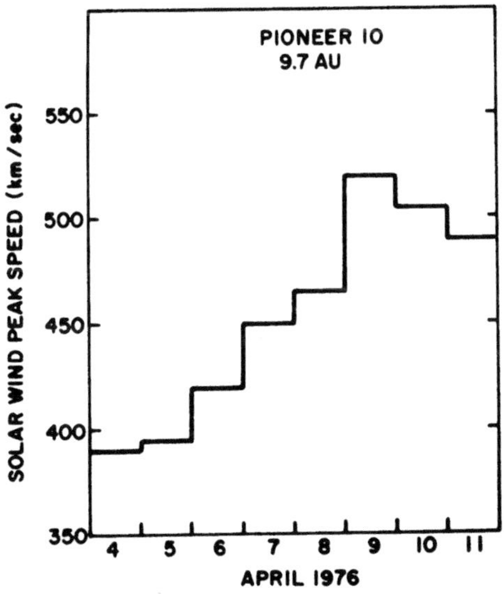

Figure 12. Pioneer 10 daily solar wind peak speed observations for 4-11 April 1976. Data courtesy of the NASA-Ames Plasma Group.

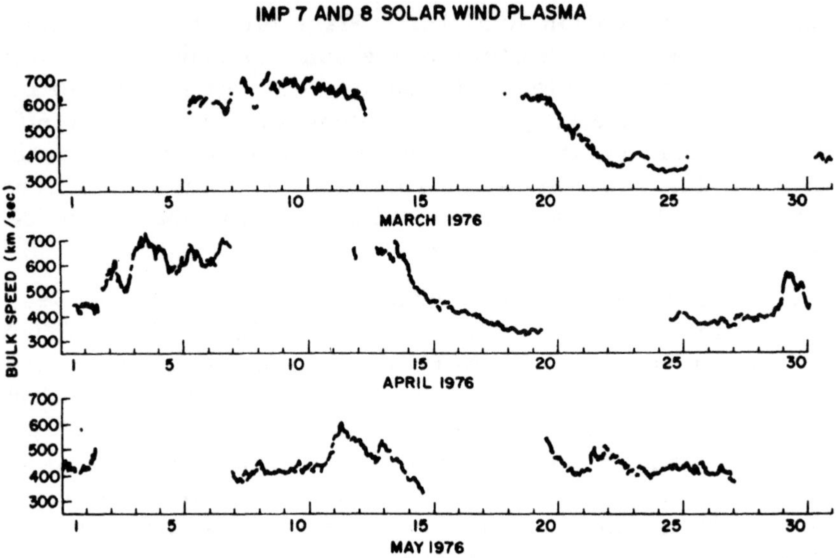

Figure 13. Solar wind bulk speed measurements observed by the earth-orbiting space craft IMP 7 and IMP 8 for the period 1 March through 31 May 1976. These data, from the MIT plasma probe, were published in Solar-Geophysical Data.

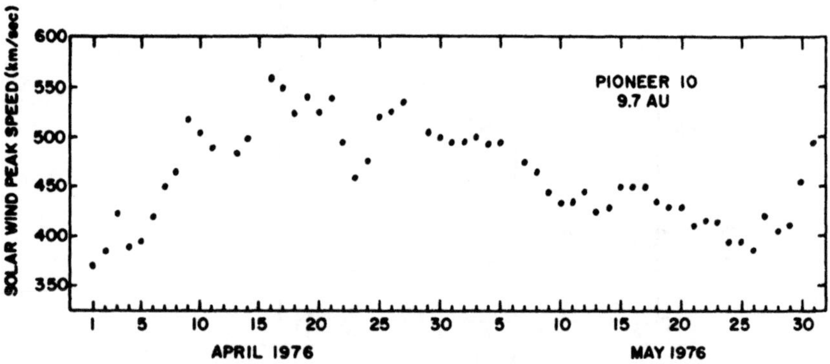

Figure 14. Pioneer 10 daily solar wind peak speed observations for the period 1 April through 31 May 1976. Data from NASA-Ames.

depending upon which Type II emission is used in the calculation. Ten days later, on 9 April 1976, an increase in the solar wind peak speed was observed at Pioneer 10 at 9.7 AU as illustrated in Figure 12. The existence of a shock passage cannot be uniquely identified because of a data gap between 1914 UT on 8 April and 0707 UT on 9 April (NASA Ames Plasma Group, private communication). When the data coverage resumed on 9 April, the rise in peak speed had occurred. Interpreting this as evidence of a shock passage results in computed average velocities of $\sim$845 km s^{-1} and $\sim$885 km s^{-1} depending upon which Type II event is attributed as the source of the shock.

The IMP 7 and IMP 8 solar wind observations during March, April, and May 1976 are shown in Figure 13. There is evidence that the high speed solar wind stream observed at the earth from 6 to 12 March, which returned again from 4 to 7 April, was also observed at 9.7 AU by the Pioneer 10 space probe. Following the method of Nolte and Roelof (1973) we would tentatively assign the solar source longitude of this high speed stream as having a width of approximately 50° between Carrington 340°-30° (i.e. 340° continuing through $360^\circ/0^\circ$ to 30° of the next solar rotation). When this source rotated past the sun-Pioneer 10 space probe line (approximately 14 days after it rotated past the sun-earth line), the plasma emission leaving the sun would eventually be detected by Pioneer 10 at 9.7 AU.

The typical daily samples of solar wind peak speed from the Pioneer 10 space probe are shown in Figure 14. The high speed solar wind stream is clearly evident in these data. Again, using the method of Nolte and Roelof (1973) we estimate that the source of the stream was at Carrington 333°-26°, strongly suggesting that the same high speed solar wind stream is being observed both at the earth and at the Pioneer 10 space probe.

Solar-Terrestrial events, 23-31 March 1976

During the period 16 through 22 March, before McMath Plage 14143 rotated onto the visible solar disk, most of the visible flare activity was from McMath Plage 14127 (Carrington Longitude 196°) with 93% of the 57 flares and subflares occurring in this region. For the three-day period that both McMath Plage 14127 and McMath Plage 14143 were on the visible hemisphere (23-25 March) 56% of the flare activity was from region 14143 on the eastern limb of the visible hemisphere, and 44% from region 14127 on the western limb. For the remaining 11 days of its transit across the visible solar hemisphere, activity in McMath Plage 14143 was responsible for 85% of the flare activity on the entire visible solar disk.

McMath Plage 14143 was observed in X-rays by the OSO 8 satellite by 12 UT on 22 March when this region was still $\sim 20^{\circ}$ behind the eastern limb of the sun. By 0450 UT on 23 March, this region was the major source of X-ray emission on the visible disk and continued to be the dominant X-ray source on the sun throughout its transit (Wolfson et al., 1977). The positions of the two principal active regions on the visible solar disk on 23 March are shown in Figure 15.

At 0837 UT on 23 March, major flare activity occurred in McMath Plage 14143 on the eastern limb of the sun just as this region was becoming visible from the earth. These flares, with onsets at 0837 UT and 0907 UT (Solar-Geophysical Data, 1976) were assigned importances of sb and sn respectively; however, the apparent minor importance of these flares may be misleading since it is possible that much of the activity was behind the limb and not visible from the earth. If these two flares were actually one solar phenomena, a portion of which could not be visible from the earth, this event would have a comprehensive flare index of 12 - the largest comprehensive flare index of STIP Interval II (Dodson and Hedeman, 1977). Akinjan et al. (1977) report that an analysis of the radio spectra observed by Soviet observatories shows that this radio event, with onset at 0841 UT on 23 March, is a

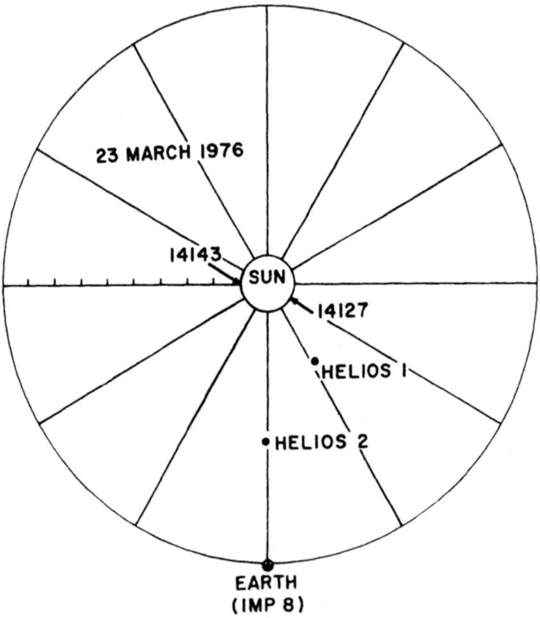

Figure 15. Positions of the Helios 1 and Helios 2 space probes relative to the sun-earth line on 23 March. The locations of McMath Plage 14143 and McMath Plage 14127 are indicated by arrows.

complex set of Type II and Type IV radio bursts. They further state that this event was one of the most intense radio bursts that they observed in 1976. The occurrence of the Type II radio burst plus the observation of the "U shaped" spectral characteristic in the radio flux maxima, both of which have been identified as associated with solar proton events (Švestka and Fritzová-Švestková, 1974; Castelli et al., 1967; Castelli and Barron, 1977) suggest that solar protons may have been released into the interplanetary medium in association with this solar flare. An impulsive, classical shaped solar proton and electron increase was observed to begin in the six-hour interval between 06 and 12 UT on 23 March on the Helios 2 space probe as shown in Figure 16. Kunow et al. (1977a) associate this solar particle event with the activity

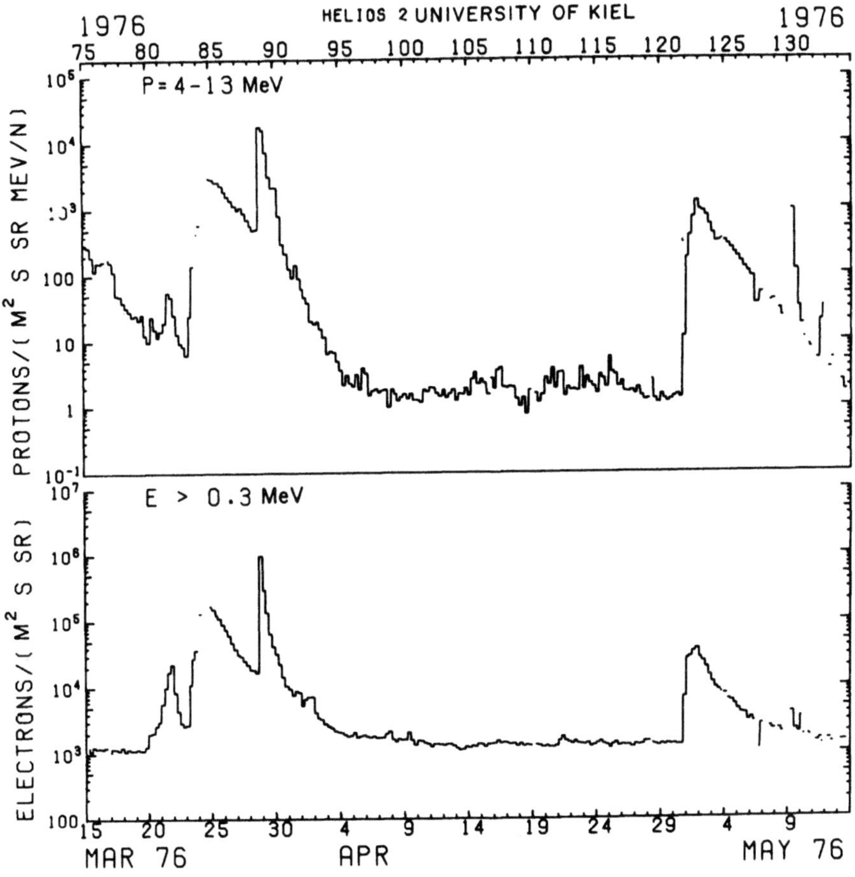

Figure 16. Six-hourly proton (4-13 MeV) and electron (> 0.3 MeV) fluxes from Helios 2 for the period 15 March-15 May 1976. Each tick mark represents one day. Data from the University of Kiel, FRG, courtesy of H. Kunow.

in McMath Plage 14143. As shown in Figure 15, the location of McMath Plage 14143 is not ideally situated for efficient solar particle detection at the earth whereas region 14127 is in a good position for efficient particle transport along the interplanetary magnetic field lines from the sun to the earth. While there was subflare activity in McMath Plage 14127 on 23 March, the significant 1N flare at 0011 UT on 24 March in this region occurs many hours after the start of the particle increase observed at the Helios 2 space probe (see Figure 16) and the Helios 1 space probe (see Figure 17). The flux observed at Helios 2 is larger than that observed at Helios 1 or at the earth. The earth orbiting IMP 7 and 8 spacecraft solar proton data published in Solar-Geophysical Data (1976) show a slowly increasing 4-12 MeV particle flux on

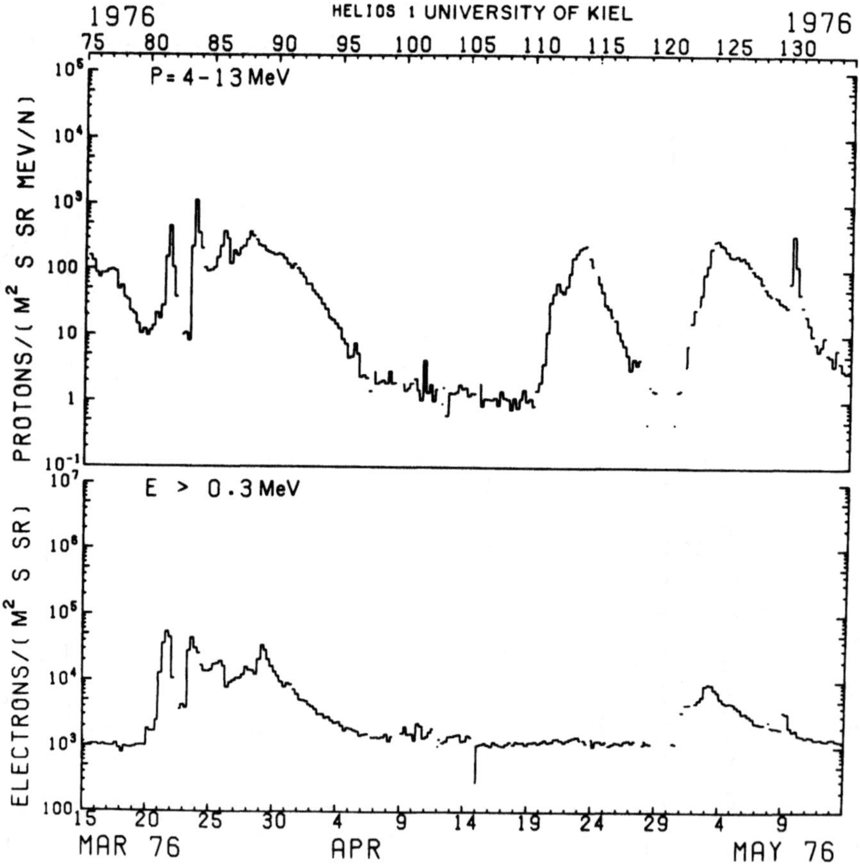

Figure 17. Six-hourly proton (4-13 MeV) and electron (> 0.3 MeV) fluxes from Helios 1 for the period 15 March-15 May 1976. Each tick mark represents one day. Data from the University of Kiel, FRG, courtesy of H. Kunow.

24 and 25 March that is about 2 orders of magnitude below the Helios 2 flux measurements and about 1 order of magnitude below the Helios 1 data. In fact the > 51 MeV proton channel on Helios 2 recorded the highest flux observed during the entire STIP Interval (Kunow et al., 1977b). The data from the Helios space probes will be a very valuable tool in ascertaining the solar coronal and interplanetary particle propagation conditions during this period.

There was a major terrestrial disturbance on 26 March with a sudden commencement at 0233 UT. The major geomagnetic storm that followed had an A_p value of 138, making it the fifth highest value of the 20th solar cycle. Intense aurora were observed at a number of mid-latitude locations. Figure 18 is an image of the auroral forms observed

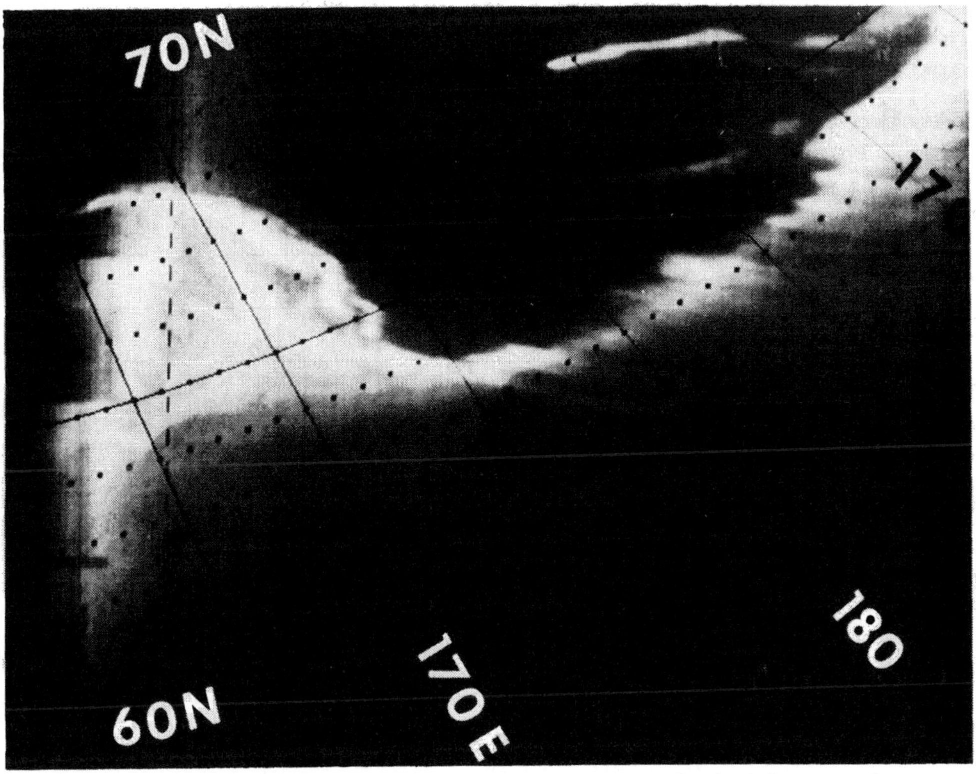

Figure 18. DMSP satellite image of the moderately bright continuous aurora plus discrete arcs and bands observed on 26 March 1976. A background geographic coordinate grid for 100 km altitude has been superimposed on the image for reference purposes. The equatorward edge of the bright continuous aurora corresponds to a corrected geomagnetic latitude of 58°. The vertical dashed line designates the satellite subtrack which passes over the equatorward edge of the aurora at 0925 UT. The satellite is in a dawn-dusk orbit; the sunset terminator appears to the left of the image.

at 0925 UT on 26 March. The intense aurora in this figure extends to 58° corrected geomagnetic latitude indicating a tremendous expansion of the earth's polar cap.

After McMath Plage 14127 rotated beyond the western limb of the sun on 25 March, McMath Plage 14143 was clearly the principal region of activity. The solar flare at $<$1905 UT on 28 March was an apparent particle producer as evidenced by the discrete impulsive increase observed on the IMP 7 and IMP 8 spacecraft at all energies (Solar-Geophysical Data, 1976). Although Helios 1 did not appear to detect a major particle increase associated with this event (see Figure 17), Helios 2 did observe a sudden impulsive increase on all energy channels (see Figure 16). The positions of the Helios 1 and Helios 2 space probes relative to the sun-earth line on 28 March are shown in Figure 19 together with the locations of McMath Plage 14143 and McMath Plage 14127.

Another major sudden commencement geomagnetic storm occurred at 0255 UT on 1 April with an A_p value of 107, and again aurora were reported at some mid-latitude locations (Solar-Geophysical Data, 1976).

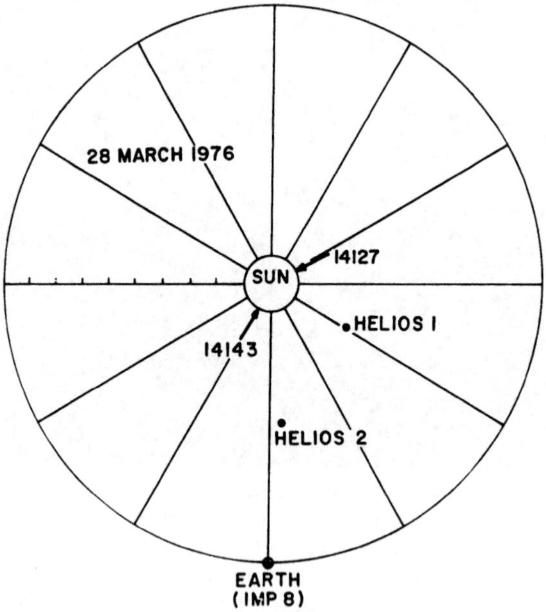

Figure 19. Positions of the Helios 1 and Helios 2 space probes relative to the sun-earth line on 28 March. The locations of McMath Plage 14143 and McMath Plage 14127 are indicated by arrows.

The 30 April solar cosmic ray event

McMath Plage 14143 rotated off the visible hemisphere on 6 April apparently diminishing in size and complexity. It returned to the eastern limb of the solar disk on 20 April 1976 at which time it was designated as McMath Plage 14179. Only five subflares were reported during its transit of the eastern hemisphere of the sun. However, on 27 April small new spots appeared, and in succeeding days they increased in number and complexity. On 29 April the frequency of subflares in McMath Plage 14179 increased markedly and continued for the next day.

The major flare from this region, at 2047 UT on 30 April, was classified as optical importance 1B accompanied by major X-ray and radio emission. This flare is of particular interest in that it was the source of a ground-level solar cosmic ray event. From examination of the neutron monitor data from 10 locations, the earliest onset of relativistic protons appears to be between 2120 and 2125 UT at Inuvik, Canada, which also had the largest increase of 12% as listed in Table 4. The relativistic particle flux in this event was quite anisotropic as illustrated by a comparison of the relative increases at Inuvik and Goose Bay, Canada, as shown in Figure 20. At lower energies there was an impulsive increase in the electron and proton flux as observed on the IMP 8 satellite (Figure 10).

Table 4

Relative Solar Proton Increase Observed by Selected Neutron Monitors for the Ground-level Event of 30 April 1976

Station	Relative Increase (%)*
Inuvik	12.0
Kerguelen Island	11.2
Tixie Bay	7.4
Alert	4.4
Thule	4.4
Goose Bay	2.7
Deep River	2.7
McMurdo	2.7

* Base line = Hourly average intensity for 2000-2100 UT on 30 April 1976.

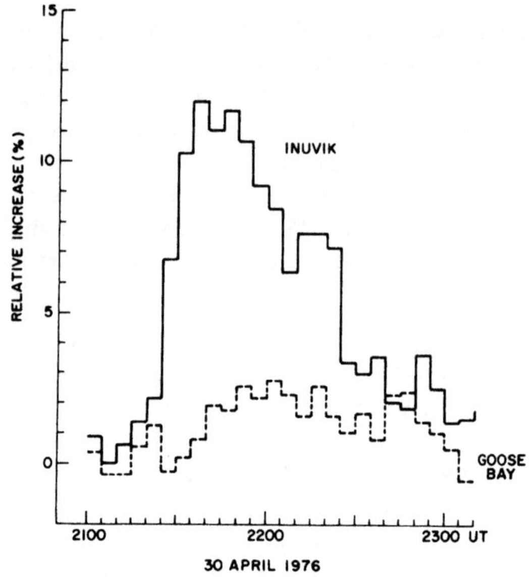

Figure 20. Relative neutron monitor increase observed at Inuvik and Goose Bay, Canada. The cutoff energy at both stations is determined by the intervening atmosphere.

Figure 21. Riometer abosrption and horizontal component of the geomagnetic field as observed at Thule, Greenland from 30 April through 5 May 1976. Figure courtesy of J. P. Turtle, Deputy for Electronic Technology, Rome Air Development Center, Hanscom AFB, Bedford, Massachusetts.

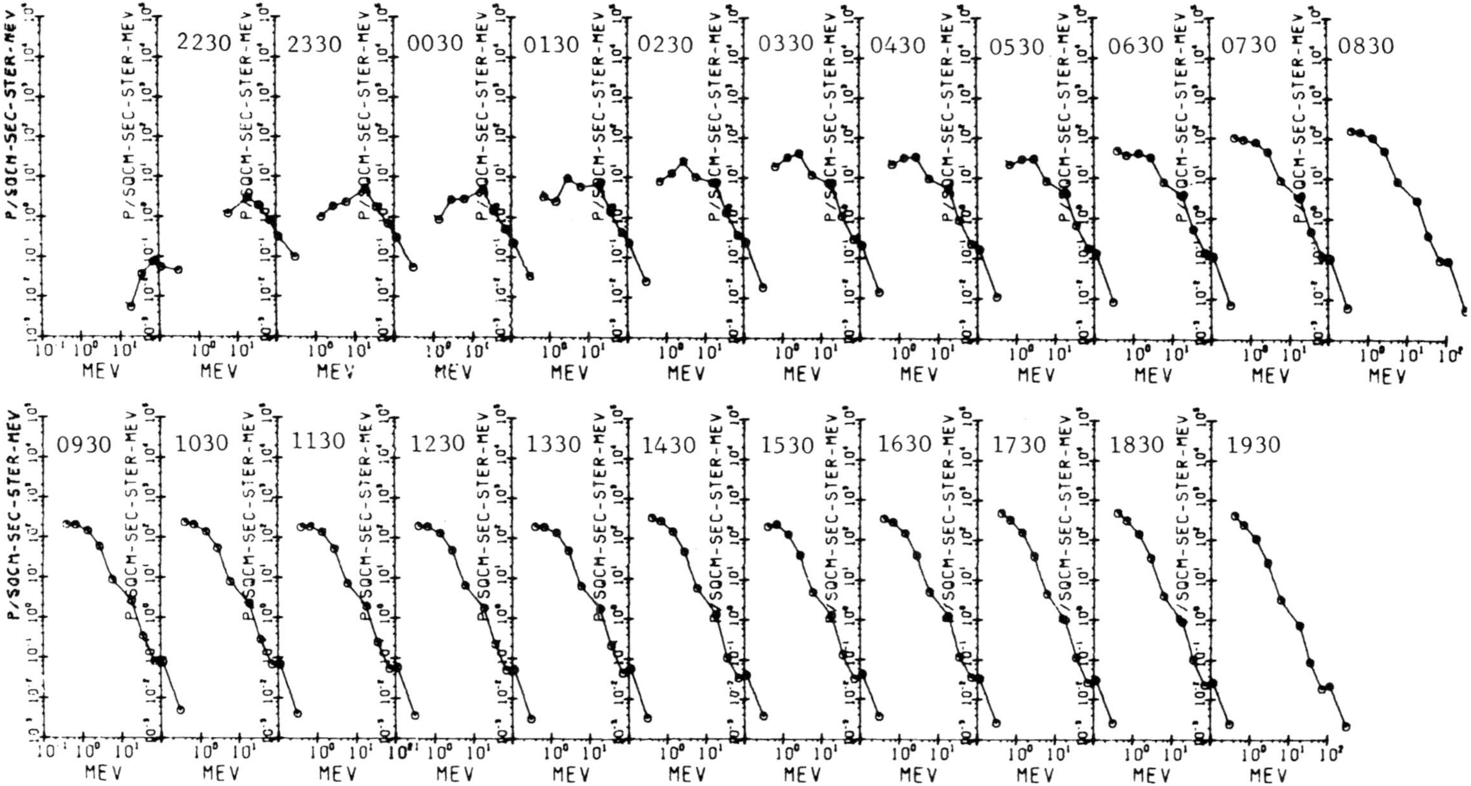

Figure 22. Solar proton spectra for late 30 April and continuing into 1 May 1976. The spectra were derived from hourly averaged data uncorrected for background from the JHU/APL experiment on IMP 8. (The apparent kink in the spectra near 100 MeV is an instrumental background effect and not a feature of the data.)

In the polar cap a maximum absorption of 2.7 dB was measured by the Thule, Greenland riometer as shown in Figure 21. The relatively modest low energy particle increase observed by IMP 8 (see Figure 10) and comparatively small riometer absorption, combined with the ground level neutron monitor increase, indicate a very hard particle spectrum. The measured differential energy spectra had a slope of approximately -2 after the flux at energies greater than 1 MeV had maximized. The velocity dependent evolution of the solar proton spectrum for the first day of the event (from 2130 UT on 30 April until 2000 UT on 1 May) as observed by the earth orbiting IMP 8 spacecraft is shown in Figure 22. A sudden commencement geomagnetic storm occurred at 1829 UT on 2 May with major disturbances for the next 24 hours, as indicated by the magnetometer record in Figure 21.

The positions of the Helios space probes on 30 April are shown in Figure 23 together with the positions of McMath Plage 14179. Although the flare position is favorably situated for efficient particle propagation from the sun to the earth, the flare location is $\sim 110^{\circ}$ to the east of the sun-Helios line. This event was observed on both the Helios 1 and Helios 2 space probes (see Figure 16 and 17), although the time-intensity

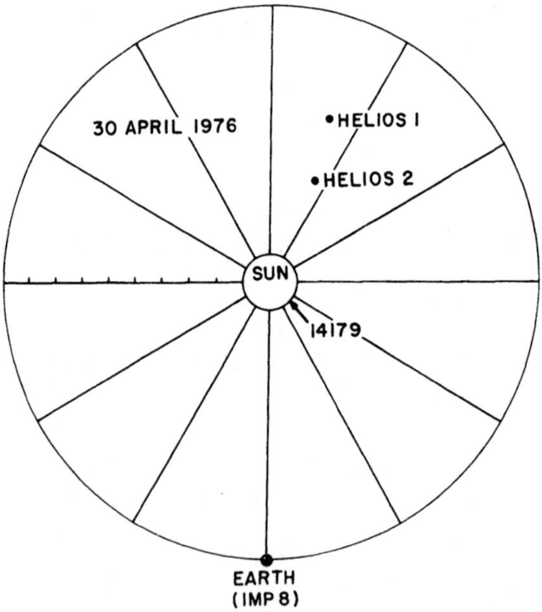

Figure 23. Positions of the Helios 1 and Helios 2 space probes relative to the sun-earth line on 30 April 1976. The location of McMath Plage 14179 is indicated by the arrow.

profile differs markedly between the two satellites (Kunow et al., 1977b). This impulsive, well isolated, relativistic solar cosmic ray event offers an unique opportunity to study solar particle propagation in the solar corona and in the interplanetary medium.

SUMMARY

STIP Intervals I and II were both during solar minimum although the solar activity that occurred during each interval was markedly different. An analysis of the solar activity, interplanetary particles, fields, and plasmas and the geomagnetic activity that occurred during STIP Interval I should result in a definition of quiet conditions during solar minimum. Conversely, similar analyses during STIP Interval II combined with the particles, fields and plasma observations of the Helios 1 and Helios 2 space probes provides interesting measurements of relatively isolated solar particle and plasma emissions during solar minimum. It is felt that the observations during these two intervals will provide an unique set of solar-terrestrial data measurements during solar minimum that may become a base line for future observations and analyses during the 21st solar cycle.

ACKNOWLEDGEMENTS

We wish to acknowledge the assistance of the following for providing data used in the preparation of this manuscript: R. E. Gold, Applied Physics Laboratory, Johns Hopkins University (IMP 7 and IMP 8 particle data); H. Kunow, University of Kiel, FRG (Helios 1 and 2 particle data); C. T. Russell, University of California, Los Angeles (for assistance in determining the Venera 9 and Venera 10 trajectories); J. P. Turtle, Deputy for Electronic Technology, RADC (Thule, Greenland riometer and magnetic field data); J. Warwick, University of Colorado Radio Astronomy Observatory (non-Io associated Jupiter radio emission data); J. Wolfe, J. Mihalov and H. Collard, NASA Ames Research Center Plasma Group (Pioneer 10 solar wind speeds); and C. J. Wolfson, Lockheed Palo Alto Research Laboratory (for providing the OSO-8 X-ray data prior to publication and for helpful discussions of the data).

REFERENCES

Akinjan, S. T., Bakunin, L. M., Chernov, G. P., Chertok, I. M., Kovalev, V. A., and Korolev, O. S.: 1977, in H. E. Coffey and J. A. McKinnon (eds.), Collected Data Reports for STIP Interval II, 20 March-5 May 1976 (UAG No. 61), World Data Center A for Solar-Terrestrial Physics, Boulder, Colorado.

Castelli, J. P., Aarons, J., and Michael, G. A.: 1967, J. Geophys. Res. 72, 5491.

Castelli, J. P., and Barron, W. R.: 1977, J. Geophys. Res. 82, 1275.

Coffey, H. E., and McKinnon, J. A. (eds.): 1977, Collected Data Reports for STIP Interval II, 20 March-5 May 1976 (UAG No. 61), World Data Center A for Solar-Terrestrial Physics, Boulder, Colorado.

Dodson, Helen W., and Hedeman, E. Ruth: 1971, An Experimental Comprehensive Flare Index and its Derivation for "Major" Flares, 1955-1969 (UAG No. 14) World Data Center A for Solar-Terrestrial Physics, Boulder, Colorado.

Dodson, Helen W., and Hedeman, E. Ruth: 1977, in H. E. Coffey and J. A. McKinnon (eds.), Collected Data Reports for STIP Interval II, 20 March-5 May 1976 (UAG No. 61), World Data Center A for Solar-Terrestrial Physics, Boulder, Colorado.

Kunow, H., Wibberenz, G., Green, G., Müller-Mellin, R., Witte, M., Hempe, H., and Fuckner, J.: 1977a, in M. A. Shea, D. F. Smart, and S. T. Wu (eds.), Contributed Papers to the Study of Travelling Interplanetary Phenomena/1977, Air Force Geophysics Laboratory, Bedford, Massachusetts (to be published).

Kunow, H., Müller-Mellin, R., Iwers, B., Witte, M., Hempe, H., Wibberenz, G., Green, G., and Fuckner, J.: 1977b, in H. E. Coffey and J. A. McKinnon (eds.), Collected Data Reports for STIP Interval II, 20 March-5 May 1976 (UAG No. 61), World Data Center A for Solar-Terrestrial Physics, Boulder, Colorado.

Nelson, G. J., and McLean, D. J.: 1977, in H. E. Coffey and J. A. McKinnon (eds.), Collected Data Reports for STIP Interval II, 20 March-5 May 1976 (UAG No. 61), World Data Center A for Solar-Terrestrial Physics, Boulder, Colorado.

Nolte, J. T., and Roelof, E. C.: 1973, Solar Phys. 33, 241.

Smerd, S. F.: 1970, Proc. Astron. Soc. Aust. 1, 305.

Solar-Geophysical Data: 1976, U. S. Department of Commerce, Boulder, Colorado.

Švestka, Z., and Fritzová-Švestková, L.: 1974, Solar Phys. 36, 417.

Wolfson, C. J., Acton, L. W., Roethig, D. T., and Smith, K. L.: 1977, in H. E. Coffey and J. A. McKinnon (eds.), Collected Data Reports for STIP Interval II, 20 March-5 May 1976 (UAG No. 61), World Data Center A for Solar-Terrestrial Physics, Boulder, Colorado.

OVERVIEW OF THE HELIOS 1 AND HELIOS 2 MISSIONS AND THEIR PARTICIPATION IN STIP INTERVALS I AND II

H. PORSCHE
Deutsche Forschungs- und Versuchsanstalt für Luft- und Raumfahrt, Oberpfaffenhofen, F.R.G.

ABSTRACT

HELIOS 1 and 2 are spacecraft with highly eccentric orbits (perihelion at 0.3095 and 0.290 AU) to investigate the innermost parts of the interplanetary space. Since the orbital velocity is varying by more than a factor of 3 between aphelion and perihelion and since the angular velocity of the spacecraft approaches half the angular velocity of the sun, the orbit properties influence the scientific objectives of the mission more than in other missions. As an introduction to the contributions of HELIOS to STIP investigations the orbital peculiarities of the two probes, their relation to one another and to the earth are discussed. Additionally some characteristics of the probes show how they are adapted to specific aspects of their research tasks. The numerous radial and magnetic spiral lineups between the two spacecraft or between one of them and the earth, and the long phases of the spacecraft near superior conjunction attain special importance.

1. INTRODUCTION

HELIOS is a joint venture of the United States (NASA) and the Federal Republic of Germany (BMFT) to investigate the innermost parts of the interplanetary space. Two spacecraft have been launched into highly elliptical orbits in the ecliptic plane. The scientific goals of HELIOS are

- to investigate the solar-interplanetary plasma between sun and earth, its particles, steady-state fields, oscillations and waves;
- to study solar and galactic cosmic rays, their mass and energy spectra and their propagation and correlation to solar X-ray events;
- to investigate the interplanetary dust, its density, distribution, particle orbits and chemical composition;
- to determine the orbit with very high precision in order to improve the observational verification of the theory of general relativity; and

M. A. Shea et al. (eds.), Study of Travelling Interplanetary Phenomena/1977, 421–429.

- to sound the solar corona by radio science methods when HELIOS is near outer conjunction.

For the implementation of their scientific tasks each of the two HELIOS probes has ten experiments on board as listed in Table I. In addition, this payload is supplemented by two passive experiments as listed in Table II.

TABLE I

ACTIVE EXPERIMENTS ON HELIOS 1 AND HELIOS 2

Experiment	Principal Investigator(s)
E 1 Plasma Particle Experiment	H. Rosenbauer, MPE Garching
E 2 Flux Gate Magnetometer	G. Musmann, Inst. f. Geophys. u. Meteorologie, Univ. Braunschweig
E 3 Flux Gate Magnetometer	F. Mariani, Lab. Plasma Spacio, Frascati, Italy N. Ness, NASA-GSFC
E 4 Search Coil Magnetometer	G. Dehmel, Inst. f. Nachrichtentechnik, Univ. Braunschweig
E 5 Plasma- and Radio-Wave Experiment	D. Gurnett, U. of Iowa, Iowa City P. Kellogg, U. of Minnesota, Minneapolis R. Weber, NASA-GSFC
E 6 Cosmic Ray Telescope	H. Kunow, Inst. f. Kernphysik, Univ. Kiel
E 7 Cosmic Ray Telescope X-Ray Monitor Gamma Burst Discriminator*	J. Trainor, NASA-GSFC
E 8 Low Energy Cosmic Ray Detector	E. Keppler, MPAe, Katlenburg-Lindau
E 9 Zodiacal Light Photometer	C. Leinert, MPA, Heidelberg-Königstuhl
E 10 Micrometeroid Analyzer	E. Grün, MPK, Heidelberg

*HELIOS 2 Only

TABLE II

PASSIVE EXPERIMENTS ASSOCIATED WITH HELIOS 1 AND HELIOS 2

Experiment	Principal Investigator
E 11 Celestial Mechanics	W. Kundt, Inst. f. Theor. Physik Univ. Hamburg W. Melbourne, JPL Pasadena
E 12 Corona Sounding	G. Levy, JPL Pasadena H. Volland, Inst. f. Astronomie, Univ. Bonn P. Edenhofer, DFVLR Oberpfaffenhofen R. Woo, JPL Pasadena

The first eight experiments listed in Table I (the plasma and cosmic ray experiments) have contributed to the investigations during the STIP Intervals. Unfortunately not all experimenters were able to attend the L. D. de Feiter Memorial Symposium in Tel Aviv to present their results at the STIP Conference.

2. THE HELIOS MISSION

The mission is characterized by two solar-orbiting spacecraft in highly eccentric orbits contained within 1 AU radial distance of the sun. Due to their high orbital velocity, each HELIOS probe travels toward the sun and back twice a year, almost an ideal orbit for gradient measurements in the interplanetary medium. The perihelia are at solar distances of 0.3095 and 0.290 AU, and the periods are 190.13 and 185.67 days respectively.

The orbital configurations between the two HELIOS space probes and the earth (including the earth-orbiting satellites such as IMP 7) are ideal for the undertaking of many experimental investigations. The orbital geometry of these two space probes is illustrated in Figure 1. HELIOS 1 was launched on 10 December 1974 from the NASA Eastern Test Range, Cape Canaveral, Florida, U.S.A. Since the earth crosses the solar equatorial plane (descending node) on 6 December and every half year later (ascending node), the apside line of HELIOS 1 is only about 4 degrees apart from the nodal line. Thus this spacecraft spends about half of its orbit time south of the solar equator and half of its time north of the equator. The nodal line crossing appears near perihelion, when the orbital velocity is at its maximum. The orbital velocity is 20.5 km s^{-1} at aphelion and 65.2 km s^{-1} at perihelion. That means the relative angular velocity between HELIOS 1 and the sun's rotation is 13.5 degrees per day at aphelion and 7.2 degrees per day at perihelion, compared to the average relative angular velocity between the sun and the earth of 13.2 degrees per day. The small relative angular velocity

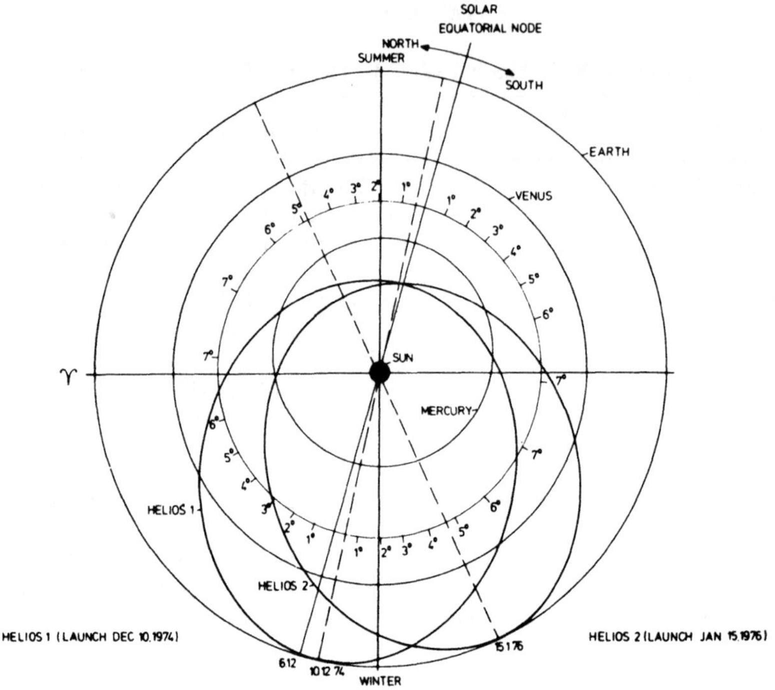

Figure 1. Orbital geometry of the two HELIOS probes. The inserted circular scale denotes the solar latitude of the ecliptic (= HELIOS orbit plane).

at perihelion is very well suited for discrimination between spatial and temporal phenomena.

Even more important than the variation of this angular velocity is the latitudinal change of the spacecraft's position during each orbit. Since the orbital velocity is very high in perihelion and the nodel line is close to the apside line, the latitudinal change is 0.9 degree per day in perihelion and more than 12 degrees (-6° to $+6^o$) within only 16 days. In comparison the earth is almost stationary. During the first perihelion passage of HELIOS 1 the earth was positioned at almost maximum southern latitude, whilst the spacecraft was crossing through its whole latitudinal range.

HELIOS 2 was launched on 15 January 1976. Its orbit geometry with respect to the sun and the earth is different from that of HELIOS 1. The angle between the solar nodal line and the apside line is approximately 40 degrees. Therefore, the orbit is no longer symmetrical with respect to the solar equator. The spacecraft stays south of the sun's equator for 134 days and north for less than 52 days per orbit. Whereas the perihelion passage of HELIOS 1 is always very close to the solar equator, the perihelion of HELIOS 2 is at a solar latitude of +4.9

degrees. Of course this is by far not a high latitude. The latitudinal changes are large prior to each perihelion; right after perihelion the change is between +4.9 and +7.15 degrees, which is small compared to the latitudinal velocity near perihelion of HELIOS 2. One can therefore consider HELIOS 2 as almost stationary with respect to solar latitude at perihelion.

An unique feature of the HELIOS mission is the relative orbital configurations of the two spacecraft, these configurations being essential to meet the objective of the HELIOS mission. A plot of the first orbit of HELIOS 2 in a sun-earth centered coordinate system is shown in Figure 2. Also illustrated are the relative positions of both HELIOS space probes with respect to the earth during STIP Interval II (15 March-15 May 1976) - a period of time when, fortunately, the sun was relatively active. Orbital positions of special interest were the radial lineups with their high importance for the investigation of the radially flowing plasma and the spiral lineups with the two spacecraft positioned on the same magnetic field line. These are the only positions that allow the gradients of velocities, densities, etc. to be measured without doubtful influences of diffusion effects and longitude or latitude dependent solar influences. It was by chance that the PIONEER 11 spacecraft was also in the same angular region of spiral lineup; thus for the first time a review of the state of the interplanetary space between its innermost parts within the orbit of Mercury to the region of Jupiter could be achieved.

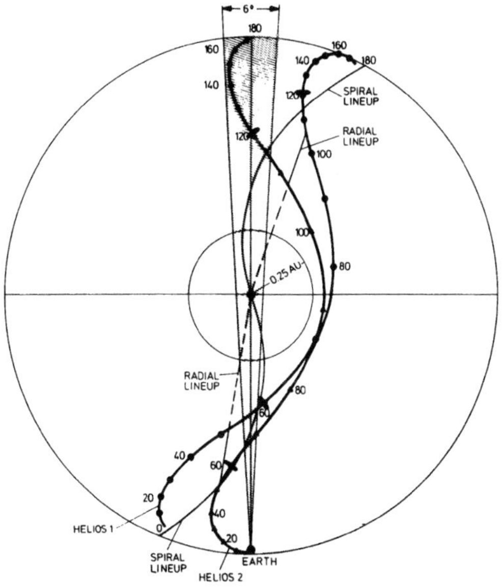

Figure 2. The orbit of HELIOS 1 and 2 in a sun-earth centered corotating coordinate system. The plot starts on 15 January 1976, the launch day of HELIOS 2. The heavy marks at day 60 and day 121 indicate the beginning and end of STIP Internal II (15 March - 15 May 1976).

The in situ research of HELIOS is supplemented by radio science investigations of the corona. Figure 2 shows that the orbit of HELIOS is ideal for these investigations. For more than two months during the first orbit of each spacecraft, the sun-earth-probe angle is less than 3 degrees. These orbital configurations were used for Faraday Rotation measurements ($k/f^2 \cdot \int n\vec{B}d\vec{s}$), for electron content ($\int nds$) and hence density measurements using the time delay of the signals, and for the investigation of the plasma-induced broadening of the bandwidth of the signals. This is the first time it has been possible to make such observations of the corona for several consecutive solar rotations. The results on coronal magnetic structure, density and its fluctuations, and solar wind characteristics have extended the measurement region of HELIOS from the entire range covering the earth's orbit to the medium and lower corona. While approaching the sun, the signal was maintained until the minimum distance between the ray path and the sun was less than 1.7 $R_\odot$ at which point the signal was lost.

3. CONTRIBUTIONS OF HELIOS TO THE STIP INTERVALS

During STIP Interval I (September-October 1975) the contributions of HELIOS were comparatively poor for two reasons. First, only one of the two spacecraft had been launched, and second, except for the perihelion phase around the 21st of September the ground station support was poor due to the low priority of the mission during that period, even though high bit rates could be transmitted, since the spacecraft was not too far from the earth.

The unique orbital configurations of the HELIOS 1 and 2 space probes were the principal reason for the selection of the dates for STIP Interval II. Due to the high bit rate and the excellent support by ground stations - almost 100% for HELIOS 2 and more than 60% for HELIOS 1 - the HELIOS contributions were successful. The high solar activity during that period favored the thorough investigations of particles and fields events and their propagation into space, and several papers from HELIOS investigators presenting results obtained from data acquired during STIP Interval II will be given during this conference.

Several weeks ago STIP Interval III (15 March-15 May 1977) ended. The dates for this interval were also selected because of the uniqueness of the orbital configurations of the two HELIOS probes and the earth. Although the sun was not as active as in STIP Interval II, we still hope to be able to again present some good results at the COSPAR conference in 1978, since the ground station support was excellent. Additionally, at the present time, we are in the last long near-superior conjunction phase of HELIOS 2 as shown in Figure 3. Investigations of the Faraday rotation and of the signal bandwidth are being conducted includin correlating analysis of the signal amplitude and bandwidth arriving at two long-distant stations.

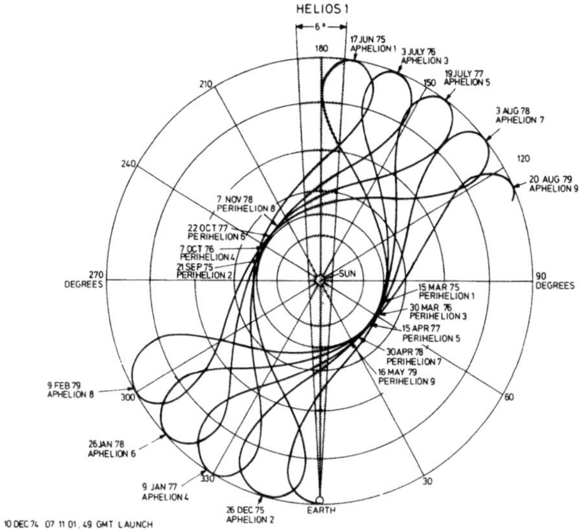

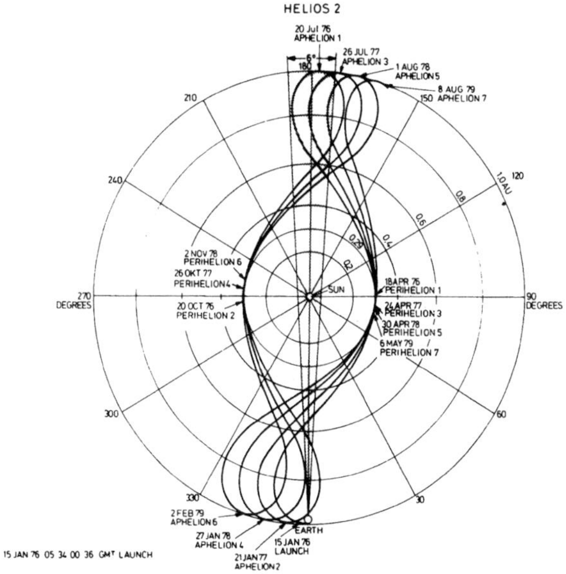

Figure 3. Orbit of HELIOS 1 and 2 in sun-earth centered corotating coordinate system.

4. FUTURE ASPECTS

In the beginning of this interplanetary project, HELIOS, there was some concern that one or both spacecraft might not survive their technical development goal of 18 months lifetime in orbit. During the initial orbit the thermal control surface experienced an ~ 5% degradation; however, the degradation rate subsequently stabilized to a lower value. This considerably lower degradation rate indicates not only that anticipated 18 month design lifetime would be achieved, but also that the lifetime of the entire mission might be considerably extended. In fact, the lifetime of the mission may approach a significant portion of a solar cycle and consequently, long term aspects as the changing solar-interplanetary relationships will be of considerable interest and importance. Plans are currently being made for another STIP Interval to be conducted late in 1977 (tentative dates 15 October-15 December) when, once again, the HELIOS space probes will both undergo perihelion. Even better research opportunities than existed during STIP Intervals I and II are expected to be available for this interval as the two VOYAGER space probes, scheduled for launch in August 1977, will be enroute to Jupiter. Thus a survey of the interplanetary space from its innermost parts to beyond the Asteroid belt and eventually as far as Jupiter might be feasible during the next two years.

5. SOME SPACECRAFT ASPECTS

A photograph of one of the two almost identical spacecraft is shown in Figure 4. Considerations of the spin stabilization and of the thermal balance led to the spoollike shape of the spacecraft as it enlarges the areas of reradiation of the heat. Overheating of the central compartment is prevented by second surface mirrors and by louvers inside the shadowed cone. Unfortunately the shape of the spacecraft and its isolating irradiated surface has induced some problems for the scientific mission. The measurements of very low energy plasma electrons (down to less than 1 eV) require a well defined low spacecraft potential, that should be of spherical or similar simple geometry. Unfortunately, the real spacecraft field is considerably different. To improve this situation a belt of conductive coating around the central compartment smoothes out some of the field disturbances, so that the measurement of those low energy electrons and of the floating potential of the spacecraft is possible. Except for a failure on HELIOS 1 when deploying one of the two branches of the antenna to measure electric field fluctuations, plasma and radio waves (it stuck and caused a shortage to the structure), no serious disturbance of the on board experiments occurred. The passive experiments, however, have been restricted to no ranging and no two-way Doppler measurements since last September. We have lost both ranging modules, and we must avoid coherent mode operation, because a detailed failure analysis has shown that in some adverse cases coherent operation is suspected to cause technical problems on board, which might kill the spacecraft.

6. CONCLUSION

HELIOS is a double mission offering numerous aspects essential for the interplanetary research in general and especially for STIP investigations. Due to the orbital characteristics of each of the two spacecraft many interesting configurations occur such as magnetic and radial lineups between the two probes and between one of the probes and the earth. If we succeed in keeping at least one of both spacecraft alive, the opportunity of conducting correlative analysis together with the data from the VOYAGERS will arise, enabling the scientific community to undertake more detailed study of the properties of matter in space between the sun and the outermost parts of the solar system.

7. ACKNOWLEDGMENT

The two HELIOS probes were built, launched and supported by joint efforts of the United States' National Aeronautics and Space Administration and by the Federal Republic of Germany's Federal Minister of Research and Technology.

Figure 4. Photograph of HELIOS 1 prior to launch.

INDEX OF AUTHORS CITED

When a name appears in the references at the end of a paper, the page number is italicized.

Aarons, J., *420*
Acton, L. W., *420*
Akinjan, S. T., 410, *419*
Alekseyev, I., 372, *387*
Alfven, H., 297, 299, *302*, 305, 306, 315, 316, *317*
Allen, C. W., 70, *79*
Allum, F. R., *41*, *340*, 345, 362, *362*, *388*
Alpers, W., 338, *340*
Altschuler, M. D., *61*, *162*, 187 *190*
Alvarez, H., 75, *79*, 80
Ananthakrishnan, S., *140*, 315, *317*
Anderson, J. D., *98*
Anderson, K. A., 36, *40*, *41*, *339*
Anderson, R. R., 71, *79*, 260, *273*
Antonov, N. M., *288*
Arens, J., 379, 384, *388*
Armstrong, J. W., 95, *97*, 102, 104, 114, *117*, 120, 131, *139*
Armstrong, T. P., *273*, *363*, 368, 369, 370, 373, 379, 380, 381, 382, 383, *387*, *388*
Arnoldy, R. L., 38, *40*
Arpigny, C., 283, *288*
Asbridge, J. R., *19*, *117*, *161*, *162*, *190*, *191*, *223*, *256*, *273*, *363*
Aubier-Giraud, M., *80*
Audouze, J., *40*
Avignon, Y. *79*
Axford, W. I., 146, 149, 153, 154, 155, 160, *161*, *162*, *163*, 164, 167, *192*, 295, *302*, 384, *388*
Axisa, F., 72, *79*

Bai, T., 36, *40*
Bakunin, L. M., *419*
Bame, S. J., *19*, *117*, *140*, 152, 153, *161*, *162*, 187, *190*, *191*, *192*, 220, 221, *223*, 225, 230, *256*; 271, *273*, 349, *363*
Banks, P. M., 173, *190*
Baoischot, A., *80*
Barber, D. J., *363*
Barnes, A., 90, *97*, 167, *190*, *191*, 299, *302*
Barnes, C. W., 155, *161*, 251, *256*
Barnstorf, H., *273*
Barouch, E., 253, *256*, 339, *339*
Barron, W. R., 411, *420*
Baumback, M. N., 75, *79*
Beard, D. B., 295, *302*
Behannon, K., *387*
Beinroth, H. J., *273*
Belcher, J. W., *98*, 124, *139*, 228, 247, *256*, 299, *302*
Bessey, R. J., 59, *60*
Beyer, M., 287, *288*
Bhandari, S. M., *140*, *317*
Biermann, L., 165, *190*, 291, 292, *302*, 306, 309, 311, 312, 316, 317, *317*
Biesbroeck, G., 288, *288*
Birmingham, T. J., 330, *339*
Biswas, S., 181, *190*
Blesing, R. G., 87, 89, *97*
Bogdanov, A. V., 280, *288*
Bohlin, J. D., 10, *18*, *19*, 149, 152, 154, *161*, 188, *190*
Bohlin, R. C., *303*
Bollea, D., 179, *190*
Bougeret, J. L., 69, 70, *79*
Bourgois, G., 120, *139*
Braginskii, S. I., 171, 173, *190*
Brandt, J. C., *99*, *193*, 292, 297, *302*, *303*, 315, *317*, *318*
Bredekamp, J. H. *190*
Bridge, H. S., 220, 221, *223*, 225

Briggs, B. H., 102, *117*
Brosowski, B., 281, *288*, 292, *302*, 306, 309, 317, *317*
Brown, J. C., 25, 26, 30, 32, 35, *40*
Brune, W. H., 294, *302*
Bukata, R. P., *363*
Bundy, R. B., *79*
Burlaga, L. F., 247, 248, 253, *256*, 339, *339*

Callahan, P. S., 92, 94, 96, *97*
Candelaria, C., *256*
Cannon, A. R., 82, *97*
Caroubalos, C., 67, *79*, *80*
Carovillano, R. L., 186, *190*
Castelli, J. P., 411, *420*
Cecchini, S., 158, *161*
Chandrasekhar, S., 308, 317, *318*
Chang, C. C., 166, *192*
Chang, H., 82, 89, 90, 91, 96, *97*
Chao, J. K., *60*
Chapman, R. D., *41*
Chapman, S., 165, 170, *190*
Chashey, I. V., *118*
Chen, G., 368, 369, 370, 372, 373, 375, 378, *388*
Chenette, D. L., 160, *162*, 230, 253, *256*
Chernikov, A. A., *288*, 294, 298, *302*, 307, 313, *318*
Chernov, G. P., *419*
Chertok, I. M., *419*
Chubb, T. A., 26, *40*
Chupp, E. L., 27, 38, *40*
Churumov, K. I., 288, *288*
Clark, T. A., *98*
Cloutier, P. A., 294, *302*
Coffey, H. E., 404, *420*
Cohen, M. H., 134, *139*
Colburn, D. S., *140*, *163*, *224*, 228, *256*, *257*
Coleman, P. J., *163*, *224*, *256*, *257*
Coles, W. A., 81, 83, 93, 96, *97*, 102, 103, 104, 105, 106, 107, *117*, 120, 121, 122, 125, 127, 128, 130, 131, 134, 135, 136, 138, 139, *139*
Colgate, S. A., 26, 27, 35, *40*
Collard, H. R., *140*, 196, *223*, *224*, 234, *256*
Combi, M. R., 294, 295, *302*
Conlon, T. F., *162*, 253, *256*
Counselman, C. C., 97, *97*
Cowsik, R., *363*
Craig, I. J. D., 59, *60*
Crockett, W. R., *118*, *140*
Croft, T., 81, *97*
Cronyn, W. M., 90, 92, *97*. 120, 131, 134, 138, 139, *139*, *140*
Crook, G. M., *273*
Cruikshank, D. P., *303*
Cuperman, S., 166, 167, 175, 176, 177, 180, 181, 186, 187, 188, 189, *190*, *191*

D'Angelo, N., 294, 298, 300, *302*, 305, 317, *318*
Daniell, R. E., Jr., 294, *302*
Datlowe, D. W., 32, *41*
Davidson, R. C., 280, *289*
Davis, L., 124, *139*, *163*, 166, *192*, *224*, 228, 247, *256*, *257*, 299, *302*
Decker, R., 373, *388*
Dehmel, G., *273*
Delsemme, A. H., 294, 295, *302*
Demenko, A. A., 288, *288*
Dennis, B. K., 36, *40*
Dennison, P. A., 87, 89, *97*, 101, 104, 105, *117*, 119, 120, 125, 131, 132, 134, *140*, *141*
Dere, K. P., 33, *40*
Dessler, A. J., 294, *302*
Detwiler, C. R., *19*
De Young, D. S., 132, *140*
Diodato, L., 215, *223*
Dobrovolskij, O. V., 315, *318*
Dobrowolny, M., 294, 298, *302*, 305, 317, *318*
Dodge, J. C., 8, *18*
Dodson, H. W., 219, 222, *223*, 402, 410, *420*
Domingo, V., 339, *339*
Donn, B. D., 308, *318*
Dryer, M., *20*, 33, *40*, 52, 53, 55, 56, 58, *60*, *61*, 62, 116,

117, *118*, 186, 187, 188, 189, *190*, *191*, *192*, *193*, 250, *256*, *257*, 274, *288*, 288, *289*, 308, 314, *318*, *319*
Duggal, S., 36, *41*, 379, *388*
Dulk, G. A., 6, 15, 18, *18*, *20*, 187, *191*
Durney, B. R., *163*
Duvall, T. L., *163*
Dyal, P., *163*, *224*, *257*

Earl, J. A., 329, 336, 338, *339*, 345, *363*
Eddington, A. S., 285, 286, *288*
Eddy, J. A., 3, *18*, *60*
Efimov, A. I., *99*
Ekers, R. D., 87, 89, 96, *97*, 102, *118*
Enskog, D., 170, *191*
Entwyler, H., *191*
Epstein, G., 67, *79*
Erickson, W. C., 89, *97*
Ershkovich, A. I., 287, *288*, 294, 298, 299, 300, *302*, 305, 307, 309, 312, 313, 314, 315, 316, *318*, *319*
Erskine, F., 120, 122, 123, 124, 125, *140*
Esposito, P. B., *98*
Evans, L. G., *79*
Eviatar, A., 175, *191*, 314, *318*

Fahr, H. J., 167, *191*
Fainberg, J., 74, 75, *79*, *80*
Fairfield, D. H., 345, *362*, *363*
Fälthammar, C. -G., 299, *302*
Fan, C. Y., *40*, *162*, *363*, *388*
Feldman, P. D., 294, *302*
Feldman, W. C., 11, *19*, *98*, *117*, *161*, *162*, 187, *190*, *191*, *223*, *256*, *273*, *363*
Fichtel, C. E., 185, *190*
Fisher, R. R., 13, *19*
Fisk, L. A., 157, 158, 160, *162*, 329, *339*
Fitzenreiter, R. J., 73, 75, 76, *79*, *80*
Fokker, A. D., 69, *79*
Formisano, V., 179, *190*, *191*
Forrest, D. J., *40*
Forslund, D. W., 175, *191*
Fowler, W. A., *40*
Frank, L. A., *224*, *273*
Fredericks, R. W., 259, *273*
Fritzová-Švestková, L., 411, *420*
Frohlich, A., *318*
Frost, K. J., 36, *40*
Fuckner, J., *339*, *340*, *420*

Garcia, C. J., *19*, *60*
Garrard, T. L., 348, *363*
Gary, S. P., 267, *273*
Geiss, J., 167, 181, *191*
Gentry, R. A., 33, *40*, 166, *191*
Gergely, T., *20*
Gleeson, L. J., 158, *162*, *163*, 356, *362*, *363*
Gloeckler, G., 37, *40*, *162*, *363*, 379, 383, *388*
Goertz, C. K., *256*, *388*
Gold, R. E., 334, *339*, 351, 354, *363*
Gold, T., 148, *162*
Goldstein, B. E., 82, 83, 89, *97*, *141*, 245, *256*
Goldstein, M. L., 158, *162*, 330, *339*
Goldstein, R. M., *98*
Gosling, J. T., 4, 6, 14, 16, 17, *18*, *19*, *21*, 35, *40*, *60*. 107, 110, *117*, *161*, *162*, 186, 187, *190*, *191*, 215, 220, 221, *223*, 225, 228, 245, *256*, *273*, *363*
Gotwols, B. L., *140*
Green, G., 338, *339*, *340*, *341*, *420*
Green, I. M., *273*
Greenstadt, E. W., 259, *273*, 274
Guigay, G., 288, *288*
Gundermann, E. J., *139*
Gurnett, D. A., 71, *79*, 260, *273*
Gurvich, A. S., 82, *97*

Haddock, F. T., 75, *79*, 80
Hallinan, T. J., 300, *303*
Hamilton, D. C., 325, 326, 327, 328, *339*, 352, 353, *363*
Han, S. M., *61*, *193*
Hansen, R. T., 13, *19*, *20*, 47, *60*, 107, 108, 110, 114, *117*, *223*
Hansen, S. F., *19*, *20*, *60*, *117*

Hardebeck, E. G., 136, *140*
Haedebeck, H. E., *139*
Harmon, J. K., 105, 110, *117*, 123, 134, 135, 136, 137, 138, 139, *139*, *140*
Harries, J. R., *362*
Harrington, J. V., 82, *98*
Harrington, R. S., *302*
Harris, D. E., 98, 120, 131, 136, *140*
Harten, A., 166, 167, 175, 180, 181, *190*
Hartle, R. E., 90, *97*, 166, 167, *190*, *191*, *192*, 280, *289*
Harvey, J. W., *98*, *162*
Harwit, M., 298, *302*
Hedeman, E. R., *23*, 402, 410, *420*
Hedgecock, P. C., *190*, 255, *256*
Heller, A. B., 309, 313, *318*
Hempe, H., *339*, *340*, *341*, *420*
Hénoux, J. C., 59, *60*
Hewish, A., 81, 87, 89, 96, *98*, 101, 104, 105, 110, *117*, *118*, 120, 121, 122, 123, 124, 125, 131, 134, *140*, 141
Hildner, E., 5, 6, 7, 12, *19*, *40*, *60*
Hirschberg, J., 56, *60*, 179, 186, *191*
Hirt, P., *191*
Hoang, S., 72, *79*, 80
Hollweg, J. V., 16, *19*, 82, *98*, 120, 134, *140*, 186, *191*
Holzer, T. E., 149, 153, *162*, 189, *191*, 194
Horan, D. M., *40*
Houminer, Z., 89, *98*, 114, *117*, 120, 122, 123, 124, 127, 128, 131, 132, 133, 134, 135, 138, 139, *140*, 156, *162*
Hovestadt, D., *40*, *162*, 348, 350, *363*, *388*
Howard, R. A., 4, *19*, *20*
Hoyle, F., 298, *302*
Hoyng, P., 27, *40*, *41*
Hudson, H. S., 23, 25, 27, *41*
Hudson, P. D., 368, 369, 372, *388*
Hundhausen, A. J., 5, 17, *19*, 52, 53, *60*, 90, *108*, 132, *140*, *141*, 149, *162*, 166, 175, 186, 187, *191*, *192*, 228, 245, 248, *256*, 260, *273*
Huntress, D., 295, *302*
Huppler, D., 301, *302*
Hurford, G. J., 348, 350, *363*
Hutcheon, I., *363*
Hyder, C. L., 299, 300, *303*, 312, 313, 314, *318*

Innanen, W. G., 359, 360, 361, *363*
Intriligator, D. S., 91, *98*, 104, 116, *118*, 138, 139, *140*, 195, 196, 198, 205, 212, 215, 216, 219, 221, 223, *224*, 225, *256*
Ioffe, Z. M., 294, 298, *303*, 305, 308, *318*
Ip, W. -H., 283, 284, 285, *288*, 293, 295, 296, 297, 299, 301, *302*, *303*, *318*
Ipavich, F. M., 262, *273*, 369, *388*
Ishimaru, A., 94, *98*, *99*
Iwers, B., *339*

Jackson, B. V., 16, *19*, 153, *162*
Jackson, J. D., 25, *40*
Jacobs, A., *318*
Jacques, S., *18*
James, J. C., 89, 90, *98*
Jockers, K., 300, 301, *303*, 306, 312, 314, 317, *318*
Jokipii, J. R., 81, 85, 90, 93, 96, *98*, 104, *118*, 120, 134, *140*, 153, 157, 160, *162*, 221, *223*, 230, 245, *256*
Jones, D. E., *163*, *224*, *257*
Jones, F. C., 330, *339*
Jordan, C., 148, *162*
Joseph, J. H., *318*

Kahler, S. W., 26, 32, *40*, *41*, 330, *339*
Kaiser, T. B., 330, 334, *339*
Kakinuma, T., 98, 102, 103, 104, 114, 116, *118*, 120, 121, 131, *140*, *141*, 149, *162*

Kalinin, A. P., *288*
Kane, S. R., 23, 38, *40*, *41*
Karpinsky, I. P., *288*
Kaufman, J. J., 102, 103, 104, *117*
Kearney, P. D., *363*
Keath, E. P., 355, *363*
Keller, H. U., 284, *288*
Kellogg, P. J., 74, 78, *79*
Kendall, W. B., *141*
Kent, M. I., 44, *60*
Keppler, E., *388*
Kerdraon, A., 69, *80*
Khazanov, B. I., *288*
Kivelson, M. G., *256*
Klecker, B., 159, *162*, *363*
Kneer, F., 59, *60*
Knight, C. A., 92, *98*
Koehler, R. L., 81, 92, *98*
Kojima, M., 103, 104, *118*, *141*
Koomen, M. J., *19*, *20*
Kopp, R. A., 148, 153, *162*, 189, *191*, 194
Korolev, O. S., *419*
Koslovsky, B., *162*
Kostyuk, N. D., 59, 60, *60*
Kovalev, V. A., *419*
Krall, N. A., 269, *273*
Kreplin, R. W., 32, 40, *41*
Krieger, A. S., *19*, *20*, *118*, 149, *162*, *163*, 187, 188, *191*, *192*
Krimigis, S. M., *140*, *273*, 325, 330, *340*, 352, *363*, 379, 381, *387*, *388*
Kropotkin, A., 372, *387*
Kruglov, Yu. M., *99*
Kundu, M., *20*
Kunow, H., 330, 334, 335, 336, *339*, *340*, *341*, 411, 413, 419, *420*
Kunstmann, J. E., 338, *340*
Kuperus, M., 59, *60*
Kurth, W. S., *79*

Lambert, D. J., 179, *192*
Landau, L. D., 306, 307, 308, *318*
Landt, J. A., 81, *98*
Lanzerotti, L. J., 325, 330, *340*, *341*, 379, 381, *388*
Lazarus, A. J., *98*
Leblanc, Y., 71, *80*
Lee, L. C., 96, *98*, 104, *118*
Lee, M. A., *340*
Leer, E., 146, *162*
Lepping, R. P., 262, *273*
Leroy, J. L., 70, *80*
Levine, R. H., 152, *162*
Levush, B., *191*
Levy, E. H., 149, 158, *162*, 379, *388*
Levy, G. S., 81, *98*
Lezniak, J. A., *256*
Lifschitz, E. M., 306, 307, 308, *318*
Lillie, C. F., 284, *288*
Lin, R. P., 23, 25, 27, 32, 36, *41*, 330, *340*, 345, *363*
Lingenfelter, R. E., 28, *41*
Little, L. T., 87, 89, 96, *97*, 102, 110, *118*, 120, 134, 135, 137, *140*
Liu, C. K., *192*
Lotova, N. A., 81, 93, *98*, 102, *117*, *118*
Lovelace, R. V. E., 90, 91, 97, *98*, 134, *140*
Lüst, R., 287, *288*, 300, *303*, 314, *317*, *318*

Maagoe, S., *117*, 125, *139*
MacQueen, R. M., 3, 4, 18, 19, *40*, 45, *60*
Magun, A., *18*
Marandino. G. E., *98*
Marochnik, L. S., 305, *318*
Marshall, F. P., 348, *363*
Martres, M. J., *79*
MaSung, L. S., 163, 330, *340*
Matheson, D. N., 120, 134, 135, 137, *140*
Matsuda, T., 56, *60*
Matyugov, S. S., *99*
McCabe, M. K., *20*
McClymont, A. N., 59, *60*
McCracken, K. G., 355, 356, *362*, *363*

McDonald, F. B., *163*, 251, *256*, *257*, 330, *340*
McIntosh, P. S., 61, *140*
McKenzie, D. L., *41*
McKenzie, J. F., 153, *162*, 309, *318*
McKibben, D. D., *224*, 356, 359, *363*
McKibben, R. B., 325, *340*
McKinnon, J. A., 404, *420*
McLean, D. J., 8, *20*, 26, *41*, 406, *420*
Mendis, D. A., 283, 284, 285, *288*, 292, 293, 294, 295, 296, 297, 299, 301, *302*, *303*, 315, *318*
Meng, C. -I., *41*
Metzler, N., 166, 177, 180, 186, *191*
Mewaldt, R. A., 341, *363*
Michael, G. A., *420*
Michels, D. J., *19*
Miggenrieder, H., *192*
Mihalov, J. D., *40*, *60*, 133, *140*, *224*, *256*
Miller, F. D., 287, *288*, 312, 313, 314, *318*
Mohler, O. C., *223*
Montgomery, M. D., 166, *192*, *363*, 368, 382, *388*
Moraal, H., 158, *162*
Moreno, G., 179, *190*, *191*, *223*
Morfill, G. E., 326, 330, *340*
Mouschovias, T. Ch., 15, 18, *19*
Muhleman, D. O., 81, 92, 96, *98*
Müller-Mellin, R., *339*, *340*, *341*, *420*
Munro, R. H., 6, 7, 12, 13, 14, *19*, *20*, *21*, *40*, *60*, 153, *162*, 187, *192*
Mutel, R. L., 92, *98*

Nakagawa, Y., *20*, 45, 46, 47, 48, 49, 51, 53, 56, 59, 60, *60*, *61*, 62, *118*, *193*, *257*
Nelson, G. J., 406, *420*
Nerney, S. F., *98*
Ness, N. F., 124, *140*, *162*, 186, *193*, 228, *256*, 308, *318*, 379, 381, *388*
Neubauer, F. M., 260, 269, *273*
Neugebauer, M., 90, 91, *98*. 134, 136, 137, 138, 139, *140*, *141*, 165, 187, *192*, 198, 216, *223*, 225, 228, *256*, *273*
Neupert, W. M., 33, 35, *41*, 42, 167, 188, *192*
Newkirk, G., *60*, *61*, 62, 71, 77, *80*, 87, 89, *98*, 149, 151, 160, *162*
Ng, C. K., 345, 346, 356, 362, *363*
Nishida, A., 167, *192*
Noble, L. M. 166, *192*
Nolte, J. T., 10, *19*, 113, *118*, 188, *192*, 337, *340*, 409, *420*
Novaco, J. C., *80*
Nowak, T., *318*
Nusinov, A. A., *288*, *302*, 307, 309, 312, *318*

O'Gallagher, J. J., *340*
Ogilvie, K. W., 166, 179, 186, *192*, *223*, 379, 384, *388*
Ohlson, J. E., *98*
Okoye, S. E., 87, 89, *98*
Ong, R. S. B., *289*, 292, 300, *303*
Oppenheimer, M., 295, *303*
Orrall, F. Q., *190*

Page, D. E., *339*
Palmeira, R. A. R., 36, *41*, *340*, *362*, 379, *388*
Palmer, I. D., 325, 329, 330, 333, 334, *362*
Palmiotto, F., *190*
Parker, E. N., 147, *162*, 165, 175, *192*, 196, *224*, 230, *256*, 368, 369, *388*
Parker, G. D., 230, *256*
Perez-de-Tejada, H., *319*
Perkins, F., 180, *192*
Pesses, M. E., 252, *256*, 385, *388*
Peterson, L. D., 32, *41*
Phillips, G. J., *117*
Pick, M., *79*
Piddington, J. H., 149, 151, *162*
Pikelner, S. B., 60, *60*
Pintêr, S., 83, *98*
Pizzo, V., *141*, 187, 188, *192*
Pneuman, G. W., 148, *162*

Poland, A. I., 6, *19*, *40*, *60*
Polenov, B. V., *288*
Pomerantz, M., 36, *41*, 379, *388*
Poquerusse, M., *79*
Potter, W. H., *79*, 80
Pramesh, R. A., *317*
Price, P. B., 348, *363*
Primbsch, J. H., *339*
Pyle, K. R., *340*

Quenby, J. J., 158, *161*, 338, *340*

Ramaty, R., 28, 36, *40*, *41*, *162*
Rankin, J. M., *97*
Rao, A. P., 136, *140*
Rao, U. R., *41*, *192*, *362*, *363*
Razdan, H., 228, *256*
Razmanov, V. M., *99*
Readhead, A. C. S., 110, *118*, 120, 134, *140*, 141
Reichman, E., *61*
Reid, G. C., 160, *162*, 384, *388*
Reinhard, R., 325, 327, 329, 333, 334, *340*, *341*, 355, 357, 358, *363*
Reynolds, R. J., *302*
Richter, A. K., *98*, *340*, 379, 384, *388*
Richter, K. D., *388*
Richter, N., 288, *288*
Rickett, B. J., 97, 102, 104, 105, 106, 107, 112, 113, 114, 116, *117*, *118*, 125, 128, 130, 131, 132, 133, *139*, *140*
Riddle, A. C., 69, 71, *80*
Robbins, D. E., 179, *191*, *192*
Robbins, M., 381, *388*
Roberts, D. E., *19*
Robertson, D. S., *98*
Robinson, R. D., *18*
Rodriques, P., 262, 263, *273*
Roelof, E. C., 131, *140*, *162*, *163*, 191, 253, *256*, *257*, 325, 330, 332, 334, 336, 337, *339*, 345, 346, 352, *363*, 410, *420*
Roethig, D. T., *420*
Roesler, F. L., *302*
Rogers, A. E. E., *98*
Roosen, R. G., *302*, *303*, *318*
Rosenau, P., *256*
Rosenbauer, H., 187, *191*, *192*
Rosenberg, R. L., 230, *256*
Ross, C. L., *19*. *40*, *60*
Rothe, E. D., 292, *302*
Rubenstein, D. M., 10, *18*
Rumsey, V. H., 83, *97*, *98*, 105, *117*, *118*, *139*
Rusch, W. V. T., *98*
Russell, C. T., 136, 138, *141*
Rust, D. M., 13, 14, *19*, *20*
Ryan, J. M., 167, *192*

Saito, K., 87, 89, *98*
Sakurai, T., 56, *60*
Salpeter, E. E., *98*, *140*
Sanderson, S. J., 267, *273*
Sarris, E. T., 260, 261, 264, 271, *273*, 347, *363*, 368, 369, 370, 372, 378, 379, 381, 382, 383, *388*
Sato, T., *98*
Sawyer, C., *117*
Scarf, F. L., 52, *61*, 166, *192*, 262, 263, 270, *273*
Schatten, K. H., 83, *98*, 149, 158, 160, *162*
Schatzman, E., 368, 369, *388*
Scherb, F., *302*
Schmahl, E. J., 5, 6, *19*
Schmidt, H. U., *302*, *317*
Schulz, M., 175, *191*
Schulze, B. M., 327, 328, 332, 333, 334, *340*
Schwenn, R., *191*, *192*
Sciambi, R. K., *40*
Scott, P. F., *140*
Scudder, J. D., *192*
Seal, R. T., *19*
Seidel, B. L., *98*
Sekanina, Z., 287, *288*
Shabanskii, V. P., 368, *388*
Shapiro, A. H., 281, *288*
Shapiro, I. I. *98*
Sharp, L. E., *98*, 120, 131, *139*, *140*
Shawhan, S. D., *140*
Sheeley, N. R., 94, *98*, *118*, *140*
Shen, W. -W., *288*, *318*
Sheridan, K. V., *18*, *41*, 187, *191*

Shibasaki, K., *98, 118, 141*
Shimizu, M., 283, *288*, 295, *303*
Shinn, D. H., *117*
Shklovskii, I. S., 3, *19*
Signorini, C., *223*
Sime, D. G., *117, 118, 140*
Simnett, G. M., 36, *41*
Simon, M., 154, *162*, 164
Simon, P., *79*
Simpson, J. A., 155, *161*, 172, 251, 253, *256*
Singer, S., 368, 379, 382, *388*
Siscoe, G. L., 156, *163*, 186, *190*, 245, 247, *256*
Skadron, G., 155, *163*
Slysh, V. I., 74, *80*
Smerd, S. F., 7, *18, 20, 21, 41*, 164, 406, *420*
Smirnov, V. N., *288*
Smith, A. M., 294, *303*
Smith, B. F., *144*
Smith, E. J., 52, *61*, 153, *163*, 196, 212, 221, *224*, 228, 230, 243, 249, 250, *256*, 257, 260, *273*, 352, *363*, 385, *388*
Smith, K. L., *420*
Smith, Z. K., *40, 60, 256*, 274
Snyder, C. W., 165, 187, *192*, 228, *256*
Soltis, C., 379, *388*
Somov, B. V., 35, *41*, 60, *61*
Sonett, C. P., *140*, 228, *256*
Southwick, R. G., *99, 193*
Spiegelglass, M., *191*
Spitzer, L., Jr., 173, *192*
Stecher, T. P., *303*
Steinberg, J. L., 14, 17, *20*, 67, 69, 71, 78, *79*, 80, *80*
Steinolfson, R. S., 40, 45, 47, 49, 56, 58, *60, 61*, 62, 116, *118*, 191, 245, 248, *256, 257*
Stelzreid, C. T., 81, *98*
Sternlieb, A., 167, 188, *191*
Stevens, G. A., *41*
Stewart, R. T., 6, 8, 13, *18, 20, 41*
Stix, T. H., 265, *273*
Stone, E. C., *341*, 348, *363*
Stone, R. G., 75, *79*
Stuhlinger, E., *60*
Sturrock, P. A., 166, *191, 192*
Suess, S. T., 95, *98*, 100, 130, *141*, 189, *192*
Sugiura, M., *192*
Sullivan, J. D., 36, *41*, 89, *98*
Suri, A. N., *40*
Svalgaard, L., 147, 152, *163*
Švestka, Z., 5, *20*, 411, *420*
Symonds, M. D., 120, 125, *140*
Syroatsky, S. I., 60, *61*

Tandberg-Hanssen, E., 5, *20*, 45, *61*, 62
Tatarskii, V. I., 82, *97*
Teegarden, B. J., 152, 160, *163*, 250, *256, 257*
Terasawa, T., 369, 372, 389, *390*
Thomas, R. J., *41*
Tidman, D. A., 269, *273*
Timothy, A. F., *19, 20, 118, 162, 163*, 187, *191, 192*
Toichi, T., 167, *192*
Tousey, R., 3, 4, *19, 20, 118, 162, 163*, 187, *191, 192*
Trainor, J. H., *163, 256, 257*
Tranger, J., *302*
Trellis, M., 70, *80*
Trotter, D. E., *190*
Trubnikov, B. A., 26, *41*
Trusov, B. P., *99*
Tuzzolino, A. J., *340*

Uchida, Y., 52, *61*
Umlauft, G., *388*
Unti, T., 134, 136, 138, *140*
Urch, I. H., 158, *163*

Vaiana, G. S., 13, *19, 20, 118, 163*, 187, *191, 192*
Vaisberg, O. L., 280, *288*
Valniček, D., 45, *61*
Van Allen, J. A., *256*, 260, 271, *273*, 359, 360, 361, *363*, 368, 369, 370, 372, 378, 379, 381, 382, 383, 384, *388*
Van Beck, H. F., *41*
Vandenberg, N. R., *98*
Van Hollebecke, M. A. I., 160, *163*, 336, *340*

Vasyliunas, V. M., 219, 221, *224*
Venkatesan, D., *340*, *341*
Vinogradov, V. A., *99*
Vitkevitch, V. V., 105, *118*, 120, *141*
Vlasov, V. I., 105, *118*, 120, *141*
Vogt, R. E., *341*, *363*
Völk, H. J., 153, 157, *163*, 330, *340*
Vollmer, O., *363*
Von Rosenwinge, T. T., *256*
Vsekhsvjatsky, S. K., 316, *318*

Waldmeier, M., 87, *98*, 150, *163*
Wallis, M. K., 280, 281, 282, 283, 286, 287, *288*, 289, 292, 295, 297, 300, *303*, 306, 309, 316, 317, *318*
Ward, B. D., 87, 89, *98*, 114, 116, *118*, 131, 132, 133, *141*
Washimi, H., *118*
Watanabe, T., 96, 98, 104, 107, 110, 114, 116, *118*, 120, 121, 122, 125, 128, 130, 131, *140*, *141*
Webb, D. F., 13, 19, *20*
Webb, S., 339, *340*
Webber, W. R., *163*, *256*, *257*, 325, 326, 327, 333, 334, *340*, *341*
Weber, E. J., 166, 167, *192*, *318*
Weber, R. R., 88, 89, 92, 93, *95*, *96*
Webster, H. F., *303*
Wegmann, R., 281, *288*, 292, *302*, 306, 309, 317, *317*
Wehinger, P., 295, 295, *303*
Weiss, A. A., *41*
Wellck, R. E., 56, 57, *60*, *61*
Wentzel, P., 151, 160, *162*
Wenzel, K., -P., *339*
Whang, Y. C., 166, *192*
Whitney, A. R., *98*
Whitney, J. D., *19*
Wibberenz, G., 160, *163*, 325, 327, 330, 333, 334, 338, 339, *340*, *341*, *420*
Wilcox, J. M., *98*, 152, 162, *163*, 186, *193*
Wild, J. P., 7, *20*, 36, *41*
Wilken, B., *388*
Wilkerson, T. D., 179, *192*
Wills, D., *140*
Wilson, D., 89, *99*
Winckler, J. R., 38, *40*, *41*
Winge, C. R., *98*
Wiseman, M., 120, 131, 132, *140*, *141*
Withbore, G. L., 187, *192*
Witte, M., 329, 336, 339, 341, *420*
Wolf, M., 312, 313, 314, *318*
Wolfe, J. H., 52, *60*, *61*, 91, *98*, 138, 139, *140*, 196, 221, *223*, *224*, 228, 230, 234, 243, *256*, *257*, 260, 352, *363*, 385, *388*
Wolff, C. L., 90, *99*, 167, *193*
Wolfson, C. J., 407, 410, *420*
Woo, R., 82, 83, 84, 85, 86, 87, 89, 90, 91, 92, 93, 94, 96, *99*
Wu, C. S., 280, *289*
Wu, S. T., 14, 17, *20*, 45, 46, 49, 51, 52, 53, 54, *60*, *61*, 62, 187, 191, *192*, *193*
Wurm, K., 285, *289*, 294, 295, *303*, 306, *318*
Wyckoff, S., 295, 296, *303*

Yakovlev, O. L., 82, *99*
Yang, F. C., *99*
Yasukawa, E., *19*, *60*
Yeh, T., 167, *193*
Yip, K. W., *99*
Yu, M., *99*

Zombeck, M. V., *19*, *20*, *60*, *192*
Zwickl, R. D., 325, 326, 327, 333, 334, *340*, *341*, 347, 348, 349, *363*

ASTROPHYSICS AND SPACE SCIENCE LIBRARY

Edited by

J. E. Blamont, R. L. F. Boyd, L. Goldberg, C. de Jager, Z. Kopal, G. H. Ludwig, R. Lüst, B. M. McCormac, H. E. Newell, L. I. Sedov, Z. Švestka, and W. de Graaff

1. C. de Jager (ed.), *The Solar Spectrum, Proceedings of the Symposium held at the University of Utrecht, 26–31 August, 1963.* 1965, XIV + 417 pp.
2. J. Ortner and H. Maseland (eds.), *Introduction to Solar Terrestrial Relations, Proceedings of the Summer School in Space Physics held in Alpbach, Austria, July 15–August 10, 1963 and Organized by the European Preparatory Commission for Space Research.* 1965, IX + 506 pp.
3. C. C. Chang and S. S. Huang (eds.), *Proceedings of the Plasma Space Science Symposium, held at the Catholic University of America, Washington, D.C., June 11–14, 1963.* 1965, IX + 377 pp.
4. Zdeněk Kopal, *An Introduction to the Study of the Moon.* 1966, XII + 464 pp.
5. B. M. McCormac (ed.), *Radiation Trapped in the Earth's Magnetic Field. Proceedings of the Advanced Study Institute, held at the Chr. Michelsen Institute, Bergen, Norway, August 16–September 3, 1965.* 1966, XII + 901 pp.
6. A. B. Underhill, *The Early Type Stars.* 1966, XII + 282 pp.
7. Jean Kovalevsky, *Introduction to Celestial Mechanics.* 1967, VIII + 427 pp.
8. Zdeněk Kopal and Constantine L. Goudas (eds.), *Measure of the Moon. Proceedings of the 2nd International Conference on Selenodesy and Lunar Topography, held in the University of Manchester, England, May 30–June 4, 1966.* 1967, XVIII + 479 pp.
9. J. G. Emming (ed.), *Electromagnetic Radiation in Space. Proceedings of the 3rd ESRO Summer School in Space Physics, held in Alpbach, Austria, from 19 July to 13 August, 1965.* 1968, VIII + 307 pp.
10. R. L. Carovillano, John F. McClay, and Henry R. Radoski (eds.), *Physics of the Magnetosphere, Based upon the Proceedings of the Conference held at Boston College, June 19–28, 1967.* 1968, X + 686 pp.
11. Syun-Ichi Akasofu, *Polar and Magnetospheric Substorms.* 1968, XVIII + 280 pp.
12. Peter M. Millman (ed.), *Meteorite Research. Proceedings of a Symposium on Meteorite Research, held in Vienna, Austria, 7–13 August, 1968.* 1969, XV + 941 pp.
13. Margherita Hack (ed.), *Mass Loss from Stars. Proceedings of the 2nd Trieste Colloquium on Astrophysics, 12–17 September, 1968.* 1969, XII + 345 pp.
14. N. D'Angelo (ed.), *Low-Frequency Waves and Irregularities in the Ionosphere. Proceedings of the 2nd ESRIN-ESLAB Symposium, held in Frascati, Italy, 23–27 September, 1968.* 1969, VII + 218 pp.
15. G. A. Partel (ed.), *Space Engineering. Proceedings of the 2nd International Conference on Space Engineering, held at the Fondazione Giorgio Cini, Isola di San Giorgio, Venice, Italy, May 7–10, 1969.* 1970, XI + 728 pp.
16. S. Fred Singer (ed.), *Manned Laboratories in Space. Second International Orbital Laboratory Symposium.* 1969, XIII + 133 pp.
17. B. M. McCormac (ed.), *Particles and Fields in the Magnetosphere. Symposium Organized by the Summer Advanced Study Institute, held at the University of California, Santa Barbara, Calif., August 4–15, 1969.* 1970, XI + 450 pp.
18. Jean-Claude Pecker, *Experimental Astronomy.* 1970, X + 105 pp.
19. V. Manno and D. E. Page (eds.), *Intercorrelated Satellite Observations related to Solar Events. Proceedings of the 3rd ESLAB/ESRIN Symposium held in Noordwijk, The Netherlands, September 16–19, 1969.* 1970, XVI + 627 pp.
20. L. Mansinha, D. E. Smylie, and A. E. Beck, *Earthquake Displacement Fields and the Rotation of the Earth, A NATO Advanced Study Institute Conference Organized by the Department of Geophysics, University of Western Ontario, London, Canada, June 22–28, 1969.* 1970, XI + 308 pp.
21. Jean-Claude Pecker, *Space Observatories.* 1970, XI + 120 pp.
22. L. N. Mavridis (ed.), *Structure and Evolution of the Galaxy. Proceedings of the NATO Advanced Study Institute, held in Athens, September 8–19, 1969.* 1971, VII + 312 pp.
23. A. Muller (ed.), *The Magellanic Clouds. A European Southern Observatory Presentation: Principal Prospects, Current Observational and Theoretical Approaches, and Prospects for Future Research, Based on the Symposium on the Magellanic Clouds, held in Santiago de Chile, March 1969, on the Occasion of the Dedication of the European Southern Observatory.* 1971, XII + 189 pp.

24. B. M. McCormac (ed.), *The Radiating Atmosphere. Proceedings of a Symposium Organized by the Summer Advanced Study Institute, held at Queen's University, Kingston, Ontario, August 3–14, 1970.* 1971, XI + 455 pp.
25. G. Fiocco (ed.), *Mesospheric Models and Related Experiments. Proceedings of the 4th ESRIN-ESLAB Symposium, held at Frascati, Italy, July 6–10, 1970.* 1971, VIII + 298 pp.
26. I. Atanasijević, *Selected Exercises in Galactic Astronomy.* 1971, XII + 144 pp.
27. C. J. Macris (ed.), *Physics of the Solar Corona. Proceedings of the NATO Advanced Study Institute on Physics of the Solar Corona, held at Cavouri-Vouliagmeni, Athens, Greece, 6–17 September 1970.* 1971, XII + 345 pp.
28. F. Delobeau, *The Environment of the Earth.* 1971, IX + 113 pp.
29. E. R. Dyer (general ed.), *Solar-Terrestrial Physics/1970. Proceedings of the International Symposium on Solar-Terrestrial Physics, held in Leningrad, U.S.S.R., 12–19 May 1970.* 1972, VIII + 938 pp.
30. V. Manno and J. Ring (eds.), *Infrared Detection Techniques for Space Research. Proceedings of the 5th ESLAB-ESRIN Symposium, held in Noordwijk, The Netherlands, June 8–11, 1971.* 1972, XII + 344 pp.
31. M. Lecar (ed.), *Gravitational N-Body Problem. Proceedings of IAU Colloquium No. 10, held in Cambridge, England, August 12–15, 1970.* 1972, XI + 441 pp.
32. B. M. McCormac (ed.), *Earth's Magnetospheric Processes. Proceedings of a Symposium Organized by the Summer Advanced Study Institute and Ninth ESRO Summer School, held in Cortina, Italy, August 30–September 10, 1971.* 1972, VIII + 417 pp.
33. Antonin Rükl, *Maps of Lunar Hemispheres.* 1972, V + 24 pp.
34. V. Kourganoff, *Introduction to the Physics of Stellar Interiors.* 1973, XI + 115 pp.
35. B. M. McCormac (ed.), *Physics and Chemistry of Upper Atmospheres. Proceedings of a Symposium Organized by the Summer Advanced Study Institute, held at the University of Orléans, France, July 31–August 11, 1972.* 1973, VIII + 389 pp.
36. J. D. Fernie (ed.), *Variable Stars in Globular Clusters and in Related Systems. Proceedings of the IAU Colloquium No. 21, held at the University of Toronto, Toronto, Canada, August 29–31, 1972.* 1973, IX + 234 pp.
37. R. J. L. Grard (ed.), *Photon and Particle Interaction with Surfaces in Space. Proceedings of the 6th ESLAB Symposium, held at Noordwijk, The Netherlands, 26–29 September, 1972.* 1973, XV + 577 pp.
38. Werner Israel (ed.), *Relativity, Astrophysics and Cosmology. Proceedings of the Summer School, held 14–26 August, 1972, at the BANFF Centre, BANFF, Alberta, Canada.* 1973, IX + 323 pp.
39. B. D. Tapley and V. Szebehely (eds.), *Recent Advances in Dynamical Astronomy. Proceedings of the NATO Advanced Study Institute in Dynamical Astronomy, held in Cortina d'Ampezzo, Italy, August 9–12, 1972.* 1973, XIII + 468 pp.
40. A. G. W. Cameron (ed.), *Cosmochemistry. Proceedings of the Symposium on Cosmochemistry, held at the Smithsonian Astrophysical Observatory, Cambridge, Mass., August 14–16, 1972.* 1973, X + 173 pp.
41. M. Golay, *Introduction to Astronomical Photometry.* 1974, IX + 364 pp.
42. D. E. Page (ed.), *Correlated Interplanetary and Magnetospheric Observations. Proceedings of the 7th ESLAB Symposium, held at Saulgau, W. Germany, 22–25 May, 1973.* 1974, XIV + 662 pp.
43. Riccardo Giacconi and Herbert Gursky (eds.), *X-Ray Astronomy.* 1974, X + 450 pp.
44. B. M. McCormac (ed.), *Magnetospheric Physics. Proceedings of the Advanced Summer Institute, held in Sheffield, U.K., August 1973.* 1974, VII + 399 pp.
45. C. B. Cosmovici (ed.), *Supernovae and Supernova Remnants. Proceedings of the International Conference on Supernovae, held in Lecce, Italy, May 7–11, 1973.* 1974, XVII + 387 pp.
46. A. P. Mitra, *Ionospheric Effects of Solar Flares.* 1974, XI + 294 pp.
47. S.-I. Akasofu, *Physics of Magnetospheric Substorms.* 1977, XVIII + 599 pp.
48. H. Gursky and R. Ruffini (eds.), *Neutron Stars, Black Holes and Binary X-Ray Sources.* 1975, XII + 441 pp.
49. Z. Švestka and P. Simon (eds.), *Catalog of Solar Particle Events 1955–1969. Prepared under the Auspices of Working Group 2 of the Inter-Union Commission on Solar-Terrestrial Physics.* 1975, IX + 428 pp.
50. Zdeněk Kopal and Robert W. Carder, *Mapping of the Moon.* 1974, VIII + 237 pp.
51. B. M. McCormac (ed.), *Atmospheres of Earth and the Planets. Proceedings of the Summer Advanced Study Institute, held at the University of Liège, Belgium, July 29–August 8, 1974.* 1975, VII + 454 pp.
52. V. Formisano (ed.), *The Magnetospheres of the Earth and Jupiter. Proceedings of the Neil Brice Memorial Symposium, held in Frascati, May 28–June 1, 1974.* 1975, XI + 485 pp.

53. R. Grant Athay, *The Solar Chromosphere and Corona: Quiet Sun.* 1976, XI + 504 pp.
54. C. de Jager and H. Nieuwenhuijzen (eds.), *Image Processing Techniques in Astronomy. Proceedings of a Conference, held in Utrecht on March 25–27, 1975*, XI + 418 pp.
55. N. C. Wickramasinghe and D. J. Morgan (eds.), *Solid State Astrophysics. Proceedings of a Symposium, held at the University College, Cardiff, Wales, 9–12 July 1974.* 1976, XII + 314 pp.
56. John Meaburn, *Detection and Spectrometry of Faint Light.* 1976, IX + 270 pp.
57. K. Knott and B. Battrick (eds.), *The Scientific Satellite Programme during the International Magnetospheric Study. Proceedings of the 10th ESLAB Symposium, held at Vienna, Austria, 10–13 June 1975.* 1976, XV + 464 pp.
58. B. M. McCormac (ed.), *Magnetospheric Particles and Fields. Proceedings of the Summer Advanced Study School, held in Graz, Austria, August 4–15, 1975.* 1976, VII + 331 pp.
59. B. S. P. Shen and M. Merker (eds.), *Spallation Nuclear Reactions and Their Applications.* 1976, VIII + 235 pp.
60. Walter S. Fitch (ed.), *Multiple Periodic Variable Stars. Proceedings of the International Astronomical Union Colloquium No. 29, Held at Budapest, Hungary, 1–5 September 1975.* 1976, XIV + 348 pp.
61. J. J. Burger, A. Pedersen, and B. Battrick (eds.), *Atmospheric Physics from Spacelab. Proceedings of the 11th ESLAB Symposium, Organized by the Space Science Department of the European Space Agency, held at Frascati, Italy, 11–14 May 1976.* 1976, XX + 409 pp.
62. J. Derral Mulholland (ed.), *Scientific Applications of Lunar Laser Ranging. Proceedings of a Symposium held in Austin, Tex., U.S.A., 8–10 June, 1976.* 1977, XVII + 302 pp.
63. Giovanni G. Fazio (ed.), *Infrared and Submillimeter Astronomy. Proceedings of a Symposium held in Philadelphia, Penn., U.S.A., 8-10 June, 1976.* 1977, X+226 pp.
64. C. Jaschek and G. A. Wilkins (eds.), *Compilation, Critical Evaluation and Distribution of Stellar Data. Proceedings of the International Astronomical Union Colloquium No. 35, held at Strasbourg, France, 19-21 August, 1976.* 1977, XIV+316 pp.
65. M. Friedjung (ed.), *Novae and Related Stars. Proceedings of an International Conference held by the Institut d'Astrophysique, Paris, France, 7-9 September, 1976.* 1977, XIV+228 pp.
66. David N. Schramm (ed.), *Supernovae. Proceedings of a Special IAU Session on Supernovae held in Grenoble, France, 1 September, 1976.* 1977, X+192 pp.
67. Jean Audouze (ed.), *CNO Isotopes in Astrophysics. Proceedings of a Special IAU Session held in Grenoble, France, 30 August, 1976.* 1977, XIII+195 pp.
68. Z. Kopal, *Dynamics of Close Binary Systems*, forthcoming.
69. A. Bruzek and C. J. Durrant (eds.), *Illustrated Glossary for Solar and Solar-Terrestrial Physics.* 1977, approx. 216 pp.
70. H. van Woerden (ed.), *Topics in Interstellar Matter.* 1977, VIII + 295 pp.

GPSR Compliance

The European Union's (EU) General Product Safety Regulation (GPSR) is a set of rules that requires consumer products to be safe and our obligations to ensure this.

If you have any concerns about our products, you can contact us on ProductSafety@springernature.com

In case Publisher is established outside the EU, the EU authorized representative is:

Springer Nature Customer Service Center GmbH
Europaplatz 3
69115 Heidelberg, Germany

Batch number: 10815974

Printed by Printforce, the Netherlands